OLD WORLD MONKEYS

Old World monkeys (Cercopithecoidea) are the most successful and diverse group of living nonhuman primates in terms of the number of species, behavioral repertoires and ecology. They have much to teach us about the processes of evolution and the principles of ecology, and are among our closest living relatives. This volume presents a broad technical account of cercopithecoid biology including molecular, behavioral and morphological approaches to phylogeny, population structure, allometry, fossil history, functional morphology, ecology, cognitive capabilities, social behavior and conservation. It will be the definitive reference on this group for professionals and graduate students in primatology, animal behavior, paleontology, morphology, systematics and physical anthropology, plus it will also be useful to senior undergraduates.

PAUL F. WHITEHEAD is Acting Collections Manager in the Division of Vertebrate Zoology at the Peabody Museum of Natural History, Yale University and Assistant Professor in the Department of Science and Mathematics at Capital Community-Technical College.

CLIFFORD J. JOLLY is Professor of Anthropology at New York University.

OLD WORLD MONKEYS

Edited by
Paul F. Whitehead and Clifford J. Jolly

CAMBRIDGE UNIVERSITY PRESS
Cambridge, New York, Melbourne, Madrid, Cape Town, Singapore, São Paulo

Cambridge University Press
The Edinburgh Building, Cambridge CB2 2RU, UK

Published in the United States of America by Cambridge University Press, New York

www.cambridge.org
Information on this title: www.cambridge.org/9780521571241

First published 2000
This digitally printed first paperback version 2006

A catalogue record for this publication is available from the British Library

Library of Congress Cataloguing in Publication data

Old world monkeys / edited by Paul F. Whitehead & Clifford J. Jolly.
p. cm.
ISBN 0 521 57124 3 (hardcover)
1. Cercopithecidae. I. Whitehead, Paul F. (Paul Frederick),
1954– . II. Jolly, Clifford J., 1939– .
QL737.P930545 2000
599.8′6–dc21 99-20192 CIP

ISBN-13 978-0-521-57124-1 hardback
ISBN-10 0-521-57124-3 hardback

ISBN-13 978-0-521-02809-7 paperback
ISBN-10 0-521-02809-4 paperback

Contents

Contributors

Brenda R. Benefit
Department of Anthropology, Southern Illinois University, Carbondale, IL 62901, USA

Fred B. Bercovitch
Caribbean Primate Research Center, University of Puerto Rico, Sabana Seca, PR 00952, Puerto Rico

Thore Bergman
Department of Anatomy and Neurobiology, Washington University School of Medicine, St. Louis, MO 63110, USA

Irwin S. Bernstein
Department of Psychology, University of Georgia, Athens, GA 30602, USA

C.M. Bocian
Ph. D. Program in Biology, CUNY Graduate School, 365 Fifth Ave, New York, NY 10016, USA

Colin A. Chapman
Department of Anthropology, University of Florida, Gainsville, FL 32611, USA

Marina Cords
Department of Anthropology, Columbia University, New York, NY 10027, USA

Todd Disotell
Department of Anthropology, New York University, 25 Waverly Place, New York, NY 10003, USA

Lynn A. Fairbanks
Neuropsychiatric Institute and Hospital Center for the Health Sciences, University of California, Los Angeles, 760 Westwood Plaza, Los Angeles, CA 90024-1759, USA

Daniel Gebo
Department of Anthropology, Northern Illinois University, DeKalb, IL 60115-2854, USA

Colin P. Groves
Dept. of Archaeology and Anthropology, Australian National University, Canberra, ACT 2600, Australia

Thomas Gundling
Department of Anthropology, William Paterson University of New Jersey, 300 Pompton Road, Wayne, NJ 07470, USA

Karen Hiiemae
Institute for Sensory Research, Syracuse University, Syracuse, NY 13244-5290, USA

Andrew Hill
Department of Anthropology, 51 Hillhouse Ave., Yale Univesity, New Haven, CT 06511, USA

Clifford J. Jolly
Department of Anthropology, New York University, New York, NY 10003, USA

Karen M. Kool
School of Biological Sciences, University of New South Wales, P.O. Box 1, Kensington, NSW 2033, Australia

Wolfgang Maier
Lehrstuhl Spezielle Zoologie, Eberhard-Karls-Universitaet, Auf der Morgenstelle 28, Germany

John F. Oates
Department of Anthropology, Hunter College, 695 Park Ave., New York, NY 10021, USA

Jane E. Phillips-Conroy
Department of Anatomy and Neurobiology, Washington University School of Medicine, St. Louis, MO 63110, USA

(which is at the base of punctuated equilibrium). The idea that speciation has proceeded through the establishment of geographical barriers (Booth, 1958) and refugia (Hamilton, 1988) is not new in the study of cercopithecoid evolution and it has precluded testing of other models of speciation (e.g. parapatric speciation by Endler, 1977).

Groves (Chapter 4) presents two cladistic analyses. One analysis is of cercopithecid genera, using the Colobidae as an outgroup. Forty-six characters are used; they are a mixture of mainly dental and cranial traits, with a small number of other features. The study supports the ideas that *Lophocebus, Gorgopithecus, Dinopithecus and Parapapio* are papionines. He suggests that *M. nemestrina* may also be a papionine, but urges caution. Groves supports the notion that *Mandrillus* and *Cercocebus* are sister-groups. His other cladistic analysis is of the *Cercopithecus*-group, with *Allenopithecus* as the outgroup; 22 cranial and dental characters are used. He affirms a suggestion that he had made previously (Groves, 1989) that vervets should be separated from *Cercopithecus*, and that the genus *Chlorocebus* should be used for green monkeys instead. This conclusion is based on three skull characters that ally vervets with patas monkeys. He also concludes that *Allenopithecus* does not belong to the Cercopithecinae.

Maier's contribution applies a comparative morphogenetic approach to a little-studied anatomical area, the ethmoidal region. Studies of the ontogeny of morphological characters have a long history in vertebrate zoology (de Beer, 1937). Martin (1990: 127) writes that, "Developmental evidence provides one of the most powerful available aids for separating primitive from derived character states and every attempt should be made to include ontogenetic information." While some regions have been repeatedly studied in primates, such as browridges, others have been comparatively neglected. The ethmoid forms an anterior portion of the cranial cavity and portions of the medial orbital walls; it is perhaps most obviously characterized by its cribiform plate. Maier identifies characters in the ethmoidal and nasal regions that distinguish cercopithecoids, including a reduced ethmoturbinal recess and posterior nasal cupula, and a lack of a sphenoidal sinus. This analysis provides an important comparative study of primates in this region.

Paleontology

Evolutionary primatology has always had an emphasis on the discovery and description of fossils because of its close ties with the study of human evolution. Physical anthropologists have sought to discover specimens and name new taxa, often in an attempt to discover hominid "ancestors." Actually, it is somewhat surprising that a field that has emphasized fossils has also wholeheartedly embraced cladism because many phylogenetic systematists (Patterson, 1981; Rosen *et al.*, 1981) de-emphasize the role of fossils in determining evolutionary relationships. At the time of OWM I, most primatologists would probably have agreed with Simpson (1961: 83) that "fossils provide the soundest basis for evolutionary classification" although Sarich had demonstrated that molecular techniques were beginning to become important.

Despite this emphasis on fossils, it has only been in recent years that cercopithecoid paleontology has been viewed as more than a sideshow to human evolution. This change has partially been due to the efforts of some particularly active paleontologists, such as Eric Delson and Meave Leakey, but also because of the variety of fossil monkeys. The Plio-Pleistocene record has been especially interesting, with discovery of the terrestrial colobine *Cercopithecoides* (Szalay and Delson, 1979), the large colobine *Paracolobus* (Birchette, 1982), the large-bodied *Theropithecus oswaldi* and new interpretations of its postcranial anatomy (Jolly, 1972; Whitehead, 1992) and evolution (Leakey, 1993), and the unique *Theropithecus brumpti* (Eck and Jablonski, 1987). There has also been increased discovery and study of Miocene cercopithecoids. Simons (1970: 104) was correct when he stated that, "Until recently the Miocene monkeys of Africa have remained largely undescribed, unnamed, or misidentified." While Miocene hominoids still receive the greatest amount of attention (Begun *et al.*, 1997), there has been progress in our understanding of Miocene cercopithecoids since von Koenigswald (1969).

Benefit's authoritative contribution (Chapter 6) demonstrates that fossils are important for the elucidation of the ecological bases of adaptive radiations, and that they still have potential for the determination of phylogeny. Benefit reviews the evidence on the phylogenetic positions of Miocene monkeys, and concludes that *Victoriapithecus* and *Prohylobates* belong to the Victoriapithecidae rather than the Cercopithecidae. Her description of the well-preserved *Victoripithecus* cranium is important in this regard. In addition, this analysis contradicts the expectation that a colobine-like skull was primitive for the cercopithecoids, an ancestral morphotype postulated by Delson (1975). Benefit argues, contrary to the idea that fossils are of little value, that "Incorporating the Miocene sister-taxa of modern hominoids and cercopithecoids into cladistic analyses is more appropriate for determining morphocline polarities within each superfamily." Benefit's analysis assumes that there is not much convergent evolution in the skull nor dentition of Old World monkeys; convergence has, however, occurred in the postcrania of different groups of cercopithecoids (Whitehead, 1992).

Benefit then conducts an extensive dental comparison to illuminate the origins and significance of bilophodonty. She states that the molars of early and middle Miocene cercopithecoids were not adapted for leaf-eating, and that the origin of bilophodonty may be related to an increase in grinding area provided by increased occlusal surfaces of the hypoconulid and entoconid. She believes that early monkeys ate hard fruits and seeds, and discusses various models of the split between hominoids and cercopithecoids. The paleoecological evidence leads her to suggest that Miocene apes inhabited forest environments, while victoriapithecids lived in "drier, more unpredictable and open environments."

Benefit's anatomical and paleocological analysis is complemented by Gundling and Hill (Chapter 7) who summarize the geological data from fossil sites in eastern Africa. Geological evidence has been treated in an equivocal manner in these days of phylogenetic systematics. It has been argued that geological sequence is "a poor guide to the polarity of characters and the order of occurrence in time need not match the order of nodes in a cladogram" (Benton and Storrs, 1996: 32). This attitude is in direct contrast to that which prevailed during OWM I, when age of appearance was used as a ranking criterion in classifications. Yet, age and geological context are important in reconstructing paleoenvironments, which is crucial if explanations of adaptive contexts are to be formulated. At the least, morphologically-based cladograms and stratigraphic data may be tested against each other.

Gundling and Hill review the chronological age and paleoecological/taphonomic reconstruction for all major cercopithecoid sites, from the early Miocene through the Pleistocene. They make the point that the pattern of evolution of genera is different from that found in the New World, because modern Old World monkey genera appear recently in the late Pliocene and Pleistocene while many modern platyrrhine genera appear in the Miocene.

Allometry

The study of allometric scaling is the analysis of change in a body part relative to change in body size. As such, it is a valuable tool to help us to understand evolutionary changes in form and function. Body size is an important variable in understanding animal biology, from the level of tissues to trophic ecology. The relationship between size and physiology has been well-documented for many years. Since the late 1970s, there has been increasing realization that size also matters for ecology (Peters, 1983) and evolution (Calder, 1984; Damuth and MacFadden, 1990). Important concepts, such as symmorphis, have originated from allometric investigations. Issues regarding the number of species present in the fossil record, as in the case of *Archaeopteryx* (Houck *et al.*, 1990), have been resolved by studies of growth series. Primates have received considerable attention because of the existence of substantial interspecific differences in body size within all families. Of course, the Hominoidea have tended to receive the most scrutiny. However, the Cercopithecoidea are also of considerable interest because they "exhibit marked diversity in adult cranial proportions and body size" (Ravosa and Profant, Chapter 9). Aside from questions dealing specifically with cercopithecoid phylogeny and function, body size should also elucidate the "role of ontogenetic and allometric factors in patterning morphological evolution."

Ravosa and Profant point out that allometric studies can help to differentiate functional convergence from close phylogenetic relationship. For example, colobines have relatively deeper mandibular bodies than do cercopithecines; but, so do large-bodied extinct papionins compared to other cercopithecines. This convergence can be functionally interpreted. In contrast, there appears to be a subfamilial difference in symphyseal robustness, with cercopithecines and colobines evolving in different directions. This type of approach was not utilized in OWM I.

Ontogenetic analyses of the cercopithecoid skull are also valuable, because they cause us to realize that "the shape of large (or older) individuals may depart significantly from that of smaller individuals only because the former have progressed farther along a growth curve that is strongely allometric" (Emerson and Bramble, 1993). Cercopithecoids have been the subject of intersexual and interspecific ontogenetic comparisons. Interspecific comparisons have implications for the study of phylogeny, since it is significant whether species show concordant or divergent allometric growth vectors. If scaling comparisons among species reveal divergent growth allometries, then this may indicate divergent adaptations; Ravosa and Profant discuss study of relative face length as a character in the formulation of cranial morphoclines, which has direct relevance to the disagreement between Delson (1975) and Benefit (Chapter 6).

Experimental functional anatomy

At the time of OWM I, experimental studies of vertebrate functional anatomy had begun on the masticatory apparatus (Crompton and Hiiemae, 1970). The experimental techniques of cineradiography and electromyography are instrumental in determining the functional significance of particular anatomical features. Sometimes, results overturn long-cherished notions that were based on non-experimental anatomical deduction (Whitehead and Larson, 1994).

We are fortunate to have a contribution from one of the pioneers of experimental functional anatomy, Karen Hiiemae (Chapter 8). Hiiemae's work demonstrates that, despite anatomical differences in the orofacial complex, cercopithecoids and mammals from other Orders have the same three stages in their feeding process. However, cercopithecoids have the ability to chew, transport and swallow food at the same time; these activities occur in series in other mammals. In addition, the movements of the tongue and jaw are not as correlated in macaques as in other studied mammals. Hiiemae then speculates that these differences between monkeys and other mammals have implications for the evolution of human speech.

Socioendocrinology, social behavior, and ecology

Behavioral studies were represented at the 1969 conference by fieldworkers and a single zoo study. By implication, contemporary, experimental laboratory studies were considered too far removed from evolutionary concerns and naturalistic behavior to be worth including – the archetypical "monkey" of comparative psychology had little or no relevance to the theme of cercopithecid diversity.

Socioendocrinology

In the present volume, there are two contributions (see Chapters 10 and 11) representing the progress that has been made towards bridging the conceptual and theoretical gap between laboratory and field. Pat Whitten (Chapter 10) reviews the developing interface between naturalistic and experimental studies of behavior–endocrine interactions in monkey behavior. Since 1970, socioecology (the study of population- or species-specific adaptations of organisms in their naturalistic setting) has evolved into behavioral ecology, which incorporates some of the insights derived from sociobiological theory, and is less content to treat species adaptation as a self-explanatory consequence of evolution. Meanwhile, behavioral endocrinology, which traditionally treated the responses of organisms to experimental manipulation as though they were universal, or at least independent of the animal's individual and phyletic history, has given rise to socioendocrinology, which interprets the interaction between the behavior and endocrine status of individuals within their social context. Whitten points out that these developments imply a convergence in methods and outlook, and looks forward to their further integration into evolutionary socioendocrinology, which will seek to understand the evolutionary origins of species- or population-specific differences, as well as individual variation in behavior–endocrine systems. Such a synthesis must rest on knowledge of how hormones work in individuals (and recognition that endocrine–behavioral relationships are likely to be both two-way and non-linear); how the individual's genome is involved in the development of its behavioral phenotype; and how individual physiology and behaviors give rise to the emergent features of a society, such as its group structure, philopatry patterns, and dominance hierarchy.

Similar themes emerge in Fred Bercovitch's (Chapter 11) consideration of the mechanisms and timing of reproductive maturation, and its inter- and intrapopulational correlates in cercopithecine monkeys. Though necessarily focused on baboons, macaques, and vervets, his review of the literature shows that interaction between rates and patterns of reproductive maturation on the one hand, and nutritional, social, and ecological factors on the other is highly complex and variable from species to species and population to population. Simplified models are poor predictors of behavior in group-living monkeys, since the choices open to individuals are limited by their social status and context. Some studies find dominant individuals of both sexes to grow faster and to reproduce earlier than their peers, but this is by no means universal. It seems that the advantage conferred by a dominant position in society does not lie in faster growth or earlier maturation *per se*, but rather in the freedom to fine-tune one's reproductive strategy in these respects. The relative freedom from social constraint enjoyed by dominant animals may provide them with a margin of time and energy that may be allocated tactically to rapid growth, or to reproductive maturation, or invested in social interaction. It any event, the complexity and subtlety of these interactions suggest that bivariate experimental designs, however appealing, are always likely to give conflicting and ambiguous results. Ultimately, interpretations will have to involve mechanisms at each of the levels of an integrated evolutionary socioendocrinology. As Whitten and Berkovitch both argue, the goal should be a synthetic behavioral primatology, in which levels of explanation, and effects working on different timescales (immediate individual reactions; individual, ontogenetic development; emergent societal features; evolving species characteristics) are seen as interacting and mutually dependent, but also kept conceptually distinct.

The emphasis upon proximal mechanisms of physiological and genetic causation evident in both Whitten's and Berkovitch's reviews, recalls the work of Solly Zuckerman, whose primary subjects were cercopithecid monkeys. For many reasons, Zuckerman's views on the causes of primate sociality were almost universally discounted, even ridiculed, in the 1960s and early 1970s. This was a time when adaptation- and ecology-based ethology provided the predominant explanatory framework for naturalistic animal behavior (Crook, 1970; Kummer, 1971). The field-based and adaptation-oriented studies of baboons by DeVore (DeVore and Hall, 1965; Hall and DeVore, 1965) and Altmann and Altmann (1970) seemed to have shown that Zuckerman's (1932) observations of hamadryas baboons, the whole basis for his theory of causation of primate society, were quite wrong, the fatally flawed product of a hopelessly un-natural situation in the London Zoo. Also downplayed was Zuckerman's insistence that explanations lacking realistic proximal causation, rooted in established neurological and endocrinological mechanisms, had little more than metaphorical value. What could not be fully appreciated either by Zuckerman or his detractors was the extent of the inter-specific and interpopulational diversity in cercopithecid behavior that must be encompassed by any comprehensive explanatory scheme. As Whitten and Berkovitch show, there is ample scope for studying variation in a family that includes *Mandrillus*, with its unique hyper-polygyny and male "fatting"; cercopithecins with tight female kingroups and peripheral males; at least two variants of baboon social organization; and the several different social styles seen in species of *Macaca*. With so much still to do, premature generalization is a temptation to be avoided.

Social behavior and ecology

Ironically, a step towards the rehabilitation of Zuckerman's observations, and a demonstration of previously under-appreciated behavioral diversity, appeared in OWM I. Kummer (1971; Kummer *et al.*, 1970), reported that wild hamadryas baboons, like Zuckerman's Monkey Hill animals, lived in a society based upon male mate-jealousy, quite different from that of yellow and anubis baboons of Kenya. Kummer also reported the existence of a zone of natural hybridization between hamadryas and anubis baboons in central Ethiopia. The existence of this natural hybrid zone sparked a discussion of the nature of species, both in baboons and in general, that remained unresolved at the end of the conference, with the authors of the taxomomic summary (Colin Groves and Dick Thorington) each favoring a different solution and nomenclature. The debate on the nature of species shows no sign of abating, and, indeed, may be intrinsically insoluble (Jolly, 1993), but continues to generate useful thoughts on the organization of nature.

Investigation of Kummer's hybrid zone also continues, and is the subject of the contribution by Jane Phillips-Conroy and her collaborators (Chapter 12). Phillips-Conroy *et al.*, document and discuss rates of dental attrition in wild baboons, and illustrate the dependence of such ontogenetic processes upon habitat features at a very local, population-specific level – a lesson that should be taken to heart by those using comparable data to interpret dietary habits among whole species or even genera of extinct primates. The fact that this study is based upon tooth-casts taken serially from wild animals exemplifies another way that laboratory and field techniques have been integrated since 1970. At that time, almost all primatologists, unlike other zoologists, strictly eschewed capture, marking, or other interaction with their subject animals, preferring a non-participant observer style. Kummer's work constituted a partial exception, combining field experiments, some on temporary captives, with traditional observation. In the further development of Kummer's project, Phillips-Conroy, Jolly, and their colleagues have developed a style of fieldwork in which hands-on methods are even more prominent; target populations of baboons (and grivet monkeys) are regularly trapped and re-trapped, as well as being subjected to naturalistic, non-intrusive observation before and after the trapping seasons. Similar methods have been gradually adopted by other fieldworkers, and are now almost routinely incorporated into programs of field study, even of highly arboreal species. Using materials that can be collected in such "hands-on" procedures, we are now able to base observational studies much more firmly upon a foundation of knowledge of the age, nutritional condition, health, population structure, and even the individual kinship, of the subject animals.

An obvious application of genetic information is in determining degrees of relatedness among individual animals interacting within a social system. Such data are needed to test propositions derived from a sociobiological explanatory framework, which emphasises the notion of natural selection as a consequence of differential individual reproduction, rather than a cause of species adaptation. The development of this theoretical orientation by Hamilton (1964), Wilson (1975), Trivers (1985), and their many followers, and its almost universal adoption as a working paradigm, undoubtedly represents the most profound change in behavioral interpretation since 1970. Moreover, cercopithecid monkeys are diurnal, live in perennial social groups, and communicate largely by sensory channels shared by humans, all of which characteristics make them ideal subjects for the kind of detailed study needed to test sociobiological predictions.

The contribution by Lynn Fairbanks (Chapter 13) belongs within this tradition. It is a review designed to empirically test the prediction, drawn from parental investment theory, that mothers will continue to invest time and effort in their juvenile and adult offspring, to the extent that such behavior does not compromise the fitness of the mother herself. Necessarily based very largely upon macaques, baboons, and vervet monkeys, often in captive or semi-natural, provisioned, populations, the study finds that the behavior of mothers fits the predictions of "sociobiological rationality" quite well. The presence of a mother, especially a high-ranking one, is certainly beneficial, and the benefit extends to the next generation; a grandmother's presence benefits her daughters' offspring. But mothers invest far more in their infant offspring (to whom the care is far more vital) than in their juvenile and adult offspring, for whom peers – especially related peers – can be a major source of support. All this makes excellent sense, and provides further confirmation of the explanatory power of the sociobiological framework.

Along with the many insights that have derived from the sociobiological interpretation of primate social behavior, however, has come the risk of a new kind of inadvertent anthropomorphism, as Irwin Bernstein points out in his critical review (Chapter 14). Such interpretations are often expressed in a terminology drawn from human behavior, used as a kind of metaphorical shorthand. An over-literal interpretation of such language may lead to unwarranted assumptions. In particular, they may lead the unwary to assume an unproven level of sophistication in intentionality and cognitive processes on the part of the non-human actors, and obscure real and evolutionarily interesting differences in cognitive ability among non-human primates. The apparently limited ability of even the "smartest" of cercopithecids to learn new and potentially useful motor patterns by imitation from conspecifics, which apparently distinguishes them sharply from great apes and humans, is a case in point.

There is still much to be learned from empirical research of a more traditional kind on naturalistic behavior in the wild. Two chapters on African colobines lie within this methodological and theoretical framework, yet nothing comparable to them could have been written at the time of OWM I, because both rely upon data painstakingly gathered over many years from ecologically diverse sites, and treated comparatively. Tom Struhsaker (Chapter 15) examines patterns of intra- and inter-specific association in

African cercopithecoids, focusing on communities that include a population of the "red" colobus (*Colobus badius*, s.l.). His conclusion is that the threat of predation, especially by large forest eagles, is the strongest predictor of polyspecific association in arboreal monkeys. However, considerable variation is introduced into the overall picture by more diverse and localized habitat factors that account for apparent exceptions, and patterns in areas where eagles pose a lesser threat. If there is a general conclusion to be drawn, it seems to be that an emergent feature of social organization such as group size and composition, the product of a long and multifactorial chain of social and ecological causation, is unlikely to be related in a simple way to any single habitat variable.

John Oates, Carolyn Bocian, and Carl Terranova (Chapter 16) also draw upon many years of fieldwork, and a rich theoretical background, as well as the relatively sophisticated tools now available for digitally-based sound analysis, in their discussion of differences in the acoustic structure of the male loud calls of various populations within the black and white *Colobus* group. Their interpretation demonstrates not only that species-groups within the cluster can be distinguished by the properties of their loud calls, but that these properties can be plausibly related to features of socioecology that differ from species to species. Finally, the behavioral complexes of which the loud calls are a component can be ordered along a polarity gradient of increasing specialization, providing a hypothesis of population history that demands testing against genetic data.

In a contribution that is a model of meticulous analysis of painstakingly gathered data, Marina Cords (Chapter 17) uses multi-year observations of *Cercopithecus mitis* in a Kenyan forest, to examine relationships of dominance among the females of a social group. For female blue monkeys, dominance is apparently a low cost, low return proposition; the subjects showed very low rates of agonisic interaction, and rank seemed to confer no measurable benefit in terms of reproduction or access to resources. Yet when they occurred, agonistic interactions indicated that the rank order among individuals remained consistent over long periods of time. The obvious inference is that the animals were retaining information that was rarely reinforced or translated into overt behavior, yet was stable and presumably important enough to be unambiguously expressed on the rare occasions that it did emerge. Such observations should inspire humility in the primatologist, reminding us that though all we can know about animal motivation comes from observation of behavior, it does not follow that observing behavior can ever tell us all that is going on in the heads of our primate subjects. Is the ability to retain a mental map of rarely expressed social relationships a mere biproduct of general cognitive ability? Or, does it have value in its own right, but only in rare circumstances, as, for example, when a group is severely stressed by a climatically-induced food-crop failure, or when it is dividing to exploit an expanding habitat?

The question of the evolutionary importance of rare behaviors arises also in the contribution by Dan Gebo and Colin Chapman (Chapter 18), which summarizes data documenting the locomotor behavior of five similarly-sized, sympatric, monkey species in an African forest. The work on simulated predation, using eagle calls, raises the question of how we should weigh the selective importance of a rarely-used but crucial locomotor pattern – such as leaping away from a predator – against a gait that is used in everyday situations, and therefore figures large in time-based locomotor profiles. The overall finding of the study – that associations between body size, substrate use, and locomotor profile, previously established in platyrrhines, are reversed in their sample – is a reminder that plausible interpretations need not be universal to be valid.

In the 1970 volume, Colin Groves called the Asian colobines "forgotten." Though the fieldwork that has appeared since that date has deprived them of the title, it is still the case that the colobines, and especially those of tropical Asia, are less well known than the cercopithecines. Because most are rare and localized in distribution, and because all are delicate and unsuited to investigation in captivity, we are unlikely ever to know the physiology, behavior, population structure, and developmental biology of the colobines in the detail that is now available for the best-known cercopithecine species. With a few exceptions (especially *S. entellus*), most colobine species are known only from one or two, purely observational, studies conducted at a few locations. Nevertheless, these studies add up to a substantial corpus of fine work, almost all of which has appeared since 1970. Yeager and Kool (Chapter 19) review this material in their synthetic treatment of naturalistic studies. Their approach is socioecological, in Whitten's sense, bringing to bear a conceptual framework derived from both sociobiology and ecology. They see social behavior as emerging from individual strategies for survival and reproduction played out upon a stage delimited by each species' habitat and physiological adaptations. Within the overall uniformity of the subfamily – distinguished especially by its common heritage of the specialized, sacculated stomach capable of digesting cellulose-rich foods such as mature leaves – there exists considerable ecological and adaptive diversity. Nevertheless, some recurrent themes do distinguish colobines from the more familiar cercopithecines; notably, a tendency towards both male and female dispersal from the natal group, and the predominance of one-male group social organization. Yeager and Kool advance reasonable, though necessarily *post hoc*, socioecological explanations for these phenomena, which should provide ample hypotheses for future fieldworkers to test.

Conservation

Whether there will remain any subjects for such studies is another matter, which brings us finally to an area of cercopithecid studies in which there is little if any progress to report. Neither this book nor its predecessor explicitly examines issues of the conservation of cercopithecid species and their habitats, but habitat destruction and declining populations are a subtext in all the naturalistic studies of extant populations. The melancholy comparison between today's situation and that of three decades ago is inescapable. In 1970, the immediate issue for cercopithecid conservation was the decimation of some species of macaques, baboons, and vervets, consumed in enormous numbers by the biomedical industry. As some skeptics had predicted (Jolly, 1966), the biomedical establishment did nothing to limit its profligacy until wild monkey populations collapsed, or threatened to do so, whereupon captive breeding programs were developed. Unfortunately, with few exceptions, these costly facilities have been built in the consumer nations, and with little or no benefit either to the wild monkey populations or to their countries of origin. Indeed, many semi-commensal populations of common species such as baboons and many macaques are probably worse off as a consequence of total bans on primate export, which have converted them from potential assets to unmitigated pests. In spite of the carnage wrought by biomedicine, however, it was still possible 30 years ago to imagine large populations of monkeys in relatively undisturbed tropical forest habitats. As Urs Rahm (1970) illustrates, ranges were still being mapped, and as late as 1988, species and subspecies unknown to science were being discovered. Logging and clearance for agriculture were a threat, but plenty of equatorial habitat remained, and newborn international movements for conservation and human population control appeared fiscally strong and politically well-supported.

It is sad to relate how far reality has fallen short of our hopes and expectations, especially during the 1990s. Most localized populations of forest cercopithecids in West Africa have now been exterminated, with whole species reduced to tiny remnants, and those of highland East Africa and much of island southeast Asia are in little better shape. In spite of declared hunting bans and National Park and forest reserve boundaries drawn on maps, the inexorable destruction of monkey populations and their habitats continues. As a result of substantial sums raised and spent on research by international conservation organizations, we know more than ever about the habitat requirements of endangered monkey species. Yet the resulting documents, often styled, with orwellian irony, "action plans", seem ever less likely to be implemented. In part, this is because of the infinitely greater resources available to those prepared to subvert conservation efforts for profit. The romantic distinction between "international" or "commercial" and "indigenous" or "traditional" exploiters of wildlife has become meaningless; all are now part of the same worldwide economic system; only the methods of destructive exploitation differ. The former have the benefit of technological sophistication and political influence; the latter, the advantage of local knowledge, patience, and sheer weight of numbers. With the development of new economic power bases, the pressure that conservationally-minded consumers could once hope to apply to commercial exploiters of tropical habitats has lost much of its force. For every exploiter who is forced to withdraw, there are two or more ready to step in. As long as indigenous governments lack the resources and the will to enforce the conservation laws already on their books, there is little hope that this situation will improve. The handful of trained and dedicated conservationists in source countries who are struggling against the tide of corruption, indifference, and ignorance stand little chance, and the conservation movement itself has constantly to resist pressure to allow its funds and efforts to be subverted along the politically expedient path of "community development", while sidestepping the fundamental problem of human population growth and consumption.

References

Altmann, S.A. & Altmann, I. (1970). Baboon ecology. *Bibl. primatol.* **12**, 1–220.

Anderson, D. (1993). A method for recognizing morphological stasis. In *Morphological Change in Quaternary Mammals of North America*, ed. R.A. Martin & A.D. Barnosky, pp. 13–23. Cambridge: Cambridge University Press.

Avise, J.C. (1994). *Molecular Markers, Natural History and Evolution.* New York: Chapman & Hall.

Barnicot, N.A. & Wade, P.T. (1970). Protein structure and the systematics of Old World monkeys. In *Old World Monkeys: Evolution, Systematics and Behavior*, ed. J.R. Napier & P.H. Napier, pp, 227–60. New York: Academic Press.

Begun, D.R., Ward, C.V. & Rose, M.D. (1997). *Function, Phylogeny, and Fossils: Miocene Hominoid Evolution.* New York: Plenum Press.

Benton, M.J. & Storrs, G.W. (1996). Diversity in the past: Comparing cladistic phylogenies and stratigraphy. In *Aspects of the Genesis and Maintenance of Biological Diversity*, ed. M.E. Hochberg, J. Clobert & R. Barbault, pp. 19–40. Oxford: Oxford University Press.

Birchette, M.G. (1982). *The Postcranial Skeleton of* Paracolobus chemeroni. Ph.D. Dissertation, Harvard University.

Booth, A.H. (1958). The Niger, the Volta and the Dahomey gap as geographical barriers. *Evolution* **12**, 48–62.

Calder, W.A., III (1984). *Size, Function and Life History*. Cambridge, MA: Harvard University Press.

Crompton, A.W. & Hiiemae, K. (1970). Molar occlusion and mandibular movements during occlusion in the American opossum. *Zool. J. Linn. Soc.* **49**, 21–47.

Crook, J.H. (1970). The socio-ecology of primates. In *Social Behavior in Birds and Mammals*, ed. J.H. Crook, pp. 103–66. New York: Academic Press.

Damuth, J. & MacFadden, B.J. (1990). *Body Size in Mammalian Paleobiology: Estimation and Biological Implications*. Cambridge: Cambridge University Press.

Darwin, C.R. (1859). *On the Origin of Species*. London: John Murray.

de Beer, G.R. (1937). *The Development of the Vertebrate Skull*. Oxford: Oxford University Press.

Delson, E. (1975). Evolutionary history of the Cercopithecidae. In *Approaches to Primate Paleobiology*, ed. F.S. Szalay, pp. 167–217. Basel: Karger.

Delson, F. & Andrews, P. (1975). Evolution and interrelationships of the Catarrhine primates. In *Phylogeny of the Primates: A Multidisciplinary Approach*, ed. W.P. Luckett & F.S. Szalay, pp. 405–46. New York: Plenum.

DeVore, I. & Hall, K.R.L. (1965). Baboon ecology. In *Primate Behavior: Field Studies of Monkeys and Apes*, ed. I.DeVore, pp. 20–52. New York: Holt, Rinehart & Winston.

Driesch, H. (1908). *The Science and Philosophy of the Organism*. London: Adam & Charles Black.

Dullemeijer, P. (1989). On the concept of integration in animal morphology. In *Fortschritte der zoologie*, vol 35: *Trends in Vertebrate Morphology*, ed. H. Splechtna & H. Hilgers, pp. 3–18. Stuttgart: Gustav Fischer Verlag.

Duncker, H.-R.C. (1985). The present situation of morphology, and its importance for biological and medical sciences. In *Fortschritte der zoologie, vol. 30: Functional Morphology in Vertebrates. Proceeding of the First International Symposium on Vertebrate Morphology, Giessen, 1983*, ed. H.-R. Duncker and G. Fleischer, pp. 3–10. Stuttgart: Gustav Fischer Verlag.

Eck, G.G. & Jablonski, N.G. (1987). The skull of *Theropithecus brumpti* compared with those of other species of the genus *Theropithecus*. In *Les faunes Plio-Pleistocènes de la Vallée de l'Omo (Ethiopie)*, Tome 3, ed. Y. Coppens & F.C. Howell, pp. 11–122. Paris: CNRS.

Eldredge, N. (1972). Systematics and evolution of *Phacops rana* (Green 1832) and *Phacops iowensis* Delo, 1935 (Trilobita) from the Middle Devonian Of North America. *Bull. Amer. Mus. Nat. Hist.* **147**, 45–114.

Eldredge, N. & Gould, S.J. (1972). Punctuated equilibria: an alternative to phyletic gradualism. In *Models in Paleobiology*, ed. T.J.M. Schopf, pp. 82–115. San Francisco: Freeman

Elliot, D.G. (1913). *A Review of the Primates*. Monograph No.1, vols. 1, 2 & 3. New York: American Museum of Natural History.

Emerson, S.B. & Bramble, D.M. (1993). Scaling, allometry, and skull design. In *The Skull*, Vol. 3: *Functional and Evolutionary Mechanisms*, ed. J. Hanken & B.K. Hall, pp. 384–421. Chicago: University of Chicago.

Endler, J.A. (1977). *Geographic Variation, Speciation and Clines*. Princeton: Princeton University Press.

Fleagle, J.G. & McGraw, W.S. (1999). Skeletal and dental morphology supports diphyletic origin of baboons and mandrills. *Proc. Natl. Acad. Sci. USA* **96**, 1157–61.

Fogleman, J.C., Danielson, P.B. & MacIntyre, R.J. (1998). The molecular basis of adaptation in *Drosophila*: The role of cytochrome P450s. In *Evolutionary Biology*, Vol. 30, ed. M.K. Hecht, R.J. MacIntyre & M.T. Clegg, pp. 15–77. New York: Plenum Press.

Forey, P.L. (1992). Fossils and cladistic analysis. In *Cladistics: A Practical Course in Systematics*, ed. P.L. Forey, C.J. Humphries, I.J. Kitching, R.W. Scotland, D.J. Siebert & D.M. Williams, pp. 124–36. Oxford: Clarendon Press.

Gans, C. (1989). Morphology, today and tomorrow. In *Fortschritte der Zoologie*, vol. 35*: Trends in Vertebrate Morphology*, ed. H. Splechtna & H. Hilgers, pp. 631–37. Stuttgart: Gustav Fischer Verlag.

Gingerich, P.D. (1993). Rates of evolution in Plio-Pleistocene mammals: Six case studies. In *Morphological Change in Quaternary Mammals of North America*, ed. R.A. Martin & A.D. Barnosky, pp. 84–106. Cambridge: Cambridge University Press.

Groves, C.P. (1989). *A Theory of Human and Primate Evolution*. Oxford: Clarendon Press.

Hall, K.R.L. & DeVore, I. (1965). Baboon social behavior. In *Primate Behavior: Field Studies of Monkeys and Apes*, ed. I. Devore, pp. 20–52. New York: Holt, Rinehart & Winston.

Hamilton, A.C. (1988). Guenon evolution and forest history. In *A Primate Radiation: Evolutionary Biology of the African Guenons*, ed. A. Gautier-Hion, F. Bourliere & J.-P. Gautier, pp. 13–34. Cambridge: Cambridge University Press.

Hamilton, W.D. (1964). The genetical theory of social behaviour. I and II. *J. Theor. Biol.* **7**, 1–52.

Harris, E.E. & Disotell, T.R. (1998). Nuclear gene trees and the phylogenetic relationships of the mangabeys (Primates: Papionini). *Mol. Biol. Evol.* **15**, 892–900.

Harris, H. (1966). Enzyme polymorphisms in man. *Proc. Roy. Soc. Lond.* B **164**, 298–310.

Hennig, W. (1950). *Grundzuge einer Theorie der phylogenetischen Systematik*. Berlin: Deutscher Zentralverlag.

Hennig, W. (1965). Phylogenetic systematics. *Ann. rev. Ent.* **10**, 97–116.

Hennig, W. (1966). *Phylogenetic Systematics* Urbana: University of Illinois Press.

Houck, M.A., Gauthier, J.A. & Strauss, R.E. (1990). Allometric scaling in the earliest fossil bird, *Archaeopteryx lithographica*. *Science* **247**, 195–8.

Hudson, R.R. (1996). Molecular population genetics of adaptation. In *Adaptation*, ed. M.R. Rose & G.V. Lauder, pp. 291–309. San Diego: Academic Press.

Jolly, C.J. (1966). Introduction to the Cercopithecoidea, with notes on their use as laboratory animals. *Symp. Zool. Soc. London* **17**, 427–57.

Jolly, C.J. (1972). The classification and natural history of *Theropithecus (Simopithecus)* (Andrews, 1916), Baboons of the African Plio-Pleistocene. *Bull. Brit. Mus. Nat. Hist. (Geol.)* **22**, 1–123.

Jolly, C.J. (1993). Species, subspecies, and baboon systematics. In *Species, Species Concepts, and Primate Evolution*, ed. W.H. Kimbel & L.B. Martin, pp. 67–107. New York: Plenum Press.

Jukes, T.H. (1966). *Molecules and Evolution*. New York: Columbia University Press.

Kimura, M. (1968). Genetic variability maintained in a finite population due to mutational production of neutral and nearly neutral isoalleles. *Genet. Res.* **11**, 247–69.

Kimura, M. & Ohta, T. (1971). *Theoretical Aspects of Population Genetics.* Princeton,NJ: Princeton University Press.

King, J.L. & Jukes, T.H. (1969). Non-Darwinian evolution. *Science* **164**, 788–98.

Krishtalka, L. (1993). Anagenetic angst: Species boundaries in Eocene primates. In *Species, Species Concepts, and Primate Evolution*, ed. W.H. Kimbel & L.B. Martin, pp. 331–44. New York: Plenum Press.

Kummer, H. (1971). *Primate Societies: Group Techniques in Ecological Adaptation*. Chicago: Aldine-Atherton.

Kummer, H., Goetz, W. & Angst, W. (1970). Cross-species modifications of social behavior in baboons. In *Old World Monkeys: Evolution, Systematics, and Behavior*, ed. J.R. Napier & P.H. Napier, pp. 351–63. New York: Academic Press.

Leakey, M. (1993). Evolution *of Theropithecus* in the Turkana Basin. In *Theropithecus: The Rise and Fall of a Primate Genus*, ed. N.G. Jablonski, pp. 85–123. Cambridge: Cambridge University Press.

Lewontin, R.C. & Hubby, J.L. (1966). A molecular approach to the study of genic heterozygosity in natural populations. II Amount of variation and degree of heterozygosity in natural populations of *Drosophila pseudoobscura. Genetics* **54**, 595–609.

Lewontin, R.C. (1974). *The Genetic Basis of Evolutionary Change.* New York: Columbia University Press.

Martin, R.D. (1990). *Primate Origins and Evolution: A Phylogenetic Perspective.* Princeton: Princeton University Press.

Mayr, E. (1942), *Systematics and the Origin Of Species.* New York: Columbia University.

Mayr, E. (1963). *Animal Species and Evolution.* Cambridge, MA: Harvard University.

Mayr, E. (1969). *Principles of Systematic Zoology*. New York: McGraw-Hill.

Mayr, E. (1974). Cladistic analysis or cladistic classification? *Z.Zool.Syst.Evolut.-forsch.* **12**, 94–128

Napier, J.R. (1970). Paleoecology and catarrhine evolution. In *Old World Monkeys: Evolution, Systematics, and Behavior*, ed. J.R. Napier & P.H. Napier, pp. 53–95. New York: Academic Press.

Napier, J.R. & Napier, P.H. (1970). *Old World Monkeys: Evolution, Systematics, and Behavior*, New York: Academic Press.

Nyhart, L.K. (1995). *Biology takes Form: Animal Morphology and the German Universities, 1800–1900.* Chicago: University of Chicago Press.

Panchen, A.L. (1992). *Classification, Evolution and the Nature of Biology*. Cambridge: Cambridge University Press.

Patterson, C. (1981). Significance of fossils in determining evolutionary relationships. *Ann. Rev. Ecol. Syst.* **12**, 195–223.

Patterson, C. (1987). Introduction. In *Molecules and Morphology in Evolution: Conflict or Compromise*, ed. C. Patterson, pp. 1–22. Cambridge: Cambridge University Press.

Peters, R.H. (1983). *The Ecological Implications of Body Size*. Cambridge: Cambridge University Press.

Rahm, U.H. (1970). Ecology, zoogeography, and systematics of some African forest monkeys. In *Old World Monkeys: Evolution, Systematics, and Behavior*, ed J.R. Napier & P.H. Napier, pp. 589–626. New York: Academic Press.

Riedl, R. (1989). Opening address and introduction. In *Fortschritte der Zoologie*, vol. 35: *Trends in Vertebrate Morphology*, ed. H. Splechtna & H. Hilgers, pp. vii–xvi. Stuttgart: Gustav. Fischer Verlag.
Rosen, D.E., Forey, P.L., Gardiner, B.G. & Patterson, C. (1981). Lungfishes, tetrapods, paleontology and plesiomorphy. *Bull. Am. Mus. Nat. Hist.* **167**, 159–267.
Sarich, V.M. (1968). Hominid origins: An immunological view. In *Perspectives on Human Evolution*, ed. S.L. Washburn & P.C. Jay, pp. 94–121. New York: Holt, Rinehart & Winston.
Sarich, V.M. (1970a). Molecular data in systematics. In *Old World Monkeys: Evolution, Systematics, and Behavior*, ed. J.P. Napier & P.H. Napier, pp. 17–24. New York: Academic Press.
Sarich, V.M. (1970b). Primate systematics with special reference to Old World monkeys. In *Old World Monkeys: Evolution, Systematics, and Behavior*, ed. J.R. Napier & P.H. Napier, pp. 175–226. New York: Academic Press.
Sarich, V.M. (1971). A molecular approach to the question of human origins. In *Background for Man: Readings in Physical Anthropology*, ed. P. Dolhinow & V. Sarich, pp. 60–81. Boston: Little, Brown & Co.
Schultz, A.H. (1970). The comparative uniformity of the Cercopithecoidea. In *Old World Monkeys: Evolution, Systematics, and Behavior*, ed. J.R. Napier & P.H. Napier, pp. 39– 51. New York: Academic Press.
Simons, E.L. (1970). The deployment and history of Old World Monkeys (Cercopithecidae, Primates). In *Old World Monkeys: Evolution, Systematics, and Behavior*, ed. J.R. Napier & P.H. Napier, pp. 97–137. New York: Academic Press
Simons, E.L. (1972). *Primate Evolution: An Introduction to Man's Place in Nature.* New York: Macmillan.
Simons, E.L. (1974). *Parapithecus grangeri* (Parapithecidae, Old World higher primates): New species from the Oligocene of Egypt and the initial differentiation of Cercopithecoidea. *Postilla* **166**, 1–12.
Simons, E.L. & Pilbeam, D.R. (1965). Preliminary revision of the Dryopithecinae (Pongidae, Anthropoidea). *Folia Primatol.* **3**, 81–152.
Simpson, G.G. (1961). *Principles of Animal Taxonomy*. New York: Columbia University.
Simpson, G.G. (1963). The meaning of taxonomic statements. In *Classification and Human Evolution*, ed. S.L. Washburn, pp. 1–31. Chicago: Aldine.
Sokal, R.R. & Sneath, P.H.A. (1963). *Principles of Numerical Taxonomy*. San Francisco: W.H. Freeman.
Strauss, E. (1999). Can mitochondrial clocks keep time? *Science* **283**, 1435–8.
Szalay, F.S. & Delson, E. (1979). *Evolutionary History of the Primates.* New York: Academic Press.
Thorington, R.W., Jr. (1970). The interpretation of data in systematics. In *Old World Monkeys: Evolution, Systematics, and Behavior*, ed. J.R. Napier & P.H. Napier, pp. 3–15. New York: Academic Press.
Trivers, R. (1985). *Social Evolution*. Menlo Park, CA: Benjamin/Cummings.
von Koenigswald, G.H.R. (1969). Miocene Cercopithecoidea and Oreopithecoidea from the Miocene of East Africa. In *Fossil Vertebrates of Africa*, Vol. 1, ed. L.S.B. Leakey, pp. 39–55. London: Academic Press.
Washburn, S.L. (1963). *Classification and Human Evolution*. Chicago: Aldine.
Washburn, S.L. & DeVore, I. (1961). The social life of baboons. *Sci. Am.* **204**, 62–71.
Whitehead, P.F. (1992). Anatomy of the forelimb in *Theropithecus oswaldi. Fifth*

North American Paleontological Convention, Special Publication No. **6**. p. 311.

Whitehead, P.F. & Larson, S.G. (1994). Shoulder motion during quadrupedal walking in *Cercopithecus aethiops*: Integration of cineradiographic and electromyographic data. *J. hum. Evol.* **26**, 525–44.

Wilson, E.O. (1975). *Sociobiology: The New Synthesis*. Cambridge, MA: Belknap.

Wright, S. (1931). Evolution in Mendelian populations. *Genetics* **16**, 97–159.

Zuckerman, S. (1932). *The Social Life Of Monkeys and Apes.* London: Kegan Paul, Trench & Trubner.

2

Molecular systematics of the Cercopithecidae

TODD R. DISOTELL

Introduction

George H.F. Nuttall (1904:4) wrote: "The persistence of the chemical blood-relationship between the various groups of animals serves to carry us back into geological times, and I believe we have but begun the work along these lines, and that it will lead to valuable results in the study of various problems of evolution." While enormous strides have been made in the field of molecular genetics, Nuttall's statement is still true. Our knowledge of the systematics of Old World monkeys, gleaned from both morphological and molecular approaches, is still poorer than many would admit.

In the 1960s, Nuttall's immunological approach was enhanced and used extensively by anthropologists and biochemists studying primates (Goodman, 1961, 1963; Sarich and Wilson, 1966, 1967). Karyological research by, for example, Chiarelli (1966) and Dutrillaux (1979) has also played a major role in primate evolutionary studies. The recognition of the potential of a "molecular clock" by Zuckerkandl and Pauling (1962) led to studies of protein sequences (Goodman *et al.*, 1987). Research on primate DNA itself, and not merely its products, began in the late 1960s and early 1970s (Kohne *et al.*, 1972; Benveniste *et al.*,1976), and was followed by restriction enzyme analyses of relationships among and between primate species and populations (Templeton, 1983; Melnick *et al.*, 1993). The mapping and sequencing of portions of primate mitochondrial and nuclear genomes has also contributed greatly to primate systematics and population genetics (reviewed in Honeycutt and Wheeler, 1989; Koop *et al.*, 1989; Melnick and Hoelzer, 1993). Eight complete catarrhine mitochondrial genomes have been sequenced to date (reviewed in Arnason *et al.*, 1998). Automated DNA sequencing and genotyping techniques using different colored fluorescent dyes have greatly enhanced our ability to collect data on an unprecedented scale (Trainor, 1990).

Analytical methods, particularly the use of computer-based algorithms, have advanced even more than molecular technology. Programs such as PAUP (Swofford 1996), HENNIG86 (Farris, 1988), MacClade (Maddison and Maddison, 1992), and PHYLIP (Felsenstein 1993) have become an integral part of all molecular systematics laboratories. Detailed treatments of analytical techniques can be found in Miyamoto and Cracraft (1991) and Swofford *et al.*, (1996).

Studies by Atchley and Fitch (1991) on the phylogeny of inbred mice and Hillis *et al*'s (1992) manipulation of a bacteriophage, have strengthened our confidence in the efficacy of the molecular approach to systematics. These demonstrations provide experimental verification of the efficacy of using molecular character data and analytical techniques in molecular systematics. Such confirmation is unique in the field of systematics, where phylogenetic hypotheses were previously only able to be compared to previously inferred phylogenies (whose accuracy is unknown).

Old World monkeys

Despite their speciosity and geographic diversity, many authors claim that cercopithecoids are not very morphologically diverse (Schultz, 1970; Szalay and Delson, 1979; Groves, 1989; Martin, 1990). Two subfamilies of Old World monkeys are universally recognized – Cercopithecinae and Colobinae. Each subfamily has developed distinctive morphological adaptations related to diet that clearly differentiate them – a specialized digestive tract in colobines and buccal pouches in cercopithecines. Beyond a basic agreement on the defining characteristics and membership of the two subfamilies (Groves, 1989), most other aspects of cercopithecoid relationships are in contention, either amongst morphologists or between morphologists and molecular researchers. Recent discussion of overall cercopithecoid systematics may be found in Szalay and Delson (1979), Napier and Napier (1985), Strasser and Delson (1987), Fleagle (1988), Groves (1989; Chapter 4), Martin (1990) and Disotell (1996).

The Cercopithecinae are the most widely studied group of Old World monkeys from a phylogenetic perspective. The subfamily has been divided into two tribes, Cercopithecini and Papionini (Szalay and Delson, 1979). Other tribes, including Cercocebini and Theropithecini (Jolly, 1966; Hill, 1974), were proposed, but were later combined into the Papionini (Jolly, 1967). Groves (1989) initially divided these monkeys into two subfamilies, the Cercopithecinae and Papioninae, though now he (Groves, Chapter 4) ranks them as tribes. He questions the placement of two genera,

Allenopithecus and *Miopithecus*, in the Cercopithecini and suggests that they may be members of the Papionini or perhaps neither clade (Groves 1989; Chapter 4). There are differences of opinion regarding the number of genera within each of the two tribes with *Allenopithecus, Cercocebus, Cercopithecus, Chlorocebus, Erythrocebus, Lophocebus, Macaca, Mandrillus, Miopithecus, Papio*, and *Theropithecus* all suggested as valid genera by various authors (e.g. Szalay and Delson, 1979; Groves, 1989).

The taxonomy and systematics of the Colobinae are as contentious. Delson (1973, 1994) sees a major split between the Asian and African colobines and has created two subtribes (without higher ranking tribes), the Presbytina, also known as Semnopithecina (Davies and Oates, 1994) and the Colobina. In contrast, Groves (1989; Chapter 4) sees the greatest difference between the Asian genus *Nasalis* (including *Simias*) and the rest of the colobines. He creates a separate subfamily, the Nasalinae, with all other colobines in the Colobinae. Various researchers place the African colobines into up to three genera (*Colobus, Procolobus*, and *Piliocolobus*) and the Asian leaf-monkeys into four to nine genera (including *Kasi, Nasalis, Presbytis, Presbytiscus, Pygathrix, Rhinopithecus, Semnopithecus, Simias, Trachypithecus*).

Cercopithecinae

The major points of disagreement regarding cercopithecine taxonomy and systematics are:

(1) Relationships within the African members of the Papionini, especially the:
  (a) relationship of *Theropithecus* to *Papio* and *Mandrillus*;
  (b) relationship of *Mandrillus* to *Papio*;
  (c) monophyly of mangabeys (*Cercocebus* and *Lophocebus*).
(2) Relationships within *Macaca* and *Papio*, including the:
  (a) relationships among the different common baboon (*Papio*) subspecies;
  (b) monophyly of macaques (*Macaca*);
  (c) interrelationships of the different macaque species groups and species.
(3) The tribal affiliations of *Allenopithecus* and *Miopithecus*.
(4) Taxonomy within the tribe Cercopithecini.
(5) The relationships of the different species groups of *Cercopithecus* (and *Erythrocebus*).

Papionini

The most widely studied group of Old World monkeys, from a genetic perspective, are the Papionini (*Macaca*, *Cercocebus*, *Lophocebus*, *Mandrillus*, *Papio*, and *Theropithecus*). For example, of the approximately 1800 cercopithecoid sequences in GenBank (Benson *et al.*, 1998) as of May 1999 all but 168 sequences are of papionins, with most of the sequences from *Macaca* (1116) and *Papio* (192).

Several reviews of papionin systematics based upon morphology have been published, including Szalay and Delson (1979), Strasser and Delson (1987), and Groves (1989; Chapter 4). Strasser and Delson (1987: 82) analyze a matrix of 37 morphological characters and derive an "apparently most parsimonious taxon phylogeny by inspection, without utilizing numerical approaches" which corresponds closely to the then current consensus phylogeny (Fig. 2.1a). Reanalysis (Disotell, 1992) of their data reveals many equally parsimonious trees (Fig. 2.1c), with almost no resolution within the Papionini (Fig. 2.1b), a result corroborated by Groves's (Chapter 4) analysis of an even larger morphological data set using computer-based parsimony algorithms. These studies demonstrate that phylogenetic analyses, with even a moderate number of taxa, must use computer-based algorithms and searches to find the most parsimonious trees. Maddison (1991: 355) enunciated this position when he stated "in answering an evolutionary question using phylogenetic analysis, the full implications of data are not revealed unless one considers all equally well supported hypotheses".

African papionins

Papionins often are defined by their 42 diploid chromosome complement. Chromosomal banding studies (Dutrillaux, 1979; Dutrillaux *et al.*, 1981, 1982; Stanyon *et al.*, 1988) reveal relatively few chromsomal changes among papionins compared to their sister tribe (Cercopithecini), which display an enomorous number of chromosomal changes, both in banding pattern and number. *Macaca* (except for *M. fascicularis*), *Lophocebus*, *Papio*, and *Theropithecus* are chromosomally indistinguishable, even with high resolution banding (Stanyon *et al.*, 1988). Dutrillaux (1979) first noted that the mangabeys, *Cercocebus* and *Lophocebus*, had major differences between their chromosomes 10 and 12. Later they found that *Cercocebus* and *Mandrillus* share a complex (three-break) rearrangement of chromosome 10 (Dutrillaux *et al.*, 1982), which they dismissed as a homoplasic or parallelly evolved trait. Additional evidence discussed

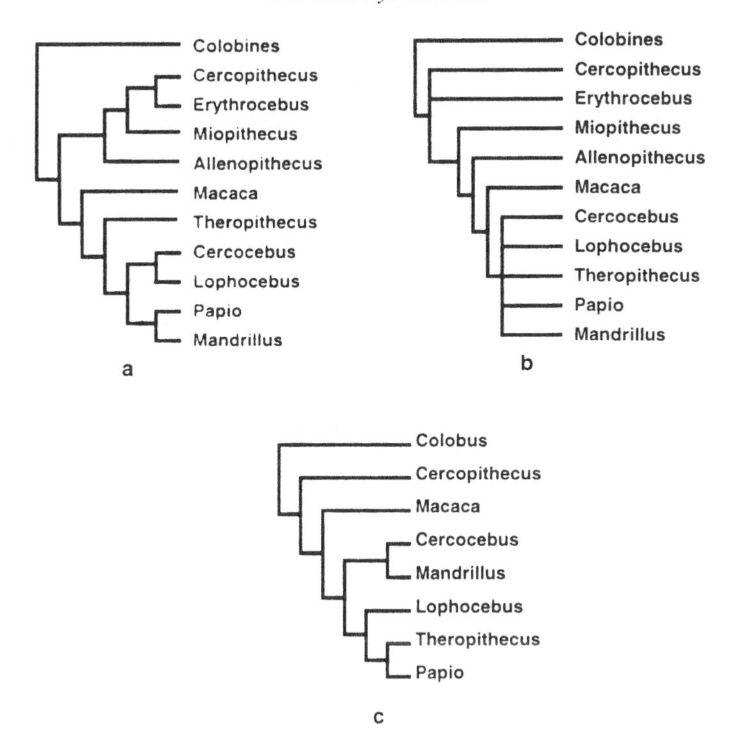

Fig. 2.1. Cladogram of cercopithecine relationships based on morphological characters from Strasser and Delson (1987). (a) Cladogram presented by Strasser and Delson (1987). Reprinted with permission, *J. hum. Evol.*, Academic Press, Harcourt Brace Jovanovich). (b) Strict consensus tree of 21 equally parsimonious trees found using PAUP with equal weighting and unordered character states. A tree with ordered characters only differed in forming an additional trichotomy consisting of *Theropithecus*, *Papio*, and *Mandrillus*. (c) Maximum parsimony tree of mitochondrial COII (cytochrome oxidase subunit II) sequences from Disotell (1992).

below reveals that this may be a synapomorphy of a *Cercocebus* plus *Mandrillus* clade.

Early protein electrophoresis studies of the hemoglobin α- and β-chains and adenylate kinase revealed identical patterns of migration for the isozymes of *Papio* and *Theropithecus*, which differed from those of *Mandrillus* (Barnicot and Wade, 1970). Sarich (1970) examined immunological distances of albumins using the microcomplement fixation technique and inferred four papionin lineages: *Macaca*, *Cercocebus*, *Mandrillus*, and a *Papio* + *Theropithecus* clade. Goodman and Moore (1971) generated additional immunological distances, using immunodiffusion and produced incongruous trees. Their *Cercocebus* (species not mentioned) sample grouped with the cercopithecin clade rather than with the other papionins.

There are several problems with this study. First, Goodman and Moore pooled their samples of *Mandrillus* and *Papio*, following the then current taxonomic practice and the belief that mandrills were most closely related to common baboons. Second, there were problems with the reciprocity of the experiments. Depending upon which anti-serum was used, quite different distances were found. For instance, when anti-*Erythrocebus* serum was used, no divergence between their *Cercocebus* sample and *Erythrocebus* or *Cercopithecus* was detected. When anti-*Papio* (a combined sample of *Papio* and *Mandrillus*) serum was used, their *Cercocebus* sample indeed was closer to the other papionins. Given these problems, the results of this study should probably be set aside.

Dene *et al.* (1976) collected additional immunodiffusion data on papionins, analyzing *Mandrillus* and *Papio* separately, which resulted in *Papio* clustering with *Theropithecus* to the exclusion of *Mandrillus*. However, *Cercocebus* and *Mandrillus* clustered with the two cercopithecins, *Erythrocebus* and *Cercopithecus*. This anomalous result may be related to the poor reciprocity between *Cercocebus* and *Mandrillus* when tested against *Erythrocebus* and *Cercopithecus* anti-sera. An alternative explanation is that of Goodman and Moore (1971), who suggest the mangabey (and mandrill) lineages may retain a more ancestral cercopithecine-like condition, at least as measured by immunodiffusion methods. Such distance-based methods do not allow us to separate ancestral and derived similarities as do character-based methods (Swofford *et al.*, 1996).

Sarich and Cronin (1976) expanded upon Sarich's earlier work on albumin microcomplement fixation studies by including more taxa and results for transferrin. Combining the distances inferred from both proteins, they found a sister-group relationship between *Papio* and *Theropithecus*. Their data also suggested that the genus *Cercocebus* (as traditionally defined) was paraphyletic (defined as a group that does not contain all of the descendants of its most recent common ancestor), with the *albigena*/*aterrimus* group (*Lophocebus*) being more closely related to baboons than to *Cercocebus*. These two conclusions were corroborated by Hewett-Emmett *et al.* (1976) who determined the amino acid sequences of the hemoglobin α- and β-chains for all papionin genera. They found that *Cercocebus* and *Mandrillus* had multiple hemoglobins, and inferred that the single hemoglobins of *Papio*, *Theropithecus*, and *Lophocebus* formed a monophyletic group to the exclusion of all *Cercocebus* and *Mandrillus* hemoglobins. Benveniste and Todaro (1976), using DNA–DNA hybridization, and Gillespie (1977) using fast-repeat high t_m sequences (FHRT), found greater similarity between *Papio* and *Theropithecus* than between

Papio and *Mandrillus* (Disotell, 1994a). A restriction enzyme analysis of a 40-kilobase region of the nuclear genome, containing the 18S and 28S ribosomal RNA sequences, places *Papio* and *Theropithecus* as sister-taxa within the Papionini (Nelkin *et al.*, 1980).

In the early 1990s, the 684 bp mitochondrial cytochrome oxidase subunit II (COII) gene was sequenced in representatives of every papionin genus, including several subspecies of *Papio* (Disotell 1992, 1994; Disotell *et al.*, 1992). Maximum parsimony (Fitch, 1971) and maximum likelihood (Felsenstein, 1981) analyses of COII sequences support a *Papio* plus *Theropithecus* clade and a *Cercocebus* plus *Mandrillus* clade (Fig. 2.1c). This suggests mangabey polyphyly (derivation from two distinct ancestral lineages), because *Lophocebus* is posited as the sister-taxon of a clade that includes gelada and common baboons. These results are corroborated by another mitochondrial sequence study (van der Kuyl *et al.*, 1995a) of a ± 390 bp portion of the 12S rRNA gene in most papionin genera. Although they did not include *Theropithecus* in their analysis, they did infer that the mangabeys are polyphyletic with *Cercocebus*, grouping with *Mandrillus*, and *Lophocebus* grouping with *Papio*. These results were confirmed by analyses of five independent nuclear loci (Harris and Disotell, 1998).

The preponderance of chromosomal and molecular evidence points to mangabey polyphyly, and a *Theropithecus* plus *Papio* clade to the exclusion of *Mandrillus*.This latter grouping has also been discussed by Cronin and Meikle (1979). Groves (1978) proposed, based on morphological data, the resurrection of the genus *Lophocebus* to correspond to the mangabey's paraphyletic status. Groves's most recent analysis (Chapter 4) also finds mangabey polyphyly, with *Cercocebus* grouping with *Mandrillus*, in his morphological analysis. Fleagle and McGraw (1999) recently found additional morphological support for a *Cercocebus–Mandrillus* clade. This group is unexpected in view of the phenetic similarity in overall morphology of the two mangabey groups and their differences compared to mandrills. Indeed, as in the case of mangabeys, *Papio* and *Mandrillus* were often considered congeners (Goodman and Moore, 1971; Delson and Napier, 1976).

Relationships within Papio *and* Macaca

Much less work has been done on papionin systematics below the genus level. Two studies of *Papio* baboon subspecies concur in their findings. Williams-Blangero *et al.* (1990) examined nine polymorphic protein loci, using protein electrophoresis, for 1164 common baboons from all five commonly defined subspecies (*Papio hamadryas hamadryas*, *P.h. anubis*, *P.h.*

ursinus, *P.h. cynocephalus*, *P.h. papio*) housed at the Southwest Foundation for Biomedical Research in Texas, USA. While they did not perform a phylogenetic analysis, for this treatment I analyzed their distance matrix (Williams-Blangero *et al.*, 1990, table V) using a Fitch–Margoliash (Fitch and Margoliash, 1967) or neighbor-joining (Saitou and Nei, 1987) analysis (PHYLIP 3.54c, Felsenstein, 1993). The resulting trees reveal that *P.h. papio* was the most divergent taxon with respect to the other four subspecies, which were not well differentiated. My examination of the mitochondrial COII gene (Disotell, 1992) found a similar result, with *P.h. papio* differentiated from the other subspecies. Both of these results stand in opposition to the more widely accepted view that the largest difference within the common baboons is between the hamadryas (*P.h. hamadryas*) and savannah baboons, which has led to partitioning of common baboons into two species (*P. hamadryas* and *P. cynocephalus*) by authors (e.g. various authors in Smuts *et al.*, 1987). Recently, Newman *et al.* (1999) proposed that *P.h. hamadryas* may be the outgroup to all *Papio* baboons, based on preliminary analyses of a 896 bp region of the mitochondrial genome. These results are, however, tentative and clearly indicate much more data need to be collected to sort out subspecific relationships of baboons.

Groves (1989; Chapter 4) notes that *Macaca* is a troubling genus, with perhaps no single synapomorphic feature uniting the 19 or so species. The most widely recognized partitioning of this speciose genus is that of Fooden (1976, 1980), who divides *Macaca* into four species groups based partially on penile morphology. These groups are the *silenus–sylvanus* group (*M. sylvanus*, *M. silenus*, *M. nemestrina*, Sulawesi species), the *fascicularis* group (*M. fascicularis*, *M. fuscata*, *M. mulatta*, *M. cyclopis*), the *sinica* group (*M. sinica*, *M. assamensis*, *M. radiata*, *M. thibetana*), and the monospecific *arctoides* group (*M. arctoides*). Female reproductive tract characters are interpreted by Fooden (1980) to suggest a closer relationship between the *silenus–sylvanus* and *fascicularis* groups. Additionaly, some researchers view the *arctoides* group as derived from or closely related to the *sinica* group (Delson, 1980; Groves, 1989).

Preliminary phylogenetic analyses of genital characters of papionin genera, based on descriptions from the literature, were found to be congruent with hypotheses derived from molecular data (Disotell, 1992, 1994). This suggests that such characters may be good indicators of phylogenetic history. Jablonski and Peng (1993) carried-out an extensive study of 177 morphological characters, including features of the skeleton, viscera, and pelage of Asian colobines and six species of macaques. They found that the four *fascicularis* group species do not form a monophyletic clade when analyzed cladistically with additive characters. One reason for incompatability, between analyses based on morphology, is best summarized by Pilbeam (1996: 159): "We face another set of problems when explaining incompatible morphological evidence (indeed, this is rarely if ever attempted) because so far we have lacked consistent and biologically sound criteria for deciding which similarities are likely to be homologies and which are not."

Macaques are the most widely studied papionin genus from a genetic perspective (Darga *et al.*, 1975; Cronin *et al.*, 1980; Melnick and Kidd, 1985; Hayasaka *et al.*, 1988; Fooden and Lanyon, 1989). Unfortunately, relatively few studies to date have examined any genetic system for all species of macaques to test Fooden's species groups or to infer relationships among the groups. Cronin and co-workers (1980) examined data for 29 protein loci in 11 macaque species using a combination of phenetic and cladistic methods, with no data for *M. sylvanus* and incomplete data for *M. silenus*. Their phylogenetic hypothesis (Fig. 2.2c) shows a four-way basal multichotomy that does not correspond very closely to Fooden's species groups. For instance, one of the lineages contains *M. mulatta*, *M. assamensis*, and *M. arctoides*, which according to Fooden are members of three different species groups. Members of the *fascicularis* group (the most widely genetically characterized macaques) are split among three of the basal lineages in the phenogram of Cronin *et al.* (1980) (Fig. 2.2c).

Melnick and Kidd (1985) pooled data from several studies to calculate gene frequencies of 9 loci in 10 macaque species to calculate tau genetic distances (Kidd and Cavalli-Sforza, 1974) and construct phylogenetic trees using a least-squares method. Their phenetic tree (Fig. 2.2a) shows all populations of species clustering together and complete concordance with Fooden's species-groups. Here, I have reanalyzed their distance matrix (Melnick and Kidd, 1985, table V) using the neighbor-joining technique (Saitou and Nei, 1987) as implemented in PHYLIP 3.54c (Felsenstein, 1993), resulting in quite a different tree (Fig. 2b). In this tree, *M. mulatta*, *M. fascicularis*, and *M. nemestrina* are paraphyletic, and show little correspondence with Fooden's species groups. Such incongruence, between the results of different analytical techniques of phylogenetic analysis using the same data, weakens our confidence in these uncorroborated hypotheses.

Fooden and Lanyon (1989) tabulated data from 22 polymorphic blood-protein loci and calculated Roger's modified genetic distances (Wright 1978). Using the Fitch–Margoliash (1967) method, they found some correspondence between the morphologically and genetically based hypotheses of macaque relationships. The three members of the *sinica* group tested (*M. sinica*, *M. radiata*, *M. assamensis*) group together in all permutations of

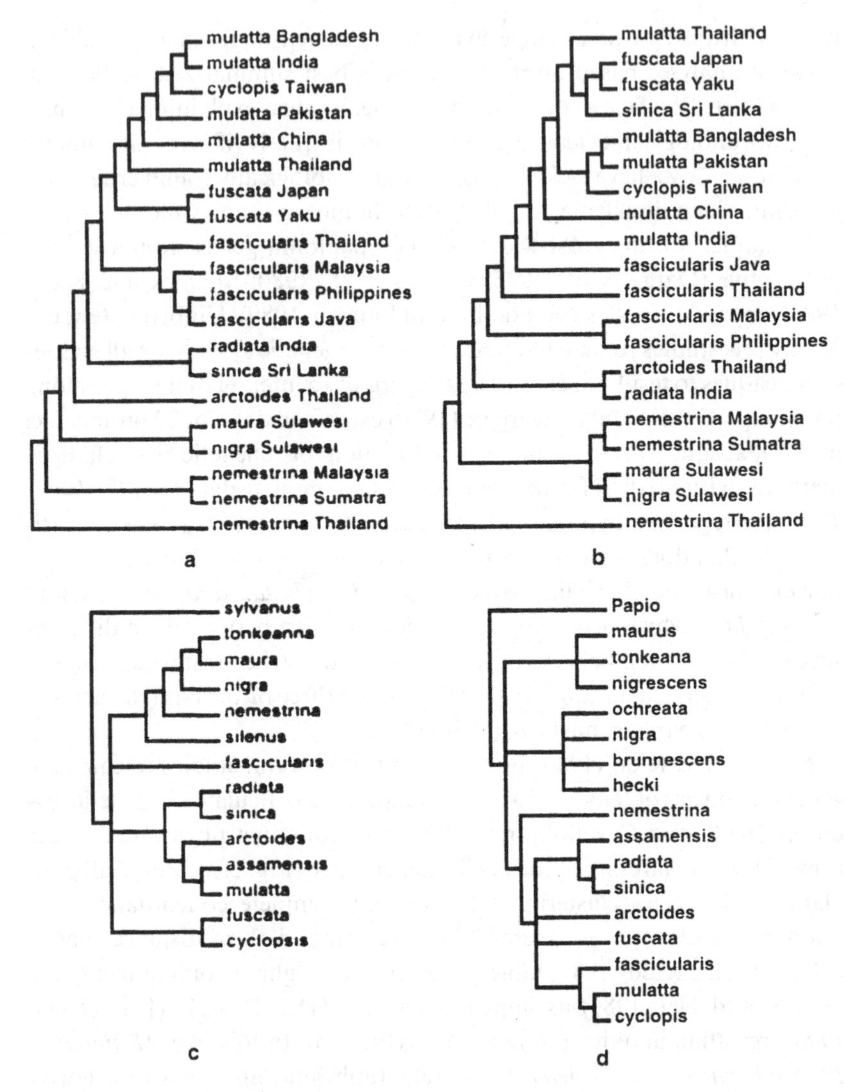

Fig. 2.2. Proposed macaque (*Macaca* spp.) phylogenies based on genetic data. (a) Phenetic tree presented by Melnick and Kidd (1985). (Reprinted with permission, *Int J. Primatol.*, Plenum Press) (b) Reanalysis of Melnick and Kidd (1985) data using neighbor-joining algorithm. (c) Cronin *et al.*, (1980) phenogram based on plasma protein and allozyme data. (Reprinted with permission of Van Nostrand Reinhold, a division of John Wiley & Sons, Inc. from *The Macaques: Studies in Ecology, Behavior and Evolution*. D.G. Lindburg, ed. Copyright 1980 by Litton Educatioal Publishing, Inc.) (d) Consensus tree of Fooden and Lanyon (1989) based on blood-protein data. (Reprinted with permission, *Am. J. Primatol.*, Wiley-Liss, Inc., a division of John Wiley & Sons, Inc.)

their analysis, with *M. arctoides* weakly joining this cluster (Fig. 2.2d). Three of the four members of the *fasicicularis* group strongly cluster together, while *M. fuscata* only weakly clusters within this clade. Interestingly, *M. nemestrina*, a member of the *silenus–sylvanus* group, is placed as the sister-taxon of all non *silenus–sylvanus* group macaques, which would make the *silenus–sylvanus* group paraphyletic (Fig. 2.2d).

Melnick *et al.* (1993) have extensively sampled several macaque species' mitochondrial genomes using restriction mapping methods. Among *fascicularis* group species, the insular *M. fuscata* and *M. cyclopis* are found to be more closely related to some *M. mulatta* populations than the *mulatta* populations are to each other. Several explanations could account for this observation. Avise (1986) notes that among various mammals, large intraspecific mtDNA differences, such as are evident in macaques (Melnick *et al.*, 1993), could lead to apparent paraphyly. Moore (1995) points-out that, while errors in phylogenetic inferences of species relationships due to ancestral polymorphism and lineage sorting are likely for taxa that split recently, analyses of mtDNA will be more efficient than analyses of nuclear loci for more anciently splitting taxa (see Hoelzer *et al.*, 1998 for more discussions).

The three most extensive DNA-based studies of macaque systematics to date all provide evidence for macaque monophyly but result in incongruent phylogenetic hypotheses within the genus. Hayasaka *et al.*'s (1996) sequence analysis of an 896 bp region of mtDNA from 13 macaque species and Morales and Melnick's (1998) restriction map analysis of a 2300 bp region of mtDNA produced incongruent phylogenetic hypotheses. Neither fully support Fooden's (1976, 1980) species group relationships. However, Tosi *et al.* (1999) sequenced approximately 3000 bp of the Y-chromosome in 18 species of macaques and found general support for Fooden's species groups as modified by Delson (1980).

The current incompatabilities between protein-based, mitochondrial, and Y-chromosome derived phylogenetic hypotheses may be due to sex-mediated differences in dispersal in these species as has been posited by Hoelzer *et al.* (1998). Final resolution of macaque systematics awaits sequence analysis of additional autosomal loci.

Cercopithecini

Only a few studies shed light on the relationships amongst the Cercopithecini either from morphological, behavioral, or genetic perspectives (Gautier-Hion *et al.*, 1988). Groves (1989) provisionally places

Miopithecus and *Allenopithecus* in the Papionini. He makes a stronger case for *Allenopithecus*, based on several shared characters with the papionin core group including female sexual swellings, molar 'flare', incisiform female lower canines, male ischial callosity midline fusion and a lack of several characters commonly found in the Cercopithecini. Remarking on *Allenopithecus*, Groves (1989:134) states, "It is therefore either a papionine with one character [lack of an M^3 hypoconulid] developed in parallel to the Cercopithecinae, or a cercopithecine with four parallelisms [female sexual swellings, molar flare, incisiform female canines, and male ischial callosity fusion] with the Papioninae."

Martin and MacLarnon (1988) present an analysis of Verheyen's (1962) cranial and dental measures of a large array of cercopithecins that were not initially analyzed phylogenetically. They convert cranial and dental metrics into Euclidean distances of residual values and performed a cluster analysis to create a dendrogram with two major clades. The first contains *Allenopithecus*, *Cercopithecus aethiops*, *C. lhoesti*, *Miopithecus*, and *Erythrocebus*. The second comprises *C. ascanius*, *C. mitis*, *C. cephus*, *C. neglectus*, *C. nictitans*, and *C. diana* (Fig. 2.3a). Gautier (1988) analyzes guenon vocalizations, and proposes two phylogenetic hypotheses and a third hypothesis in which he combines his results with those of Dutrillaux *et al.* (1988), Ruvolo (1988), and Martin and MacLarnon (1988). Gautier makes his own judgements when the different hypotheses varied (Fig. 2.3b). For this review, I used PAUP to generate a majority rule consensus tree for the 11 taxa in common examined by Martin and MacLarnon's (1988) analyses, in which they produced two morphologically-derived cladograms, and Gautier (1988) who generated two alternate vocally derived cladograms. The consensus tree reveals that these analyses only agree in clustering *Erythrocebus*, *Miopithecus*, and *C. lhoesti* in a clade, with the other eight taxa in a basal multichotomy (Fig. 2.3c). Thus, our current state of knowledge regarding cercopithecin relationships from non-genetic data is rather poor (Disotell, 1996).

Cercopithecins are very diverse karyotypically, varying in diploid number from 48 to 72. Dutrillaux (1988) and Ledbetter (1981) have both suggested explicit phylogenetic hypotheses, based on inferred patterns of chromosomal change as deduced from analysis of banding patterns. Ledbetter (1981), using *Macaca mulatta* as an outgroup, proposes that the first two taxa to diverge are *Allenopithecus* and *Miopithecus* respectively, followed by a clade containing *Erythrocebus*, *Cercopithecus aethiops*, *C. sabaeus*, and *C. lhoesti* (Fig. 2.4a). Next to diverge are *C. diana* and *C. neglectus*. This leaves two clades: one consisting of *C. cephus*, *C. ascanius*, and *C. petaurista* , and another consisting of *C. nictitans*, *C. mitis*, *C. wolfi*,

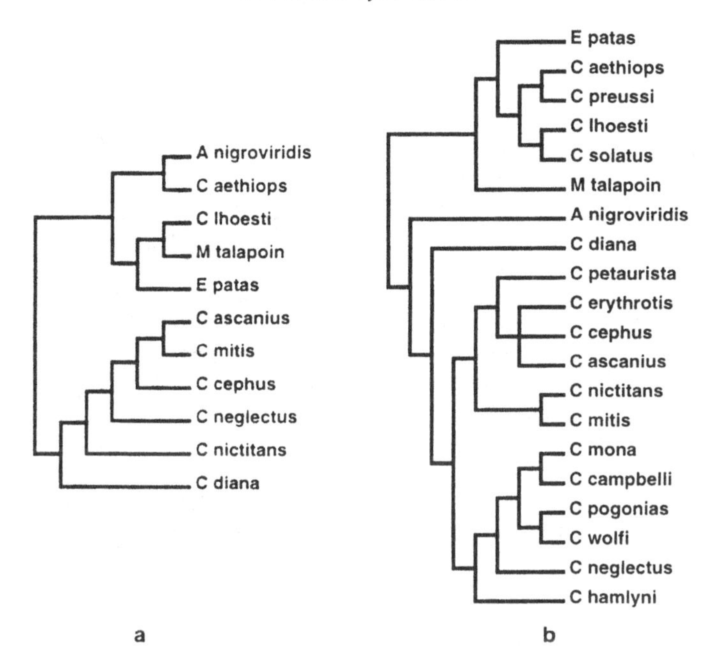

Fig. 2.3. Proposed guenon (*Allenopithecus*, *Cercopithecus*, *Miopithecus*, *Erythrocebus*) phylogenies based on morphological and behavioral data. (a) Consensus cladogram presented by Martin and MacLarnon (1988) based on cranial and dental metrics. (Reprinted with permission, Cambridge University Press). (b) Consensus cladogram presented by Gauthier (1988) based on vocalization data. (Reprinted with permission, Cambridge University Press). (c) Majority-rule consensus cladogram of Martin and MacLarnon's (1988) two hypotheses and Gauthier's (1988) two hypotheses.

and *C. pogonius*. Dutrillaux *et al.* (1988) presents a tree with major differences from that of Ledbetter. Dutrillaux posits a basal *Allenopithecus*, which is the sister-taxon of two clades. One consists of *Erythrocebus*, *Cercopithecus aethiops*, *C. sabaeus*, *C. lhoesti*, *Miopithecus* (plus three taxa not examined by Ledbetter – *C. pygerythrus*, *C. preussi*, and *C. solatus*). The remaining cercopithecins are in the other clade (Fig. 2.4b).

My reanalysis of Dutrillaux's data, using maximum parsimony (equal weighting and uniform transformation costs with PAUP), finds six equally parisomonious trees that have a majority rule consensus cladogram which does not differ significantly from that suggested by Dutrillaux *et al.* (1988). The strict consenus cladogram between Dutrillaux *et al.* (1988) and Ledbetter (1981) is shown in Figure 2.4c. Its most notable feature is the clade containing *Erythrocebus*, *Cercopithecus aethiops*, *C. sabaeus*, and *C. lhoesti*. Sarich's (1970) earlier work, using microcomplement fixation of

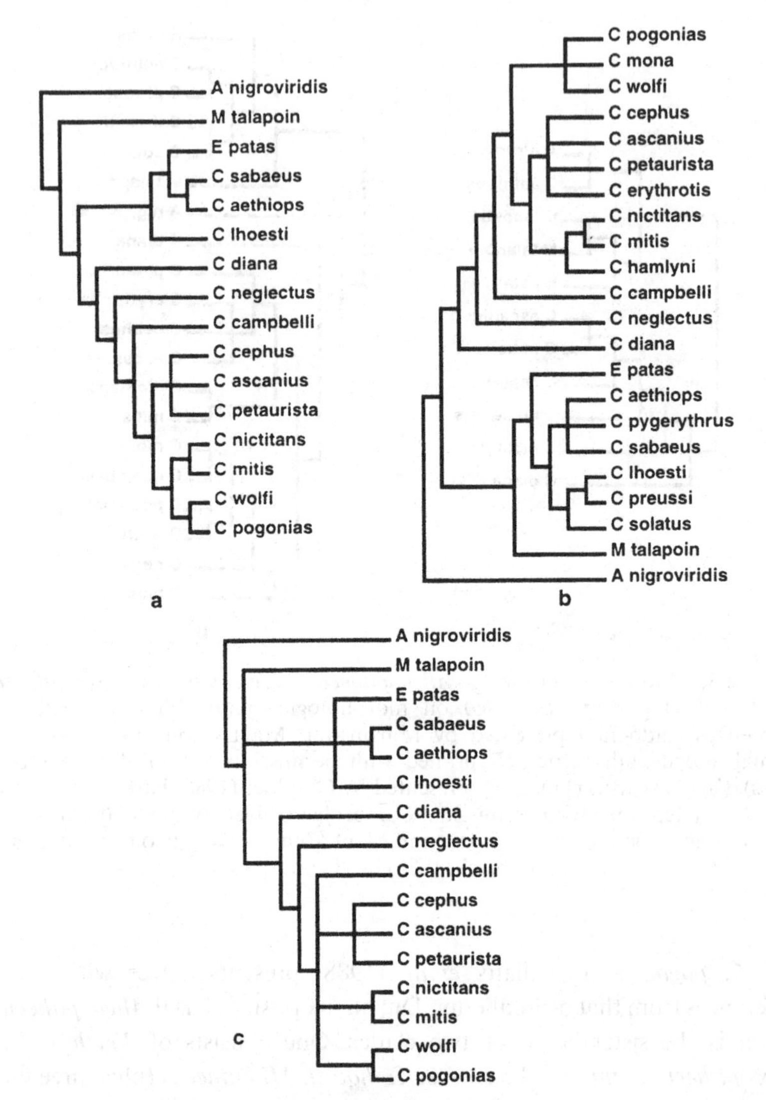

Fig. 2.4. Chromosomal phylogenetic hypotheses of the guenons (*Allenopithecus*, *Cercopithecus*, *Miopithecus*, *Erythrocebus*). (a) Ledbetter (1981). (Reprinted with permission, Dr. D.H. Ledbetter). (b) Dutrillaux *et al.* (1988). (Reprinted with permission, Cambridge University Press). (c) Strict consensus tree of a and b.

albumins, also found *C. aethiops* and *Erythrocebus patas* closer to each other than *C. aethiops* to the other species of *Cercopithecus* tested (*C. diana, C. cephus*, and *C. mona*). If these trees approximate the true pattern of cladogenesis, and cercopithecin taxonomy is to reflect cladistic relationships, then either *Erythrocebus* must be subsumed in *Cercopithecus*, or as

Groves proposes (1989; see Chapter 4) the taxon *Chlorocebus* must be ressurrected to include *C. aethiops*. Groves (see Chapter 4), however, does not support the close relationship of *C. lhoesti* to a *Chlorocebus* (*C. aethiops*) clade.

The largest study of cercopithecin genetics to date is that of Ruvolo (1982, 1988), in which she examined 14 genetic loci in 18 species using protein electrophoresis. While protein electrophoresis can detect approximately 85–90% of amino acid replacements (Lewontin, 1995), the majority of nucleotide substitutions may remain undetected at a given loci. Ruvolo (1988) presents two-distance based phylogenetic trees, using Nei's (1972) and Manhattan (Wright, 1978) distances analyzed with UPGMA and the Distance Wagner (Farris, 1972) techniques respectively. These trees differ from each other quite substantially, again demonstrating the importance of understanding and choosing the proper analytical technique to analyze systematic data. Ruvolo also presents a cladistically-derived hypothesis based on outgroup comparisons and character analysis, which differs significantly from her distance based trees. Her concluding consensus tree (Fig. 2.5a) has *Allenopithecus* and *Erythrocebus* as the first two taxa to branch off respectively, followed by a three way spit between *Miopithecus*, *Cercopithecus aethiops*, and the rest of the *Cercopithecus* species. Amongst the latter, *C. hamlyni* and *C. lhoesti* are the first to diverge (Fig. 2.5a). Ruvolo (1988: 139) concludes by stating that "the chromosomal phylogenies conflict with each other as well as with the electrophoretic consensus tree in the placement of the relatively primitive species." Coupled with the evidence just presented on morphologically and behaviorally derived hypotheses, this generalization could be expanded to most members of the entire tribe when all analyses to date are taken into account.

Van der Kuyl *et al.* (1995a) recently sequenced a ±390 bp region of the mitochondrial 12S rRNA gene in 15 cercopithecins and many other catarrhines. Differing from the chromosomally- derived hypotheses, they posit *C. neglectus* as the most basal cercopithecin and group *Erythrocebus* with the non-*aethiops* guenons (Fig. 2.5b). The only agreement between the van der Kuyl *et al.* (1995a) tree and the consensus tree of Ruvolo (1988) is in the grouping of *C. cephus*, *C. ascanius*, *C. mitis*, and *C. nictitans* into a common lineage. Rather unexpectedly, their analysis places *Miopithecus* as the most basal member of the entire Cercopithecinae. However, van der Kuyl *et al.* (1995b) note that portions of the mitochondrial genome, especially those containing the 12S rRNA molecule, have been duplicated and transferred to the nuclear genome several times. If their earlier study (van der Kuyl *et*

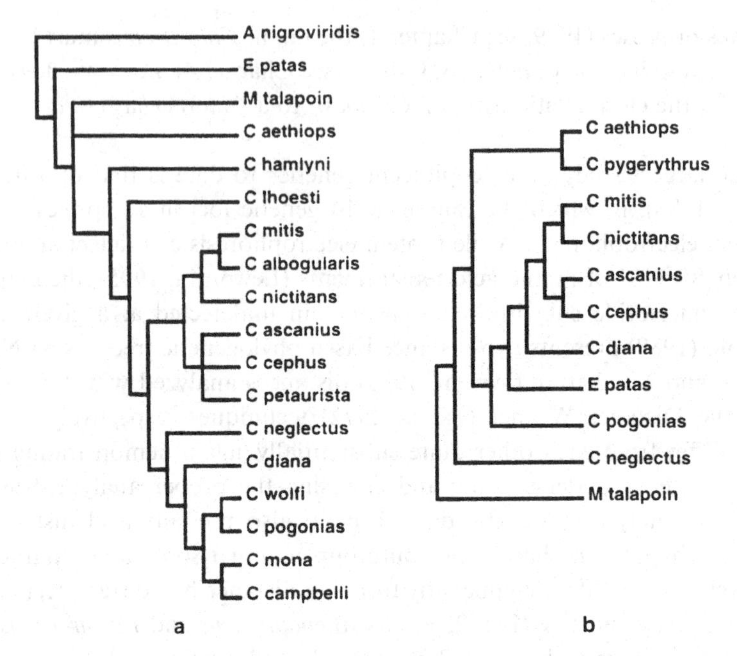

Fig. 2.5. Molecular phylogenetic hypotheses of the guenons (*Allenopithecus*, *Cercopithecus*, *Miopithecus*, *Erythrocebus*). (a) Ruvolo consensus (1988). (Reprinted with permission, Cambridge University Press). (b) vander Kuyl *et al.* (1995a). (Reprinted with permission, *J. Mol. Evol.*, Springer-Verlag Inc.)

al., 1995a) of the *Miopithecus* sequence is one of these nuclear counterparts, their intial placement of *Miopithecus* so far away from the rest of the guenons may be explained (though this position is not supported by the authors). Given the paucity of data regarding *Miopithecus*, and its low chromosomal number ($2N = 54$) compared with most of the cercopithecins, this hypothesis of a sister-group relationship to the Cercopithecinae may need to be considered. Collura and Stewart (1995) also note this phenomenon of mitochondrial pseudogenes and caution us in the interpretation of results using known duplicated regions. Much more study needs to be made of the occurrence of such duplicates.

It seems relatively clear that the *aethiops*-group of guenons are more closely related to *Erythrocebus*.Thus, Grove's resurrection of *Chlorocebus* seems sensible. *Miopithecus* may either be a basal cercopithecin, or perhaps a basal cercopithecine, the genetic data are equivocal. The position of *Allenopithecus* is impossible to predict based upon current genetic and chromosomal data (Disotell, 1996).

Colobinae

The major disagreements amongst researchers regarding colobine taxonomy and systematics are:

(1) The monophyly of the Asian colobines.
(2) Diversity among the African forms (one, two, or three genera).
(3) The interrelationships among species of *Colobus*.
(4) The number of Asian genera (four to eight).
(5) The relationship of the proboscis monkey (*Nasalis*) and the simakobu (*Simias*).
(6) The relationships of snub-nosed monkeys (*Rhinopithecus*) to douc monkeys (*Pygathrix*).
(7) The relationships among and between leaf-monkeys (*Presbytis*), langurs (*Trachypithecus*), and Hanuman or gray langurs (*Semnopithecus*).

Szalay and Delson (1979) split the Colobinae into two subtribes, the Asian Presbytina and the African Colobina. Within the Presbytina, they group *Presbytis* and *Pygathrix* together to the exclusion of *Nasalis*. Strasser and Delson (1987) place *Pygathrix* closer to *Nasalis* than to *Presbytis*, and include several other genera in their study (Fig. 2.6a). Reanalysis with PAUP (Swofford, 1996) of this data set (which did not include character states for *Rhinopithecus* and *Simias*), using character states ordered as in Strasser and Delson (1987), results in a very different phylogenetic hypothesis (Fig. 2.6b). *Nasalis* is posited as the basal taxon and *Presbytis* is the sister-taxon to the African species in this reanalysis. Again, this discrepancy stresses the importance of examining all possible equally parsimonious topologies (Maddison, 1991). Groves (1989; Chapter 4) does not split the colobines into African and Asian clades, but rather into a long-faced clade (including *Nasalis)* and a short-faced clade containing the other colobine species; *Trachypithecus*, *Semnopithecus*, and *Presbytis* have uncertain affinties. Peng *et al.* (1993) examine 14 variables measured on skulls of five Asian colobine genera (including *Presbyticus*, also known as *Rhinopithecus avunculus*) and derive the phenogram shown in Fig. 2.6c using cluster analysis (SPSS/PC). The phenogram is basically congruent with the tree presented by Strasser and Delson (1987) for Asian forms.

Only 95 DNA sequences for all colobines are found in GenBank as of May 1999. Thus, very little can be said about the molecular systematics of the Colobinae. Few genetic studies shed light on aspects of their evolutionary history. The many papers of Goodman's group examined only one

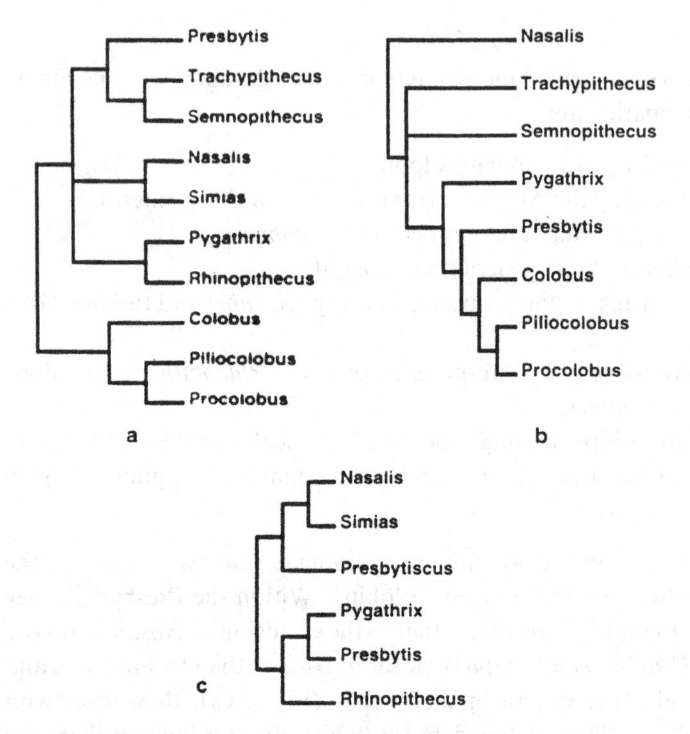

Fig. 2.6. Morphologically-based hypotheses of the colobines. (a) Strasser and Delson (1987). (Reprinted with permission, *J. hum. Evol.*, Academic Press, Harcourt Brace Jovanovich). (b) Reanalysis of Strasser and Delson (1987) data with PAUP and character states ordered as in a, no data presented for *Simias* and *Rhinopithecus* (c) Peng *et al.* (1993), *Presbytiscus* = (*Rhinopithecus avunculus*). (Reprinted with permission, *Folia Primatol.*, S. Karger, Inc.)

individual of *Presbytis* and/or of *Colobus* and thus shed little light on colobine systematics (Goodman *et al.*, 1982). Dutrillaux *et al.* (1986) examined chromosomes of only four colobines: *Colobus angolensis*, *C. guereza*, *C. polykomos*, and *Presbytis* (*Trachypithecus*) *cristata*. They grouped *C. polykomos* with *C. guereza* followed by *C. angolensis* in one clade, with *Presbytis* (*Trachypithecus*) *cristata* as an outgroup. This conclusion corresponds with the tree suggested by Oates and Trocco (1983), based on vocalizations, for the species common to the two studies (Fig. 2.7c).

Sarich (1970) discusses the phylogenetic relationships of several colobine species inferred from albumin microcomplement fixation experiments, and finds *Nasalis* and *Presbytis* most similar amongst the species studied. *Pygathrix* either groups with the African *Colobus* or in a basal three-way split with *Colobus* and a *Nasalis*/*Presbytis* clade. Later refinements of this work, using genetic distances derived from both albumin and transferrin

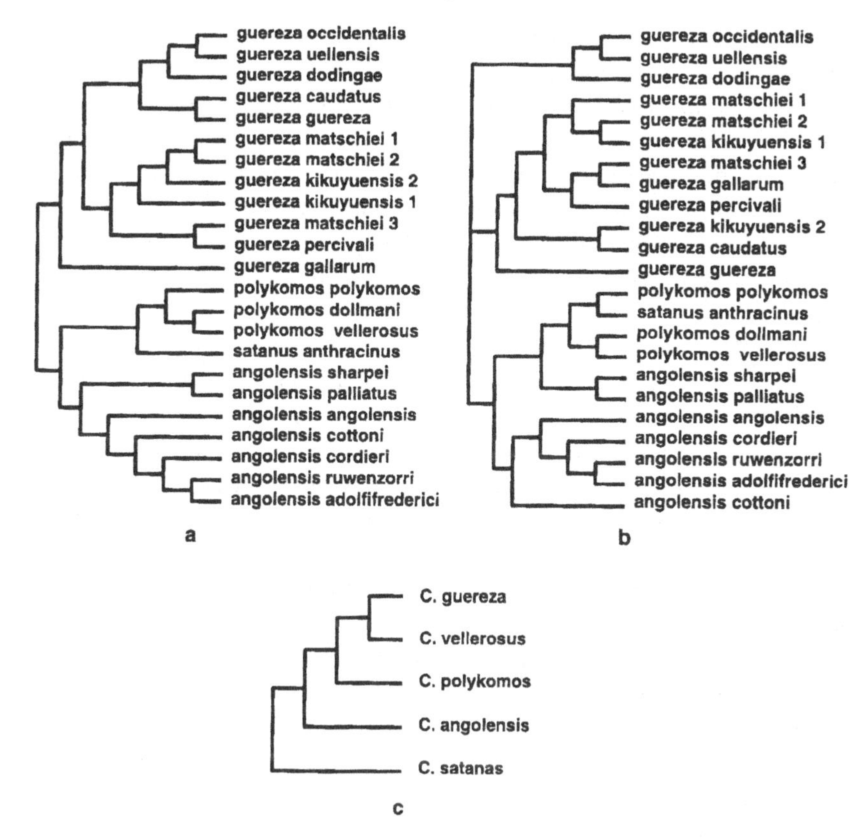

Fig. 2.7. Distance based analyses of craniometric data presented by Hull (1979). Trees modified for simplicity by removing subspecific replicates found to be monophyletic in both analyses. (a) Ward's method dendrogram for males on corrected Mahalanobois' distance (from Hull, 1979). (Reprinted with permission, *Am. J. phys. Anthrop.*, Wiley-Liss, Inc., a division of John Wiley & Sons, Inc.) (b) Neighbor joining analysis of same data. (c) Tree presented by Oates and Trocco (1983) based on vocalization data. (Reprinted with permission, *Folia Primatol.*, S. Karger, Inc.)

(Sarich and Cronin, 1976), find a basal multichotomy among *Colobus*, *Presbytis*, and *Pygathrix*. Such basal multichotomies are common in these distance-based studies of albumin and transferrin whenever the species have diverged more than around five million years or earlier (e.g. Cronin and Sarich, 1976; Cronin *et al.*, 1980).

Two works in progress are of interest here. Data collected by myself (Disotell, 1992) and Ruvolo (pers. com.) on the mitochondrial COII gene suggest that *Nasalis* clusters with other Asian colobines to the exclusion of African *Colobus*. This finding does not support Groves's views discussed above. Messier and Stewart (1997) concurs with an African/Asian split based on mitochondrial cytochrome b sequences and those of the nuclear encoded lysozyme locus. Analysis of their preliminary data (Stewart, pers. comm.) yields a monophyletic Asian clade with four deep lineages comprised of *Nasalis*, *Rhinopithecus/Pygathrix*, *Semnopithecus entellus/vetulus*, and *Trachypithecus francoisi/obscurus/cristatus*, which (as of this writing) cannot be more clearly differentiated. Zhang and Ryder (1998) have analyzed a 424 bp fragment of the mitochondrial cytochrome b gene from 10 Asian and one African colobine genera. They also support the monophyly of Asian assemblage and weakly cluster *Nasalis* with *Rhinopithecus* and *Pygathrix*.

Hull (1979) analyzed 76 cranial measurements from 1072 African colobine skulls, using distance techniques, to generate a phenogram for numerous populations and subspecies of black-and-white *Colobus*. Dendrograms (Fig. 2.7a) were created using Ward's (1963) Error Sum method of Mahalanobis distance matrices calculated separately for the two sexes. He found monophyletic *Colobus guereza*, *C. angolensis*, and *C. polykomos* clades. *C. satanas* is the sister-taxon of the *polykomos* lineage. Since Ward's (1963) method is not a standard method used in phylogenetic analyses, and has been criticized by Sneath and Sokal (1973), I reanalyzed Hull's distance matrices using the neighbor-joining method (Saitou and Nei, 1987) and found a very different topology (Fig. 2.7b) with *C. guereza*, *C. angolensis*, and *C. polykomos* paraphyletic. Such disparate results from the different analytical techniques used to analyze the same data weaken our confidence in this approach. Oates and Trocco (1983) present a cladogram (Fig. 2.7c) of black-and-white colobines, based on an analysis of loud call variation, which differs from Hull's original analysis and my reanalysis of his data in almost every respect (compare Figures 2.7a,b, and c).

Conclusions

The molecular phylogeny of papionins (Fig. 2.1c) is the most strongly corroborated, given the number of independent analyses that support the view put forth here. Macaque species level and common baboon subspecific phylogeny require further molecular scrutiny. *Miopithecus* may tenuously be placed with the Cercopithecini, or as a basal member of the whole cercopithecine lineage. *Allenopithecus*'s position within the cercopithecines is even more problematic, although I would lean against its papionin status (the classical views can't all be wrong after all). The resurrection of the genus genus *Chlorocebus* makes sense, along with its sister-taxon status to *Erythrocebus*, which is supported by Grove's morphological analysis (Chapter 4). Regarding colobines, all that molecular anthropologists can suggest is that an African/Asian split is likely and that *Rhinopithecus* and *Pygathrix* are sister-taxa.

It must be kept in mind that any phylogenetic tree inferred from a single molecule or locus is actually a 'gene-tree' and not necessarily a species tree (Pamilo and Nei, 1988). As can be seen in the above discussion, several phylogenies inferred from molecular data are incongruent with those derived from morphological or behavioral analyses and also with those derived from other molecular studies. When only a few loci have been analyzed for any group, it is therefore not surprising that they lead to different phylogenetic hypotheses. In such cases, until more independent data are assessed, it is best to entertain several or all of the hypotheses when using the phylogeny to analyze further a lineage. Until analyses of multiple independent data sets begin to agree with each other (e.g. see Harris and Disotell, 1998), any phylogenetic hypotheses must be considered provisional.

Clearly, much work needs to be done regarding even basic aspects of cercopithecoid systematics. While most of molecular phylogenetic research on Old World monkeys has focused on members of the tribe Papionini, an increasing number of researchers are examining the lesser known (from a genetic perspective) guenons, their probable near relatives *Miopithecus* and *Allenopithecus*, and the subfamily Colobinae. More work needs to be done on speciose genera such as *Macaca*, *Cercopithecus*, and *Presbytis*, which should provide us with additional insights into the processes of speciation.

Approaches using restriction maps and sequences of the mitochondrial genome have proven very useful and still show great promise due to the maternal inheritance and fast rate of change of this molecule. Moore (1995) points-out that these properties lead to an effective population size that is one-fourth that of nuclear-encoded autosomal genes, making mitochondrially-based trees more likely to represent accurately species trees under many circumstances (see Hoelzer *et al.*, 1998 for further discussion). However, since the mitochondrial genome is basically a single linkage-group, we still must examine multiple independent nuclearly-encoded loci in order to provide corroboration for phylogenetic hypotheses. Thus, the continued searching for, and sequencing of, independent nuclear loci will be critical (Disotell, 1994b). Microsatellite analyses are proving increasingly useful in examining the population genetic structure and history within species (Bowcock *et al.*, 1994; Morin *et al.*, 1994). Perhaps more difficult, but no less worthy, would be further research into the mechanisms and processes of chromosomal evolution.

Issues of phylogeny and population genetics are also an increasingly important aspect of epidemiology of primate infectious diseases (Gao *et al.*, 1999) conservation research, as many of these species are endangered or threatened (Oates, 1986; Mittermeier and Cheney, 1987). Technological advances now allow the extraction of genetic materials from museum specimens, hair and feces, allowing noninvasive sampling approaches (Pääbo, 1989; Vigilant *et al.*, 1989; Constable *et al.*, 1995). Such studies could occupy many research groups, in close collaboration with field researchers and local authorities, and need to be carried out soon before habitat destruction and extinction make them impossible. These systematic and population genetic studies will be crucial to further advances in many areas of research involving Old World monkeys.

References

Arnason, U., Gullberg, A. & Janke, A. (1998). Molecular timing of primate divergences as estimated by two nonprimate calibration points. *J. Mol. Evol.* **47**, 718–27.

Atchley, W.R. & Fitch, W.M. (1991). Gene trees and the origins of inbred strains of mice. *Science* **254**, 554–8.

Avise, J.C. (1986). Mitochondrial DNA and the evolutionary genetics of higher animals. *Philos. Trans. R. Soc. Lond. (Biol.)* **312**, 325–42.

Barnicot, N.A. & Wade, P.T. (1970). Protein structure and the systematics of Old World monkeys. In *Old World Monkeys: Evolution, Systematics, and Behavior*, ed. J.R. Napier & P.H. Napier, pp. 227–60. New York: Academic Press.

Benson, D.A., Boguski, M.S., Lipman, D.J., Ostell, J., Ouellette, B.F. (1998). Gen Bank. *Nucleic Acids Res.* **26**, 1–7.

Benveniste, R.E. & Todaro, G.J. (1976). Evolution of type C viral genes: evidence for an Asian origin of man. *Nature* **261**, 101–8.

Bowcock, A.M., Ruiz-Linares, A., Tomfohrde, J., Minch, E., Kidd, J.R. & Cavalli-Sforza, L.L. (1994). High resolution of human evolutionary trees with polymorphic microsatellites. *Nature* **368**, 455–7.

Chiarelli, B. (1966). Caryology and taxonomy of the catarrhine monkeys. *Am. J. phys. Anthrop.* **24**, 155–69.

Collura, R.V. & Stewart C.B. (1995). Insertions and duplications of mitochondrial DNA in the nuclear genomes of catarrhine primates. *Nature* **378**, 485–9.

Constable, J.J., Packer, C., Collins, D.A. & Pusey, A.E. (1995). Nuclear DNA from primate dung. *Nature* **373**, 393.

Cronin, J.E., Cann, R. & Sarich, V.M. (1980). Molecular evolution and systematics of the genus *Macaca*. In *The Macaques: Studies in Ecology, Behavior, and Evolution*, ed. D.G. Lindburg, pp. 31–51. New York: Van Nostrand Reinhold.

Cronin, J.E. & Meikle, W.E. (1979).The phylogenetic position of *Theropithecus*: congruence among molecular, morphological, and paleotological evidence. *Syst. Zool.* **28**, 259–69.

Cronin, J.E. & Sarich, V.M. (1976). Molecular evidence for the dual origin of the mangabeys among Old World monkeys. *Nature* **260**, 700–2.

Darga, L.L., Goodman, M., Weiss, M.L., Moore, G.W., Prychodko, W., Dene, H.,

Tashian, R. & Koen, A. (1975). Molecular systematics and clinal variation in macaques. In *Isozymes IV: Genetics and Evolution*, ed. C.L. Markert, pp. 797–812. New York: Academic Press.

Davies, G. & Oates, J.F. (1994). *Colobine Monkeys: Their Ecology, Behaviour and Evolution*. Cambridge: Cambridge University Press.

Delson, E. (1973). *Fossil Colobine Monkeys of the Circum-Mediterranean Region and the Evolutionary History of the Cercopithecidae (Primates, Mammalia)*. PhD thesis, Columbia University.

Delson, E. (1980). Fossil macaques, phyletic relationships and a scenario of deployment. In *The Macaques: Studies in Ecology, Behavior, and Evolution*, ed. D.G. Lindburg, pp. 10–30. New York: Van Nostrand Reinhold.

Delson, E. (1994). Evolutionary history of the colobine monkeys in paleoenvironmental perspective. In *Colobine Monkeys: Their Ecology, Behavior and Evolution*, ed. A.G. Davies & J.F. Oates, pp. 11–44. Cambridge: Cambridge University Press.

Delson, E. & Napier, P.H. (1976). Request for determination of the generic names of the baboon and mandrill (Mammalia: Primates, Cercopithecoidea). *Bull. Zool. Nomen.* **33**, 46–60.

Dene, H.T., Goodman, M. & Prychodko, W. (1976). Immunodiffusion evidence on the phylogeny of the primates. In *Molecular Anthropology*, ed. M. Goodman & R.E. Tashian, pp. 171–95. New York: Plenum.

Disotell, T.R. (1992). *Molecular Evolution of the Papionini (Primates: Cercopithecinae)*. PhD thesis, Harvard University.

Disotell, T.R. (1994a). Generic level relationships of the Papionini (Cercopithecoidea). *Am. J. phys. Anthrop.* **94**, 47–57.

Disotell, T. (1996). The phylogeny of Old World monkeys. *Evol. Anthropol.* **5**, 18–24.

Disotell, T.R., Honeycutt, R.L. & Ruvolo, M. (1992). Mitochondrial DNA phylogeny of the Old World monkey tribe Papionini. *Mol. Biol. Evol.* **9**, 1–13.

Dutrillaux, B. (1979). Chromosomal evolution in primates: tentative phylogeny from *Microcebus murinus* (Prosimian) to Man. *Hum. Genet.* **48**, 251–314.

Dutrillaux, B., Muleris, M. &Couturier, J. (1988). Chromsomal evolution of Cercopithecinae. In *A Primate Radiation: Evolutionary Biology of the African Guenons*, ed. A. Gautier-Hion, F., Bourliere, J.-P., Gautier & J. Kingdon, pp. 150–9. Cambridge: Cambridge University Press.

Dutrillaux, B., Couturier, J. & Viegas-Pequignot, E. (1981). Chromosomal evolution in primates. *Chromosomes Today* **7**, 176–91.

Dutrillaux, B., Couturier, J., Muleris, M., Lombard, M. & Chauvier, G. (1982). Chromosomal phylogeny of forty-two species or subspecies of cercopithecoids (Primates, Catarrhini). *Ann. Génét.* **25**, 96–109.

Dutrillaux, B., Couturier, J., Muleris, M., Rumpler, Y. & Viégas-Pequignot, E. (1986). Relations chromosomiques entre sous-orders et infra-ordres et schéma évolutif général des primates. *Mammalia* **50**, 108–21.

Farris, J.S. (1972). Estimating phylogenetic trees from distance matrices. *Am. Nat.* **106**, 645–68.

Farris, J.S. (1988). *HENNIG86: Program and Documentation*. New York: Department of Ecology and Evolution, State University of New York at Stony Brook.

Felsenstein, J. (1981). A likelihood approach to character weighting and what it tells us about parsimony and compatibility. *Biol. J. Linn. Soc.* **16**, 183–96.

Felsenstein, J. (1993). *PHYLIP: Phylogenetic Inference Package, Version 3.5:*

Program and Documentation. Seattle: Department of Genetics, University of Washington.
Fitch, W.M. (1971). Toward defining the course of evolution: Minimal change for a specific tree topology. *Syst. Zool.* **20**, 406–16.
Fitch, W.M. & Margoliash, E. (1967). The construction of phylogenetic trees – a general applicable method utilizing estimates of the mutation distance obtained from cyctochrome c sequences. *Science* **155**, 279–84.
Fleagle, J.G. (1988). *Primate Adaptation and Evolution*. New York: Academic Press.
Fleagle, J.G. & McGraw, W.S. (1999). Skeletal and dental morphology supports diphyletic origin of baboons and mandrills. *Proc. Natl. Acad. Sci. USA* **96**, 1157–61.
Fooden, J. (1976). Provisional classification and key to living species of macaques. *Folia Primatol.* **25**, 225–36.
Fooden, J. (1980). Classification and distribution of living macaques (*Macaca* Lacepede 1799). In *The Macaques: Studies in Ecology, Behavior, and Evolution*, ed. D.G. Lindburg, pp. 1–9. New York: Van Nostrand Reinhold.
Fooden, J. & Lanyon, S.M. (1989). Blood-protein allele frequencies and phylogenetic relationships in *Macaca*: A review. *Am. J. Primatol.* **17**, 209–41.
Gao, F., Bailes, E., Robertson, D.L. *et al.* (1999). Origin of HIV-1 in the chimpanzee *Pan troglodytes troglodytes. Nature* **397**, 436–41.
Gautier, J.-P. (1988). Interspecific affinities among guenons as deduced from vocalizations. In *A Primate Radiation: Evolutionary Biology of the African Guenons*, ed. A. Gautier-Hion, F., Bourliere, J.-P. Gautier & J. Kingdon, pp. 194–226. Cambridge: Cambridge University Press.
Gautier-Hion, A., Bourliere, F., Gautier, J.-P. & Kingdon, J. (1988). *A Primate Radiation: Evolutionary Biology of the African Guenons*. Cambridge:Cambridge University Press.
Gillispie, D. (1977). Newly evolved repeated DNA sequences in primates. *Science* **196**, 889–91.
Goodman, M. (1961). The role of immunological differences in the phyletic development of human behavior. *Hum. Biol.* **33**, 131–62.
Goodman, M. (1963). Serological analysis of the systematics of recent hominoids. *Hum. Biol.* **35**, 377–436.
Goodman, M., Miyamoto, M.M. & Czelusniak, J. (1987). Pattern and process in vertebrate phylogeny revealed by coevolution of molecules and morphology. In *Molecules and Morphology in Evolution: Conflict or Compromise?*, ed. C.P. Patterson, pp. 141–76. Cambridge: Cambridge University Press.
Goodman, M. & Moore, G.W. (1971). Immunodiffusion systematics of the primates. I. The Catarrhine. *Syst. Zool.* **20**, 19–62.
Goodman, M., Romero-Herrera, A., Dene, H., Czelusniak, J. & Tashian, R.E. (1982). Amino acid sequence evidence on the phylogeny of primates and other eutherians. In *Macromolecular Sequences in Systematics and Evolutionary Biology*, ed. M. Goodman, pp. 115–91. New York: Plenum.
Groves, C.P. (1978). Phylogenetic and population systematics of the mangabeys (Primates: Cercopithecoidea). *Primates* **19**, 1–34.
Groves, C.P. (1989). *A Theory of Human and Primate Evolution*. Oxford: Clarendon Press.
Harris, E.E. & Disotell, T.R. (1998). Nuclear gene trees and the phylogenetic relationships of the mangabeys (Primates: Papionini). *Mol. Biol. Evol.* **5**, 626–44.

Hayasaka, K., Gojobori, T. & Horai, S. (1988). Molecular phylogeny and evolution of primate mitochondrial DNA. *Mol. Biol. Evol.* **5**, 626–44.
Hayasaka, K., Fujii, K. & Horai, S. (1996). Molecular phylogeny of macaques: implications of nucleotide sequences from an 896-base pair region of mitochondrial DNA. *Mol. Biol. Evol.* **13**, 1044–53.
Hewett-Emmett, D., Cook, C.N. & Barnicot, N.A. (1976). Old World monkey hemoglobins: deciphering phylogeny from complex patterns of molecular evolution. In *Molecular Anthropology*, ed. M. Goodman & R.E. Tashian, pp. 357–405. New York: Plenum.
Hill, W.C.O. (1974). *Primates: Comparative Anatomy and Taxonomy*, vol. VII: *Cynopithecina*. Edinburgh: Edinburgh University Press.
Hillis, D.M., Bull, J.J., White, M.E., Badgett, M.R. & Molineux, I.J. (1992). Experimental phylogenetics: generation of a known phylogeny. *Science* **255**, 589–92.
Hoelzer, G.A. Wallman, J. & Melnick, D.J. (1998). The effects of social structure, geographical structure and population size on the evolution of mitochondrial DNA: II. Molecular clocks and the lineage sorting period. *J. Mol. Evol.* **47**, 21–31.
Honeycutt, R.L. & Wheeler, W. (1989). Mitochondrial DNA: variation in humans and primates. In *DNA Systematics: Human and Higher Primates*, ed. S.K. Dutta & W. Winter, pp. 91–129. Boca Raton: CRC Press.
Hull, D.B. (1979). A craniometric study of the black and white *Colobus* Illiger 1811 (Primates: Cercopithecoidea). *Am. J. phys. Anthrop.* **51**, 163–82.
Jablonski, N.G., and Peng, Y.-Z. (1993). The phylogenetic relationships and classification of the doucs and snub-nosed langurs of China and Vietnam. *Folia Primatol.* **60**, 36–55.
Jolly, C.J. (1966). Introduction to the cercopithecoidea, with notes on their use as laboratory animals. *Symp. Zool. Soc. Lond.* **17**, 427–57.
Jolly, C.J. (1967). The evolution of baboons. In *The Baboon in Medical Research*, vol. II, ed. H. Vagtbog, pp. 427–57. Austin: University of Texas Press.
Kidd, K.K. & Cavalli-Sforza, L.L. (1974). The role of genetic drift in the differentiation of Icelandic and Norwegian cattle. *Evolution* **28**, 381–95.
Kohne, D.E., Chiscon, J.A. & Hoyer, B.H. (1972). Evolution of primate DNA sequences. *J. hum. Evol.* **1**, 627–44.
Koop, B.F., Tagle, D.A., Goodman, M. & Slightom, J.L. (1989). A molecular view of primate phylogeny and important systematic and evolutionary questions. *Mol. Biol. Evol.* **6**, 580–612.
Ledbetter, D.H. (1981). *Chromosomal Evolution and Speciation in the Genus* Cercopithecus *(Primates, Cercopithecinae)*. PhD dissertation, University of Texas, Austin.
Lewontin, R.C. (1995). *Human Diversity*. New York: Scientific American Library.
Maddison, D.R. (1991). The discovery and importance of multiple islands of most-parsimonious trees. *Syst. Zool.* **40**, 315–28.
Maddison, W.P. & Maddison, D.R. (1992). *MacClade: Interactive Analysis of Phylogeny and Character Evolution, Version 3.00*. Sunderland, MA: Sinauer Associates.
Martin, R.D. (1990). *Primate Origins and Evolution: A Phylogenetic Reconstruction*. Princeton: Princeton University Press.
Martin, R.D. & MacLarnon, A.M. (1988). Quantitative comparisons of the skull and teeth in guenons. In *A Primate Radiation: Evolutionary Biology of the African Guenons*, ed. A. Gautier-Hion, F., Bourliere, J.-P., Gautier & J. Kingdon, pp. 160–83. Cambridge: Cambridge University Press.

Melnick, D.J. & Hoelzer, G.A. (1993). What is mtDNA good for in the study of primate evolution? *Evol. Anthropol.* **2**, 1–10.
Melnick, D.J. & Kidd, K.K. (1985). Genetic and evolutionary relationships among Asian macaques. *Int. J. Primatol.* **6**, 123–60.
Melnick, D.J., Hoelzer, G.A., Absher, R. & Ashley, M.V. (1993). MtDNA diversity in rhesus monkeys reveals overestimates of divergence time and paraphyly with neighboring species. *Mol. Biol. Evol.* **10**, 282–95.
Messier, W. & Stewart, C.B. (1997). Episodic adaptive evolution of lysozyme in the primates. *Nature* **385**, 151–4.
Mittermeier, R.A. & Cheney, D.L. (1987). Conservation of primates and their habitats. In *Primate Societies*, ed. B.B. Smuts, D.L. Cheney, R.M. Seyfarth, R.W. Wrangham & T.T. Struhsaker, pp. 477–90. Chicago: University of Chicago Press.
Miyamoto, M.M. & Cracraft, J. (1991). *Phylogenetic Analysis of DNA Sequences.* New York: Oxford University Press.
Moore, W.S. (1995). Inferring phylogenies from mtDNA variation: mitochondrial-gene trees versus nuclear-gene trees. *Evolution* **49**, 718–26.
Morales, J.C. & Melnick, D.J. (1998). Phylogenetic relationships of the macaques (Cercopithecoida: *Macaca*), as revealed by high resolution restriction site mapping of mitochondrial ribosomal genes. *J. hum. Evol.* **34**, 1–23.
Morin, P.A., Moore, J.J., Chakraborty, R., Jin, L., Goodall, J. & Woodruff, D.S. (1994). Kin selection, social structure, gene flow, and the evolution of chimpanzees. *Science* **265**, 1193–201.
Napier, J.R. & Napier, P.H. (1985). *The Natural History of the Primates.* Cambridge: MIT Press.
Nei, M. (1972). Genetic distance between populations. *Am. Nat.* **106**, 283–92.
Nelkin, B., Strayer, D. & Vogelstein, B. (1980). Divergence of primate ribosomal RNA genes as assayed by restriction enzyme analysis. *Gene* **11**, 89–96.
Newman, T.K., Jolly, C.J. & Rogers, J. (1999). Mitochondrial DNA sequence variation in baboons (Papio hamadryas). *Am. J. phys. Anthrop.* (Supp.) **28**, 210.
Nuttal, G.H.F. (1904). *Blood Immunity and Blood Relationships.* Cambridge: The University Press.
Oates, J.F. (1986). *Action Plan for African Primate Conservation: 1986–1990.* Gland: IUCN ISSC Primate Specialist Group.
Oates, J.F. & Trocco, T.F. (1983). Taxonomy and phylogeny of black-and-white colobus monkeys: inferences from an analysis of loud call variation. *Folia Primatol.* **40**, 83–113.
Pääbo, S. (1989). Ancient DNA: extraction, characterization, molecular cloning, and enzymatic amplification. *Proc. Natl. Acad. Sci. USA.* **86**, 1939–43.
Pamilo, P. & Nei, M. (1988). Relationships between between gene trees and species trees. *Mol. Biol. Evol.* **5**, 568–83.
Peng, Y.-Z., Pan, R.-L. & Jablonski, N.G. (1993). Classification and evolution of Asian colobines. *Folia Primatol.* **60**, 106–17.
Pilbeam, D.R. (1996). Genetic and morphological records of the Hominoidea and hominid origins: A synthesis. *Mol. Phylogenet. Evol.* **5**, 155–68.
Ruvolo, M. (1982). *Genetic Evolution in the African Guenon Monkeys (Primates, Cercopithecinae).* PhD thesis, Harvard University.
Ruvolo, M. (1988). Genetic evolution in the African guenons. In *A Primate Radiation: Evolutionary Biology of the African Guenons*, ed. A. Gautier-Hion, F. Bourliere, J.-P. Gautier & J. Kingdon, pp. 127–49. Cambridge: Cambridge University Press.

Saitou, N. & Nei, M. (1987). The neighbor-joining method: a new method for reconstructing phylogenetic trees. *Mol. Biol. Evol.* **4**, 406–25.

Sarich, V.M. (1970). Primate systematics with special reference to Old World monkeys: a protein perspective. In *Old World Monkeys: Evolution, Systematics, and Behavior*, ed. J.R. Napier & P.H. Napier, pp. 175–226. New York: Academic Press.

Sarich, V.M. & Cronin, J. (1976). Molecular systematics of the primates. In *Molecular Anthropology*, ed. M. Goodman & R.E. Tashian, pp. 141–70. New York: Plenum.

Sarich, V.M. & Wilson, A.C. (1966). Quantitative immunochemistry and the evolution of primate albumins. *Science* **154**, 1563–6.

Sarich, V.M. & Wilson, A.C. (1967). Immunological time scale for hominid evolution. *Science* **158**, 1200–3.

Schultz, A.H. (1970). The comparative uniformity of the Cercopithecoidea. In *Old World Monkeys: Evolution, Systematics, and Behavior*, ed. J.R. Napier & P.H. Napier, pp. 39–52. New York: Academic Press.

Smuts, B.B., Cheney, D.L., Seyfarth, R.M., Wrangham, R.W. & Struhsaker, T.T. (1987). *Primate Societies*. Chicago: University of Chicago Press.

Sneath, P. & Sokal, R.R. (1973). *Numerical Taxonomy*. San Francisco: W.H. Freeman.

Stanyon, R.C, Fantini, C., Camperio-Ciani, A., Chiarelli, B. & Ardito, G. (1988). Banded karyotypes of 20 Papionini species reveal no necessary correlation with speciation. *Am. J. Primatol.* **16**, 3–17.

Strasser, E. & Delson, E. (1987). Cladistic analysis of cercopithecid relationships. *J. hum. Evol.* **16**, 81–99.

Swofford, D.L. (1996). *PAUP*: Phylogenetic Analysis Using Parsimony (and Other Methods)*, version 4.0. Sunderland, MA: Sinaver Associates.

Swofford, D.L., Olsen, G.J., Waddell, P.J. & Hillis, D.M. (1996). Phylogeny inference. In *Molecular Systematics*, 2nd edn., ed. D.M. Hillis, C. Moritz & B.K. Mable, pp. 407–514. Sunderland, MA: Sinauer Associates.

Szalay, F. & Delson, E. (1979). *Evolutionary History of the Primates*. New York: Academic Press.

Templeton, A.R. (1983). Phylogenetic inference from restriction endonuclease cleavage site maps with particular reference to the evolution of human and apes. *Evolution* **37**, 221–4.

Tosi, A.J., Morales, J.C., Melnick, D.J. (1999). Y-chromosome phylogeny of the macaques (Cercopithecidae: *Macaca*). *Am. J. phys. Anthrop.* (Suppl.) **28**, 266.

Trainor, G.L. (1990). DNA sequencing, automation, and the human genome. *Anal. Chem.* **62**, 418–26.

van der Kuyl, A.C., Kuiken, C.L., Dekker, J.T. & Goudsmit, J. (1995a). Phylogeny of African monkeys based upon mitochondrial 12S rRNA sequences. *J. Mol. Evol.* **40**, 173–80.

van der Kuyl, A.C., Kuiken, C.L., Dekker, J.T., Perizonius, W.R.K. & Goudsmit, J. (1995b). Nuclear counterparts of the cytoplasmic mitochondrial 12S rRNA gene: a problem of ancient DNA and molecular phylogenies. *J. Mol. Evol.* **40**, 652–7.

Verheyen, W.N. (1962).Contribution à la craniologie comparée des primates. *Ann. Mus. r. Afr. cent., Sciences Zoologiques, série 8*, **105**, 1–256.

Vigilant, L., Pennington, R., Harpending, H., Kocher, T. & Wilson, A.C. (1989). Mitochondrial DNA sequences in single hairs from a southern African population. *Proc. Natl. Acad. Sci. USA.* **86**, 9350–4.

Ward, J.H. Jr. (1963). Hierarchical grouping to optimize an objective function. *J. Am. Stat. Assoc.* **58**, 236–44.

Williams-Blangero, S., Vandberg, J.L., Blanjero, J., Konigsberg, L. & Dyke, B. (1990). Genetic differentiation between baboon subspecies: relevance for biomedical research. *Am. J. Primatol.* **20**, 67–8.

Wright, S. (1978). *Evolution and the Genetics of Populations*. Chicago: University of Chicago Press.

Zhang, Y.-P. & Ryder, O.A. (1998). Mitochondrial cytochrome b gene sequences of Old World monkeys: with special reference on the evolution of Asian colobines. *Primates* **39**, 39–49.

Zuckerdandl, E. & Pauling, L. (1962). Molecular disease, evolution, and genetic heterogeneity. In *Horizons in Biochemistry*, ed. M. Kasha & N. Pullman, pp. 189–225. New York: Academic Press.

3

Molecular genetic variation and population structure in *Papio* baboons

JEFFREY ROGERS

Introduction: why study the genetics of baboons?

Baboons (*Papio hamadryas*) have been the subject of much scientific investigation. The ecology, behavior, demography, anatomy and physiology of baboons have been examined in detail. This scientific attention has also included a number of studies of specific genes, proteins and non-coding DNA sequences. Genetic research on baboons has usually addressed one of the following three topics. First, some studies have focused on the structure and/or function of specific baboon genes or proteins. Second, some investigations have used baboons as animal models in analyses of genetic factors in common human diseases such as athersclerosis, osteoporosis and hypertension (VandeBerg and Williams-Blangero, 1996). Third, many studies have used genetic data as tools to investigate other aspects of baboon biology, such as comparing reproductive success among competing males or assessing the phylogenetic relationships among *Papio* and other primate genera.

In the case of baboons, but not all nonhuman primates, the third type of genetic study has been the most common. Some studies have examined proteins or DNA sequences in baboons in order to answer questions concerning social behavior or demography. For example, genetic variation among individual baboons has been used to ascertain paternity for a series of infants and to measure male reproductive success (Altmann *et al.*, 1996). Quantification of genetic variation within natural populations can also lead to insights regarding demography. The amount of variability segregating within a natural population is influenced by a number of processes, including the history of changes in the size of the population, the pattern of mating within the population, and the frequency of migration into the population from outside. Estimates of the amount of genetic variation in any given population can be used to infer various aspects of population demography and history that cannot be observed directly (Rogers and Kidd, 1996). Finally, genetic data can be used to reconstruct the phylogenetic relationships among baboons, other cercopithecines, and more distantly related taxa (Disotell, 1996).

Empirical studies of genetic differences among baboons began in the early 1960s (Buettner-Janusch, 1963; Barnicot *et al.*, 1965). Over the years, researchers have identified a number of polymorphic genetic markers in the genus *Papio*, and some of these marker systems have been analyzed in natural populations. The two goals of this chapter are to summarize previously published information about protein and DNA sequence variation among baboons and to present new data and analyses of DNA polymorphisms. It will not be possible to be comprehensive in the coverage of past studies or in the discussion of the evolutionary implications of currently available data. This chapter is not intended to be a definitive analysis of baboon population genetics, since there are significant gaps in our knowledge, and definitive conclusions are not possible at this time.

Given the long-standing debate concerning the taxonomy of baboons, a brief comment regarding classification and nomenclature is warranted. While it is clear that *Papio* baboons exhibit a high degree of morphological, behavioral and ecological diversity relative to most other primate species, there is now a substantial amount of evidence indicating that the various forms or types of baboons (excluding geladas) are best considered subspecies, as opposed to full species. Jolly (1993) has provided an excellent review of this controversial subject. There may be some justification for recognizing seven or more distinct subspecies (Jolly, 1993). In this chapter, I will use the classification based on five subspecies (hamadryas or sacred, olive, yellow, chacma, and Guinea baboons) and refer to all of them as *Papio hamadryas*. Subspecies designations are therefore *P. h. hamadryas, P. h. anubis, P. h. cynocephalus, P. h. ursinus, and P. h. papio*, respectively.

The early genetic markers: protein polymorphisms

The first studies of baboon genetics involved the analysis of expressed proteins using a set of methods collectively known as protein electrophoresis (Buettner-Janusch, 1963; Barnicot *et al.*, 1965). These methods are not able to detect all the amino acid variation in a sample of proteins, but variability in several different proteins was soon observed among baboons. Jolly and Brett (1973) evaluated the known baboon genetic polymorphisms and discussed the application of those polymorphic markers to questions of baboon population biology. However, at the time of their review, only eight polymorphisms had been described in baboons and most provided little information because they had few alleles and low heterozygosity. Furthermore, only one natural population had been investigated for genetic variability (Buettner-Janusch and Olivier, 1970). As a result, Jolly and Brett looked primarily to the future and concluded that the study of polymorphic protein markers in natural populations presented the possibility of generating novel data for analyses of baboon taxonomy, population demography and social behavior. They further suggested that field studies of genetic structure should be undertaken in a number of localities to initiate analyses of broad patterns of variability.

Despite the initial interest in these newly discovered genetic markers, protein polymorphisms have not provided a substantially improved understanding of the social behavior, demography or geographic differentiation of baboons. One problem is the generally low level of heterozygosity of these markers. Most have only two or three alleles, and therefore provide little information about genetic differences among individuals in any given population of interest. The second problem is that protein polymorphisms are often time consuming and expensive to analyze. This is not true of all such polymorphisms, but it has discouraged large-scale genotyping.

Despite these problems, a substantial amount is known about protein polymorphisms in *Papio hamadryas*. The most thorough review of this literature (VandeBerg, 1992) lists 31 autosomal and X-linked loci that exhibit variation in one or more baboon subspecies. VandeBerg includes all the genetic markers known in baboons at that time, and presents both the numbers of alleles per locus and subspecies differences. Although the total number of genetic markers increased during the period between 1973 and 1992, these markers were still not sufficient for a reliable quantitative analysis of baboon population genetic structure. Among the 31 polymorphic protein systems listed, 20 had no more than two alleles within any one baboon subspecies, and no more than three alleles throughout all the baboons investigated. Eight loci are known to be polymorphic in only one subspecies. In one case (GPI), variability exists between subspecies rather than within them, as *P.h. papio* appears to be fixed for an allele not found in any other subspecies (Williams-Blangero *et al.*, 1990).

Another class of protein polymorphism has been described among baboons. Immunoglobulin allotypes consist of antigens present on the immunoglobulin IgG molecules that circulate in serum. Two classes of allotypes have been studied in baboons, Inv allotypes that occur on IgG light chains and Gm allotypes that occur on the IgG heavy chains (Steinberg, 1969). These genetic markers are detected through immunological reactions. Immunoglobulin markers are generally more polymorphic than other protein markers (Steinberg *et al.*, 1977; Olivier *et al.*, 1986), but they are not inherited as simple co-dominant systems and therefore are more difficult to interpret. The number of immunoglobulin systems is also small; only two types (Gm and Inv) have been reported for baboons, though other related systems are known in humans. This approach has provided valuable information, but it too had limited impact on general questions of baboon population genetics.

Restriction fragment length polymorphisms

Beginning in the early 1980s, the study of genetic variation within mammalian species was tremendously improved by the discovery of restriction fragment length polymorphisms or RFLPs (Botstein *et al.*, 1980). This approach to identifying and routinely assaying differences among indivduals in DNA sequence proved to be highly successful. The number of known human RFLPs increased dramatically from 159 in 1983 (Human Gene Mapping Conference 7, 1984) to more than 3200 in 1991 (Human Gene Mapping Conference 11, 1991). Using molecular genetic methods and DNA probes developed for studies of human variation, the RFLP approach was been used by several investigators to identify DNA polymorphisms in baboons. Hixson *et al.* (1988, 1989b, 1993) found that several baboon apolipoprotein loci exhibit this kind of variation. Restriction site polymorphisms among captive baboons in multigenerational pedigrees were analyzed to estimate the effects of specific genes on physiological variation (Kammerer *et al.*, 1993) and to conduct genetic linkage studies of baboon chromosomes (Kammerer *et al.*,1992, Rogers *et al.*, 1995a). This work clearly established that human DNA clones could be used as probes to detect DNA polymorphism among baboons, and that this approach was an efficient strategy for identifying variation relevant to biomedical research. Table 3.1 lists known RFLP systems for *Papio* baboons.

RFLPs have also been used as genetic markers in analyses of natural baboon populations. These studies have shown that wild baboons are highly polymorphic at the nucleotide level. Among the baboons of Mikumi National Park, Tanzania, Rogers and Kidd (1993) found 14 polymorphisms in five loci by screening six different restriction enzymes in a sample of 27 individuals. Nucleotide heterozygosity for this population was estimated to be 0.0033 (Rogers and Kidd, 1996). This level of variability is higher than estimates of nucleotide variability in human populations. Nucleotide diversity (π) was estimated to be 0.0053 for the Mikumi

Table 3.1. *Restriction fragment length polymorphisms in baboons*

Locus	Polymorphic Enzymes	Reference
AT3	PvuII, BglII, EcoRI, HindIII	Rogers & Kidd (1993)
HEXB	PvuII, BglII, TaqI, HindIII	Rogers & Kidd (1993)
APOB	PvuII, BglII, TaqI	Rogers & Kidd (1993)
VIM	BglII	Rogers & Kidd (1993)
REN	(a) BglII, TaqI	Rogers & Kidd (1993)
	(b) MspI, HindIII	Perelygina *et al.* (1994)
IGF1	EcoRI	Perelygina *et al.* (1995a)
IGF1R	TaqI, MspI	Perelygina *et al.* (1995a)
PDGFRβ	BglII, PvuII, TaqI, PstI, EcoRI, MspI	Perelygina *et al.* (1995b)
LCAT	PvuII	Rainwater *et al.* (1992)
CETP	MspI	Cole *et al.* (1996a)
LIPC	TaqI	Cole *et al.* (1996b)
LDLR	AvaII	Hixson *et al.* (1989b)
APOA1	PstI	Hixson *et al.* (1988)
APOA2	BglII	Rogers & Hixson (unpub. data)

baboons, while Li and Sadler (1991) estimated diversity to be 0.0011 for humans. A test of difference between means indicates that the Mikumi baboons have significantly higher diversity than humans ($p<0.0012$). It should be noted that this is a comparison between baboons sampled in a single small area of central Tanzania and humans drawn from mixed European populations, not a single local population of humans. Baboons drawn from an equivalently sized area might show even more variation.

This high level of DNA sequence variability in the baboons is consistent with other studies that have compared the amount of DNA polymorphism in humans and nonhuman primates. Ruano *et al.* (1992) showed that for a short segment of a single nuclear locus (HOX2B also called HOXB6), chimpanzees and gorillas have more intraspecies variation than do humans. Studies of mitochondrial DNA (mtDNA) have shown that chimpanzees (Morin *et al.*, 1994), gorillas (Ruvolo *et al.*, 1994) and several species of macaques (Melnick and Hoelzer, 1993) exhibit higher levels of intraspecific mtDNA variation than humans. It appears that humans are unusual among primates in having low levels of DNA sequence polymorphism, probably because the human population has expanded rapidly from a small effective size during the last 100,000–200,000 years (Rogers and Jorde, 1995).

Microsatellite polymorphisms

Weber and May (1989) and Litt and Luty (1989) described a new class of DNA polymorphisms that has had a major impact on research in human genetics. These researchers described the first microsatellite polymorphisms, – also called simple sequence repeat (SSR) or short tandem repeat (STR) polymorphisms. This class of nuclear DNA sequences consists of tandem repeats of two to five base pairs. These repeats are highly susceptible to mutations that add or remove repeats. As a result, many microsatellite loci have accumulated a large number of alleles that differ only in the number of tandem repeats (Weber, 1990). Polymorphisms of this type have become the predominant genetic markers used for linkage mapping in the human genome (Dib *et al.*, 1996). They have also been used to investigate the history of genetic differentiation among human populations (e.g. Bowcock *et al.* 1994, Jorde *et al.* 1995). Over 8000 STR loci have been described in humans (Dib *et al.*, 1996; see also the Genome Database, Baltimore, MD, http://gdbwww.gdb.org/), most with heterozygosities >0.70.

Given the large amount of information concerning human genetic variation generated through the study of microsatellite polymorphisms, it is reasonable to infer this approach would also be valuable for other primates. Inoue and Takenaka (1993) reported three microsatellite (CA dinucleotide repeat) polymorphisms in baboons: one 3-allele system, and two 2-allele systems. The loci described by Inoue and Takanaka (1993) were cloned from *Macaca fuscata*, primers were synthesized to complement the macaque DNA sequence, and the primers were later tested in baboons. This approach, the cloning of new microsatellite loci, can be used to develop this type of DNA marker in any primate species. The three markers reported by Inoue and Takanaka (1993) do not seem to be highly polymorphic in baboons, but only a small number of individuals were studied and it is possible additional alleles would be observed if more baboons were screened.

As noted above, thousands of microsatellites are known from the human genome, with primer sequences readily available in publications or in computer databases. An alternative to cloning new microsatellite loci from any nonhuman primate species of interest is to test human polymerase chain reaction (PCR) primers to determine whether they will amplify the homologous locus in the species under study. If the human primers do amplify the locus cleanly enough to be useful (i.e. without substantial numbers of spurious PCR products that obscure the desired product), then the locus can be tested for polymorphism in the nonhuman primate. This strategy has been used to identify polymorphic microsatellites in baboons (Rogers *et al.*,

1995a; Altmann *et al.*, 1996), and other primate species (Morin *et al.*, 1994; Rogers *et al.*, 1995b).

We have employed several human microsatellite loci in studies of genetic linkage mapping in baboons. Our long-term goal is to construct a complete linkage map of the baboon genome, with an average spacing between loci of 10–15 centiMorgans. The majority of markers in this map will be microsatellites originally cloned from the human genome. In the first report describing this effort, five dinucleotide repeat loci that are located on human chromosome 1q (D1S306, D1S194, D1S104, D1S215, D1S158) were assayed in a multigenerational pedigree of captive baboons (Rogers *et al.*, 1995a). The average number of alleles observed per locus was 9.0, and the average heterozygosity per locus was 0.74. The five loci are all linked in baboons, with the order and approximate recombination distances apparently conserved between humans and baboons. More recently, Perelygin *et al.* (1996) have examined eight microsatellites known to map to human chromsome 18. The average number of alleles in this set was 8.9 and the average heterozygosity 0.65. As in humans, these loci are syntenic in baboons and their physical order along the chromosome is apparently the same in humans and baboons.

Human microsatellites are also being used in studies of natural populations of baboons. Figure 3.1 presents the allele frequencies for two human microsatellite loci typed in the free-ranging yellow baboons of Mikumi National Park, Tanzania. D8S165 consists of a dinucleotide repeat first recognized to be polymorphic in a panel of olive baboons from the Southwest Foundation (Rogers and Witte, unpub. data). A randomly selected sample of Mikumi yellow baboons exhibits a total of 10 alleles at this locus. Figure 3.1 also presents allele frequencies for D2S144, a tetranucleotide repeat, that has 9 alleles in these animals. The estimated heterozyosity in the Mikumi population is 0.803 for D2S144, and 0.836 for D8S165. Figure 3.2 illustrates the variation in D2S144.

Altmann *et al.* (1996) investigated 10 human microsatellites in the baboons of Amboseli National Park, Kenya. The 10 DNA polymorphisms, along with two protein markers, were typed in 76 individuals. There were an average of 4.9 alleles per microsatellite locus. Altmann and her colleagues used the genotype information to determine paternity for a series of infants and to calculate kinship coefficients within and between matrilines. St. George *et al.* (1995) studied variability in three microsatellites among 30 free-ranging individuals sampled from each of two yellow baboons populations – Mikumi National Park, Tanzania and the Tana River Primate Reserve, Kenya. We found the average heterozygosity across the three loci to be 0.75 in Mikumi and 0.69 in the Tana population. These studies

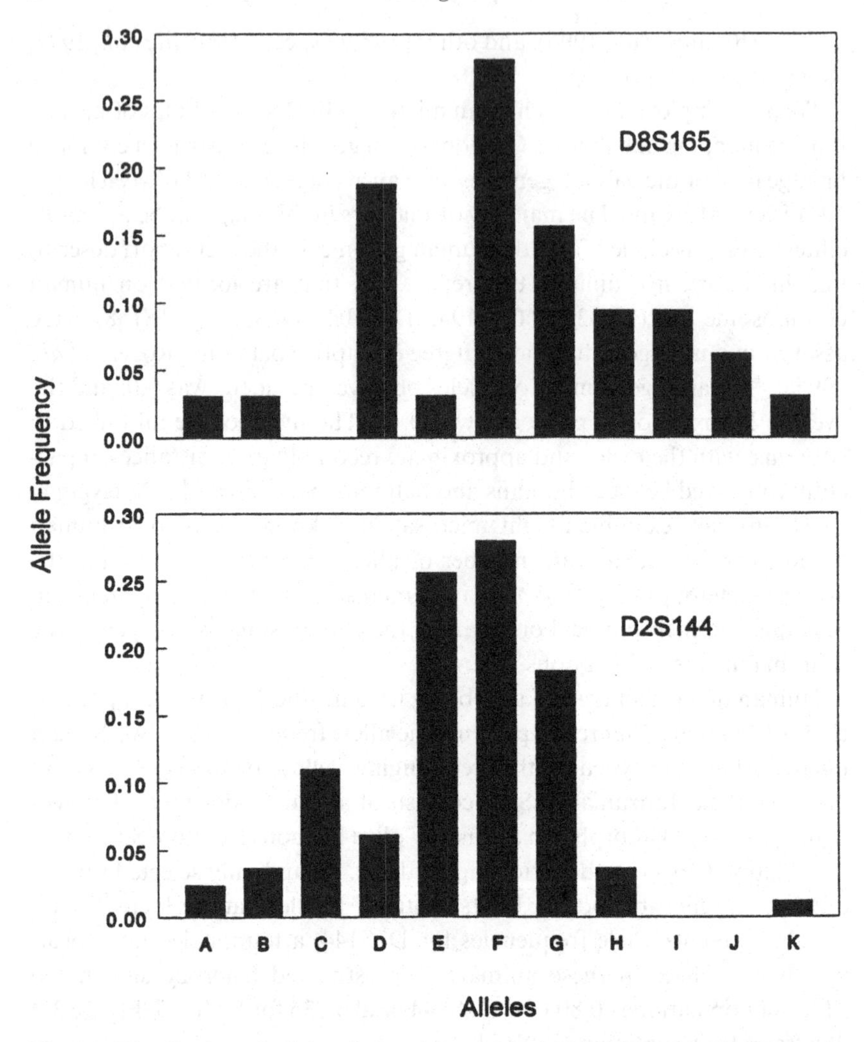

Fig. 3.1. Distribution of alleles observed for two loci, D8S165 and D2S144, among yellow baboons. Allele C for D8S165 was not observed in this group of individuals, but may be found in a larger sample. The same is true for alleles I and J for D2S144.

demonstrate that human microsatellite loci will be more valuable for analyses of reproductive behavior, demography and genetic differentiation among wild baboons than protein polymorphisms or RFLPs because they have more alleles per locus and higher heterozygosity.

Given that the number of known microsatellite polymorphisms in nonhuman primates is increasing steadily, this approach is likely to dramatically improve our ability to quantify population genetic structure in many pri-

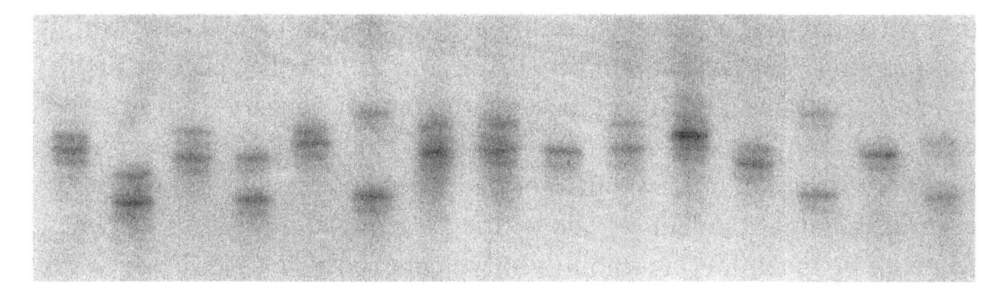

Fig. 3.2. Autoradiograph showing microsatellite variation at locus D2S144. Each vertical lane represents the polymerase chain reactin (PCR) amplification product of one baboon. Bands in each lane differ in position because of the differences in the length of the PCR product. Longer products have more copies of the tandem repeat unit.

mates including *Papio* (see Bruford and Wayne, 1993). There are also other advantages to this approach. It may not be necessary to obtain DNA from blood samples in order to investigate a set of microsatellites, if hair or feces are available (Morin *et al.*, 1993). Once useful primers have been identified and the proper PCR reaction conditions determined, the amplification and genotyping of microsatellites is much more rapid than typing of RFLPs. Though the use of human PCR primers may be an efficient strategy for hominoids and Old World monkeys, it may in other cases be necessary to clone new loci from the species of interest, as done by Inoue and Takanaka (1993). Cloning of new microsatellite loci from nonhuman species will be especially important when studying platyrrhines or strepsirhines, where human PCR primers are much less likely to function well.

Other types of DNA polymorphism

The baboon microsatellite polymorphisms described above are dinucleotide and tetranucleotide repeats, and probably occur in DNA segments that do not code for functional proteins or RNA molecules. The presence of a variable number of 2- or 4-base pair repeats would almost certainly disrupt the function of a protein- or RNA-coding gene. It is reasonable to assume that most or all the known baboon microsatellites occur in the intron sequences of genes, or in the intervening sequences between functional genes. Two tandem repeat polymorphisms in baboons are known to occur in coding regions, but these are not classic microsatellites. The D4 dopamine receptor gene (DRD4) contains a protein-coding repeat that is polymorphic (Livak *et al.*, 1995). This 48 base pair segment codes for a 16 amino acid motif in the DRD4 protein, and occurs a variable number of times in different alleles. Several primate species, including baboons, exhibit variation in DRD4 repeat number (Livak *et al.*, 1995). The baboon apo(a) gene also contains a polymorphic series of tandem repeats, each over 300 base pairs long (Hixson *et al.*, 1989a; White *et al.*, 1994). This variation is similar to variation observed in apo(a) among humans. This type of polymorphism, larger tandem repeats within protein-coding regions of genes, seems to be much less common than typical di-, tri-, and tetranucleotide repeat microsatellites.

All the genetic polymorphisms discussed above involve loci in the nuclear genome. The analysis of variability in mitochondrial DNA within species has also become an important aspect of primate population genetics. Mitochondrial DNA can be used for phylogenetic analyses among closely related species (Melnick and Hoelzer, 1993; Horai *et al.*, 1995), among subspecies (Morin *et al.*, 1994) and among recently separated populations within a species (Cann *et al.*, 1987; Jorde *et al.*, 1995). Little is known about variability of mtDNA within *Papio*. Newman and co-workers (1996, 1997) have surveyed mtDNA haplotypes in the Awash hybrid zone where hamadryas and olive baboons interbreed. More work on the variability and distribution of mtDNA haplotypes, within and between subspecies of baboons, would provide an indication of matrilineal relationships among extant baboon populations and complement studies of protein and microsatellite markers (see Melnick and Holezer, 1993).

Population genetic structure

More than 90 different genetic polymorphisms have been described in *Papio*. Our understanding of the demography and evolutionary biology of baboons would be substantially improved if a significant number of highly informative genetic markers were analyzed in a series of natural populations distributed across the geographic range and taxonomic diversity of *Papio*. Allele frequency information for 30–40 or more highly polymorphic loci would permit quantitative analysis of the amount of genetic variation in specific local populations and the degree to which that variation is shared with other populations of the same and different subspecies. Reliable quantitative descriptions of the amount and pattern of genetic differences among social groups within a given small area would facilitate inferences about the pattern and genetic consequences of male migration among social groups. Analysis of the relationships among geographic distance, subspecies classification and genetic differentiation among populations will allow reconstruction of the longer-term evolutionary history of genetic change. This type of information would eventually generate an overall picture of baboon population genetic structure. Such a picture is developing for human populations (Cavalli-Sforza *et al.*, 1994; Bowcock *et al.*, 1994), and it is leading to important insights regarding human evolution. The state of knowledge regarding the population genetic structure of macaque species, especially rhesus, is also better than the situation for baboons. Extensive data describing the distribution of protein and mtDNA variation within and among macaque species are available (Melnick, 1988; Melnick and Hoelzer, 1993, 1996).

However, few of the available polymorphisms have been investigated in more than one or two natural populations of baboons. Each individual polymorphism is subject to substantial random fluctuations, so a significant number of independent loci is required before a reliable overall picture is produced. Consequently, the patterns of genetic variation within or between local baboon populations cannot now be described in reliable quantitative terms. (In this context, a local population or locality is considered to be all the baboons found in an area no larger than about 600–800 square kilometers. See discussion of inbreeding effective population size below.) Some population structure data are available, and therefore some preliminary descriptions are possible. But the following discussion of baboon population genetic structure is based on minimal information. Any general conclusion must be tested against significant amounts of additional data incorporating more genetic markers and more local populations.

Variability within local populations

The most basic parameter to be estimated is the amount of genetic variation within a local population. Several studies have surveyed multiple protein markers among baboons from a single locality (e.g. McDermid *et al.*, 1973; Shotake, 1981; Sanders, 1981). Generally, protein polymorphism is observed in 25–50% of the markers tested. Unfortunately, this is not a random sample of genetic loci, since investigators have tended to assay systems that have been informative in past studies of other populations. Estimation of the overall level of variability requires a random sampling of loci. One study of DNA polymorphism was undertaken as a random assessment of variation (Rogers and Kidd, 1993, 1996). Five loci were screened for RFLP variation in the yellow baboons of Mikumi National Park, and the average heterozygosity was found to be 0.0033. This level of variability is higher than nucleotide variability estimated for human populations. At this point, no other baboon population has been surveyed in this manner.

Estimation of inbreeding effective population size allows one to infer aspects of the evolutionary history and dynamics of animal populations (Hartl and Clark, 1989). Neutral genetic variation (i.e. variation that has essentially no influence on the fitness of an individual) will accumulate in populations as a result of new mutations or the introduction of new variants through gene flow from other conspecific populations. Random genetic drift will gradually eliminate this variation, and the balance between the introduction of new variation and loss of existing variation determines the amount of variability observed in a population at any given time. Larger populations generally retain more variation than smaller ones, and hence the amount of variation present provides some indication of the size of a given population over recent evolutionary time. In addition, the pattern of mating within a population can influence the rate at which variation is lost. Unequal reproductive success among individuals can also influence the amount of variability retained. Since the boundaries of natural baboon populations are usually difficult or impossible to establish through field observations, and the long-term patterns of mating and reproductive success are also difficult to determine, analyses of the effective size of populations based on estimates of the current level of genetic variation can lead to new information about population structure and dynamics. Knowing the inbreeding effective population size in a specific region allows the investigator to infer aspects of a population's history, demography and evolutionary relationships with other populations.

The RFLPs identified in the Mikumi baboons are likely to be functionally neutral, because most polymorphisms will be found in the non-coding regions of genes or intervening sequences. As a result, the observed level of DNA polymorphism can be used to estimate the inbreeding effective population size N_e. Rogers and Kidd (1996) estimated N_e for the Mikumi baboons to be approximately 14,200 and inferred that the majority of adult males disperse less than 20 kilometers when migrating between groups. In Mikumi, baboon troops range in size from about 20 individuals to over 100, but most have between 50 and 90 individuals (Rogers, 1989, unpub. obs.). Home ranges for troops are as large as 80–100 square kilometers. Mikumi Park is an area of about 3300 square kilometers, and it is likely there are between 150 and 250 baboon troops in the park. If the average group is about 70 individuals, with 20–30 adults, then the number of breeding adult baboons is about 4000–6000. Since this number is lower than the calculated N_e, this implies that the entire baboon population of the park behaves as a single interconnected genetic population, and that other baboon groups outside the park are also part of the same interbreeding population.

However, this analysis depends on several assumptions, including the assumption that population density has remained roughly stable for many generations. If the density of baboons was significantly different in the past, then this estimate of the area covered by the "genetic population" or deme may be incorrect.

A number of studies have analyzed allele frequency differences among social groups in a local baboon population. Brett *et al.* (1976) investigated the ABO blood group polymorphism in 11 troops from the hybrid zone in the Awash Park, Ethiopia. They found that allele frequencies varied significantly among neighboring groups. Byles and Sanders (1981) determined ABO allele frequencies among five troops of Kenyan olive baboons, and observed significant differentiation among troops thought to exchange males and therefore undergo reciprocal gene flow. Similar patterns occurred with two other polymorphisms (Sanders, 1981). Studies of immunoglobulin allotypes in olive baboons (Steinberg *et al.*, 1977; Coppenhaver and Olivier, 1986; Olivier *et al.*, 1986) also found substantial inter-troop differences within localities. Shotake (1981) studied 10 protein polymorphisms in the same hamadryas–anubis hybrid zone investigated by Brett *et al.* (1976). This analysis found significant genetic differentiation among nearby troops, and Shotake was able to show that allele frequency differences are smaller among troops from the same subspecies than across subspecies. Groups containing a mixture of olive and hamadryas baboons were intermediate. Studies of protein and DNA restriction site polymorphisms in the yellow baboons of Mikumi National Park, Tanzania indicate that F_{ST}, a measure of genetic variance across groups, is 0.049 among complete social groups and 0.057 if only natal animals are included (Rogers, 1989). Like the previous studies, this analysis indicates substantial differentiation among groups.

How do we explain significant genetic differences among social groups that regularly exchange members? Inter-troop migration by males should tend to reduce allele frequency differences among groups. But the number of breeding adults in individual baboon groups can be small, and the number of breeding males can be very small (see Altmann *et al.*, 1996). In such circumstances, random genetic drift is a powerful force producing random changes in allele frequencies within any one group. The variance effective size of social groups is a statistic that predicts the probable rate of random genetic drift within individual groups (Hartl and Clark, 1989). The smaller the variance effective size of a group, the larger the random fluctuations in allele frequencies. It seems likely, given the observed pattern of significant inter-group differentiation, that the variance effective size of baboon social groups is quite small. If we assume that all male yellow baboons disperse from their natal social groups, the variance effective size of those groups can be estimated from

$$F_{ST} = \frac{1}{(4Nm + 1)}$$

where m is the migration rate per generation (here 0.5) and N is the variance effective group size (Hartl and Clark, 1989). Using $F_{ST} = 0.049$ (Rogers, 1989), and solving for N leads to the conclusion that the variance effective size of Mikumi baboon troops is 9.7. This predicts that the average baboon troop will undergo genetic drift at a rate equivalent to that of a breeding group with about 15 adult females and three adult males, assuming no variance in reproductive success among males or among females The observed average census size of groups is larger than 9.7, suggesting that real baboon groups deviate from the theoretical model. For example, it is likely that all adults do not have equal reproductive success. When variance effective size is low, allele frequencies are likely to fluctuate rapidly from generation to generation in each troop within a population. Any one troop can be considered a temporary combination of alleles, the result of chance fluctuations due to migration and the stochastic process of Mendelian inheritance. The total amount of genetic variation found in a given social group is a function of its size and the inbreeding effective size of the local population. But the allele frequencies in any one group are likely to change rapidly, and hence diverge from neighboring groups.

Variability among localities, regions, or subspecies

Only a few genetic markers have been analyzed in several baboon populations distributed over large geographic distances. The ABO blood group system has been typed in more populations than any other polymorphism (Weiner and Moor-Jankowski, 1969; McDermid *et al.*, 1973; Downing *et al.*, 1975; Brett *et al.*, 1976; Byles and Sanders 1981). There is substantial variation in observed allele frequencies across these studies, which collectively have sampled all five subspecies of *Papio*. But no statistical analysis of genetic structure is warranted, because any one marker is subject to both random genetic drift in individual populations and sampling error in the estimation of allele frequncies. To calculate meaningful estimates of the overall differentiation across populations within and between subspecies would require a larger number of loci, all typed on the same populations. The protein polymorphism in transferrin has also been typed in several localities (Buettner-Janusch and Olivier, 1970; McDermid *et al.*, 1973; Shotake, 1981; Rogers, 1989), but this is still too little information. These markers suggest that there may be genetic variability both among localities within a subspecies and between subspecies, but no detailed analyses would be meaningful.

Olivier and his colleagues have described variability in immunoglobulin allotypes across several populations of olive baboons (Steinberg *et al.*, 1977; Olivier *et al.*, 1986). Comparisons of the frequencies of Gm allotypes indicate that substantial differentiation occurs among localities separated by 50–100 kilometers. However, Olivier and his colleagues also found significant allotype frequency differences among baboon troops from the same locality, i.e., troops separated by no more than 20 kilometers. The Gm allotype analyses suggest that the genetic composition of local baboon populations, i.e., troops found within an area of less than a few hundred square kilometers, can show significant genetic differentiation. Furthermore, genetic differences between localities much farther apart geographically are not necessarily greater than differences within localities. The studies of Gm allotypes by Olivier and his colleagues present systematic analysis of the same genetic markers in hundreds of baboons from more than two dozen troops representing about 10 localities. However, basing conclusions regarding baboon population structure on these loci alone is highly problematic, and these populations were not typed for ABO or transferrin variation.

In general, protein polymorphisms are distributed across broad geographic areas that include several research localities and subspecies. Williams-Blangero *et al.* (1990) found that five of the nine protein polymorphisms they studied were shared by at least two subspecies. Their measure of genetic distance among subspecies shows that Guinea baboons are apparently the most distinct. Seven of the ten pairwise comparisons between subspecies exhibited statistically significant differences in alleles frequencies. Six out of ten protein polymorphisms are shared by hamadryas and olive baboons in the Awash hybrid zone (Shotake, 1981). This sharing of allelic variation across subspecies may result from gene flow among them. Alternatively, the polymorphisms may predate the divergence of the subspecies.

However, some populations do not exhibit variation for specific genetic markers that are broadly distributed throughout the species. For example, Sanders (1981) surveyed several hundred olive baboons from Gilgil, Kenya and found no variation in transferrin. This protein is polymorphic in several baboon populations drawn from various subspecies (Buettner-Janusch and Olivier, 1970; McDermid *et al.*, 1973; Rogers, 1989). The Gilgil baboons seem to have lost their variability in transferrin, but remain polymorphic for other markers (Sanders, 1981). Presumably this is the result of fixation of one transferrin allele by genetic drift or selection.

The studies of protein variability in populations of *Papio* baboons suggest that: (a) many polymorphisms are shared across broad geographic ranges and include several subspecies; and (b) there are substantial differences in allele frequencies among localities, which in some instances involved the fixation of one allele in a given population (or set of populations). However, we must recognize that both the number of highly informative protein markers and the number of localities sampled is low. The electrophoretic analysis of protein variation has provided only a preliminary understanding of the genetic structure of baboon populations. DNA markers have not yet been surveyed broadly across natural populations.

Summary

This chapter can be summarized as follows:

(1) Baboons exhibit substantial amounts of intra-specific genetic variation, both at the protein level and at the DNA nucleotide level.
(2) The inbreeding effective population size of one population of yellow baboons has been estimated, on the basis of RFLP data from five loci, to be about 14,200.
(3) Allele frequencies often vary substantially among social groups within a small area. The variance effective size for one series of social groups was estimated, using protein and RFLP data, to be about 10. This estimate is also based on a small sample of loci, and could change significantly once additional data are available.
(4) Allele frequencies for protein polymorphisms often differ significantly between subspecies, and between regions within *Papio hamadryas anubis*, but this observation is based on too few loci to permit accurate quantification.
(5) Little is known about the details of the broad geographic distribution of genetic variation in baboon populations, either across large areas within a subspecies, or between subspecies. Additional studies including many more highly polymorphic markers, such as microsatellites, would greatly improve our understanding of baboon genetic diversity and population genetic structure.

References

Altmann, J., Alberts, S.C., Haines, S.A., Dubach, J., Muruthi, P., Coote, T., Geffen, E., Cheesman, D.J., Mututua, R.S., Saiyalel, S.N., Wayne, R.K., Lacy, R.C. & Bruford, M.W. (1996). Behavior predicts genetic structure in a wild primate group. *Proc. Nat. Acad. Sci. USA* **93**, 5797–801.

Barnicot, N., Jolly, C.J., Huehns, E.R. & Dance, N. (1965). Red cell and serum protein variants in baboons. In *The Baboon in Medical Research*, vol. I, ed. H. Vagtborg, pp. 323–38. Austin, TX: University of Texas Press.

Botstein, D., White, R.L., Skolnick, M. & Davis, R.W. (1980). Construction of a genetic linkage map in man using restriction fragment length polymorphisms. *Amer. J. Hum. Genet.* **32**, 314–31.

Bowcock, A.M., Ruiz-Linares, A., Tomfohrde, J., Minch, E., Kidd, J.R. & Cavalli-Sforza, L.L. (1994). High resolution of human evolutionary trees with polymorphic microsatellites. *Nature* **368**, 455–7.

Brett, F.L., Jolly, C.J., Socha, W. & Weiner, A.S. (1976). Human-like ABO blood groups in wild Ethiopian baboons. *Yrbk. Phys. Anthrop.* **20**, 276–89.

Bruford, M.W. & Wayne, R.K. (1993). Microsatellites and their application to population genetic studies. *Curr. Opin. Genet. Develop.* **3**, 939–43.

Buettner-Janusch, J. (1963). Hemoglobins and transferrins of baboons. *Folia Primatol.* **1**, 73–87.

Buettner-Janusch, J. & Olivier, T.J. (1970). Distribution of transferrin phenotypes in selected troops of Kenya baboons. *Am. J. phys. Anthrop.* **33**, 303–6.

Byles, R.H. & Sanders, M.F. (1981). Intertroop variation in the frequencies of ABO alleles in a population of olive baboons. *Inter. J. Primat.* **2**, 35–46.

Cann, R.L., Stoneking, M. & Wilson, A.C. (1987). Mitochondrial DNA and human evolution. *Nature* **325**, 31–6.

Cavalli-Sforza, L.L., Menozzi, P. & Piazza, A. (1994). *The History and Geography of Human Genes*. Princeton: Princeton University Press.

Cole, S.A., Birnbaum, S. & Hixson, J.E. (1996a) MspI RFLP at the CETP locus in baboons. *Anim. Genet.* **27**, 63.

Cole, S.A., Presley, L. & Hixson, J.E. (1996b). TaqI RFLP at the LIPC locus in baboons. *Anim. Genet.* **27**, 63.

Coppenhaver, D.H.& Olivier, T.J. (1986). Immunoglobulin allotypes of Kenyan olive baboons: troop frequencies, linkage disequilibria and comparisons with other studies. *Intern. J. Primat.* **7**, 335–50.

Dib, C., Faure, S., Fizames, C., Samson, D., Drouot, N., Vignal, A., Millasseau, P., Marc, S., Hazan, J., Seboun, E., Lathrop, M., Gyapay, G., Morissette, J. & Weissenbach, J. (1996). A comprehensive genetic map of the human genome based on 5,264 microsatellites. *Nature* **380**, 152–4.

Disotell, T.R. (1996). The phylogeny of Old World monkeys. *Evol. Anthrop.* **5**, 18–24.

Downing, H.J., Burgers, L.E. & Getliffe, F.M. (1975). ABO blood groups of two subspecies of chacma baboons (*Papio ursinus*) in South Africa. *J. Med. Primat.* **4**, 103–7.

Hartl, D.L. & Clark, A.G. (1989). *Principles of Population Genetics*, 2nd edn. Sunderland, MA: Sinuaer Press.

Hixson, J.E., Borenstein, S., Cox, L.A., Rainwater, D.L. & VandeBerg, J.L. (1988). The baboon gene for apolipoprotein A-I: characterization of a cDNA clone and identification of DNA polymorphisms for genetic studies of cholesterol metabolism. *Gene* **74**, 483–90.

Hixson, J.E., M.L. Britten, G.S. Manis & Rainwater, D.L. (1989a).

Apolipoprotein(a) glycoprotein isoforms result from size differences in apo(a) mRNA in baboons. *J. Biol. Chem.* **264**, 6013–16.

Hixson, J.E., Kammerer, C.M., Cox, L.A. &. Mott, G.E. (1989b). Identification of LDL receptor gene marker associated with altered levels of LDL cholesterol and apolipoprotein B in baboons. *Arteriosclerosis* **9**, 829–35.

Hixson, J.E., Kammerer, C.M., Mott, G.E., Britten, M.L., Birnbaum, S., Powers, P.K. & VandeBerg, J.L. (1993). Baboon apolipoprotein A-IV. *J. Biol. Chem.* **268**, 15667–73.

Horai, S., Hayasaka, K., Kondo, R., Tsugane, K. & Takahata, N. (1995). Recent African origin of modern humans revealed by complete sequences of hominoid mitochondrial DNA. *Proc. Nat. Acad. Sci. USA* **92,** 532–6.

Human Gene Mapping Conference 7 (1984). *Cytogenet. Cell Genet.* **37,** 1–666.

Human Gene Mapping Conference 11 (1991). *Cytogenet. Cell Genet.* **58**, 1–2200.

Inoue, M. & Takenaka, O. (1993). Japanese macaque microsatellite PCR primers for paternity testing. *Primates* **34**, 37–45.

Jolly, C.J. (1993). Species, subspecies and baboon systematics. In *Species, Species Concepts and Primate Evolution*, ed. W.H. Kimbel & L.B. Martin, pp. 67–107. New York: Plenum Press.

Jolly, C.J. & Brett, F.L. (1973). Genetic markers and baboon biology. *J. Med. Primatol.* **2**, 85–99.

Jorde, L.B., Bamshad, M.J., Watkins, W.S., Zenger, R., Fraley, A.E., Krakowiak, P.A., Carpenter, K.D., Soodyall, H., Jenkins, T. & Rogers, A.R. (1995). Origins and affinities of modern humans: a comparison of mitochondrial and nuclear genetic data. *Amer. J. Hum. Genet.* **57**, 523–38.

Kammerer, C.M., Hixson, J.E., Aivaliotis, M.J., Porter, P.A. & VandeBerg, J.L. (1992). Linkage heterogeneity between the C3 and LDLR and the APOA4 and APOA1 loci in baboons. *Genomics* **14**, 43–8.

Kammerer, C.M., Hixson, J.E. & Mott, G.E. (1993). A DNA polymorphism for lecithin:cholesterol acyltransferase (LCAT) is associated with high density lipoprotein cholesterol concentrations in baboons. *Atherosclerosis* **98**, 153–63.

Li, W.H. & Sadler, L.A. (1991). Low nucleotide diversity in man. *Genetics* **129**, 513–23.

Litt, M. & Luty, J.A. (1989). A hypervariable microsatellite revealed by *in vitro* amplification of a dinucleotide repeat within the cardiac muscle actin gene. *Amer. J. Hum. Genet.* **44,** 397–401.

Livak, K.J., Rogers, J. & Lichter, J.B. (1995). Variability of dopamine D4 receptor (DRD4) gene sequence within and among nonhuman primate species. *Proc. Nat. Acad. Sci. USA* **92**, 427–31.

McDermid, E.M., Vos, G.H. & Downing, H.J. (1973). Blood groups, red cell enzymes and serum proteins of baboons and vervets. *Folia Primat.* **19**, 312–26.

Melnick, D.J. (1988). The genetic structure of a primate species: rhesus macaques and other cercopithecine monkeys. *Inter. J. Primat.* **9,** 195–231.

Melnick, D.J. & Hoelzer, G.A. (1993). What is mtDNA good for in the study of primate evolution. *Evol. Anthrop.* **2**, 2–10.

Melnick, D.J. & Hoelzer, G.A. (1996). The population genetic consequences of macaque social organisation and behaviour. In *The Evolution and Ecology of Macaque Societies*. ed. J.E. Fa & D.G. Lindburg, pp. 413–43. Cambridge: Cambridge University Press.

Morin, P.A., Moore, J.J., Chakraborty, R., Jin, L., Goodall, J. & Woodruff, D.S. (1994). Kin selection, social structure, gene flow and the evolution of chimpanzees. *Science* **265**, 1193–201.

Morin, P.A., Wallis, J., Moore, J.J., Chakraborty, R. & Woodruff, D.S. (1993). Non-invasive sampling and DNA amplification for paternity exclusion, community structure and phylogeography in wild chimpanzees. *Primates* **34**, 347–56.

Newman, T.K., Jolly, C.J., Disotell,T.R., Phillips-Conroy, J. & Brett, F.L. (1996). Spatial population structure in a baboon hybrid zone, central Ethiopia, based on mtDNA variation. *Am. J. phys. Anthrop. (Suppl.)* **22,** 177.

Newman, T.K. & Disotell, T.R. (1997). The distribution of mtDNA markers in 1973 and its implications for mechanisms of hybridization in the Awash. *Am. J. phys. Anthrop. (Suppl.)* **24**, 177–8.

Olivier, T.J., Coppenhaver, D.H. & Steinberg, A.G. (1986). Distribution of immunoglobulin allotypes among local populations of Kenyan olive baboons. *Am. J. phys. Anthrop.* **70**, 29–38.

Perelygin, A.A., Kammerer, C.M., Stowell, N.C. & Rogers, J. (1996). Conservation of human chromosome 18 in baboons (*Papio hamadryas*): a linkage map of eight human microsatellites. *Cytogenet. Cell Genet* **75**, 207–9.

Perelygina, L.M., Kammerer, C.M. & Henkel, R.D. (1995a). RFLPs at the baboon insulin-like growth factor 1 (IGF1) and the IGF1 receptor (IGF1R) loci. *Anim. Genet.* **26,** 280.

Perelygina, L.M., Kammerer, C.M. & Henkel, R.D. (1995b). An RFLP map of the baboon platelet-derived growth factor recertor gene. *Mammal. Genome* **6**, 373–5.

Perelygina, L.M., Kammerer, C.M. & Henkel, R.D. (1994). RFLPs in the baboon renin locus. *Anim. Genet.* **25,** 198.

Rainwater, D.L., Blangero, J., Hixson, J.E., Birnbaum, S., Mott, G.E. & VandeBerg, J.L. (1992). A DNA polymorphism for LCAT is associated with altered LCAT activity and high density lipoprotein size distributions in baboons. *Arteriosclerosis and Thrombosis* **12**, 682–90.

Rogers, J. (1989). *Genetic Structure and Microevolution in a Population of Tanzanian Yellow Baboons*. PhD dissertation, Yale University.

Rogers, A.R. and Jorde, L.B. (1995). Genetic evidence on the origin of modern humans. *Hum. Biol.* **67**, 1–36.

Rogers, J. & Kidd, K.K. (1993). Nuclear DNA polymorphisms in a wild population of yellow baboons (*Papio hamadryas cynocephalus*) from Mikumi National Park, Tanzania. *Amer. J. phys. Anthrop.* **90**, 477–86.

Rogers, J. & Kidd, K.K. (1996). Nucleotide polymorphism, effective population size and dispersal distances in the yellow baboons (*Papio hamadryas cynocephalus*) of Mikumi National Park, Tanzania. *Amer. J. Primatol.* **38**, 157–68.

Rogers, J., Witte, S.M., Kammerer, C.M., Hixson, J.E. & MacCluer, J.W. (1995a). Linkage mapping in *Papio* baboons: conservation of a syntenic group of six markers on human chromosome 1. *Genomics* **28,** 251–4.

Rogers, J., Witte, S.M. & Slifer, M.A. (1995b). Five new microsatellite DNA polymorphisms in squirrel monkeys. *Amer. J. Primat.* **36,** 151. (Abstract.)

Ruano, G., Rogers, J., Ferguson-Smith, A.C. & Kidd, K.K. (1992). DNA sequence polymorphism within hominoid species exceeds the number of phylogenetically informative characters for a HOX2 locus. *Mol. Biol. Evol.* **9,** 575–86.

Ruvolo, M., Pan, D., Zehr, S., Goldberg, T., Disotell, T.R. & von Dornum, M. (1994). Gene trees and hominoid phylogeny. *Proc. Nat. Acad. Sci USA* **91**, 8900–4.

Sanders, M.F. (1981). *Genetic Diversity in a Population of Kenyan Olive Baboons* (Papio anubis). PhD dissertation, University of California, Los Angeles.

Shotake, T. (1981). Population genetical study of natural hybridization between *Papio anubis* and *P. hamadryas*. *Primates* **22**, 285–308.

St. George, D., Rogers, J., Witte, S.M., Turner, T.R., Weiss, M.L., Phillips-Conroy, J., Phillips, R. & Smith, E.O. (1995). Microsatellite polymorphisms in two wild populations of yellow baboons (*Papio hamadryas cynocephalus*). *Am. J. phys. Anthrop.* (Suppl.) **20**, 203. (Abstract.)

Steinberg, A.G. (1969). Globulin polymorphisms in man. *Ann. Rev. Genetics* **3,** 25–62.

Steinberg, A.G., Olivier, T.J. & Buettner-Janusch, J. (1977). Gm and Inv studies on baboons, *Papio cynocephalus*: Analysis of serum samples from Kenya, Ethiopia and South Africa. *Am. J. phys. Anthrop.* **47**, 21–30.

VandeBerg, J.L. (1992). Biochemical markers and restriction fragment length polymorphisms in baboons: their power for paternity exclusion. In *Paternity in Primates: Genetic Tests and Theories*, ed. R.D. Martin, A.F. Dixson & E.J. Wickings, pp. 18–31. Basel:Karger.

VandeBerg, J.L. & Williams-Blangero, S. (1996). Strategies for using nonhuman primates in genetic research on multifactorial diseases. *Lab. Anim. Sci.* **46,** 146–51.

Weber, J. (1990). Informativeness of human $(dC\text{-}dA)_n - (dG\text{-}dT)_n$ polymorphisms. *Genomics* **7**, 524–30.

Weber, J. & May, P.E. (1989). Abundant class of human DNA polymorphisms which can be typed using the polymerase chain reaction. *Amer. J. Hum. Genet.* **44,** 388–96.

White, A.L., Hixson, J.E., Rainwater, D.L. & Lanford, R.E. (1994). Molecular basis for null lipoprotein(a) phenotypes and the influence of apolipoprotein(a) size on plasma lipoprotein(a) level in the baboon. *J. Biol. Chem.* **269**, 9060–6.

Wiener, A.S. & Moor-Jankowski, J. (1969). The ABO blood groups of baboons. *Amer. J. phys. Anthrop.* **30**, 117–22.

Williams-Blangero, S., VandeBerg, J.L., Blangero, J., Konigsberg, L. & Dyke, B. (1990). Genetic differentiation between baboon subspecies: Relevance for biomedical research. *Amer. J. Primat.* **20,** 67–81.

4

The phylogeny of the Cercopithecoidea

COLIN P. GROVES

Introduction

It is time for a fresh look at the phylogeny of the Old World monkeys. There have been recent attempts to analyse the group by molecular or chromosomal means, and it is time that "traditional" methods of analysis caught up, holding-out the prospect of new macroscopic diagnostic tools. The purpose of this chapter is to examine Cercopithecoid phylogeny by cladistic analysis, using primarily skull characters, but including other sources of evidence where these are available. The results should be regarded as preliminary, in that characters were taken from the literature and tested on specimens, rather than new characters being sought. However, a few new insights have emerged.

Traditionally, the Cercopithecoidea have been contained within a single family, Cercopithecidae, with two subfamilies, Colobinae and Cercopithecinae. The work of Benefit (1993) has established a second family, Victoriapithecidae, to include plesiomorphic early Miocene fossils. The Victoriapithecidae will not be further considered in the present review, which is concerned with reconstructing the phylogeny of extant cercopithecoids and their immediate fossil relatives.

Hill (1966, and elsewhere) supported upgrading the two extant subfamilies to family level. The differences are not extreme, but highly characteristic; adopting full family status for them would give more taxonomic flexibility within each to recognise fine degrees of relationship among genera (Groves,1989). It is clear that the two form a clade with respect to the Victoriapithecidae. It has been argued (Groves 1989) that a plesiomorphic fossil group should not have a new, higher category erected for it, but should be given the same rank as the highest-level living taxa, or (if of limited taxonomic diversity) classed as a plesion. In this way, classifications have a chance to remain tolerably stable and not be upset every time a new, more divergent fossil is discovered.

Within the Cercopithecinae (=Cercopithecidae of Hill and Groves), it is usual to recognize two divisions (generally ranked as tribes), the Cercopithecini and the Papionini, although some genera may sit uneasily within either of these groups – for example, *Theropithecus* and *Miopithecus*, according to Jolly (1966), or *Macaca*, according to Groves (1989). Within the Colobinae (=Colobidae of Hill and Groves), there is not even this degree of consensus. Strasser and Delson (1987) divide them (as Colobinae) geographically, and at no more than subtribal level, whereas Groves (1989) splits *Nasalis* from the rest, recognizing the division at the level of subfamily.

It is evident that little can be taken for granted in Cercopithecoid taxonomy. In what follows, therefore, the following usages will be maintained: Cercopithecidae, the "omnivore" Old World monkeys, i.e. the Cercopithecoidea exclusive of the Colobidae and Victoriapithecidae, vernacular: cercopithecids; Cercopithecinae, the *Cercopithecus*/*Erythrocebus* group (usually given tribal rank as Cercopithecini), vernacular: cercopithecines; Papioninae, the *Papio*/*Cercocebus* group (usually given tribal rank as Papionini), vernacular: papionines.

The cercopithecid/colobid split

The Colobidae are clearly monophyletic; they have a complex, and evidently highly derived, stomach morphology, including a fermentation chamber and a reticular groove. The Cercopithecidae lack all trace of this; their soft tissue specialization is the possession of cheek pouches (Murray, 1972). Skeletal differences are harder to find.

One skeletal difference that is traditionally cited is the position of the lacrimal fossa within the lacrimal bone or across the lacrimo-maxillary suture (Strasser and Delson, 1987). This difference simply does not hold, being polymorphic even within species in both families (Benefit and McCrossin, 1993).

Another commonly cited difference is the greater interorbital width of the Colobidae, which Verheyen (1962) expressed as a percentage of glabella-prosthion length. There is, however, considerable overlap between the two families – the cercopithecid range is 7.3–16.8 and the colobid range is 13.0–33.0. *Nasalis* has the lowest index among the Colobidae, and some *Cercopithecus* species and *Macaca sylvanus* have particularly high indices among the Cercopithecidae; these account for the overlap.

The listing of cranial differences among the Cercopithecoidea by Olivier *et al.* (1955) is very instructive – there is as much difference between

Cercopithecus, Macaca and *Papio* as there is between these three and the Colobidae. Among the characters listed by these authors, a supraorbital notch does not occur in the Colobidae and usually does occur in the Cercopithecidae, but in a few individuals (especially of *Cercopithecus*, where it may be a polymorphism) it is absent. The lower margin of the orbit is said to be at or above the level of the upper margin of the pyriform aperture in Cercopithecidae, below it in Colobidae. However, *Nasalis*, not considered by Olivier *et al.* (1955), sorts with the Cercopithecidae in this indicator of facial elongation (Vogel, 1966). There is said to be only one mental foramen on each side in Colobidae, more than one in Cercopithecidae, but this is subject to variation. According to Ravosa and Shea (1994), the colobid skull tends to be more airorhynch (the probable derived condition) than the cercopithecid skull, although individual ranges overlap. The similarities of *Nasalis* to the Cercopithecidae seem best explained as primitive retentions (Groves unpub. data), though the skull is airorhynch in the typically colobid fashion (Ravosa and Shea, 1994).

The dentition is diagnostic. The molar cusps are much higher and sharper in Colobidae, the indicators of bilophodonty (mid-tooth waisting, transverse cresting) are better developed, and molar trigonids are shorter in Colobidae. The anterior lower premolar is much less sectorial in Colobidae; this is rather difficult to quantify, as the width:length index overlaps between the two families (Delson, 1975).

Stephan *et al.* (1988) report different values of the encephalization index (on the Tenrecinae baseline): in Colobidae, 5.78–7.19; in Cercopithecidae, 7.39–11.2. This would appear to be a readily quantifiable difference, with no overlaps between the two families. The Cercopithecidae appear to show a derived condition, but caution must be exercised in interpreting this as evidence for their monophyly, since the Hominoidea and some Platyrrhini show high indices in parallel.

Strasser (1988, 1994) has documented differences in pedal morphology between the two families. These relate to the paraxonic condition of the Colobidae and the mostly mesaxonic condition of the Cercopithecidae. The situation is more complex than this, and both families are evidently derived in opposite directions, although many of the features of the Cercopithecidae are similar to those of the Hominoidea, and so perhaps primitive.

In summary, the monophyly of Colobidae seems unquestionable, while a question mark must hang over that of Cercopithecidae, although it does seem probable. In this study, it will be assumed that the Cercopithecidae are monophyletic with respect to the Colobidae. For reasons given above, it is convenient to recognise the two divisions at family level.

Divisions within the Colobidae

Phylogeny within the Colobidae is analysed by Groves (unpub. data), using characters of the skull, postcranial skeleton, dentition, soft anatomy and external appearance. The provisional scheme of Strasser and Delson (1987), dividing an African subtribe Colobina from an (admittedly plesiomorphic) Asian subtribe Presbytina, is not supported. Instead, a monophyletic short-faced clade emerges, contrasted with the long-faced *Nasalis, Dolichopithecus* and *Libypithecus* (which do not, or not certainly, together form a monophyletic group). The African fossil genera *Paracolobus* and *Rhinocolobus* cluster together, but their association with the main short-faced clade is insecure.

Within the short-faced clade, the African genera *Colobus* and *Procolobus* (including *Piliocolobus*) cluster together, and are joined by the poorly known fossil genus *Microcolobus*. There is a non-African clade consisting of *Pygathrix, Rhinopithecus* and the late Miocene fossil *Mesopithecus* (in agreement with Jablonski and Peng, 1993). The other genera – *Presbytis, Semnopithecus, Trachypithecus* and the fossil African *Cercopithecoides* – do not group consistently with any other clade or with each other. The monophyly of *Trachypithecus* is itself so far unverified, and ideally a new analysis should be undertaken in which this genus is split into its component parts, but the datasets for different species are incomplete and in some cases overlap little.

On this basis, the taxonomy of the Colobidae would be as follows:

Plesion *Dolichopithecus*
Plesion *Libypithecus*
Subfamily Nasalinae
  Nasalis

Subfamily Colobinae
  Incertae cedis: *Presbytis*
    Trachypithecus
    Semnopithecus
    Cercopithecoides
  Tribe Rhinopithecini
    Pygathrix

Rhinopithecus
Mesopithecus
Tribe Colobini
Colobus
Procolobus
Microcolobus

Divisions within the Cercopithecidae

The questions to be asked in the Cercopithecidae are:

- What is the status of *Allenopithecus* and *Miopithecus*? Do they truly belong in the *Cercopithecus* group, or might they be part of the Papionine clade, as tentatively proposed by Groves (1989) in the case of *Allenopithecus*? Alternatively, might they (or one of them) even be a sister group to *Cercopithecus* plus Papionines?
- Is *Macaca* monophyletic?
- Is *Macaca*, or one or more of its constituents, part of the Papionine clade, or does it occupy some other position in the phylogeny?
- Does *Mandrillus* form a monophyletic group with *Papio*, as usually envisaged, or is it related to *Cercocebus*?
- Does *Theropithecus* form a monophyletic group with *Papio*?
- Is *Cercopithecus aethiops* closer to *Erythrocebus* than it is to other species of *Cercopithecus*, and does it therefore warrant its own genus name?

The questioning of some assumptions about the Cercopithecidae is necessary because they have never been properly tested. The reasons for suspecting that testing is in order will now be discussed.

The position of Allenopithecus *and* Miopithecus

Verheyen (1962) regarded *Allenopithecus* as a subgenus of *Cercopithecus*, but drew attention to its numerous distinctive cranial characters (he regarded *Erythrocebus* in the same way). On the other hand, he found nothing especially distinctive about the skull of *Miopithecus*, so did not recognize it at even subgeneric level in *Cercopithecus*.

Jolly (1966) first drew attention to the fact that the talapoin, unlike other members of Verheyen's subgenus *Cercopithecus*, has periodic sexual swellings in the female. He not only resurrected the genus *Miopithecus* for it, but even placed it *incertae cedis* among the Cercopithecinae

(=Cercopithecidae of the present study). His decision reflects uncertainty whether this character state links it to the Papioninae (his Cercocebini) or was acquired in parallel (as it presumably has been in certain colobids, and indeed in the chimpanzee).

Hill (1966) found that *Allenopithecus* is also highly distinctive and recognized it as a full genus. He noted that it has sexual swellings in the female, as in *Miopithecus*, and cited its general macaque-like body build and the form of its molar teeth. In the terminology of Delson (1975), it has molar flare and is the only member of the *Cercopithecus* group to do so.

Groves (1978) listed some characters distinguishing the *Cercopithecus*-like and baboon-like groups, and later (1989), revising his assessment of polarities somewhat, even proposed that *Allenopithecus* might be a Papionine.

The question of macaque monophyly

The genus *Macaca* is very hard to define. These are nondescript brown monkeys (even then, some of them are black!), lacking the defining features of other Old World monkey genera, such as maxillary and mandibular fossae or excessive facial elongation. Some have periodic sexual swellings, others do not; some have shortened tails, others do not; some have hypoconulids on the posterior lower molars, others do not. A number of fossils are placed in *Macaca*, but by virtue of lacking any of the derived states of other genera, not by possession of any distinctive features of the genus itself.

One species seems especially different from the rest – *Macaca sylvanus*. The only non-Asian species, it resembles the baboon-group in having a degree of molar flare, and ischial callosities that consistently fuse across the midline in the male. The possibility that these are genuine synapomorphies with the baboon-group needs to be tested.

Ideally, the monophyly of the other species should be investigated as well. But the cranio-dental differences between the recognised species are still in need of definition, and the possibility that some of the recognised species-groups might not be monophyletic units makes it very hard to know what operational taxonomic units to test. Under these circumstances, I have chosen to test only *Macaca sylvanus* and *M. nemestrina*, using the latter as proxy for "the typical macaque".

The phyletic position of the macaques

Laying *M. sylvanus* aside for the moment, the relationships of macaques in general are quite unclear. They are commonly included in the Papionine group, but whether this is valid depends on the polarity of the character states defining the latter – hypoconulids, sexual swellings, molar flare, and the like. These questions about the macaques raise further questions about the status of the otherwise apparently well-defined Papionine group as a whole.

The relationships of Mandrillus

Szalay and Delson (1979) put the mandrills in *Papio*, but other authors have raised questions about whether they and the true baboons really are sister taxa. Dutrillaux *et al.* (1980) and Stanyon *et al.* (1988) report that *Mandrillus* shares with *Cercocebus* (*sensu stricto*, i.e. not including *Lophocebus*) a complex reorganization of chromosome 10, and extra heterochromatin on chromosome 12, which *Papio* does not share. Disotell *et al.* (1992) found that mtDNA relationships of *Mandrillus* are with *Cercocebus*, rather than *Papio*. Groves (1978), in differentiating *Lophocebus* from *Cercocebus*, took the relationship of *Mandrillus* and *Papio* for granted, so the distribution of skull characters has not yet been investigated.

The relationships of Theropithecus

Jolly (1966) separated *Theropithecus* at tribal level, as Theropithecini (to be equal with Cercopithecini and "Cercocebini" (correctly Papionini)). Palaeontologists find no difficulty in identifying the molar teeth of *Theropithecus*, and in tracing them back into the Pliocene (Jablonski, 1993). There would seem to be considerable time depth to the separation of *Theropithecus* from its nearest relatives. Geneticists insist on a close relationship between *Theropithecus* and *Papio* (Disotell *et al.* 1992), which is not inconsistent with the fossil data (Cronin and Meikle, 1979).

The morphological and genetic findings are not necessarily in conflict. *Papio* and *Theropithecus* could be sister taxa, yet have separated in the Pliocene. It is necessary to test whether they are sister taxa or whether some other genus is closer to one or other.

The relationships of Cercopithecus aethiops

Groves (1989), largely on the basis of the cranial studies of Verheyen (1962), using many of the characters listed below, raised the possibility that *Cercopithecus aethiops* might be more closely related to *Erythrocebus patas* than to the other *Cercopithecus* species. If this is so, then we could either unite it generically with *Erythrocebus*, award it a separate genus, or put *Erythrocebus* back into *Cercopithecus* as was done by Verheyen (1962).

The genus *Erythrocebus* is widely recognized, and to sink it into *Cercopithecus* might be unwise. The species *aethiops* was assigned to a genus *Chlorocebus*, which long antedates *Erythrocebus*, so to unite the two generically would be even more confusing. Groves (1989) favoured splitting the two species both from *Cercopithecus* and from each other, reserving *Chlorocebus* for *C. aethiops* only.

The preferred interpretation of the chromosomal data by Dutrillaux *et al.*(1980) unites not only *C. aethiops* and *E. patas*, but also *C. lhoesti* and *C.(=Miopithecus) talapoin*, on a clade separate from the other members of *Cercopithecus*. We are brought back to the first problem – the relationships of *Miopithecus*!

Cladistic analysis of the Cercopithecidae

Two cladistic analyses of the Cercopithecidae were undertaken – one of genus-level taxa, the other of the *Cercopithecus*-group alone. In the first case, the assumption was made that Cercopithecidae and Colobidae are monophyletic, so that Colobidae can be used as a suitable outgroup. In the second case, *Allenopithecus* was used as the outgroup (see justification below). The OTUs (operational taxonomic units) were species–groups, which are comparatively uncontroversial in *Cercopithecus*.

Material and methods

For the genus-level analysis, craniodental and postcranial (Strasser, 1994) characters were used, with a few soft-tissue characters (Murray, 1972; pers. obs.). After an initial examination of illustrations of craniodental characters (Elliot, 1913), a preliminary list of characters was tested on a sample (*n* usually >5 per OTU) of modern skulls in the Natural History Museum, London. In most cases, preliminary characters were rejected as subject to too much individual variation, but a number remained for the analysis. Fossil taxa could not be checked beyond descriptions and illustrations in the literature (Delson, 1975; Freedman, 1957, 1960, 1965, 1976; Szalay and Delson, 1979). The final dataset had 46 characters. The characters are described in Table 4.1, and the dataset is given in Fig. 4.1a. Characters were not differentially weighted; multiple-state characters were Ordered, unless otherwise noted.

For the cladistic analysis of the *Cercopithecus* group, only craniodental characters were used. These were initially taken from Verheyen (1962), then checked on skulls in the Natural History Museum, London (and, in most cases, accepted). No fossil taxa were used, and one distinctive living species, *Cercopithecus dryas* (=*salongo*), was not included as no adult skulls were available. The final datset had 22 characters. The characters are described in Table 4.2, and the dataset is given in Fig. 4.1b.

In each case, the dataset was run in PAUP, version 3.1.1 (Swofford, 1989), Heuristic search, then transferred to MacClade, version 2.1 (Maddison and Maddison, 1989) for further examination.

The PAUP results were first bootstrapped (100 replicates) to assess the consistency of support for each clade. To test for significance, Permutation Tail Probability (PTP) Tests were run, followed by Topology-Dependant (T-PTP) Tests for selected nodes.

As originally designed, the PTP (Faith and Cranston, 1991) and T-PTP (Faith, 1991) tests applied within-character randomization to the in-group but not to the outgroup(s). Trueman (Faith and Trueman, 1996) has argued that this may bias the test. A more appropriate T-PTP test would take as its test statistic the minimum of support for three tests: rejection of the null hypothesis from Faith–T-PTP; and two alternative ways for estimating the length difference from minimum-length trees for the original and randomized data, (1) with outgroup taxa randomized and (2) with outgroups excluded. In the present instance, all three T-PTP results are reported for each test, with interpretation based on Trueman's suggested minimum-of-support aproach.

Cercopithecidae

Results

Fifty-six shortest trees (length 130, Consistency Index 0.38) were found. The strict and 50% majority-rule trees have very little structure (Fig. 4.2). In the strict consensus tree, a clade uniting *Mandrillus* and *Cercocebus* is the only one that appears. In the 50% majority-rule tree, this clade joins in succession *Papio*, then *Theropithecus* and *Paradolichopithecus*, then *Macaca nemestrina* and *Procynocephalus*. The only other clade is the one uniting *Miopithecus* and *Cercopithecus*; *Allenopithecus* does not join them.

Using MacClade, some of the patterns which would have been predicted from the literature were recreated. The linking of *Theropithecus* in a clade with *Papio* increases tree length to 133. Restoring the most parsimonious

Table 4.1. *Character list for cladistic analysis of Cercopithecidae*

Presumed derived states of each character are given
Characters are all unordered

1 Inion lower than Frankfurt Plane
2 Coronal suture goes straight across, no V-shaped backward "kink"
3 Vault sutures complex, serrate
4 Temporal lines converge from posterior edge of torus (state 1), from lateral ends (2)
5 Lateral orbital pillars diverge inferiorly
6 Interorbital space narrowed, averaging $<11\%$ of glabella-prosthion distance
7 Malar occupies less than half of lower margin of orbit
8 Interorbital pillar drops perpendicularly, not sloping antero-inferiorly
9 Orbital height = (state 1) or < (state 2) orbit-to-alveolar distance
10 Nasion region narrow, ridge-like
11 Nasal tips pointed, not blunted
12 Pyriform aperture higher, at least $2\times$ its width
13 Pyriform aperture not lozenge-shaped
14 Supraorbital torus flattened, not projecting
15 Supraorbital notch always present, deep
16 Mandibular fossa present
17 Suborbital fossa present
18 Inferior orbital margin behind superior
19 External auditory meatus reaches lateral cranial outline
20 External auditory meatus inferior border in a V
21 Posttympanic process elongated
22 Tympanic not fused with postglenoid: separated (state 1), widely separated (2)
23 Postglenoid process shortened
24 Postglenoid process mediolaterally restricted
25 Choanal sides parallel, not widely divergent posteriorly
26 Medial pterygoids diverge at posterior ends (1), strongly so (2)
27 Posterior edge of vomer incised
28 Vomer inflated where it meets presphenoid
29 Median crest on palate
30 Palatine canals troughlike, not rounded and flush with palate surface
31 Posterior margin of palate not posterior to third molars
32 Upper and lower mandibular margins diverge anteriorly
33 No molar flare
34 No hypoconulid on third lower molar
35 Female canines slender and elongated
36 Upper lateral incisor not pointed and narrow
37 Upper lateral incisor slopes medially
38 Incisor row length less than molar row
39 Incisor row much enlarged, its width approx. = (1) or > (2) molar row length
40 Cheek pouches present, larger: pouch area:bodyweight ratio <0.30(state 1), >0.30(2)
41 Asymmetrical laryngeal sac
42 No sexual swelling
43 Oestrus cycle lengthened, >30 days
44 Ischial callosities of male fused across midline
45 Pedal Ray I lengthened
46 Phalanx 1 shortened relative to Metatarsal 1

Table 4.2. *Character list for cladistic analysis of* Cercopithecus-*group*

Presumed derived states of each character are given
Characters are all unordered

1	Frontals inflated (not flat)
2	Supraorbital torus reduced
3	Temporal lines follow posterior margin of torus
4	Pterygoid fossa shallow
5	Styloid reduced, blunt
6	External auditory meatus with no V-notch inferiorly
7	Orbit angular (not rounded)
8	Sphenomaxillary fissure has a lateral branch
9	Nasal tips concave or straight (not convex)
10	Internasal suture closes at adulthood (not remaining open)
11	Interorbital pillar constricted at f-m suture
12	Nasal bases don't reach orbital margin level
13	Premaxilla broadened, its suture bowed out
14	Pyriform aperture oval (not angular)
15	Overbite (not edge-to-edge bite)
16	Choana low, wide (not high, narrow)
17	Upper incisors form a unitary bite-row
18	Posterior palatine notches shallower (not deeply indented)
19	Digastric scars reduced (not expanded to whole inferior symphyseal surface)
20	Inferior mandibular margin curved (not straight)
21	Ascending ramus slopes back (not upright)
22	Inferior orbit border level with or in front of superior in Frankfurt plane

structure, but moving *Allenopithecus* to the cercopithecine clade, increases the length to 135. Leaving it on the papionine lineage and moving *Lophocebus* up to the baboon/mandrill clade likewise gives a length of 135 if *Lophocebus* is on the *Papio/Theropithecus* twig, but only 133 if it is sister to a clade containing *Papio/Theropithecus* and *Mandrillus/Cercocebus*. Finally, restoring the original structure but putting the two macaques together gives a tree of length 139.

In view of these results, it did not seem worthwhile to apply T-PTP tests or to compute Bremer support (Källersjö *et al.*, 1992) for any groups on these trees. A bootstrap analysis gave 91% replicability to the *Mandrillus/Cercocebus* clade, but only 36% to the clade uniting them with *Papio*; the *Cercopithecus/Miopithecus* and *M. nemestrina/Procynocephalus* clades both showed 45% bootstrap values. Paul Whitehead (pers. comm.) queries whether the strength of support for a *Mandrillus/Cercocebus* association could be in part an artefact of the lack of close associations between other taxa; this could well be the case, although the existence of the clade itself is clearly real.

```
#NEXUS
[Cercopithecidae phylogeny]

begin data;
    dimensions ntax=16 nchar=46;
    format missing=? symbols="012";
;

matrix
outgroup              0000000000000000000000000000000000000000000000
M_sylvanus            1100111010011100010011010101001100010102007100
M_nemestrina          1000100001011010000012000110010010010002102010
Papio                 1000100111010011101012001110100100010001001101
Mandrillus            1011100021010001101010001010100100001017002110
Theropithecus         1000111111010011111011000110000100010101702001
Cercocebus            0011110021110101101110001000000000001011101111
Lophocebus            1100100001011001101001110100001100000002100101
Allenopithecus        1110010010100010000110010111111001000007707100
Miopithecus           11000110000011000010011100000001111110002102000
Cercopithecus         011000101000111000000010100000001011100002010000
Procynocephalus       ???????0???1???00?????????????0100000 1?10???????
Paradolichopithecus   0??0?0?0???1?1?110?????????????010001010???????
Dinopithecus          0??000?11001101000 0??10000?0?00000?1010???????
Parapapio             00000100100111010?1011110?10?0110001?10???????
Gorgopithecus         ???001?11001?0001???????????????00???10???????
;
end;

#NEXUS
[Cercopithecus and related genera]

begin data;
    dimensions ntax=11 nchar=23;
    format missing=? symbols="012";
    ;

matrix
Allenopithecus        00000000000000000000000
Miopithecus           10010011001100000200010
Erythrocebus          00001200101010002011000
C_aethiops            00001201001011102011111
C_mitis               11110111011001101201111
C_lhoesti             11010210010000002201100
C_hamlyni             01010210010000002201110
C_neglectus           11110111111001020201111
C_diana               11110111011101122101100
C_cephus              11110011011101111201101
C_mona                11010011011101111201101
;
end;
```

Fig. 4.1. Data matrices for PAUP analyses: (top) Cercopithecoidea, (bottom) *Cercopithecus*-group.

Table 4.3. *T-PTP results for* Cercopithecus*-group tree (Fig. 4.2); 100 randomisations in each case*

Test:	Clade 1 monophyly	Clade 2 monophyly	Clade 3 monophyly	Clade 4 non-monophyly	Clade 5 non-monophyly
Faith TPTP					
extra steps: observed data	3	2	1	4	9
randomized data	1 to −7	0 to −10	0 to −11	10 to −1	7 to −1
p	<0.01	<0.01	<0.01	NS	<0.01
All-randomized					
TPTP extra steps: observed data	6	6	0	6	9
randomized data	1 to −8	0 to −11	0 to −11	10 to −1	7 to −2
p	<0.01	<0.01	<0.03	NS	<0.01
No-outgroup					
T-PTP extra steps: observed data	6	6	0	6	6
randomized data	2 to −7	2 to −7	1 to −8	7 to −1	8 to −1
p	<0.01	<0.01	<0.06	<0.09	<0.11

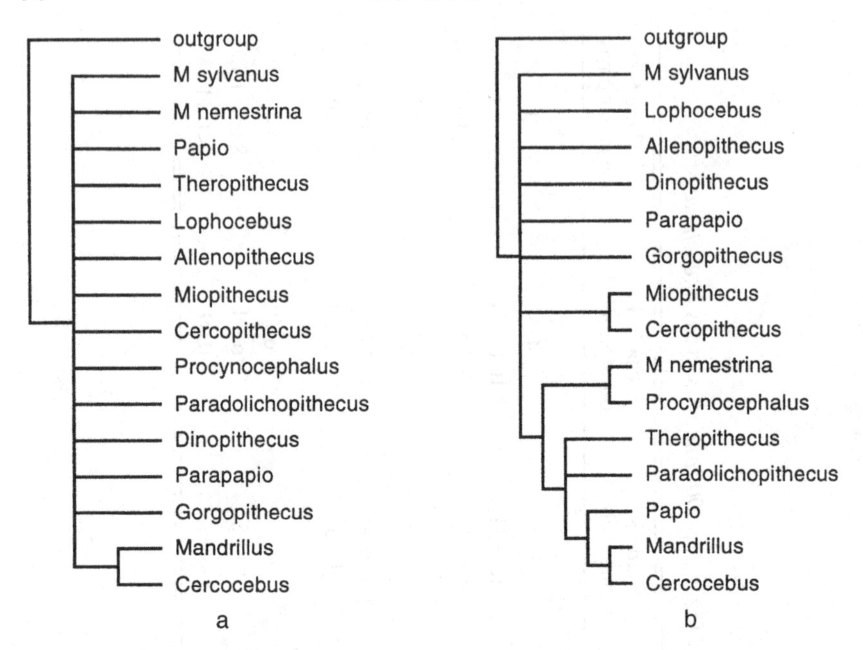

Fig. 4.2. (a) Strict and (b) 50% majority-rule consensus trees for Cercopithecidae, using characters listed in Table 4.1. Length 130, Consistency Index 0.38.

Discussion

That the two macaques refuse to associate, even on the 50% majority-rule tree, is interesting. However, this is simply a lack of evidence for their monophyly rather than evidence against it. That *Mandrillus* and *Cercocebus* associate so clearly, given the general lack of structure in the trees, is also interesting and in good agreement with molecular findings.

The consistency index for the tree is low (0.38 for the shortest tree), as is bootstrap replicability. Many of the clades can be defined only by character changes which are reversed higher up, or are parallel to changes elsewhere:

(1) *Macaca nemestrina* and *Procynocephalus* share one uniquely derived state: palatine canals troughlike.

(2) The non-*M. nemestrina*/*Procynocephalus* clade has four derived conditions: alveolar and inferior margins of mandible diverge anteriorly; ischial callosities of male joined across the midline (reversed in *Theropithecus*); pyriform aperture not lozenge-shaped; cheek-pouches reduced. These last two are both reversals from the basal cercopithecid condition.

(3) *Papio, Mandrillus* and *Cercocebus* share one uniquely derived condition

not found in *Theropithecus* or *Paradolichopithecus*:choanal walls parallel.

(4) *Mandrillus* and *Cercocebus* share a number of derived character states, nearly all unique to the clade: temporal lines diverge from posterior margin of supraorbital torus; orbital height less than orbit-to-molar distance; tympanic tube widely separated from postglenoid; medial pterygoids do not diverge posteriorly (reversal); I2 slopes medialward; incisor row much enlarged; pedal phalanx lengthened (reversal).

(5) Finally, *Cercopithecus* and *Miopithecus* share three derived conditions, only one of them unique:shape of female canines; loss of molar flare (parallel to *M. nemestrina*); loss of M3 hypoconulid (parallel to *Allenopithecus*).

Conclusions

There is no evidence that *Allenopithecus* is part of the cercopithecine group; equally, it does not associate with the papionine group. There is no evidence that *Lophocebus* is part of the papionine group (and the same applies to the African fossil genera *Dinopithecus, Gorgopithecus* and *Parapapio*). On the other hand, an association of *Macaca nemestrina* (but not *M. sylvanus*!) with the papionine clade is weakly supported.

Before one can support the dismemberment of the genus *Macaca*, the other species must be examined to see whether they belong with *M. nemestrina* or form one or more further clades. It is surely the case that *Procynocephalus* will be sunk into whatever genus ends up containing *Macaca nemestrina*.

The case for sister-group relationship between *Mandrillus* and *Cercocebus* is strong, but in the absence of any good indications of higher-level relationships among the other genera, it is impossible to recognize this by a special suprageneric grouping. Combining them at the generic level (*Cercocebus* E.Geoffroy, 1812, having priority over *Mandrillus* Ritgen, 1824) would cause confusion, and one may doubt whether the world is quite ready for *Cercocebus sphinx* and *Cercocebus leucophaeus.*

Cercopithecus

Results

A single tree, of length 53 and Consistency Index of 0.509, resulted from the PAUP analysis (Fig. 4.3). The first split was between an *aethiops/patas* clade and the rest; then successively *Miopithecus talapoin* split off, then

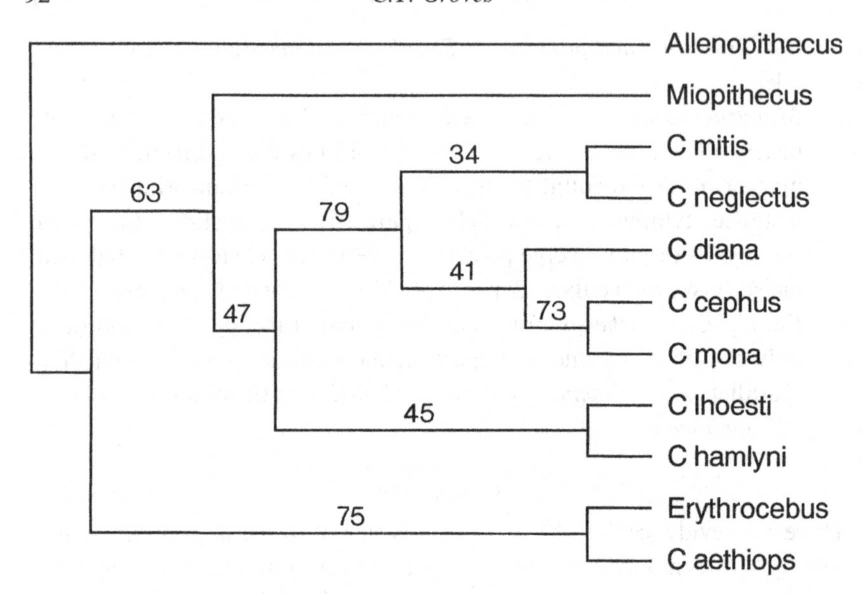

Fig. 4.3. Shortest tree for *Cercopithecus*-group. Length 53; Consistency Index 0.509.

C.hamlyni and *lhoesti*, then a *mitis/neglectus* clade separated from a *diana(cephus/mona)* clade.

T-PTP tests on the four leading hypotheses of relationship are shown in Fig. 4.4. The null hypotheses that clade 1 (*Erythrocebus* plus *C.aethiops*) or clade 2 (*Miopithecus* plus other *Cercopithecus*) appear by chance alone are rejected by all three variants of the test ($p<0.01$). The null hypothesis that clade 3 (other *Cercopithecus*) appears by chance alone is rejected by Standard (Faith, 1991) and All-randomised T-PTP ($p<0.01$), but not by the No-outgroups variant($p<0.06$). The null hypothesis of monophyly of taxon-group 4 (*Cercopithecus*, including *C.aethiops*), however, cannot be rejected.

Support for clades 1 and 2 but not for 3 and not against 4 suggests that an alternative tree which has *Miopithecus* as sister to *Erythrocebus*, and with the root placed within *Cercopithecus* s.l., could not be rejected in these data.

Bootstrap replication was much stronger than for the Cercopithecidae tree. The *patas/aethiops* clade has 75% replicability, and the non-*patas/aethiops* clade 63%. The non-*talapoin* clade has only 47% replicability. The non-*hamlyni/lhoesti* clade appears in 79% of replicates, and the *cephus/mona* link appears in 73%.

Giving *talapoin* sister-group status to the others, in line with its possession (retention?) of female sexual swellings and papionine-like behavioral

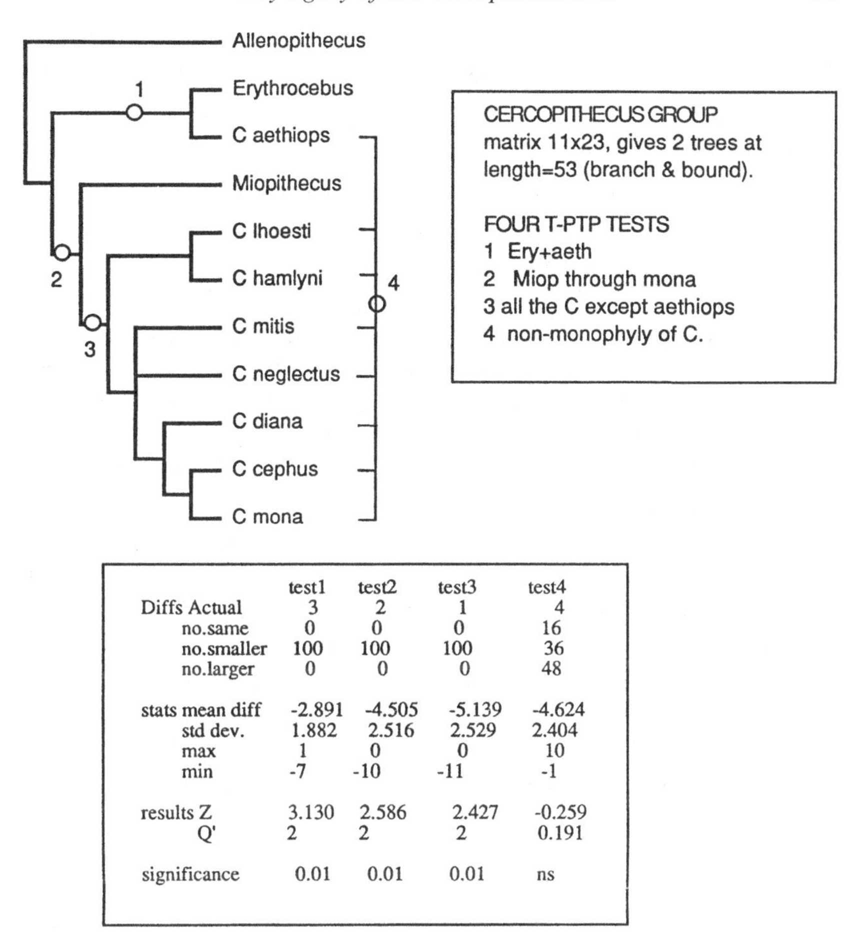

		test1	test2	test3	test4
Diffs	Actual	3	2	1	4
	no.same	0	0	0	16
	no.smaller	100	100	100	36
	no.larger	0	0	0	48
stats	mean diff	-2.891	-4.505	-5.139	-4.624
	std dev.	1.882	2.516	2.529	2.404
	max	1	0	0	10
	min	-7	-10	-11	-1
results	Z	3.130	2.586	2.427	-0.259
	Q'	2	2	2	0.191
significance		0.01	0.01	0.01	ns

Fig. 4.4. Numbered clades in the *Cercopithecus*-group tree, with T-PTP results as indicated.

features (Rowell, 1972), increases the length to 56, and reduces the consistency index to 0.48. Recreating the "chromosome tree" of Dutrillaux *et al.* (1980), with *patas, aethiops, talapoin* and *lhoesti* versus the rest, increases tree length to 64, and reduces consistency index to 0.42 (other branches manipulated to give the most parsimonious result).

Discussion

Allenopithecus nigroviridis was chosen as outgroup because it has often been included in the Cercopithecini, even in the genus *Cercopithecus*, but is so clearly outside the group that the first (Cercopithecidae) analysis places it preferentially as a papionine.

The rationale for using only craniodental characters was to see whether the assumptions that result in the talapoin being given sister status to the rest would be duplicated in an analysis which did not incorporate them. The result, at least superficially, was negative.

(1) The *aethiops/patas* association is characterized by three unique changes: orbits angular; upper incisors form a single straight bite; inferior margin of zygomata curved down, not straight.
(2) The other group (clade 2) is characterized by four changes, of which only one is subject to homoplasy: supraorbital torus reduced (reversed in *C. hamlyni*); external auditory meatus rounded below, not V-shaped; nasal bones concave at free ends; inferior orbit margin in front of superior.
(3) The non-*Miopithecus* group (clade 3) has the following three derived conditions, two of them unique:temporal lines follow posterior margins of supraorbital torus medialwards; pyriform aperture margins not angled; inferior border of suborbital plane curved.
(4) The association between *C. hamlyni* and *C. lhoesti* is supported by only one, not unique, derived state: no overbite (reversal).
(5) The non-*hamlyni/lhoesti* clade has two homoplastic changes: styloid process reduced (reversed in *C. mona*); posterior palatine notches reduced (parallelled by *aethiops*).
(6) *C. diana* and the *C. mona/cephus* clade are associated by two derived conditions, neither unique: choanae high, narrow (parallelled by *M. talapoin*); inferior margin of mandible curved (parallelled by *C. neglectus*).
(7) The *C. mona/cephus* clade is supported by a single derived state, not unique: sphenomaxillary fissure in orbit has no lateral branch (parallelled by *M. talapoin*).

The really convincing splits are the one between the *aethiops/patas* clade and the rest, and between the main *Cercopithecus* group and *C. hamlyni* and *C. lhoesti* (which may or may not be associated together). The inclusion of the talapoin within "true" *Cercopithecus* may be partly dependent on size; supraorbital torus reduction, reduced nasals and sloping orbital apertures (perhaps connected with gnathic reduction), rounded meatus (perhaps related to supramastoid crest diminution) are characters likely to appear in small-sized species.

Conclusions

These results do not support the chromosome-based hypotheses of either Dutrillaux *et al.* (1980), in which *C. lhoesti* belongs in an *aethiops/patas/tal-*

apoin clade, or Ponsa *et al.* (1994), in which *C. aethiops* and a *Miopithecus/Erythrocebus* clade are associated with different groups of other *Cercopithecus* species but not with each other. However, there is considerable homoplasy in chromosome evolution, and these interpretations were offered simply as the most parsimonious ones.

Agreement with the cladogram of Ruvolo (1988) is not bad. This author considered it more likely that talapoin and *aethiops* were part of a "true *Cercopithecus*" clade separate from *patas*, but this was in the nature of a "best solution" rather than an only possible one. Ruvolo placed *C. hamlyni* and *lhoesti* as the first species to separate from true *Cercopithecus*, though preferably on a common lineage.

The cladogram of Gautier (1988), based on vocalizations, is not too dissimilar from the present one, given that his clade uniting *patas, talapoin, aethiops* and *lhoesti* was characterized mainly by character loss. On the other hand, in Gautier's cladogram, no divergent position is awarded to *hamlyni*.

The taxonomic change which seems inescapable from these results is the separation of *C. aethiops* from *Cercopithecus* sensu stricto. As discussed above and in Groves (1989), we have three options: to include both it and the patas monkey in *Cercopithecus*, awarding them each subgeneric status; to separate it and the patas from *Cercopithecus*, and place them in a genus apart, for which the prior available name would be *Chlorocebus* not *Erythrocebus*; or to separate them, but award them separate genera. The names for vervet and patas monkeys would then be, respectively: *Cercopithecus (Chlorocebus) aethiops* and *Cercopithecus (Erythrocebus) patas; Chlorocebus aethiops* and *Chlorocebus patas; Chlorocebus aethiops* and *Erythrocebus patas.*

The first option recalls Verheyen's (1962) solution, but as the generic name *Erythrocebus* is well-known it would cause confusion. The second would introduce a completely new nomenclature, and result in utter confusion and complete dismay; the differences between patas and vervet are considerable (limb elongation, digital shortening, hair saturation, facial depigmentation, elongated vomer, angulated occiput in *E. patas*; early closure of internasal suture, curvature of inferior zygomatic border, reduced post-palatine notches and digastric scars, higher chromosome number in *C. aethiops*). The third would result in the introduction of one new concept but would preserve a well-known one. I recommend the third option.

What is Cercopithecus dryas/salongo?

See Kuroda *et al.* (1985) for details of this species. It is a thoroughly distinct species, not part of any of the well-known species–groups of the genus. It appears to lack the character states of the *Chlorocebus/Erythrocebus* group, and to possess those of true *Cercopithecus*: orbit not oval, upper incisors do not form a single unified bite surface, nasals not pointed at tips, small supraorbital torus. In addition the limbs are black, a character usual in *Cercopithecus* but not seen in other cercopithecines. On the other hand, it has a lozenge-shaped pyriform aperture; the species is remarkably un-sexually dimorphic. While a member of *Cercopithecus*, it seems to be a very divergent species, possibly the sister to all other species.

Conclusions and summary: taxonomy of Cercopithecidae

This chapter has examined the phylogeny of the Cercopithecidae, and come to certain unexpected conclusions. As not all of these are secure, but need further study, it is inadvisable to change the taxonomy too radically; the following changes, however, seem inescapable:

- *Procynocephalus* is a synonym of *Macaca* (*nemestrina*-group).
- *Allenopithecus* does not belong to the Cercopithecinae.
- *Chlorocebus* should be recognized for the species/superspecies *aethiops*.

A great deal of further work must be done, and new characters must be sought. Further investigation is necessary to confirm or refute the hypotheses that *Macaca sylvanus* is generically distinct from other macaques, that *Macaca* (even without *M. sylvanus*) is polyphyletic, that *Lophocebus* is the most divergent papionine (perhaps not a papionine at all), and that *Miopithecus* should be sunk into *Cercopithecus*. The phylogeny of the Old World monkeys is still very largely unexplored territory.

Acknowledgements

I thank Paula Jenkins for permission to examine skulls of Old World monkeys in the Natural History Museum, London; Doug Brandon-Jones, Nina Jablonski and Beth Strasser for helpful discussions; and Paul Whitehead and Cliff Jolly for comments on the manuscript. I am especially grateful to John Trueman for running T-PTP tests on my data, and for numerous discussions of the value and significance of probability testing in cladistic analysis.

References

Benefit, B.R. (1993). The permanent dentition and phylogenetic position of *Victoriapithecus* from Maboko Island, Kenya. *J. Hum. Evol.* **25**, 83–172.

Benefit, B.R. & McCrossin, M.L. (1993). The lacrimal fossa of Cercopithecoidea, with special reference to cladistic analysis of Old World Monkey relationships. *Folia Primatol.* **60**, 133–45.

Cronin, J.E. & Meikle, W.E. (1979). The phyletic position of *Theropithecus*: congruence among molecular, morphological, and paleontological evidence. *Syst. Zool.* **28**, 259–69.

Delson, E.(1975). Evolutionary history of the Cercopithecidae. In *Approaches to Primate Biology*, ed. F.S. Szalay, *Contrib. Primatol.* **5**, 167–217.

Disotell, T.R., Honeycutt, R.L. & Ruvolo, M. (1992). Mitochondrial phylogeny of the Old-World monkey tribe Papionini. *Mol. Biol. Evol.* **9**, 1–13.

Dutrillaux, B., Couturier, J. & Chauvier, G. (1980). Chromosomal evolution of 19 species or sub-species of Cercopithecinae. *Ann. Génét.* **23**, 133–43.

Elliot, D.G. (1913). *A Review of the Primates*, 3 vols. New York: American Museum of Natural History.

Faith, D.P. (1991). Cladistic permutation tests for monophyly and nonmonophyly. *Syst. Zool.* **40**, 366–75.

Faith, D.P. & P.S. Cranston (1991). Could a cladogram this short have arisen by chance alone? *Cladistics* **7**, 1–28.

Faith, D.P. & Trueman, J.W.H. (1996). When the Topology-Dependent Permutation Test (T-PTP) for monophyly returns significant support for monophyly, should that be equated with (a) rejecting a null hypothesis of nonmonophyly, (b) rejecting a null hypothesis of 'no structure', (c) failing to falsify a hypothesis of monophyly, or (d) none of the above? *System. Biol.* **45**, 580–6.

Freedman, L. (1957). Fossil Cercopithecoidea of South Africa. *Ann. Transv. Mus.* **23**, 121–257.

Freedman, L. (1960). Some new Cercopithecoid specimens from Makapansgat, South Africa. *Palaeont. Afr.* **7**, 7–45.

Freedman, L. (1965). Fossil and subfossil primates from the limestone deposits at Taung, Bolt's Farm and Witkrans, South Africa. *Palaeont. Afr.* **9**, 19–48.

Freedman, L. (1976). South African fossil Cercopithecoidea: a re-assessment including a description of new material from Makapansgat, Sterkfontein and Taung. *J. hum. Evol.* **5**, 297–315.

Gautier, J-P. (1988). Interspecific affinities among guenons as deduced from vocalizations. In *A Primate Radiation: Evolutionary Biology of the African Guenons*, ed. A. Gautier-Hion, F. Bourlière, J.-P. Gautier & J. Kingdon, pp.194–226. Cambridge: Cambridge University Press.

Groves, C.P. (1978). Phylogenetic and population systematics of the mangabeys (Primates: Cercopithecoidea). *Primates* **19**, 1–34.

Groves, C.P. (1989). *A Theory of Human and Primate Evolution*. Oxford: Oxford University Press.

Hill, W.C.O. (1966). *Catarrhini, Cercopithecoidea, Cercopithecinae. The Primates: Comparative Anatomy and Taxonomy*, 6. Edinburgh: Edinburgh University Press.

Jablonski, N. (Ed.) (1993). *Theropithecus: The Rise and Fall of a Primate Genus*. Cambridge: Cambridge University Press.

Jablonski, N. & Peng Y-Z. (1993). The phylogenetic relationships and classification of the Doucs and Snub-nosed Langurs of China and Vietnam. *Folia Primatol.* **60**, 36–55.

Jolly, C.J. (1966). Introduction to the Cercopithecoidea with notes on their use as laboratory animals. *Symp. Zool. Soc. Lond.* **17**, 427–57.

Källersjö, M., Farris, J.S., Kluge, A.G. & Bult, C. (1992). Skewness and permutation. *Cladistics* **8**, 275–87.

Kuroda, S., Kano, T. & Muhindo, K. (1985). Further information on the new monkey species, *Cercopithecus salongo* Thys van den Audenaerde 1977. *Primates* **26**, 325–33.

Maddison, W.P. & Maddison, D.R. (1992). MacClade, version 2.1. Sunderland, MA: Sinauer Associates.

Murray, P. (1972). The role of cheek pouches in Cercopithecine monkey adaptive strategy. In *Primate Functional Morphology and Evolution*, ed. R.H. Tuttle, pp.151–94. The Hague: Mouton.

Olivier, G., Libersa, C. & Fenart, R. (1955). Le crâne du semnopithèque. *Mammalia* **19**, 1–292.

Ponsa, M., Egozcue, J. & Garcia, M. (1994). The phylogeny of guenons using PAUP analysis of cytogenetic characters. *Hum. Genet.* **88**, 387–92.

Ravosa, M.J. & Shea, B.T. (1994). Pattern in craniofacial biology: evidence from the Old World Monkeys (Cercopithecidae). *Int. J. Primatol.* **15**, 801–22.

Rowell, T.E. (1972). *Social Behaviour of Monkeys*. London: Penguin Books.

Ruvolo, M. (1988). Genetic evolution in the African guenons. In *A Primate Radiation: Evolutionary Biology of the African Guenons*, ed. A. Gautier-Hion, F. Bourlière, J.-P. Gautier & J. Kingdon, pp.127–39. Cambridge: Cambridge University Press.

Stanyon, R., Fantini, C., Camperio-Ciani, A., Chiarelli, B. & Ardito, G. (1988). Banded karyotypes of 20 Papionini species reveal no necessary correlation with speciation. *Amer. J. Primatol.* **16**, 3–17.

Stephan, H., Baron, G. & Frahm, H.D. (1988). Comparative size of brains and brain components. In *Comparative Primate Biology*. 4. *Neurosciences*, ed. H.D. Steklis & J. Erwin, pp.1–38. New York: Alan R. Liss.

Strasser, E. (1988). Pedal evidence for the origin and diversification of cercopithecid clades. *J. hum. Evol.* **17**, 225–45.

Strasser, E. (1994). Relative development of the hallux and pedal digital formulae in Cercopithecidae. *J. hum. Evol.* **26**, 413–40.

Strasser, E. & Delson, E. (1987). Cladistic analysis of cercopithecid relationships. *J. hum. Evol.* **16**, 81–99.

Swofford, D. (1989). PAUP: Phylogenetic Analysis Using Parsimony, version 3.1.1. Champaign: Illinois Natural History Survey.

Szalay, F.S. & Delson, E. (1979). *Evolutionary History of the Primates*. New York: Academic Press.

Verheyen, W.N. (1962). Contribution à la craniologie comparée des Primates. *Ann. K. Mus. Midden-Afrika, 8o, Zool. Wetensch.* **105**, 1–253.

Vogel, C. (1966). Morphologische Studien am Gesichtsschädel catarrhiner Primaten. *Bibl. Primatol.* No. 4. Basel: Karger.

5

Ontogeny of the nasal capsule in cercopithecoids: a contribution to the comparative and evolutionary morphology of catarrhines

WOLFGANG MAIER

Introduction

Members of a species are adequately characterized when all stages of the individual life are understood. All ontogenetic stages have to be investigated as to their canalizing or constraining effects on evolutionary transformations. Comparative life history studies show potential for elucidating microevolutionary processes (Stearns, 1992). Although comparative morphogenetic studies have a long tradition (Garstang, 1922; DeBeer, 1937; Bonner, 1982; Maier, 1993a), they have not yet proven useful for systematics on middle and higher taxonomic levels.

Ontogenetic stages were included by Hennig (1966) as "semaphoronts". It is sometimes difficult to define stages that are really comparable in more than a few details, because of heterochronic changes in developmental processes. Heterochrony is a descriptive, not causal, concept that needs adaptational explanations for shifts in the developmental program. The shifts themselves might be a valuable source of systematic and adaptational information. Heuristically, ontogenetic studies have proven to be valuable in craniology. Craniogenetic studies have led to a deeper understanding of the morphology of the vertebrate skull (Gaupp, 1906; DeBeer, 1937; Starck, 1967; Novacek, 1993).

The present study applies the comparative morphogenetic approach to the ethmoidal region of anthropoid primates and identifies characteristic features in the ethmoidal and nasal regions of cercopithecoids.

The ethmoidal region is a good example of heterochronic processes. In eutherian mammals, its structural differentiation – as compared to the braincase and some sensory organs – is retarded during intra-uterine development, and it continues to grow and change into postnatal life (Augier, 1931; Starck, 1967). This is functionally understandable when one considers lactation. Although growth and differentiation of nasal and facial structures is continued into postnatal life, most published studies of primates are based on stages too young to exhibit species-specific structures. An exception is the study of *Propithecus* by Starck (1962), who pointed-out the importance of studying the late ontogeny of the ethmoidal region. In late fetal and early postnatal stages, the ethmoidal skeleton is greatly modified by absorption, and morphological evaluation of the remaining structures becomes difficult.

The primate ethmoidal region is comparatively unknown. Because it is hidden beneath the facial elements, and is fragile in the adult skull, it has played almost no role in primate paleontology. In fetuses, it is prominent and can easily be distinguished from the exocranial components of the facial skeleton. During ontogeny, especially the later stages, the cartilaginous nasal capsule of the fetus becomes profoundly transformed, and only a few parts, such as the external nose and nasal floor, remain cartilaginous into adulthood. Other parts are resorbed, and still others are transformed by enchondral ossification. Though the ethmoid is usually called a "replacement bone", the ethmoidal structures do not simply replace cartilages, but are also shaped by "appositional ossification" and remodelling of bony structures.

The cartilaginous nasal capsule plays an important role in the ontogeny of the facial skull. By interstitial and appositional growth, the cartilaginous structures expand and position the overlying membranous bones (Scott, 1967), thus functioning as a "Stemmkoerper". At the same time, they play direct biomechanical roles. At earlier ontogenetic stages, the nasal cartilages are an integral component of the architecture of the fetal facial skull, only gradually replaced by bony tissue (Maier, 1987). Much of the height and length of the skeletal muzzle of cercopithecoids develops by postnatal sutural growth and structural reorganization. The endocranial cartilages may therefore not be helpful in elucidating these later phases of growth and differentiation.

Materials and methods

Microscopic studies of ontogenetic stages are the only suitable methods for analysis of the structural transformations. Preparations of cleaned skulls are not sufficient and may even be misleading (Hershkovitz, 1977). Even the studies of Seydel (1891), Kollman and Papin (1925), Cave and Haines (1940) and Cave (1967, 1973) show deficiencies because of neglect of earlier ontogenetic stages. Hill (1966) illustrates parasagittal sections of various primate heads, and Geist (1933) depicts a section of the head of *Macaca mulatta*. Frets (1913, 1914) and Maier (1980,1983, 1987,1993b, 1997) deal with some specific aspects of the nasal region of primates, based on histological serial sections.

The present study is based largely upon serial sections of three papionin fetuses: *Macaca fascicularis* (Crown–Rump–Length (CRL): 61 mm), *Theropithecus gelada* (CRL: 88 mm), and *Papio anubis* (CRL: 115 mm). Additional serial sections of *Macaca fascicularis* (CRL: 105 mm), *Cercopithecus aethiops* (CRL: 60 mm), *Trachypithecus vetulus (senex)* (CRL unknown), and *Nasalis larvatus* (CRL: 100 mm) were also consulted. Cross-sections of the nasal region were drawn with the camera lucida, and the description is based on the complete sectional series. Other taxa, including colobines, have been examined, but no differences in the principal anatomical conditions were observed. Sagittally-sectioned adult heads of *Papio*, *Theropithecus*, *Cercopithecus*, and *Presbytis* were also studied. A number of cleaned infant and juvenile skulls of various primate taxa have been examined.

Living hominoids, the sister-group of the cercopithecoids (Andrews, 1985,1988; Andrews and Martin, 1987; Strasser and Delson, 1987), are plesiomorphic in many aspects of their nasal skeleton (Maier, 1993b), while they are derived in others (Maier, 1997). The hylobatids assume a key position for elucidating character states of the last common ancestor of hominoids and cercopithecoids. Sectional series of fetal stages of *Hylobates pileatus* (CRL: 100 mm, Head–Length (HL): 39 mm), *H. moloch* (HL: 51 mm), *Pongo pygmaeus* (CRL: 170 mm), *Gorilla gorilla* (CRL and HL unknown), *Pan troglodytes* (CRL: 71 mm; CRL: 80 mm, preservation poor), and *Homo sapiens* (HL: 26.5 mm, 62.5 mm) were available for comparisons.

Platyrrhines are considered to be the sister group of catarrhines. I have chosen the following taxa: *Callimico goeldii* (CRL: 62 mm), *Saimiri sciureus* (CRL: 110 mm), and *Callicebus moloch* (CRL: 61 mm).

Occasional remarks on the nasal cavity of some strepsirhine prosimians are included in order to indicate the morphological status of anthropoids in general.

Anatomy of the fetal nasal skeleton of cercopithecoids

External nose

Catarrhine primates derive their name from the close position of their nasal openings. Platyrrhiny is plesiomorphic, although the nasal cupula may be secondarily broadened in some neotropical monkeys (Maier, 1980). Catarrhiny does not mean the possession of a narrow interorbital septum, but rather the narrowing of the medial lamella of the cupular cartilage. This results in the nares becoming closely approximated and oriented more anteriorly. Cercopithecoid nares are slit-like and open anterolaterally (Wen, 1930; Schultz, 1935,1956; Kingdon, 1974; Maier, 1980). Whereas the position of their nares is apomorphic, their shape remains plesiomorphic and reminiscent of the strepsirhine nasal structure (Hofer, 1977).

In fetal *Macaca* and *Theropithecus* (Ma.2–1–2, Th.7–2–2, respectively. See Figs. 5.2 and 5.3), the nares are directed laterally and their openings are framed by a medial and lateral lamella of the cupular cartilage. *Papio* (Pa.26–2–2) differs in having a relatively long, tubelike cupular cartilage. *Nasalis* shows a very specialized structure of the external nasal cartilages.

The outer nose of hominoids is not basically different from this pattern, especially in fetal stages (Schultz, 1956). Gibbons look similar to cercopithecoids (Schultz, 1969). *Pan* and *Gorilla* show a secondary expansion of the cupular cartilages and connective tissue pads during pre- and postnatal growth. The medial lamella of the cupular cartilage is pronounced in human fetuses (Reinbach, 1963). Reinbach reports the existence of a small superior alar process in an older human fetus, but this seems to be the only record in catarrhines.

Parasagittal section

Figure 5.1a shows the parasagittal section of a preserved specimen of an adult male chacma baboon (*Papio ursinus*), Fig. 5.1b the section of a cleaned skull of the same species. Both figures illustrate the simplified internal relief of the nasal cavity, reflecting the pronounced microsmatism of these primates. Relevant details of the nasal capsule will be analyzed later in this paper, with the aid of histological cross-sections of fetal cercopithecoid material (see Figs. 5.2–5.4).

The epithelial lining of the nasal capsule reveals a prominent, but isolated concha in the vestibule of the nose (marginoturbinal). It is well-separated from the narrow conchae of the middle portion of the nasal cavity (maxilloturbinal, mxt; nasoturbinal, nat). The anterior nasal cavity is connected with the buccal cavity by two wide nasopalatine ducts. The nasoturbinal continues into the anterior lamella of the ethmoturbinal I (et I). The two other lamellae of the ethmoturbinals (posterior et I and et II) are oriented in a vertical position. This is a plesiomorphic arrangement, because they show a horizontal orientation in hominoids. Fetal gibbons are intermediate, because their ethmoidal conchae have an oblique position. Behind ethmoturbinal I, the entrance to a shallow lateral recess is found. In some specialized ceboids (cf. *Cebus*, *Saimiri*), the orientation of the ethmoturbinals is horizontal due to an extreme narrowing of the dorsal nasal cavity (Frets, 1913,1914; Cave, 1967; Hershkovitz, 1977; Maier, 1983, 1993b). Since callitrichids, *Callicebus*, and others show a more primitive pattern, this must be convergently developed to hominoids (see below). Informative median sections of various primates are depicted in the classical works of Zuckerkandl (1887) and Seydel (1891), which confirm this conclusion.

Marginoturbinal

The marginoturbinal is typical of therian mammals, including "prosimians" and platyrrhines. The cartilage of the marginoturbinal is part of the nasal roof that is rolled inward (Figs. 5.2–5.4). Rostrally, it begins at the lateral margin of the external nasal opening (naris, na), and commonly continues into the atrioturbinal and into the maxilloturbinal. It is peculiar to the cercopithecoids, however, that the marginoturbinal is an isolated concha, not posteriorly followed by an atrioturbinal, and therefore lacking a direct connection with the maxilloturbinal. Maier (1980) wrongly stated that a marginoturbinal is generally missing in catarrhines. In hylobatids, "pongids", and hominids a faint marginoturbinal is usually found, at least as a fetal structure. Though the retention of a marginoturbinal is *per se* a plesiomorphic character, its isolated position in cercopithecoids is apomorphic. This is also documented in the plates of Seydel (1891). Cave and Haines (1940, Fig.9) show a small marginoturbinal in a chimpanzee.

Atrioturbinal and anterior transverse lamina

In most therian mammals, the anterior transverse lamina consists of a cartilage pillar supporting the nasal cupula and the atrioturbinal against the paraseptal cartilages of the nasal floor. In "prosimians" and platyrrhines, the marginoturbinal continues through a well-developed atrioturbinal, supported by a complete anterior transverse lamina, directly into the maxilloturbinal. In the taxa studied, both structures are present in adult specimens as well (Maier, 1980). In catarrhines, the lamina usually is interrupted and the atrioturbinal is vestigeal. The absence of a well-developed atrioturbinal and the reduction of a complete anterior transverse lamina, seem to be functionally correlated. Maier (1980) has speculated about the loss of this pillar. Due to the evolution of incisal biting and mobile lips, the nasal cupula may have lost its function as a mechanosensitive rostrum.

The lack of a complete anterior transverse lamina and of an atrioturbinal – at least in adults – may be considered a synapomorphy of the catarrhine primates (Maier 1980), but in fetal *Theropithecus* (Th.14–2–2, Fig. 5.3) and *Papio* (Pa. 43–2–1, 46–2–1, Fig. 5.4), an anterior transverse lamina

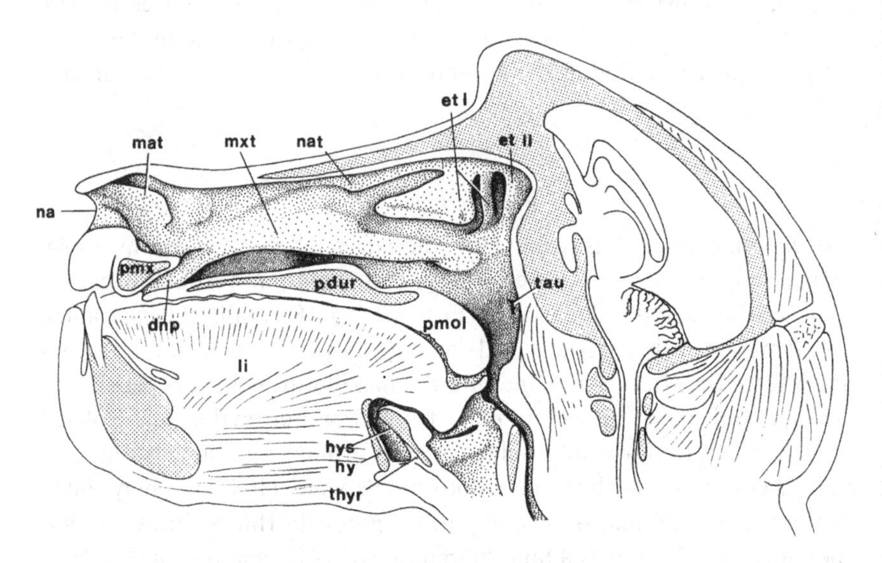

Fig. 5.1a. Longitudinal section of the head of an adult male chacma baboon (*Papio ursinus*). The structures of the air passage are stippled. Note the isolated marginoturbinal (mat) and the persisting nasopalatine duct (dnp). The larynx develops an asymmetric subhyoidean diverticulum.

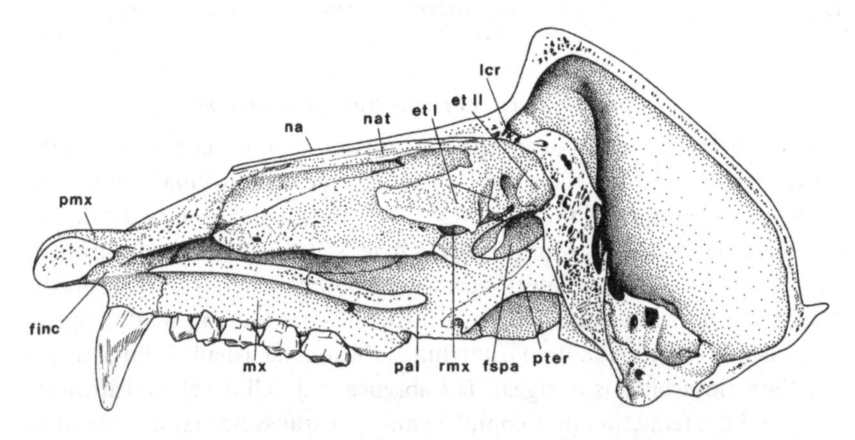

Fig. 5.1b. Longitudinal section of the skull of an adult male chacma baboon (not the same specimen as in Fig. 5.1a; in both specimens, the incisors are missing).

appears to be almost complete. The same holds true for *Trachypithecus*, although some signs of early resorption are visible (Maier, 1997). These genera seem to show vestiges of soft-tissue atrioturbinals, which do not continue posteriorly into the maxilloturbinals.

In the fetal *Hylobates*, the anterior transverse lamina is nearly complete, but its margino- and atrioturbinals are vestigeal (Maier, 1997). In *Gorilla*, there is a broad and almost complete transverse lamina, connected with a well-developed lateroventral cartilage. It is vestigeal in *Pan*, *Pongo*, and *Homo*. The turbinals at the nasal entrance are not well-developed in hominoids; this seems to be a synapomorphy of this group. The transitional occurrence of a more or less complete anterior transverse lamina in fetuses, and its absence in adults, is a synapomorphy of catarrhines.

Key for Figs. 5.1–5.4
aet – ethmoidal artery
aor – ala orbitalis
asph – alisphenoid
at – atrioturbinal
bol – olfactory bulb
cdn – cartilage of nasopalatine duct
cn – nasal cupola
cnp – posterior nasal cupola
con – orbitonasal commissure
cpa – palatine cartilage
cps – paraseptal cartilage
crs – crista semicircularis
dnl – nasolacrimal duct
dnp – nasopalatine duct
etI – ethmoturbinal I
etII – ethmoturbinal II
fep – epiphanial foramen
fet – ethmoidal foramen
finc – incisal foramen
fr – frontal
fspa – sphenopalatine foramen
hsl – semilunar hiatus
hy – hyoid
hys – hyoidal diverticulum of larynx
la – lacrimal
lao – antorbital lamina
lcr – cribrosal lamina
li – tongue
ls – lacrimal sac
lta – anterior transversal lamina
mat – marginoturbinal
mor – orbital muscle
mx – maxillary
mxt – maxilloturbinal
nar – nares
nas – nasal
nat – nasoturbinal
ne – ethmoidal nerve
nmx – maxillary nerve
nopt – optic nerve
npd – nasopalatine duct (Steno)
pal – palatine
pdur – hard palate
peo – ethmoidal process of orbitosphenoid
plv – ventrolateral process
pmol – soft palate
pmx – premaxillary
pnd – nasopharyngeal duct
por – periorbital
ppn – paranasal process
ppp – posterior paraseptal process
pss – supraseptal planum
pter – pterygoid
pun – uncinate process
ret – ethmoturbinal recess
rmx – maxillary recess
sio – interorbital septum
sme – mesethmoidal spine
sn – nasal septum
spe – mesethmoidal spine
tau – auditory tube
thyr – thyroid
vo – vomer

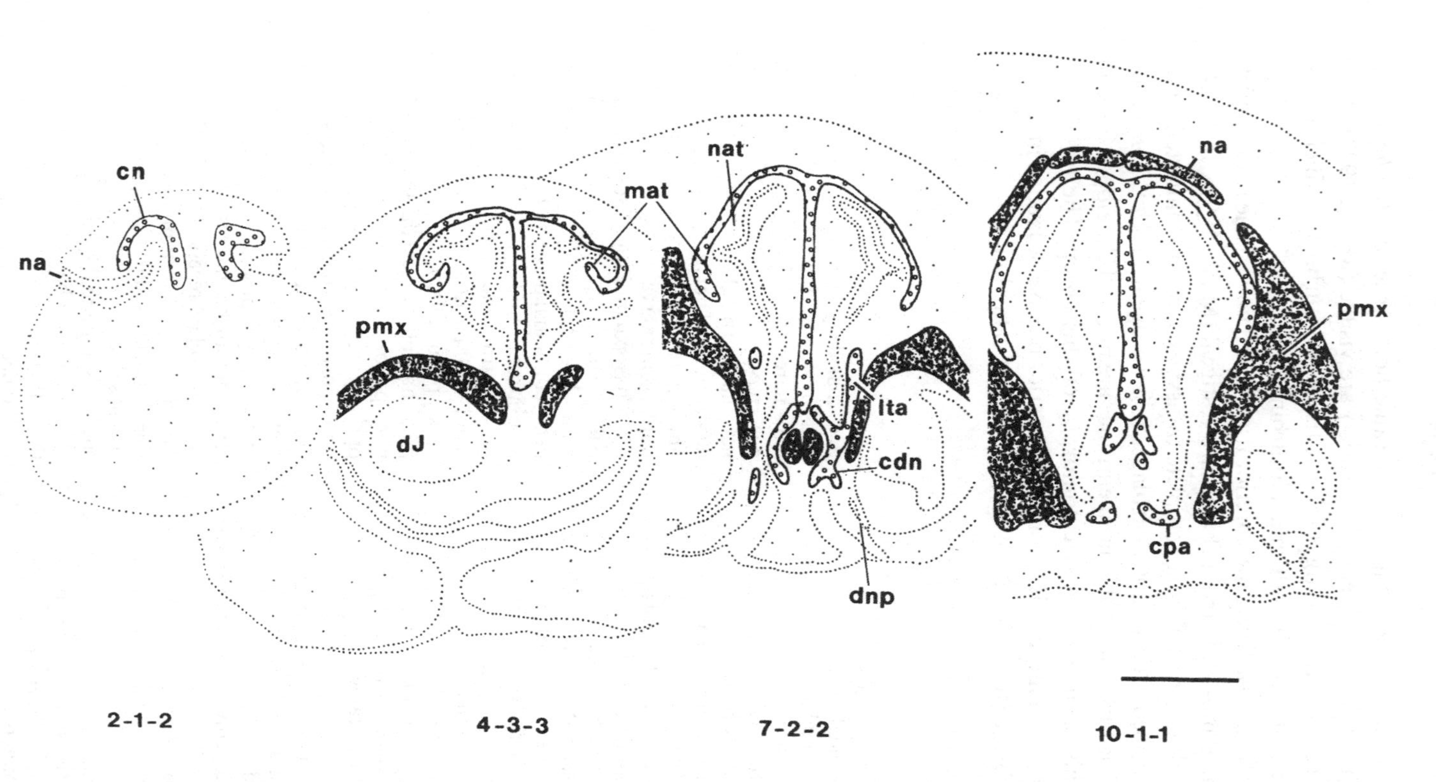

Fig. 5.2. Semi-diagrammatic cross-sections of the nasal region of a fetus of *Macaca fascicularis* (CRL 55 mm). The numbers indicate the position of the section within the histological series housed at the Department of Zoology at Tubingen, Germany; in the text, reference to this series of pictures is made with the prefix Ma. The bar-length is always 1 mm. For key see Fig. 5.1.

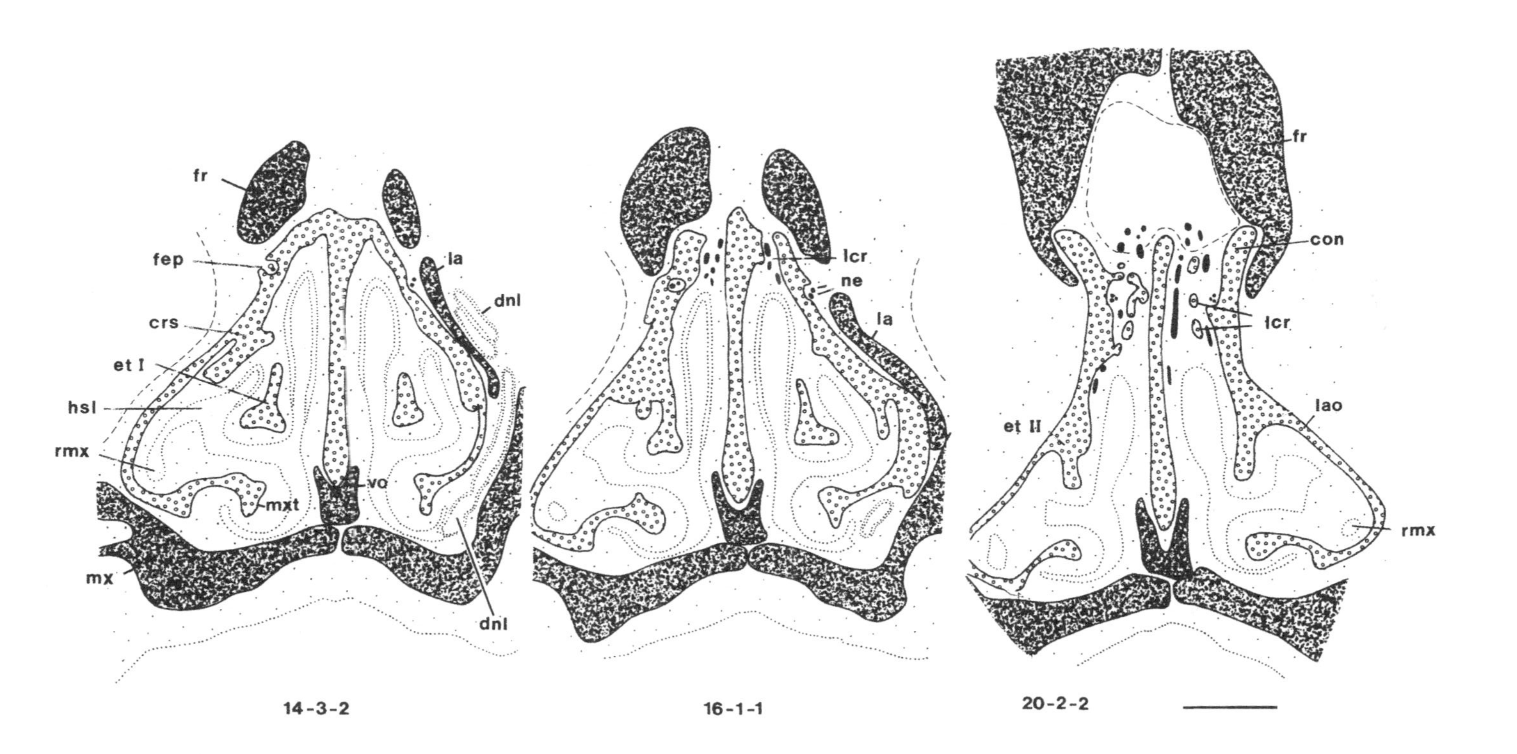
fr
fep
crs
et I
hsl
rmx
mxt
vo
mx
la
dnl
dnl
14-3-2
lcr
ne
la
16-1-1
fr
con
lcr
et II
lao
rmx
20-2-2

Fig. 5.2. (*cont.*)

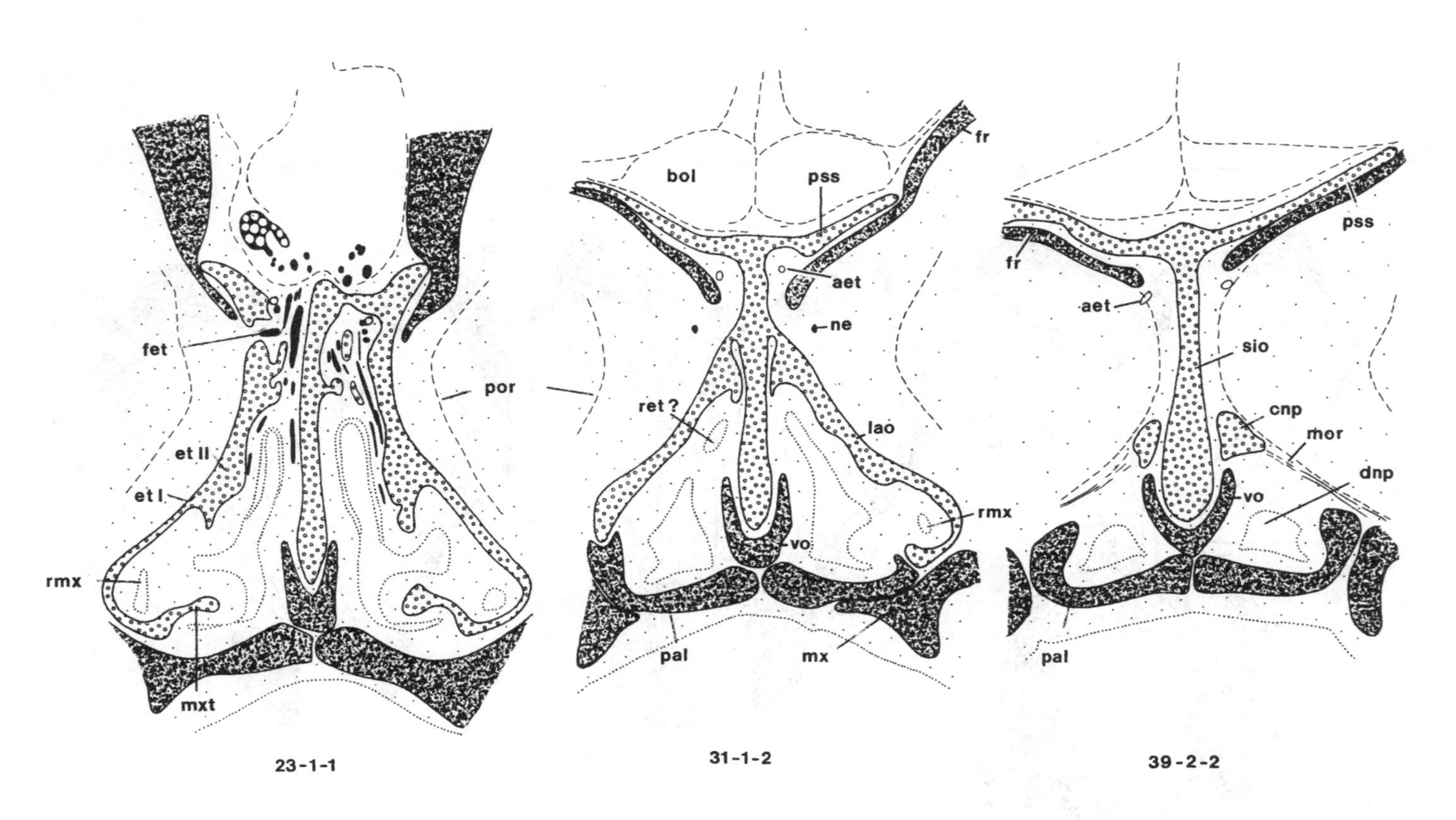
fet
por
et II
et I
rmx
mxt
23-1-1
bol
pss
fr
aet
ne
ret ?
lao
rmx
vo
pal
mx
31-1-2
pss
fr
aet
sio
cnp
mor
dnp
vo
pal
39-2-2

Fig. 5.2. (*cont.*)

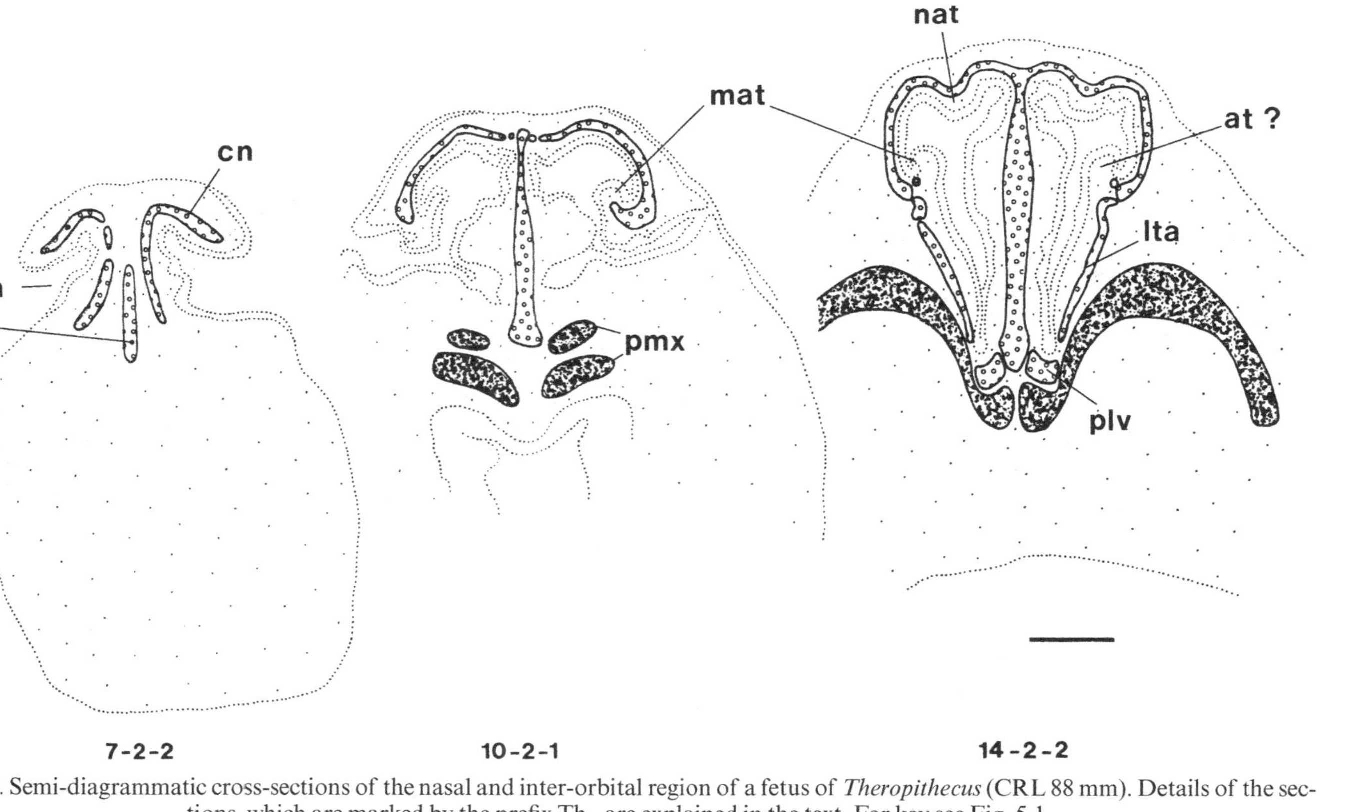

Fig. 5.3. Semi-diagrammatic cross-sections of the nasal and inter-orbital region of a fetus of *Theropithecus* (CRL 88 mm). Details of the sections, which are marked by the prefix Th., are explained in the text. For key see Fig. 5.1.

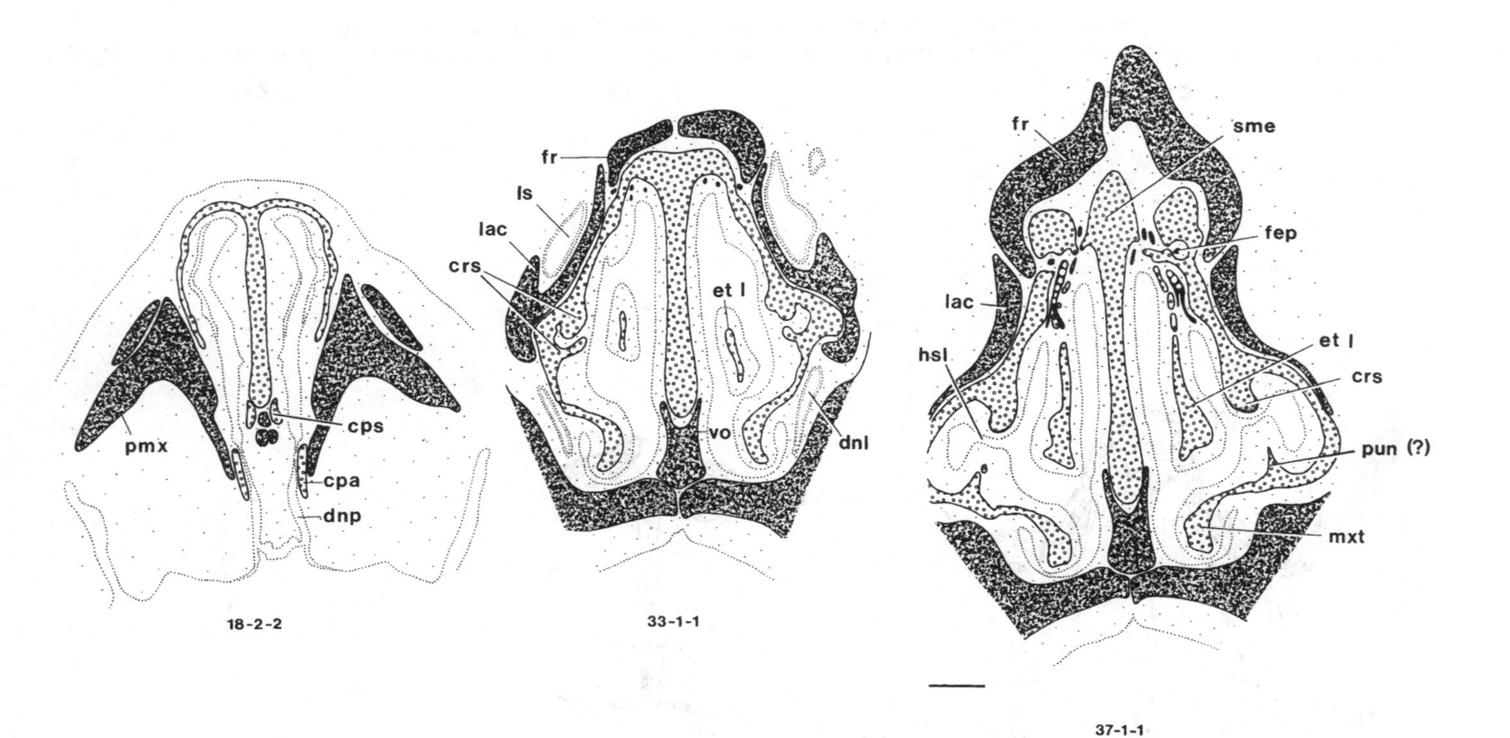
pmx
cps
cpa
dnp
18-2-2
fr
ls
lac
crs
et l
vo
dnl
33-1-1
fr
sme
fep
lac
hsl
et l
crs
pun (?)
mxt
37-1-1

Fig. 5.3. (*cont.*)

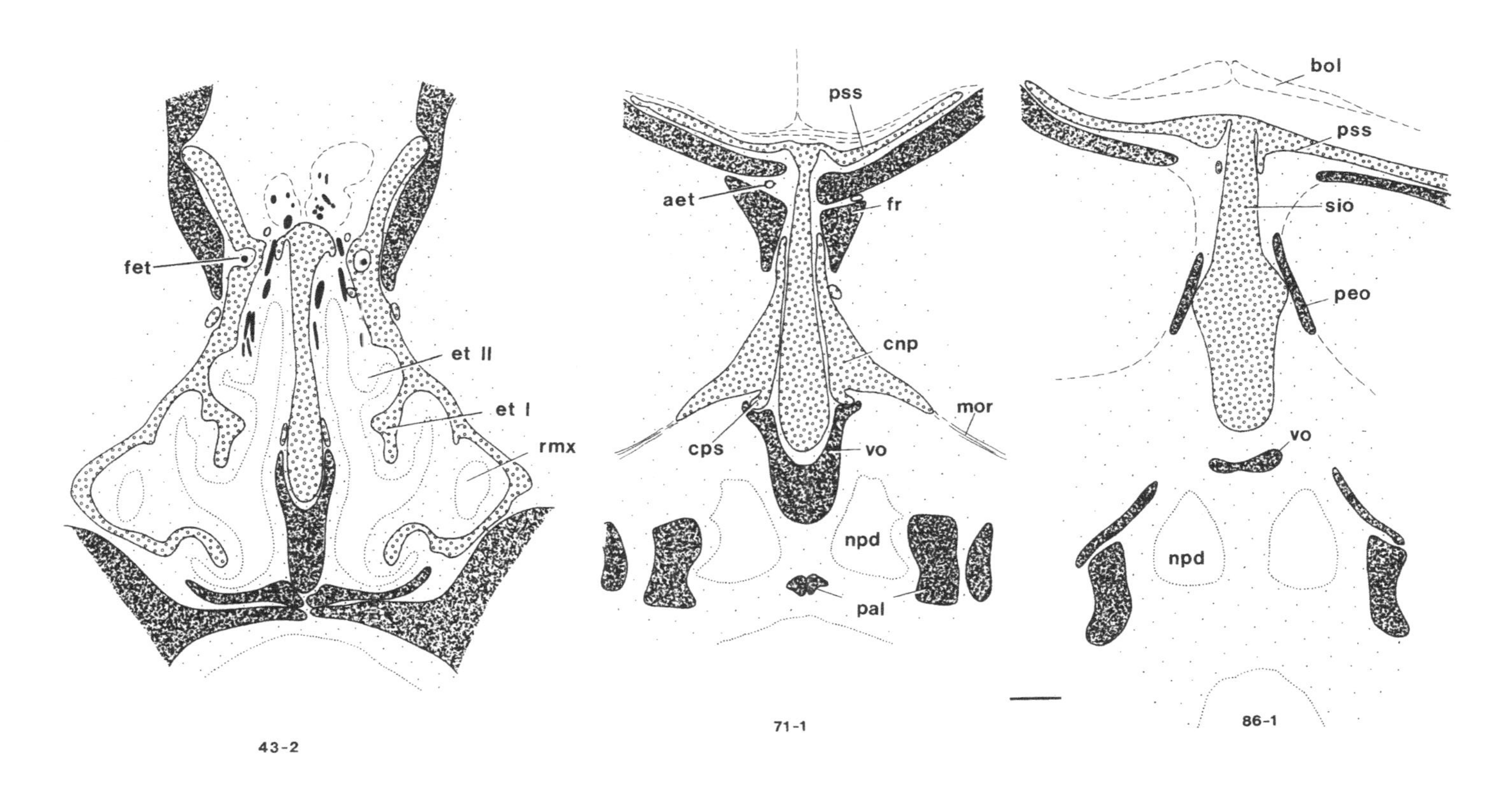
fet
et II
et I
rmx
43-2
pss
aet
fr
cnp
mor
cps
vo
npd
pal
71-1
bol
pss
sio
peo
vo
npd
86-1

Fig. 5.3. (*cont.*)

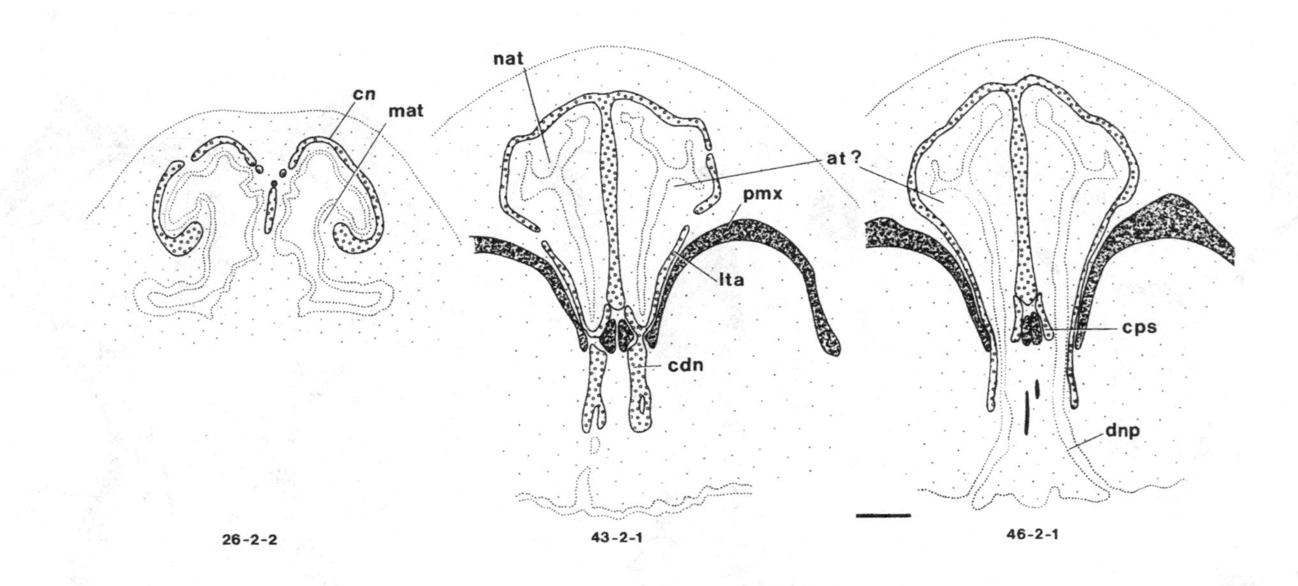

Fig. 5.4 Semi-diagrammatic cross-section of the nasal and interorbital region of a fetus of *Papio anubis* (CRL 115 mm). Details of the sections, which are marked by the prefix Pa., are explained in the text. Note the compound structure of the inter- orbital septum. For key see Fig. 5.1.

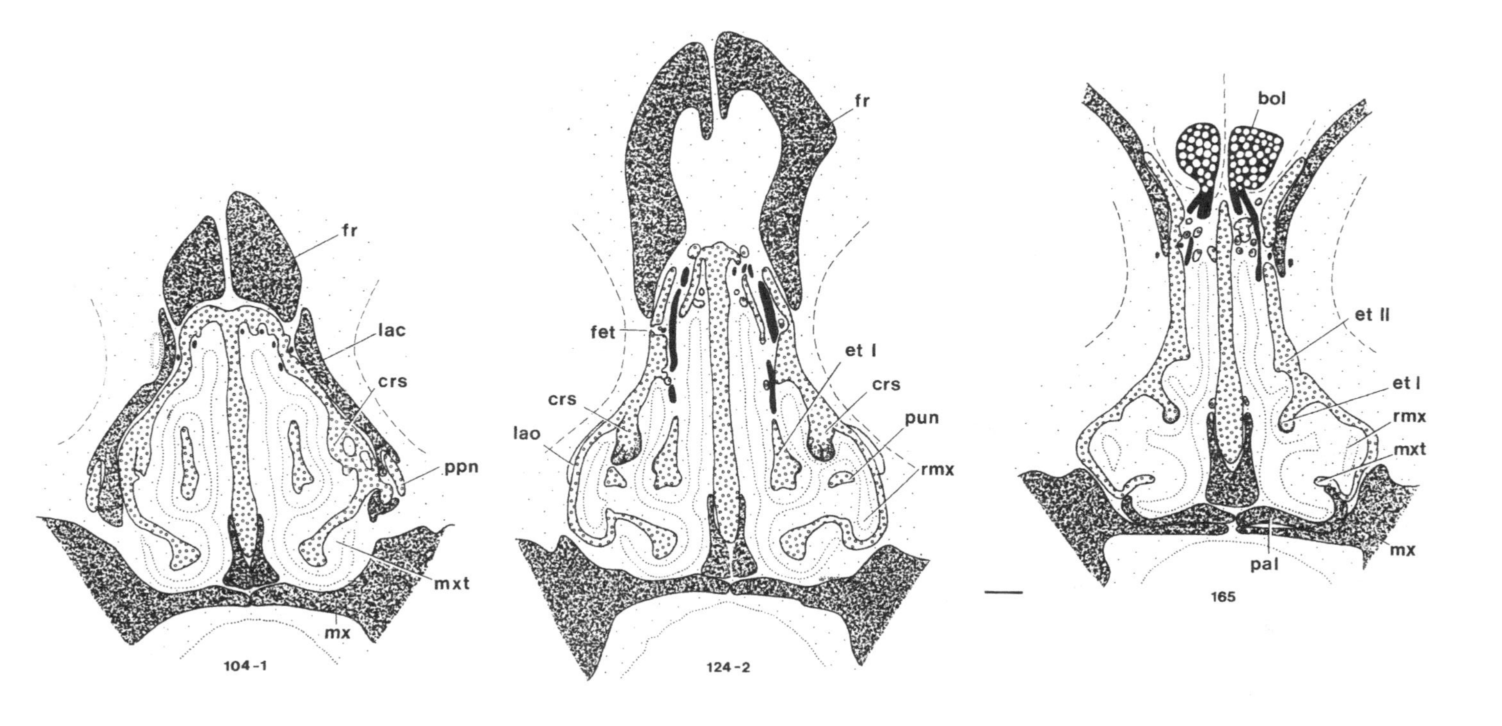
fr
lac
crs
ppn
mxt
mx
104-1
fr
fet
et I
crs
crs
lao
pun
rmx
124-2
bol
et II
et I
rmx
mxt
mx
pal
165

Fig. 5.4. (*cont.*)

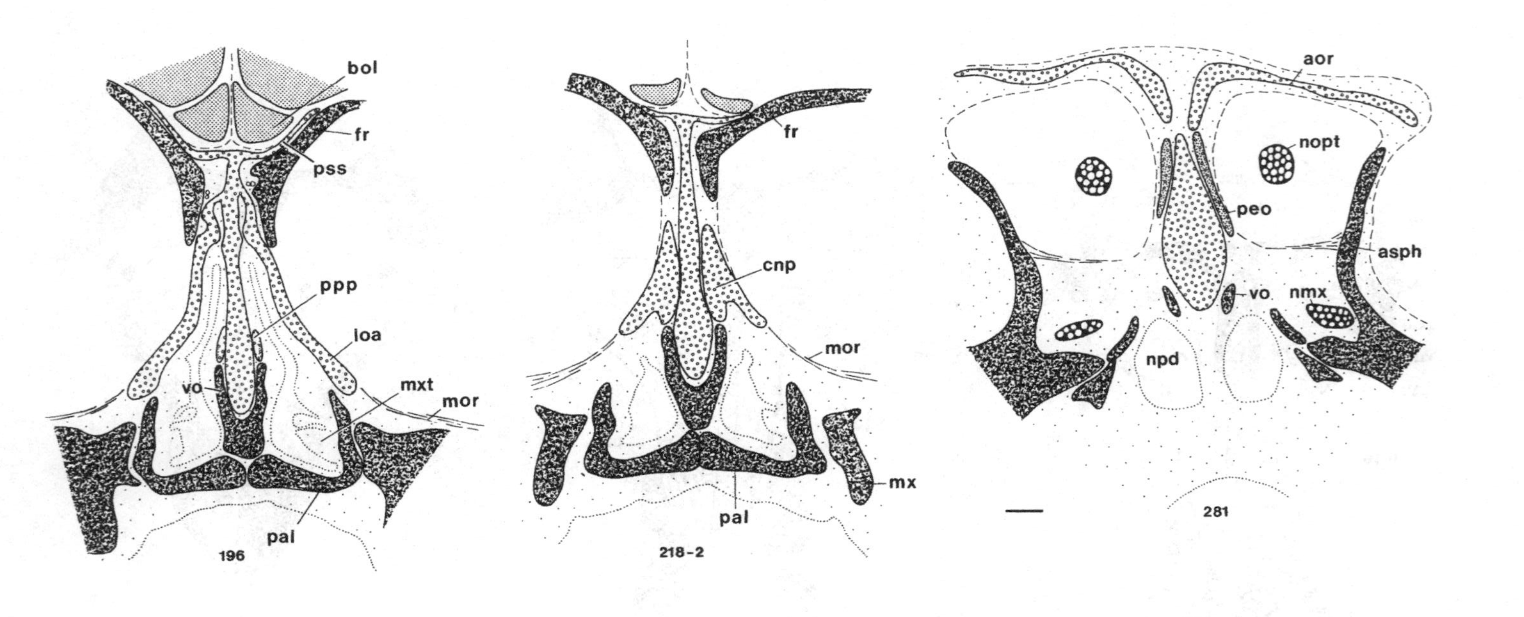
bol
fr
pss
ppp
loa
vo
mxt
mor
pal
196
fr
cnp
mor
mx
pal
218-2
aor
nopt
peo
asph
vo
nmx
npd
281

Fig. 5.4. (*cont.*)

Nasoturbinal

A nasoturbinal is present in all three specimens of cercopithecoid fetuses, although in variable grades of differentiation. In the fetal macaque, it appears to be confined to the anterior portion of the nasal cavity without showing any skeletal support (Ma.7–2–2, Fig. 5.2). In juvenile heads of *Macaca* and *Cercopithecus*, a short nasoturbinal is restricted to the anterior nasal region as well. In *Theropithecus*, an outer groove of the anterior nasal roof forms a low crest at the inside, which lends some support to the connective tissue of this concha (Th.14–2–2, Fig. 5.3). An immature female gelada does not show any nasoturbinal. In the adult *Papio*, the nasoturbinal becomes more prominent posteriorly (Fig. 5.1a) and continues into the anterior root of the first ethmoturbinal but not into the semicircular crest, as typical for many primitive mammals and "prosimians". In the nasal skeleton of the adult, the nasoturbinal is marked only by a low bony crest running along the nasofrontal suture. The fetus of *Papio* shows a low nasoturbinal continuing into the first ethmoturbinal. The figures of Seydel (1891) depict nasoturbinals.

Fetuses of *Hylobates* reveal only a tiny remnant of a nasoturbinal, but it seems to be well-developed in adult specimens (Cave and Haines, 1940). In a fetal *Pan*, the nasoturbinal is a very small swelling of the epithelium in the roof of the anterior nasal cavity. In the *Gorilla* fetus, the nasoturbinal is pronounced even in the vestibular region. Seydel (1891) and Cave and Haines (1940) confirm its presence in the large apes. Reinbach (1963) depicts a well-developed nasoturbinal in an advanced human fetus, which disappears underneath the semicircular crest (uncinate process, Reinbach, 1963). It is less clear in my series. In human anatomy, the "agger nasi" is considered a vestige of the nasoturbinal (Seydel, 1891; Warwick and Williams, 1973). The nasoturbinal is sometimes missing in platyrrhines, which is the result of parallel reduction. Thus, strong reduction of this concha appears to be a widespread trend within anthropoids.

Nasal floor cartilages

Maier (1980) described the complicated and complete cartilage structures around the nasopalatine (Stenson's) duct in an adult *Cercopithecus talapoin*. Nasal floor cartilages of a fetal macaque were documented by Frets (1914), but his peculiar nomenclature renders interpretation difficult. Adult cercopithecoids exhibit a pair of funnel-shaped nasopalatine ducts (Fig.

5.1a), which run through the incisal foramina of the hard palate. The incisal foramen is bisected by the medial bony bridge formed by the medial palatine processes of the premaxilla. All three depicted fetuses present the same structural configuration; the others that were examined also coincide. The lateroventral process and the lower portion of the anterior transverse lamina combine to form the cartilage of the nasopalatine duct, which surrounds the nasopalatine duct anteriorly (Ma.7–2–2, Fig. 5.2; Pa.43–2–1, 5.4). A peculiar feature is a very broad lateral lamella of cartilage, which covers the duct from the side. In the baboon, this cartilage is in direct connection with the side wall of the nasal capsule (Pa.46–2–1, Fig. 5.4). An independent palatine cartilage frames the incisal foramen from the lateral and posterior side (Ma.10–1–1, 5.2).

The fully-developed nasopalatine duct of cercopithecoids is an unexpected feature whose function remains obscure, especially as no vestige of a vomeronasal organ (Jacobson's organ) has been observed. The duct is greatly modified in hominoids. In gibbons, it persists into adult life (Maier, 1997) and there are vestiges of a cartilage nasopalatine duct. In an advanced *Gorilla* fetus, a narrow epithelial duct still connects nasal and buccal cavity (Maier, 1991, 1997). The fetuses of *Pan* show funnel-shaped nasopalatine ducts, which do not break through the oral epithelium. They are accompanied by well-differentiated cartilages (Maier, 1997). In *Pongo*, only a vestige of a nasopalatine duct is developed; the duct itself may be represented by an epithelial cyst, which is situated within the incisal foramen (Maier, 1997). Maier (1991, 1997) depicts remnants of a nasopalatine duct in the form of solid epithelial strands in a late fetus of *Homo*. Well-differentiated relics of nasal floor cartilages are described in older fetuses of *Homo* by Reinbach (1963) and Maier (1991).

The vomeronasal organ has disappeared in cercopithecoids, hylobatids, and the great apes, but it is present as a transitory relic in *Homo*. The vestige of the organ in a young fetus of *Pan* looks very similar to humans (Starck, 1960). It may be that early embryos of hominoids are not known well enough, but its complete absence in cercopithecoids is abundantly documented.

Tarsiers and platyrrhines still have functioning Jacobson's organs, and the nasal floor cartilages show very variable shapes (Frets, 1914; Maier, 1980; Starck, 1984). The microscopic anatomy of the sensory epithelium in these haplorhines shows signs of simplification, whereas it seems to be progressively developed in most strepsirhines (Schilling, 1970; Stephan *et al.*, 1982).

Lateral recess, semicircular crest, and uncinate process

Traditional studies (Gaupp, 1906; Voit, 1909; Reinbach, 1952; Zeller, 1987; Maier, 1993a) subdivide the nasal cavity into anterior, posterior, and lateral compartments. The last two are considered to be additions through early mammalian evolution (Starck, 1967; Kuhn, 1971; Zeller, 1989). In fetal skulls of primitive mammals, the lateral recess is anteriorly bordered by a sharp rim of cartilage, the semicircular crest. After ossification, it may persist as a fragile uncinate process, which partly frames the semilunate hiatus anteriorly and ventrally.

In the macaque fetus (Ma.14–3–2, Fig. 5.2), a semicircular crest is present, but it does not form a free uncinate process. In a much older fetus of *M. fascicularis*, the process and the maxillary recess are almost non-existant. In juvenile skulls of rhesus macaques, a bony uncinate process has always been observed. In *Papio*, there is a typical uncinate process both as cartilage and as bony structure (Pa.124–2, Fig. 5.4). *Theropithecus* shows a pronounced semicircular crest, but the uncinate process appears to be fused with the root of the maxilloturbinal (Th.37–1–1, Fig. 5.3). The homology of this crest remains doubtful because a small, isolated cartilage is also present. In cercopithecoids, the semicircular crest is covered by the anterior portion of the first ethmoturbinal, whereas it lies well in front of this concha in "prosimians". This altered relationship mirrors the secondary diminution of the structures of the lateral recess. In *Nasalis*, there is no indication of a semicircular crest; its maxillary recess is formed as a groove, but it is not separated as a closed epithelial duct. This coincides with Seydel (1891), who noticed a shallow depression in the place of the maxillary recess in an adult *Nasalis*. In a fetus of *Trachypithecus*, both the uncinate process and the maxillary recessus are present in a very rudimentary state. There appears to be intra-familiar variation in this structural system. The maxillary sinus, which develops from the recess, is relatively spacious in *Cercopithecus*, whereas it is tiny in the other taxa (Seydel, 1891).

In *Hylobates*, there is a narrow, ventrally-oriented lateral chamber with a maxillary recess. The faint semicircular crest, which shows no free uncinate process, is hidden underneath the first ethmoturbinal. In *H. moloch*, it is anteriorly continuous with a low nasoturbinal. This reduction of the semicircular crest, as well as the lateral prominence of the pars lateralis, seems to be autapomorphic for gibbons. Adult gibbons have variable conditions at the semilunate entrance to the lateral recess, but a maxillary sinus is invariably present (Cave and Haines, 1940).

Of the large hominoids, *Pongo* is reported to have an aberrant entrance to the lateral recess. It is shifted far posteriorly, and its round opening is said to have no uncinate process. The maxillary sinus is extremely expanded and tends to replace the sphenoidal sinus (Cave and Haines, 1940). My fetal orang shows a fairly normal semicircular crest with a free uncinate process, although both are displaced posteriorly. Humans and African apes have well-developed – though variable – uncinate processes and lateral recesses in fetal and adult stages. In the latter, maxillary sinuses are formed by secondary pneumatization of the recesses (Weinert, 1926; Cave and Haines, 1940).

Although the dorsal parts of the nasal cavity are strongly compressed in most platyrrhines, all of them show a very prominent semicircular crest with a free uncinate process. The lateral recess with its maxillary extension is well-developed in fetal stages. The uncinate process may be confluent with the nasoturbinal.

Paranasal sinuses

Paranasal sinuses are secondary extensions of recesses of the nasal cavity into the adjacent bony elements. After resorption of the cartilages, the nasal epithelium expands by the process of pneumatization into mechanically less important structures. Thereby, a reduction of weight and saving of bony substance is achieved (Weidenreich, 1924). Pneumatization seems to increase with body size, but phylogenetic factors are involved as well. Paranasal sinuses of anthropoid primates have been described in great detail (Seydel, 1891; Paulli, 1900; Cave and Haines, 1940; Cave, 1967; Hershkovitz, 1977).

In primitive mammals, the lateral recess is divided into upper (frontal) and lower (maxillary) portions, separated posteriorly by a horizontal lamina (frontomaxillary septum). This arrangement is found in most studied strepsirhines, where even frontoturbinals are retained (Paulli, 1900; Kollman and Papin, 1925; Maier, 1993b). In anthropoids, the frontal recess is confined to a small space next to the narrow entrance, whereas the maxillary recess forms a narrow epithelial duct filling the interior of the maxillary process of the nasal capsule (Figs. 5.2–5.4). An upper posterior recess, containing frontoturbinals, cannot be distinguished in anthropoids, and the complete loss of this structural unit is a synapomorphy of the Anthropoidea.

In contrast to hominoids, pneumatization is almost absent in cercopithecoids. Even the large cercopithecoids have almost no sinuses. Suppression of paranasal sinuses is a synapomorphy of cercopithecoids. The maxillary recess forms in fetal life, but does not – or not significantly – expand subsequently. Paranasal sinuses do not play a role in the construction of the bony muzzle or of the orbital tori. Sphenoidal sinuses are also absent; this feature is probably connected with the complete reduction of the fetal ethmoturbinal recess (Maier, 1993). Nevertheless, a small pocket-like lateral recess (comprising both frontal and maxillary recesses) persists, its entrance hidden behind the anterior lamina of the first ethmoturbinal. In a cleaned skull, it is just visible below the "medial concha" (Fig. 5.1b).

In the adult chacma baboon, a wide, oval opening (hiatus) leads into the spacious lateral recess. In a subadult gelada, there is a slit-like semilunar hiatus, the upper margin of which is somewhat bulbous, reminiscent of the human ethmoidal bulla. In the small crania of *Cercopithecus* and *Presbytis*, narrow and oblique slits open into very small lateral recesses. The anatomical details of this region will be clarified by serial sectioning of adult specimens. Judging from human anatomy, individual variation may be considerable. It appears that systematic and size-dependent factors could be involved in shaping this regressive region of the nasal cavity of cercopithecoids.

Ethmoturbinals

The upper conchae of human anatomy are homologous to the ethmoturbinals of other mammals. In cercopithecoids, the number of conchae seems to depend on absolute size: small monkeys usually show one pair; female baboons two pairs; and male baboons three pairs. This generalization is based on only a few specimens. Which of the conchae of primates are homologous with those of primitive mammals and "prosimians" on the one hand, and those of hominoids on the other? Studies of many fetal mammals have revealed a fairly similar "bauplan" of the nasal cavity (Maier, 1993a,b). Among strepsirhine prosimians, *Daubentonia* possesses four ethmoturbinals, while the other investigated taxa have three (Kollmann and Papin, 1925). Published information is complicated to interpret, because the first ethmoturbinal generally has two lamellae, and interturbinals and epiturbinals may also occur. The homology of the ethmoidal bulla is one of the main problems of nasal morphology (Seydel, 1891).

When compared to "prosimians", cercopithecoids exhibit reduction of the ethmoturbinal system, but retain a serial arrangement of the ethmoturbinals. This is a plesiomorphic trait (Fig. 5.1) that contrasts with the apomorphic condition in hominoids, which includes a ventrodorsal series of horizontally-orientated conchae. The narrowing of the cribriform plate causes the fixation points of the turbinals to be shifted well down to the lateral wall, except for *Theropithecus* where the ethmoturbinal I is mainly fixed to the cribral plate (Th.37–1–1, Fig. 5.3). In the other taxa, portions of this turbinal may indirectly reach the cribral plate. In *Trachypithecus*, the ethmoturbinal I is fixed to the middle of the lateral wall and shows no connection with the cribral plate. Colobines seem to be more derived than cercopithecines in this respect, despite their relatively wide interorbital pillar (Vogel, 1966). The regression-index of the olfactory bulb is lowest in *Colobus* as well (Stephan, 1975). From my material, it is not clear whether the second ethmoturbinal is the posterior lamella of the first ethmoturbinal, or whether it is homologous with the second. I tend to accept the first alternative. The identification of a third one would depend on this decision. Seydel (1891) recorded two ethmoturbinals in a *Papio* and three in a *Mandrillus*.

The cribriform plate becomes complicated, and its trabeculae are not situated at a single level, because of the narrowing of the subcerebral roof of the nasal capsule. I suspect that the "squeezed" cribriform plate of cercopithecoids includes vestiges of ethmoturbinals (see below). The structures of the posterior roof of the nasal capsule become difficult to understand, because of later ossification.

In *Hylobates*, the most dorsal portion of the nasal cavity appears to be compressed between the two frontals, and the anterior portion of its cribral plate looks narrowed. More posteriorly, the upper nasal cavity appears spacious enough to allow parts of the first two ethmoturbinals to be directly fixed to the ventral side of the cribral plate. Due to its compressed state, the dorsal parts of the nasal capsule are similar to cercopithecoids and especially to *Theropithecus*, with which it shares the existence of a prominent mesethmoidal spine (Th.37–1–1, Fig. 5.3). The gibbon possesses a third ethmoturbinal, which is fixed to the side wall of the nasal capsule (antorbital lamina). In adult specimens, Seydel (1891) observed a first ethmoturbinal and a vestigeal second.

The explanation of the horizontal orientation of the upper nasal conchae of the hominoids is an old morphological problem. However, comparative anatomy of hominoids and their ontogenesis show transitions (Seydel, 1891). Both fetal and adult gibbons show obliquely fixed ethmoturbinals, and the orang appears also a little plesiomorphic, as do fetal humans. The differences between the apomorphic hominoid and the plesiomorphic anthropoid states are not categorical. The individual variation is enormous, as is known from human anatomy. Seydel (1891) maintains that the development of ethmoidal sinuses (cellulae) is the result of a secondary broadening of the interorbital pillar in humans and the African apes. Whereas Seydel claimed that these cellulae are confined to humans, Cave and Haines (1940) describe "anterior and posterior ethmoidal sinuses" in *Gorilla* and *Pan*. The investigation of fossil hominoids should help to clarify these questions.

Primitive platyrrhines show a reduced number of ethmoturbinals, but their arrangement is mostly vertical. In some cebids, they are horizontal. The larger platyrrhines appear to have paranasal sinuses (Hill, 1957; Cave, 1967). The ethmoidal structures of platyrrhines appear to have undergone a structural radiation, but the basic pattern helps to understand better the anthropoid "groundplan".

Maxillary process, orbital lamina, and maxilloturbinal

Parts of the side wall of the cartilaginous nasal capsule are exposed at the medial wall of the orbit. By enchondral ossification, this plate later becomes the orbital lamina of the ethmoid, which constitutes a considerable portion of the medial wall of the orbit in cercopithecoids. There is no formation of ethmodial sinuses (cellulae) in cercopithecoids, gibbons nor orangs.

At the lower edge of the orbital wall of the nasal capsule, the cartilage turns inward to become the maxilloturbinal. The free edge, sometimes called maxillary process, rests on the maxilla anteriorly and on the ascending process of the palatine posteriorly. It becomes a simple, obliquely-oriented cartilage plate in its most posterior portion (Ma.31–2–2, Fig. 5.2; Pa.196, fig 5.4). Maier (1987) argues that this oblique plate plays a mechanical role in the architecture of the fetal skull. By separate ossification and by partial resorption of the lateral wall of the nasal capsule, the maxilloturbinal ("inferior concha") appears as an isolated and distinct element of the ethmoid, mainly fixed to a low crest of the maxilla. This distinction, which stems from anatomy of adult humans, does not make sense in an ontogenetic perspective, and this concha is part of the ethmoidal bone. The soft tissue maxilloturbinal can be followed into the anterior parts of the nasopharyngeal ducts(cf. Ma.31–2–2, Fig. 5.2; Th.71–1, Fig. 5.3; Pa.196, 218, Fig. 5.4). It may use parts of the ascending process of the palatine as a support (Th.71–1, Fig. 5.3; Pa.218–2, Fig. 5.4), as can be seen in the adult skull (Fig. 5.1b).

Whereas the paranasal process of the pars lateralis of the nasal capsule is almost missing in *Hylobates*, the maxillary process is very prominent and extends far posteriorly. It rests on the dorsal sides of both the maxillary and the palatine. Only the most posterior part of the maxillary process consists of a solid cartilage plate. This lateroventral portion of the nasal capsule is more similar to the plesiomorphic condition in platyrrhines (Maier, 1993b) than to the state in cercopithecoids, where reduction has further proceeded.

Cribral plate, supraseptal plane

The cribral plate appears to be vaulted dorsally by lateral compression of the frontals, and its ventral side is probably fused both to the ethmoturbinals and to the side wall of the nasal capsule. It is relatively wide in *Theropithecus* (Th.37–1– 1, Fig. 5.3), but more narrow in *Macaca* (Ma.16–1–1, 20–2–2, Fig. 5.2) and *Papio* (Pa.124–2, 165, Fig. 5.4). Colobines do not appear more spacious in this region, in spite of their greater interorbital breadth (Vogel, 1966). The area of the cribral plate is framed by prominent cartilage rims (precribrosal border and paracribrosal border), which are fixed to the surrounding bones of the cranial vault. In *Theropithecus*, the precribrosal border is a prominent spine of cartilage, the mesethmoidal spine (Th.37–1–1, Fig. 5.3). Such a spine is present in a number of strepsirhines (*Lemur*, *Lepilemur*, *Propithecus*, *Indri*) and other mammals. It is prominent in *Hylobates*, but is otherwise not known from catarrhines. I have observed it in young fetuses of *Alouatta* and *Ateles*, but not in *Callimico* and the callitrichids. The notion of Seydel (1891) that the smaller species of cercopithecoids do not possess a true cribral plate can be refuted by this study, and by carefully cleaned juvenile skulls.

The cribral plate of *Hylobates* is displaced upward and partly fused with the side wall of the nasal capsule and the first ethmoturbinal, much as in cercopithecoids. It is likely that a number of the cribral foramina are of a secondary nature.

The cribral plate of the large apes and of humans is wider and more horizontally-oriented than in all other catarrhines. The question arises whether this is a plesiomorphic condition or whether it is a secondary expansion due to size increase, providing absolutely more space at the dorsal roof of the nasal capsule. Increase in brain size is suggested as being responsible for such a secondary widening (Seydel, 1891). The course of the ethmoidal nerves and vessels does not indicate a secondarily modified situation (see below). The ventral insertion of the medial nasal concha (ethmoturbinal I) may be the result of a more narrow nasal cavity in ancestors. Computer tomographs of muzzles of middle Tertiary fossil hominoids may help to clarify this question.

In most primitive mammals, the cribral plate is followed by a more or less extensive unperforated portion (infracribral lamina). In most taxa, this solid roof of the posterior cupula forms part of the floor of the anterior cranial cavity. However, in all observed cercopithecoids, the posterior cupula, which recedes ventrally, is excluded from the cranial cavity by a supraseptal planum. This planum appears to be formed by a medial fusion of the orbitonasal commissures. Sometimes, isolated supraseptal cartilages become integrated into this planum, which posteriorly continues into the wings of the orbital alae or orbitosphenoids respectively. In late ontogenetic stages and in all juveniles, the orbital processes of the frontals meet in the midline, and the planum has disappeared – apart from anteromedial processes of the orbital wings. This resuls in the secondary exclusion of a tiny compartment of the supracribral space from the cranial cavity. In cercopithecoids, the paired ethmoidal arteries (and sometimes branches of the ethmoidal nerve) run anteriorly through this "subplanal" space (Ma.31–1–2, Fig. 5.2; Th.71–1, Fig. 5.3; Pa.196, Fig. 5.4).

The term supraseptal planum appears to have been coined by Reinbach (1963) for a late human fetus. Fig.10A in Maier (1993b) depicts this structure in an advanced fetus of *Homo*. There exists no such supraseptal planum in *Hylobates*; the orbital processes of the frontals rest on the compressed cartilage of the dorsal part of the posterior cupula. The posterior cupula containing the ethmoidal recessus is, in contrast to platyrrhines (Maier, 1993b), shifted ventrally. In hylobatids, the distance between the small cribral plate and the orbital wings is very large, and the orbitonasal commissures disappear early. The young fetus of *Pan* shows a perfect supraseptal plane, reminiscent of cercopithecoids (Maier, 1993b, fig. 10B). However, the "subplanar space" does not contain the posterior ethmoidal nerve and artery as in platyrrhines and cercopithecoids. Therefore, it may by a convergently developed structure. A fetal *Gorilla* shows a complicated, double layered planum above the posterior end of its cribral plate (Maier, 1993b, fig. 10C). *Pongo* is very distinctive in this structure; there exists a wide infracribrosal lamina (cf. Maier, 1993b, fig. 10D), more like that of some platyrrhines than other known catarrhines.

Maier (1983) depicts a supraseptal planum for *Saimiri* and a similar structure in *Cebus* (Maier, 1993b, fig. 7B). In adult skulls of these species, one can find sphenethmoidal ossifications lying on top of the suture between the frontals. I also find a well-developed supraseptal planum in *Callithrix jacchus* (CRL 31 mm). In a *Callicebus moloch* (CRL 61 mm), it is missing because the septum projects into the cranial cavity. *Callimico* remains plesiomorphic by retaining a broad and flat infracribral plate reaching the orbital region (Maier, 1993b, fig. 8A). Therefore, it is difficult to decide whether a supraseptal planum belongs to the "groundplan" of the platyrrhine ancestor. It seems likely that such a structure developed several times independently as a consequence of the reduction of the posterior nasal capsule.

Posterior cupula, ethmoturbinal recess, and sphenoidal sinus

The reduction of the posterior nasal cavity (ethmoturbinal recess) and its cartilaginous wall (posterior cupula) in all anthropoids has resulted in parts of the ancestral nasal septum becoming secondarily exposed as an interorbital septum (see below). In fetal stages of cercopithecoids, the posterior parts of the nasal capsule are retained as solid cartilage plates (maxillary process), lying at the sides of the interorbital septum and supporting the interorbital pillar. The most posterior end of the cupula is a simple cartilaginous process resting on top of the wings of the vomer, but it may also serve as a buttress for the orbital processes of the frontals in older specimens (Ma.39–2–2, Fig. 5.2; Th.71–1, Fig. 5.3; Pa.218–2, Fig. 5.4). The medial process of the cupula, which directly contacts the vomer, is probably homologous to the posterior paraseptal cartilage. A posterior transverse lamina, seen in platyrrhines and hominoids, is absent in cercopithecoids (Maier, 1993b). This complete loss may be considered as synapomorphy of this taxon. Occasionally, very shallow excavations of the nasal epithelium are found, which may or may not be remnants of the posterior ethmoturbinal recesses (Ma.31–1–2, Fig. 5.2).

The lateral edge of the cupular cartilage serves as origin of the smooth orbital muscle, which spans the inferior orbital fissure. The postorbital septum and the smooth orbital muscle (Mueller's muscle; cf. Warwick and Williams, 1973), as parts of the periorbita, probably constitute the structural matrix for the bony postorbital closure. This preadaptive condition was also considered by Cartmill (1980). In the adult skulls of cercopithecoids, the posterior orbital plate of the osseous ethmoid still shows a sharp posterolateral edge for the insertion of the orbital muscle.

Due to the expansion of the orbitosphenoid processes (see below), which completely cover the posterior cupula, the orbital muscle of gibbons has lost its connection with this posterior portion of the orbital plate. It is present in extant pongids and hominids. The posterior cupula in platyrrhines and hominoids has been discussed (Maier, 1987, 1993b). A young fetus of *Hylobates* shows a fairly well-developed posterior cupula underneath the orbitosphenoidal process, containing a slender ethmoidal recess.

It is more similar to *Homo*, *Pan*, and *Gorilla* than to *Pongo*, which may be derived in this feature. The tiny ethmoturbinal recess of platyrrhines, as well as of hominoids, appears to be the place of origin of the sphenoidal sinus. Both structures are missing in cercopithecoids, autapomorphically.

Ethmoidal process of the orbitosphenoid, interorbital septum

The anterior (preoptic) root of the orbital ala begins to ossify as the orbitosphenoid in later fetal stages. This ossification not only invades the trabecular plate to form the presphenoid, but it also develops a plate of appositional bone growing rostrally at each side of the trabecular plate or interorbital septum. Frick (1954) and Starck (1967) call it the ethmoidal process of the orbitosphenoid. In "prosimians" and hominoids, as in other mammals, these orbitosphenoidal processes enclose the posterior nasal cupula, which is later resorbed. Where developed, the ethmoturbinal recess may invade the pre- and orbitosphenoid as the sphenoidal sinus (Maier, 1993b).

I can document several stages of development of the interorbital septum *sensu lato*. In the young fetus of *Macaca*, no ossification occurs at the orbital ala. In a fetal *M. fascicularis*, the preoptical roots have just begun to ossify. In the *Theropithecus*, the development of the ethmoidal process has just begun (Th.86–1, Fig. 5.3), and in the *Papio*, the bony laminae have grown forward for some distance (Pa.281, Fig. 5.4). In cleaned skulls of late fetal and perinatal *Macaca mulatta*, the thin bony plates of the orbitosphenoid have grown forward, closely apposing the cartilaginous or fusing with the ossifying interorbital septum *sensu stricto*. In somewhat older juvenile skulls of *Macaca*, the orbitosphenoids have reached the frontal as well as the orbital plate of the ethmoid. This results in the bony interorbital septum of cercopithecoids becoming a compound trilaminar structure, consisting of the median septum *sensu stricto* and of the two adjacent ethmoidal plates of the orbitosphenoid.

The orbitosphenoid processes are a characteristic feature of strepsirhine skulls, but they remain short because of the posterior extension of the nasal capsule (Maier, 1993b, fig.5). An interorbital septum is found in earlier fetal stages of strepsirhines (Henkel, 1928), but later on it is covered by the orbitosphenoid (Starck, 1962; Muehlenkamp, 1993). In the platyrrhine skulls, we find a three-layered interorbital septum (Maier, 1983, figs.6,7).

No interorbital septum is present in later fetuses of the large hominoids, but there is a well-developed posterior nasal capsule (orbital plate), extending in front of or below the optic foramen. In these hominoids, the orbitosphenoid is a significant element of the most posterior portion of the medial orbital wall. In *Hylobates*, the posterior cupula is much shorter, and the ethmoidal plates of the orbitosphenoid are distinctly anteriorly elongated. Even in the young fetus, the orbitosphenoid processes come to lie at the sides of the posterior cupulae. Morphologically, the posterior cupula of gibbons is intermediate between the hominids and the cercopithecoids. *Pongo* shows a derived posterior cupula, without a floor formed by the posterior transverse lamina, but containing a shallow ethmoidal recess. The formation of the posterior nasal septum is complicated and aberrant in this species (Maier, 1993b). The most posterior parts of the nasal cupulae are massive cartilage plates lying at both sides of the ossified presphenoid. These plates have overgrown the true septum, forming a supraseptal mass reminiscent of *Homo* and *Gorilla*. The large hominoids do not possess protrusive and overlapping ethmoidal processes of the orbitosphenoids, probably due to the posterior extension of the posterior cupula. The ethmoidal recesses give rise to the sphenoidal sinuses of subadult and adult animals, which extend into the sphenoid bone by pneumatization. The formation of paranasal sinuses in *Pongo* is different from the other large hominoids (Cave and Haines, 1940).

Platyrrhines show considerable reduction of the posterior cupula, but the ethmoidal recess retains a dorsal position between the orbital processes of the frontals. Their secondary interorbital septum tends to develop ventral to the posterior cupula and not dorsal to it, as in catarrhines (Frets 1914; Maier, 1993b). However, there exist considerable differences among platyrrhines.

In cercopithecoids, the anterior roots of the orbital ala bulge dorsally and lie on top of the low trabecular plate. The preoptic roots, together with the orbitosphenoids, become posterior parts of the interorbital septum (Fig. 5.4). The ossifying preoptic roots soon begin to fuse in the midline above the trabecular plate. The trabecular plate remains cartilaginous at first, but then begins to ossify enchondrally to the presphenoid *sensu stricto*. Thus, three heterogeneous skeletal elements later fuse to the presphenoid *sensu lato* and to a compound interorbital septum.

Discussion and conclusions

Given the presumed monophyly of extant anthropoids, what does the nasal region tell us about their common ancestor? We assume that it was somewhat more macrosmatic than catarrhines, and that its external nasal region was reminiscent of the strepsirhine condition, much as in platyrrhines. It possessed a broader internarial area, a continuous margino–atrio–maxilloturbinal system, and a functioning vomeronasal organ. There was considerable convergence of the eyeballs and of the orbits, as seen in the skulls of fossil oligopithecines, parapithecines and propliopithecines (Simons, 1989,1990). Accentuation of stereoscopy led to a shortening of the snout, and resulted in a narrowing of the upper portions of the nasal cavity, as well as in some degree of postorbital closure. This caused considerable reduction of the lateral recesses of the nasal cavity, of which the frontoturbinals had been lost. However, the lateral cartilages persist as strong plates contributing to the fetal construction of the palatal arch (paranasal process, maxillary process). The cribriform plate became reduced in size and laterally narrowed. Closure of the posterior cribral foramina, accompanying the gradual reduction of the ethmoturbinal recesses, may have led to the formation of the supraseptal planum covering the posterior course of the ethmoidal artery. The orbitonasal fissure became narrowed to a small ethmoidal foramen. The ethmoidal processes of the orbitosphenoid contributed to strengthen the secondarily exposed interorbital septum. The lack of any significant pneumatization is a peculiar characteristic of cercopithecoids, in which the reduction of the posterior nasal structures proceeded furthest among catarrhines. Reduction of the posterior nasal structures followed different paths in platyrrhines, where dorsal portions of the posterior cupula and its recess are preserved as vestiges.

The shortening and heightening of the facial skull in early anthropoids (Maier, 1993b) provided the structural basis for stronger incisal biting. Spatulate incisors tend to close the median diastema and thus interrupt the philtrum, which primitively connected the rhinarium with the oral opening of the nasopalatine duct and the vomeronasal organ. Median fusion and mobility of the upper lip and haplorhiny were the results of such changes.

In fetuses, as in strepsirhines, the nasal capsule constitutes a broad interorbital pillar. This medial pillar is suited to transmit the pressure forces of the tongue, which also play an important role during suckling (Herring, 1985). Tooth eruption and masticatory activity changes the distribution of forces. Milk incisors, which appear first, cause pressure forces close to the midline. Milk molars shift the biting forces laterally to the alveolar processes and zygomatic roots (Maier, 1987). During the juvenile period, the exocranial bones of the muzzle and of the orbit gradually take over the masticatory forces, whereas the ethmoidal structures tend to become relieved and disappear or remain as fragile supports of the nasal conchae. In all major catarrhine groups, the interorbital pillar remains relatively broad in some taxa (callitrichids, some cebids, hominoids, and colobines).

Adaptive pressure led to extreme narrowing of the interorbital region in ceboids and cercopithecoids (Vogel, 1966; Maier, 1983). The details of these intra-group changes are not yet adequately studied.

Biomechanics of primate skulls, let alone fetal stages, have not yet been analyzed in detail (Preuschoft *et al.*, 1985). However, it seems obvious that the cartilaginous nasal capsule constitutes an integral part of the growing facial skull. Considering the reduced olfactory functions, it is informative to see which parts are retained for mechanical reasons. It is mainly the side walls of the lateral and posterior parts of the capsule that maintain their morphogenetic functions.

Quantitative analysis of the olfactory bulb indicates that cercopithecids show a regression index (0.1) that is double that of hominids (Stephan, 1975). *Miopithecus* and *Colobus* appear to be distinctly lower than *Macaca* and *Cercopithecus*. Baboons and geladas were not included in that analysis. My fetus of *Macaca* shows that the olfactory epithelium is more expanded than in humans. The internal structures of the cercopithecoid nasal cavity are simplified, but the plesiomorphic primate "bauplan" is still clearly discernible. The baboon-like elongation of the muzzle affects only the anterior portion of the nasal capsule, whereas the conservative middle portion remains relatively unaffected.

The essential characteristics of the nasal structures of cercopithecoids may be summarized as follows. The anterior part tends to become enlarged with the facial skull. The posterior part is extremely reduced, although it has not disappeared completely. In these modified parts, we find most of the apomorphic features. The retention of a well-developed marginoturbinal in the vestibulum, and the persisting nasopalatine ducts, are the most remarkable features of the anterior nasal region. The ethmoturbinal recess and the posterior nasal cupula are even more reduced, and no sphenoidal sinus is developed. The middle portion, which is connected with reduced olfactory functions, is relatively small and simple, but the structural arrangement is plesiomorphic. The lateral part of the nasal capsule as well as the anterior two ethmoturbinals are preserved. However, the elements of the lateral portion are diminutive and the lateral recess does not give rise to any remarkable paranasal sinuses, except for a small maxillary sinus. The diminution of the lateral recess appears to influence the course of the ethmoidal nerves and vessels. The area of the cribral plate is narrowed between the orbital processes of the frontals; the cribral plate appears vaulted and ventrally fused with the upper parts of the ethmoturbinals. The reduction of the posterior nasal cupula secondarily exposes an interorbital septum, which is strengthened by two laterally posited ethmoidal processes of the orbitosphenoid. The interorbital septum is thus a trilaminar structure, which is dorsally supplemented by the upward vaulted orbital roots.

Cercopithecoids share a considerable number of plesiomorphies with the platyrrhines, especially at the nasal floor. In other traits, especially those connected with the reduction of the posterior portion, they look more derived than the hominoids. Gibbons show many similarities with cercopithecoids, but in some cases it is likely that common specializations are due to convergence. In many features of the nasal region, cercopithecoids are more distant to the common catarrhine ancestor than hominoids.

References

Andrews, P. (1985). Family group systematics and evolution among catarrhine primates. In *Ancestors: The Hard Evidence*, ed. E. Delson, pp. 14–22. New York: Alan R. Liss.

Andrews, P. (1988). A phylogenetic analysis of the Primates. In *The Phylogeny and Classification of the Tetrapods*, ed. M.J. Benton, pp. 143–75. Oxford: Clarendon Press.

Andrews, P. & Martin, L. (1987). Cladistic relationships of extant and fossil hominoids. *J. hum. Evol.* **16**, 101–18.

Augier, M. (1931). Squelette cephalique. In *Traits d'Anatomie Humaine*, vol. I part 4. Paris: F. Poirier & Charpy.

Bonner, J.T. (1982). *Evolution and Development*. Berlin: Springer.

Cartmill, M. (1980). Morphology, function, and evolution of the anthropoid postorbital septum. In *Evolutionary Biology of Primates and Continental Drift*, ed. R.L. Ciochon and A.B. Chiarelli, pp.243–74. New York: Plenum.

Cave, A.J.E. (1967). Observations on the platyrrhine nasal fossa. *Am. J. phys. Anthrop.* **26**, 277–88.

Cave, A.J.E. (1973). The primate nasal fossa. *Biol. J. Linn. Soc.* **5**, 377–87.

Cave, A.J.E. & Haines,R.W. (1940). The paranasal sinuses of the anthropoid apes. *J. Anat.* **74**, 493–523.

DeBeer, G. (1937). *The Development of the Vertebrate Skull*. Oxford: Oxford University Press.

Frets, G.P. (1913). Beiträge zur vergleichenden Anatomie und Embryologie der Nase der Primaten. II Die Regio ethmoidalis des Primordialcraniums mit Deckknochen von einigen platyrrhinen Affen. *Morph. Jahrbuch* **45**, 557–726.

Frets, G.P. (1914). Bcitrage zur vergleichenden Anatomie und Embryologie der Nase der Primaten. III. Die Regio ethmoidalis des Primordialcraniums mit Deckknochen von einigen Catarrhinen, Prosimiae und dem Menschen. *Morph. Jahrb.* **48**, 238–79.

Frick, H. (1954). *Die Entwicklung und Morphologie des Chondrocranium von Myotis Kaup*. Stuttgart: Thieme.

Garstang,W. (1922). The theory of recapitulation: A critical re-statement of the biogenetic law. *Linn.J. Zool.* **35**, 81–101.

Gaupp, E. (1906). Die Entwickelung des Kopfskelettes. In *Handbuch der vergleichenden und experimentellen Entwickelunglehre des Wirbeltiere*, ed. O. Hertwig, pp. 573–873. Jena: Fischer.

Geist, F.D. (1933). Nasal cavity, larynx, mouth and pharynx. In *The Anatomy of the Rhesus Monkey*, ed. C.G. Hartman & W.L. Straus, pp.189–209. Baltimore: Williams and Wilkins. (Reprint by Hafner Publ., New York, 1961.)

Henckel, K.O. (1928). Studien uber das Primordialcranium und die Stammesgeschichte der Primaten. *Morph.Jahrbuch* **49**, 105–78.

Hennig, W. (1966). *Phylogenetic Systematics*. Urbana, IL: University of Illinois.

Herring, S.W. (1985). Postnatal development of masticatory muscle functions. *Fortschritte Zool.* **30**, 213–15.

Hershkovitz, P. (1977). *Living New World Monkeys (Platyrrhini)*, vol. I. Chicago: University of Chicago.

Hill, W.C.O. (1957). *Primates. Comparative Anatomy and Taxonomy*. vol. III: *Pithecoidea (Platyrrhini)*. Edinburgh: University Press, Edinburgh.

Hill, W.C.O. (1966). *Primates. Comparative Anatomy and Taxonomy*, vol. VI *Cercopithecidae*. Edinburgh: Edinburgh University Press.

Hofer, H. (1977). The anatomical relations of the ductus vomeronasalis and the occurrence of taste buds in the papilla palatina of *Nycticebus coucang* (Primates, Prosimiae). With remarks on strepsirhinism. *Morph. Jahrb.* **123**, 836–56.

Kingdon, J. (1974). *East African Mammals*, vol. I. Chicago: University of Chicago.

Kollman, M. & Papin, L. (1925). Etudes sur les lemuriens. Anatomie comparée des fosses nasales et de leurs annexes. *Archs Morph. gén. exp.* **22**,1–58.

Kuhn, H.-J. (1971). Die Entwicklung und Morphologie des Schadels von *Tachyglossus aculeatus*. *Abh. senckenb. naturforsch. Ges.* **528**, 1–192.

Maier, W. (1980). Nasal structures in Old and New World primates. In *Evolutionary Biology of New World Monkeys and Continental Drift*, ed. R.L.Ciochon & A.B.Chiarelli, pp. 219–41. New York: Plenum.

Maier, W. (1983). Morphology of the interorbital region of *Saimiri sciureus*. *Folia primatol.* **41**, 277–303.

Maier, W. (1987). Functional principles of the growing skull of primates as shown by the posterior cupula of the nasal capsule. In *Définition et Origines de l' Homme*, ed. M. Sakka, pp.199–207. Paris: CNRS.

Maier, W. (1991). Aspects of ontogenetic development of nasal and facial skeletons in primates. In *Craniofacial Abnormalities and Clefts of the Upper Lip*, ed. G. Pfeifer, pp.115–24. Stuttgart: Thieme.

Maier, W. (1993a). Cranial morphology of the therian common ancestor, as suggested by the adaptations of neonate marsupials. In *Mammal Phylogeny*, ed. F.S. Szalay, M.J. Novacek, & M.C. McKenna, pp. 165–81. New York: Springer.

Maier, W. (1993b). Zur evolutiven und funktionellen Morphologie des Gesichtsschaedels der Primaten. *Z. Morph. Anthrop.* **79**, 279–99.

Maier, W. (1997). The nasopalatine ducts and the nasal floor cartilages in catarrhine primates. *Z. Morph. Anthrop.* **81**, 289–300.

Muehlenkamp, I. (1993). Beitrag zur ontogenetischen Entwicklung der Regio ethmoidalis von *Lemur catta*. Medical Dissertation, Frankfurt.

Novacek, M.J. (1993). Patterns of diversity in the mammalian skull. In *The Skull*, vol. II. ed. J. Hanken & B.K. Hall, pp. 438–545. Chicago: University of Chicago.

Paulli, S. (1900). Uber die Pneumaticitaet des Schaedels bei den Saeugethieren. III. Ueber die Morphologie des Siebbeins und die der Pneumaticitaet bei den Insectivoren, Hyracoideen, Chiropteren, Carnivoren, Pinnipedien,

Edentaten, Rodentiern, Prosimiern und Primaten, nebst einer zusammenfassenden Uebersicht ueber die Morphologie des Siebbeins und die der Pneumaticitaet des Schaedels bei den Saeugethieren. *Morph. Jb.* **28**, 483–564.

Preuschoft, H., Demes, B., Meyer, M. & Baer, H.F. (1985). Die biomechanischen Prinzipien im Oberkiefer von langschnauzigen Wirbeltieren. *Z. Morph. Anthrop.* **76**, 1–24.

Reinbach. W. (1952). Zur Entwicklung des Primordialcraniums von *Dasypus novemcinctus* Linne. I. *Z. Morph. Anthrop.* **44**, 375–444.

Reinbach, W. (1963). Das Cranium eines menschlichen Feten von 93 mm Sch.-St.-Lg. *Z. Anat. Entwickl. gesch.* **124**, 1–50.

Schilling, A. (1970). L'organe de Jacobson du lémurien malgache *Microcebus murinus* (Miller 1777). *Mém. Mus. Nat. Hist. Nat*, N.S., A **61**, 203–80.

Schultz, A.H. (1935). The nasal cartilages in higher primates. *Am. J. Phys. Anthrop.* **20**, 205–12.

Schultz, A.H. (1956). Postembryonic age changes. In *Primatologia*, vol. 1, ed. H. Hofer, A.H. Schultz & D. Starck, pp. 887–964. Basel: Karger.

Schultz, A.H. (1969). *The Life of Primates*. London: Weidenfeld and Nicolson.

Scott, J.H. (1967). *Dentofacial Development and Growth*. London: Pergamon Press.

Seydel, O. (1891). Ueber die Nasenhoehle der hoeheren Saeugetiere und des Menschen. *Morph. Jahrb.* **17**, 44–99.

Simons, E.L. (1989). Description of two genera and species of Late Eocene Anthropoidea from Egypt. *Proc. Natl. Acad. Sci. USA* **86**, 9956–60.

Simons, E.L. (1990). Discovery of the oldest known anthropoidean skull from the paleogene of Egypt. *Science* **247**, 1507–9.

Starck, D. (1960). Das Cranium eines Schimpansenfetus (*Pan troglodytes*, Blumenbach 1799) von 71 mm SchStlg., nebst Bemerkungen ueber die Koerperform von Schimpansenfeten. *Morph. Jahrb.* **100**, 559–647.

Starck, D. (1962). Das Cranium von *Propithecus* spec. (Prosimiae, Lemuriformes, Indriidae). *Bibl. primat.* **1**,163–96.

Starck, D. (1967). Le crâne des mammifères. In *Traité de Zoologie*, vol. 16, ed. P.-P. Grassé, pp. 405–549. Paris: Masson.

Starck, D. (1984). The nasal cavity and nasal skeleton of *Tarsius*. In *Biology of Tarsiers*, ed. C. Niemitz, pp. 275–90. Stuttgart: Fischer.

Stearns, S. (1992). *The Evolution of Life Histories*. Oxford: Oxford University Press.

Stephan, H. (1975). Allocortex. In *Handbuch der mikroskopischen. Anatomie des Menschen*, W. Bargmann, ed., vol.4, part 9, pp. 1–998. Berlin: Springer.

Stephan, H., Frahm, H. & Baron, G. (1982). Comparison of brain structure volumes in Insectivora and Primates. II. Accessory olfactory bulb (AOB). *J. Hirnforsch* **23**, 575–91.

Strasser, E. & Delson, E. (1987). Cladistic analysis of cercopithecid relationships. *J. hum. Evol.* **16**, 81–99.

Vogel, C. (1966). Morphologische Studien am Gesichtsschädel catarrhiner Primaten. *Bibl. Primatol.* **4**, 1–226.

Voit, M. (1909). Das Primordialcranium des Kaninchens unter Beruecksichtigung der Deckknochen. *Anat. Hefte* **38**, 425–616.

Warwick, R. & Williams, P.L. ed. (1973). *Gray's Anatomy*, 35th edn. London: Longman.

Weidenreich, F. (1924). Uber die pneumatischen Nebenraeume des Kopfes. *Z. Anat. Entwickl. gesch.* **72**, 55–93.

Weinert, E. (1926). Die Ausbildung der Stirnhoehlen als stammesgeschichtliches Merkmal. *Z. Morph. Anthrop.* **25**, 243–357.
Wen, I.C. (1930). Ontogeny and phylogeny of the nasal cartilages in primates. *Contrib. Embryol.* **130**, 111–34.
Zeller, U. (1987). Morphogenesis of the mammalian skull with special reference to *Tupaia*. In *Morphogenesis of the Mammalian Skull*, ed. H.-J. Kuhn & U. Zeller, pp.17–50. Hamburg: Parey.
Zeller, U. (1989). Die Entwicklung und Morphologie des Schaedels von *Ornithorhynchus anatinus*. *Abh. senckenb. naturforsch. Ges.* **545**, 1–188.
Zuckerkandl, E. (1887). *Das periphere Geruchsorgan der Säugetiere. Eine vergleichend-anatomische Studie*. Stuttgart: Enke.

6

Old World monkey origins and diversification: an evolutionary study of diet and dentition

BRENDA R. BENEFIT

Introduction

The question of whether the earliest cercopithecoids were adapted for folivory or frugivory has implications for understanding the divergence of Old World monkeys and apes. Because the molars of all modern cercopithecoid monkeys are bilophodont, and most mammals with lophodont dentition eat leaves, the origin of Old World monkeys is commonly associated with a trend toward the inclusion of more leaves in their annual diets than in those of primitive apes and basal catarrhines (Jolly, 1970; Napier, 1970; Simons, 1970; Delson, 1975a,b, 1979; Andrews, 1981; Temerin and Cant, 1983; Andrews and Aiello, 1984).

The first suggestion that the earliest monkeys may not have been folivorous, but instead were highly frugivorous, came from an analysis of shear crest lengths (predominantly the lengths of cusp margins) on the lower second molars of the middle Miocene *Victoriapithecus* (Kay 1975, 1977a). Kay (1975, 1978, 1984) and Maier (1977a,b) proposed that cercopithecoid bilophodonty evolved as a consequence of selection for an efficient grinding mechanism: lophs act as guides for interlocking cusps and basins during occlusion; the size of the entoconid grinding facet is expanded; and the functional life of the crown is lengthened by increasing crown height.

Some proponents of the analogy-based scenario argued that two species existed within the *Victoriapithecus* sample, one interpreted to be a frugivorous cercopithecine, and the other a more folivorous colobine based on its supposedly longer shear crests (Delson, 1975a,b, 1979; Simons and Delson, 1978; Szalay and Delson, 1979). The more frugivorous species was depicted as eating leaves facultatively. Temerin and Cant (1983) reasoned that, without the ability to consume leaves, contemporaneous Miocene apes evolved suspensory adaptations for traveling greater distances in search of fruits than did the partly folivorous Old World monkeys. In spite of Kay's work and the fact that, at the time, no postcrania of Miocene apes (excluding *Oreopithecus*) were known to possess suspensory adaptations, Temerin and Cant's (1983) scenario was widely accepted. The Old World monkey/ape divergence seemed easy to understand in terms of apes having derived locomotor ability and monkeys having specialized diets.

In this chapter, the origin of Old World monkeys and their molar bilophodonty is viewed from the perspective of new fossils of *Victoriapithecus* and the early Miocene cercopithecoid *Prohylobates*. *Victoriapithecus* discoveries clarify interpretation of its phylogenetic position and lead to a better understanding of the morphology and adaptations of the earliest cercopithecoids. Based on functional comparisons of the victoriapithecids with informative new fossils of Miocene apes, the nature of the cercopithecoid/hominoid divergence is reconsidered. The emergence of colobine and cercopithecine monkeys, which was closely linked to differentiation of their teeth and crania, is discussed based on analyses of the fossil evidence as well as information from neontological comparisons. Inferences about the dietary habits of fossil catarrhines are drawn from the measurement of molar features such as shear crest lengths (including subsets of all eight shear crests, enabling moderately worn and fragmentary fossils to be analyzed), degree of cusp relief, and degree of cusp proximity or flare, which are functionally related to diet among extant cercopithecoids (Kay, 1977a,b,c, 1978, 1981, 1984; Kay and Hylander, 1978; Kay and Covert, 1984; Benefit, 1985, 1987, 1990; Benefit and McCrossin, 1990).

Phylogenetic position of the earliest known Old World monkeys

Fossil history

Sarich and Cronin (1976) estimate that Old World monkeys and apes diverged from a common ancestor as recently as 20–25 million years ago (Ma), but little is known about the morphology of cercopithecoids prior to 15 Ma. The oldest known cercopithecoid remains, an isolated upper second molar, canine, proximal radius and ulna from the 19 Ma site of Napak, resemble middle Miocene *Victoriapithecus* and may belong to this genus (Bishop, 1964; Pilbeam and Walker, 1968; Pickford *et al.*, 1986). The skeleton of *Victoriapithecus*, from 15 Ma deposits at Maboko Island, is the best documented of all fossil cercopithecoids. It is represented by 2500 specimens, including all skeletal elements except the trapezoid, centrale, and a few ribs and vertebrae (Von Koenigswald, 1969; Senut, 1986; Harrison, 1987, 1989; Benefit, 1993, 1994; Benefit and McCrossin, 1991, 1993, 1997;

McCrossin and Benefit, 1992; Strasser, 1997). Craniodental and postcranial variability indicate that only one species, *V. macinnesi*, occurs at Maboko (Benefit, 1987, 1993, 1994; Harrison, 1989; *contra* Von Koenigswald, 1969 and Delson, 1975a,b). *Prohylobates* is known by only four partial mandibles of *P. tandyi* from early Miocene deposits at Wadi Moghara, a single mandibular fragment of *P. simonsi* from middle Miocene localities at Gebel Zelten (Simons, 1969, 1994; Delson, 1979; Pickford, 1987), and approximately 19 specimens including isolated teeth, partial mandibles, a partial maxilla and pisiform from the 17 Ma site of Buluk (Leakey, 1985; Fleagle *et al.*, 1997). How well the appearance and lifeways of these 19–15 million year old monkeys represent those of the earliest cercopithecoids depends on their relationship to the modern subfamilies.

Characters of Victoriapithecidae

The cercopithecoid affinities of known early and middle Miocene monkeys are indicated by their high mandibular genial pit and constriction of molar width between mesial and distal aspects of the crown (Simons, 1969). In addition, *Victoriapithecus* holds in common with extant cercopithecoids: a derived absence of a maxillary sinus; absence of a lingual pillar on upper incisors; upper male canines with a deep sulcus continuing onto the root; short extension of enamel onto the mesiobuccal root of P^3 and P^4; upper molars that are nearly as long as wide and which have an expanded hypocone that directly opposes (and is nearly as large as) the metacone; tall and sexually dimorphic extension of enamel onto the mesiobuccal root of P^3; bilophodont lower molars; strong medial trochlear keel on the distal humerus; and ischial callosities (Table 6.1; Von Koenigswald, 1969; Strasser and Delson, 1987; Benefit, 1985, 1987; Harrrison, 1987, 1989; Benefit and McCrossin, 1991, 1993; McCrossin and Benefit, 1992).

That *Victoriapithecus* and *Prohylobates* belong to a family of Old World monkeys (Victoriapithecidae), distinct from that to which colobines and cercopithecines belong (Cercopithecidae), is strongly indicated by their dental morphology (Table 6.2; Von Koenigswald, 1969; Benefit, 1993, 1994).Victoriapithecids share only two derived dental features exclusively with Cercopithecinae (i.e. curved distal margins of I^1 and di^1) and only one derived trait with Colobinae (i.e. high di^2 crown height) (Tables 6.3 and 6.4; Benefit, 1993, 1994). In contrast, *Victoriapithecus* possesses at least 16 dental traits found consistently among basal catarrhines (propliopithecids and pliopithecids), Miocene hominoids (including both small- and large-bodied forms that lack an entepicondylar foramen and possess a Y-5 molar

Table 6.1. *Distribution of dental features shared by victoriapithecids and cercopithecids among modern and fossil catarrhines*

	Derived dental features shared by victoriapithecids and cercopithecids				
Anterior Dentition	Miocene hominoid	Modern hominoid	*Victoriapithecus*	Cercopithecinae	Colobinae
Enamel height at cervix upper and lower I1–C1, di^1–dc^1, and dc_1	even on all sides	even	uneven	uneven	uneven
Lingual pillar on I^1	present	present	absent	absent	absent
Mesial sulcus continues onto male C^1 root	no	no	yes	yes	yes
Crown height relative to width di^1	varied	low CH 110%W	moderate CH 140%W	moderate	moderate
Root cross-section upper and lower I1–C1, di^1–dc^1, and dc_1	cylindrical	cylindrical	compressed	compressed	compressed
di_2	4-sided	4-sided	3-sided	3-sided	3-sided
Length of mesial and distal crests from cusp tip of di^2, dc^1 and dc_1	equal	equal	M>D	M>D	M>D
Number of cusps and presence of trigon basin dp^3	2, no	2, no	4, yes	4, yes	4, yes
Hypocone position relative to metacone dp^4	distal	distal	opposite	opposite	opposite
Protolophid dp_4	oblique	oblique	transverse	transverse	transverse
dp^4 hypocone size	small	small	moderate	moderate	moderate
dp^4 continuous lingual margin	no	no	yes	yes	yes
dp_4 proximity of distal cusps	apart	apart	close	close	close
dp_4 metaconid position relative to protoconid	distal	distal	opposite	opposite	opposite
Upper and lower Molars					
quadrate arrangement of cusps	no	no	yes	yes	yes

Source: (Based on data in Benefit, 1993, 1994)

pattern) and modern hominoids, but not among living and fossil cercopithecids (Table 6.2). Cercopithecines and colobines share an additional 14 traits not found in *Victoriapithecus* (Table 6.2).

The distinctive cranial morphology of Victoriapithecidae confirms that it is the sister-taxon of Cercopithecidae (Benefit and McCrossin, 1991, 1993, 1997). The skull of *Victoriapithecus* closely resembles the basal catarrhine *Aegyptopithecus* (Simons, 1987) and many Miocene hominoid genera (especially *Afropithecus*) (Leakey and Leakey, 1986; Leakey *et al.*, 1991; Leakey and Walker, 1997), *Sivapithecus* (Pilbeam, 1982), and to some extent *Dryopithecus* (Moya Sola and Kohler, 1993), in having: (1) a lower neurocranium relative to length and width (unknown for Miocene apes); (2) more airorhynchous hafting of the neurocranium and face, with less flexed basicranium and longer midcranial region as measured from postglenoid process to M^3 (unknown for Miocene apes); (3) steep and linear facial profile; (4) frontal trigon formed in part by the anterior convergence of temporal lines and presence of supraorbital costae; (5) taller than wide orbits; (6) orbits and zygomatic that are angled dorsally relative to the Frankfurt plane, so that the top of the orbit is posterior to its inferior margin instead of perpendicular to the anatomical plane; (7) deep malar region of the zygomatic; and (8) shallow palate (Fig. 6.1, Benefit, 1995; Benefit and McCrossin, 1997). Because these features are shared by propliopithecids, Miocene hominoids, and Miocene monkeys, they were primitive for catarrhines and retained by *Victoriapithecus* (Fig. 6.2; Benefit and McCrossin 1991, 1993, 1997). Consequently, at least eight cranial features shared by Colobinae and Cercopithecinae are derived.

Other cranial differences between victoriapithecids and modern cercopithecids include: mandibular coronoid process that extends above the condyle (shared exclusively with *Aegyptopithecus* among catarrhines); longer-than-wide neurocranium (shared with the fossil colobine *Libypithecus*); and pronounced sagittal and nuchal crests (shared with *Libypithecus* and fossil *Theropithecus*).These features may have been primitive for cercopithecoids, but the limited distribution of the former two traits, and potential homoplasy of the latter, make this hypothesis difficult to test.

Cercopithecoid characters

Since cranial and dental evidence indicates that victoriapithecids are the sister-taxon of cercopithecids, it is parsimonious to assume that features shared by it and one or the other modern subfamily should be primitive for

Table 6.2. *Distribution of dental features distinguishing victoriapithecids among modern and fossil catarrhines*

	Dental features distinguishing victoriapithecids from cercopithecids				
Anterior dentition	Miocene hominoid	Modern hominoid	*Victoriapithecus*	Cercopithecinae	Colobinae
Crown height relative to width dc_1	moderate–low	moderate–low	moderate–low	high	moderate–high
Crown elongation di^1	varied	varied	great as in hylobatids W 50–60%L	moderate W 70–80% L	moderate W 60–85% L
dc1	moderate–low	low	moderate W70%L	great–moderate	great–moderate
Crown height relative to width dc_1	moderate–low	moderate–low	moderate–low	high	moderate–high
Shape – elongation					
dp^3	W>=L	W>L	W=L	W<L (elongated)	W<L (elongated)
dp^4	W>L	W>=L	W>=L	W<L	W<L
dp_3	less	less	less	great	great
dp_4	moderate	less	less	great	great
P^3	W/L>130	W/L>130	W/L 126	W/L 105–110	W/L 116
P^4	W/L>140	W/L>140	W/L 139	W/L 120–124	W/L 132
P_3			OcL/W 112	OcL/W 126–145	OcL/W 121
P_4 orientation	oblique	straight	oblique	straight	straight
Number of cusps					
dp_4	5	5	5	4	4
Bilophodont					
dp^3	no	no	no	yes	yes
dp^4	no	no	no	yes	yes
dp_4	no	no	variable	yes	yes
Crista obliqua					
dp^4	yes	yes	yes	no	no

Hypocone position relative to metacone dp^3	–	–	distal	opposite	opposite
Cusp relief dp_3	moderate–high	moderate–high	moderate–high	low	moderate
Occlusal shelf lengths dp^4	M > D	M > D	M < D	M > D	M > D
Occlusal shelf widths P^3			wide	moderate	moderate
protolophid dp_3	varied	varied	absent	present	present
dp^3 mesial loph position	–	–	mesial to cusp tips	between cusp tips	between cusp tips
dp^3 distal shelf isolated from trigon basin by a crest	no	no	no	yes	yes
dp_3 hypoconid and entoconid size	minute	minute	moderate	larger	larger
dp_3 protoconid height	tall	varied	tall	low	moderate
dp_3 metaconid height (MH) relative to entoconid height (EH)	MH >> EH	MH >> EH	MH > EH	MH >= EH	MH >= EH
Crown H/honing facet H female P_3	94–100%	nearly equal	83%	56–64%	56%
dp_4 hypoconulid size	moderate	moderate	small	absent	absent
Upper Molars					
Crista obliqua	present	present	varied (present 91%, M^1, 71% M^2, 33% M^3)	absent	absent
Lower Molars					
M1/M2 hypoconulid	present	present	varied (present on 86% M_1s and 25% M_2s)	absent	absent

Source: (Based on data in Benefit, 1993, 1994)

Table 6.3. *Distribution of dental features shared by cercopithecines and victoriapithecids among modern and fossil catarrhines*

	Primitive cercopithecoid dental features shared by cercopithecines and victoriapithecids				
Anterior dentition	Miocene hominoid	Modern hominoid	*Victoriapithecus*	Cercopithecinae	Colobinae
I^1 Root length and curvature	varied	varied	long and curved distobuccally	long and curved distobuccally	relatively short and straight
I^2 position relative to I^1	varied	varied	I^1 anterior to I^2	I^1 anterior to I^2	I^1 and I^2 set equally forward
Crown elongation di_1	moderate	great-moderate	low W80%L	moderate–low	moderate, except low in *Colobus*
Distal margin shape					
I^1	straight	straight	curved	curved	straight
di^1	straight	straight	curved	curved in Papionini straight in Cercopithecini	straight
Premolars					
Occlusal shelf lengths dp_4	moderate	moderate	moderate	moderate	short
Occlusal shelf widths					
P^4			wide	moderate–wide	moderate
Upper molars					
Flare (index of mesial and distal cusp proximity relative to crown width)	varied	low	highest	high	low
Cusp relief	varied	moderate–low	low	low	high
Shear crest lengths	varied	moderate–low	low	low	high
Lower molars					
Flare	varied	low	highest	high	low
Cusp relief	varied	low	low	low	high
M1/M2 mesial width (MW) relative to distal width (DW)	MW = DW	MW = DW	MW = DW	MW = DW	MW < DW

Source: (Based on data in Benefit, 1993, 1994)

Table 6.4. *Distribution of dental features shared by colobines and victoriapithecids among modern and fossil catarrhines*

	Primitive cercopithecoid dental features shared by colobines and victoriapithecids				
Anterior dentition	Miocene hominoid	Modern hominoid	*Victoriapithecus*	Cercopithecinae	Colobinae
Crown height relative to width di^2	low (except *Simiolus*)	low	high	moderate to high	high
Cusp relief dp_4	varied	low	moderate	low	moderate
Premolars					
P^4 paracone height			>crown L	<crown L	>crown L
Upper and Lower Molars					
Mesial shelf length					
M^1			short	long	short
M_1–M_3	short		short	long	short

Source: (Based on data in Benefit, 1993, 1994)

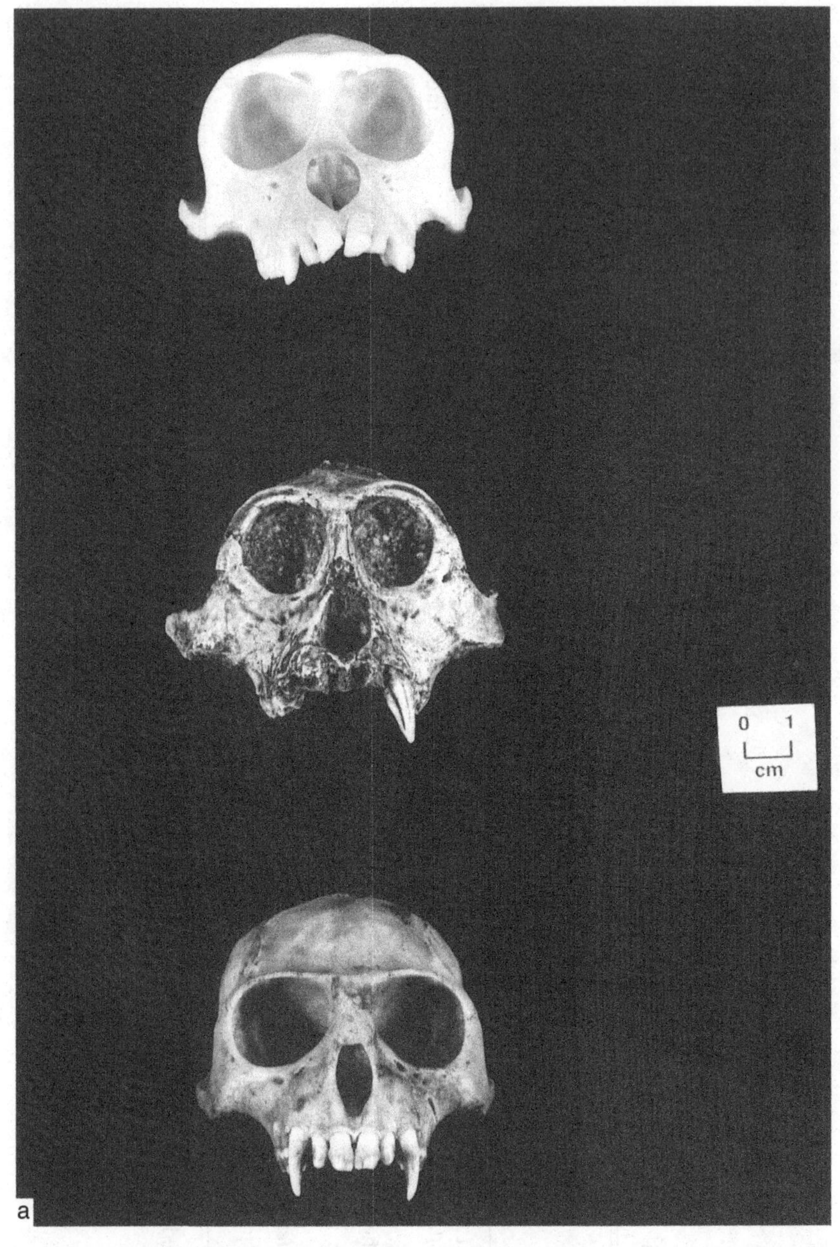

Fig. 6.1. Male *Victoriapithecus* cranium KNM-MB 29100 (center), compared with female *Macaca fascicularis* (above) and female *Colobus guereza* (below), facial view (left) and left lateral view (right).

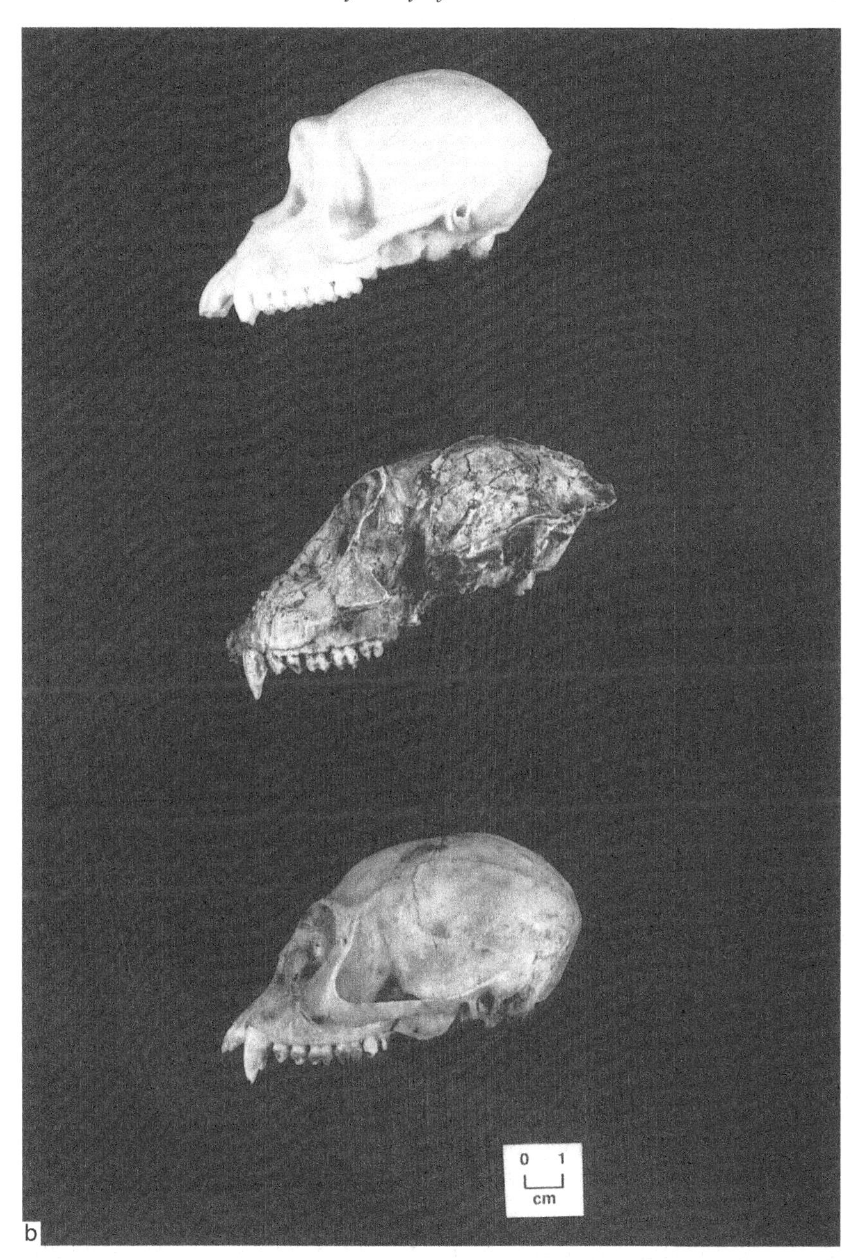

Fig. 6.1. (*cont.*)

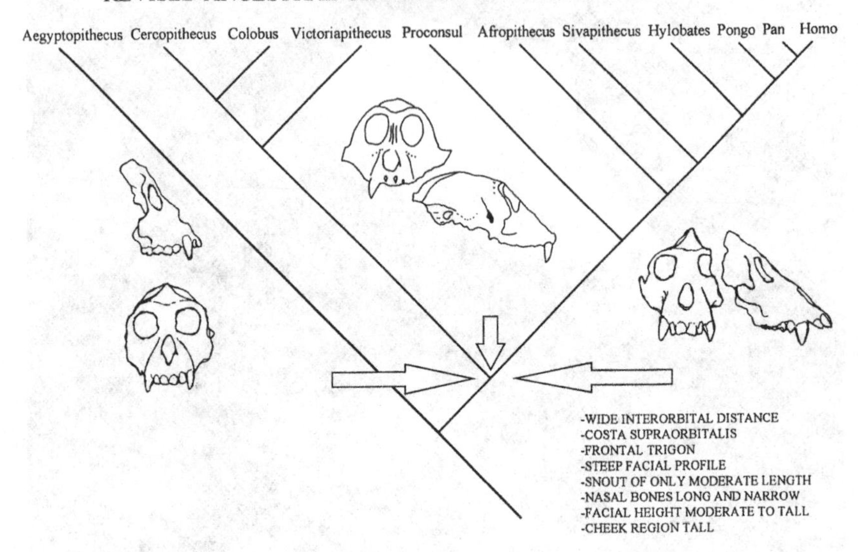

Fig. 6.2. Ancestral catarrhine cranial morphotype based on consideration of fossil and neontological evidence.

cercopithecoids. This view presupposes that incidences of convergent evolution were extremely rare during the evolutionary history of Old World monkeys. Accordingly, facial and dental features shared by *Victoriapithecus* and Cercopithecinae should be primitive for the superfamily. These include: a narrow interorbital septum; long and narrow nasal bones; low and narrow nasal aperture; moderately long muzzle; long and anteriorly tapering premaxilla with I^1s positioned anterior to I^2s; degree of elongation, root length and distal margin shape of I^1 and di^1; moderate to wide length of dp_4 and P^4 mesial shelves; lower molars with nearly equal mesial and distal crown widths; both upper and lower molars with low crown relief; low shear crest lengths; and a high degree of flare (Table 6.3). Reconstruction of these features as primitive is supported by their occurrence among *Libypithecus, Rhinocolobus*, some *Paracolobus*, and modern *Nasalis*, and all except a narrow interorbital septum among propliopithecids and the Miocene hominoids *Afropithecus* (Benefit and McCrossin, 1991, 1993, 1997). The only colobine-like features expected to be primitive for cercopithecoids are a wide palate, tall di^2 crown height, moderate dp_4 cusp relief, and tall P^4 paracone, shared exclusively by colobines and *Victoriapithecus.* Although short molar mesial shelves are shared by *Victoriapithecus* and Colobinae, the longer cercopithecine-like condition is found in *Prohylobates*, making the polarity of this trait equivocal.

Others argue that high and rounded crania with relatively smooth frontals, such as those shared by the archaic catarrhine *Pliopithecus*, modern colobines and gibbons were primitive for catarrhines and cercopithecoids (Vogel, 1966, 1968; Harrison, 1987; Strasser and Delson, 1987). According to this view, the 30 derived dental features shared by colobines and cercopithecines, and the eight cranial features shared by *Victoriapithecus*, *Aegyptopithecus*, *Afropithecus*, and early Pliocene African colobines, evolved convergently. This argument is far less parsimonious than the one suggested above, and it makes several assumptions that are unfounded. It treats pliopithecids as closer to the last common ancestor of Old World monkeys and apes than propliopithecids although the former are less likely ancestors of modern catarrhines because of their dental specializations, middle Miocene age, and Eurasian distribution (Szalay and Delson, 1979). It uses modern hominoids as an outgroup for resolving whether the colobine or cercopithecine cranial morphology is primitive, although hominoids are a cranially diverse group and it is unclear whether short-snouted gibbons with rounded frontals, orang-utans with *supraorbital costae*, or African apes with brow ridges represent the primitive condition. Assuming that hylobatid crania are primitive, because they branched-off before other modern ape clades, fails to recognize that many aspects of their skeleton may be more derived than that of other apes (Groves, 1989; McCrossin and Benefit, 1994; Benefit and McCrossin, 1995). Incorporating the Miocene sister-taxa of modern hominoids and cercopithecoids into cladistic analyses is more appropriate for determining morphocline polarities within each superfamily (Benefit and McCrossin, 1991, 1993, 1995, 1997; McCrossin and Benefit, 1994, 1997).

Ancestral cercopithecoid diet and the emergence of bilophodonty

Shared characters

Several cranial and dental features shared by *Victoriapithecus* and cercopithecines are related to their common reliance on incisal biting during food preparation: a longer and lower neurocranium, which achieves a posteriorly inclined orientation of the temporal muscle; a low and relatively narrow nasal aperture, which does not restrict the size of the I^1 roots; narrow interorbital septum; a long and anteriorly tapering premaxilla; moderate snout length; heteromorphic upper incisors; and large, distally curved and procumbent I^1s (Vogel, 1966, 1968). The strongly implanted upper central incisors are adapted for biting into fruits or seeds.

Molar similarities between victoriapithecids and cercopithecines reflect a shared diet of fruit. Relative to colobine molars, cercopithecine molars have: thicker enamel; a closer approximation of cusp tips and loph(id)s leading to greater crown flare (Fig. 6.3) and reduced size of the central basin; lower occlusal relief above, and taller crown height below, the base of the median lingual notch (Fig. 6.4); shorter shear crest lengths (Fig. 6.5); mesiodistally longer mesial shelf; and nearly equal mesial and distal crown widths (Delson, 1973; Kay, 1977a, b; Benefit, 1987, 1993). The first three of these features are strongly correlated with frugivory among extant cercopithecids (Kay, 1975, 1977a,b, 1978; Benefit, 1987, 1993; Benefit and McCrossin, 1990). The close proximity of cercopithecine molar cusps ensures that occluding cusp tips and occlusal surfaces contact each other during chewing. Lingual cusps of upper molars and buccal cusps of lower molars wear flat rapidly. As flat and thick enamel rims form around circles of exposed dentine, their already low cusp height decreases, lophid length increases, and crushing surfaces increase in area. Molars of *Victoriapithecus* and *Prohylobates* are "hyper-cercopithecine-like" in having even shorter shear crests, lower cusp relief, and greater molar flare with more closely approximated cusps, than those of extant cercopithecines, including highly frugivorous and seed-eating mangabeys and macaques (Figs. 6.3–6.5). No features correlated with folivory are found on the molars of early and middle Miocene cercopithecoids, although *Victoriapithecus* shares short mesial shelf length with Colobinae.

Dietary reconstruction

The proportions of fruits and leaves eaten annually by victoriapithecine monkeys were estimated by regressions expressing the relationship between shear crest length, flare/cusp proximity, occlusal relief, and the average proportion of fruits and leaves consumed annually by a series of extant cercopithecoids (Table 6.1; Benefit, 1987, 1990; Benefit and McCrossin, 1990). On the basis of lower second molar shear quotients, the annual diet of *Victoriapithecus* from Maboko is estimated to have consisted of 79% fruits and 7% leaves. *Prohylobates* from Buluk was observed to have a slightly higher shear quotient (−4.6) than *Victoriapithecus* (−6.9) and an annual diet of 74% fruits and 13% leaves. Because of their advanced stage of wear, shear could not be evaluated on molars of *P. tandyi* and *P. simonsi*. However, the degree of lingual cusp relief on both north African species, indicates that they consumed about 84% fruits annually and were as highly frugivorous as their eastern African relatives.

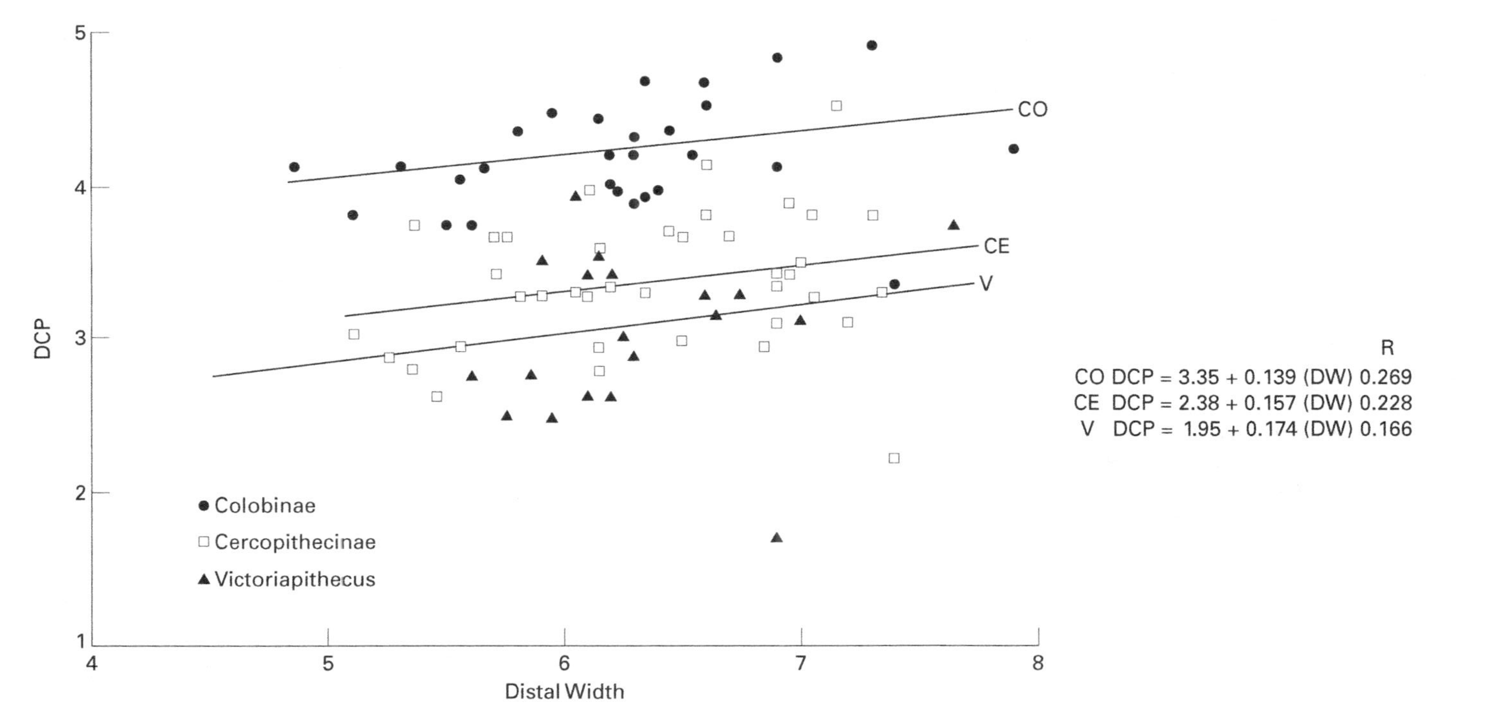

Fig. 6.3. Bivariate plot of lower second molar DCP (distal cusp proximity measured between tips of the hypoconid and entoconid) and distal width (maximum crown width measured between sides of the distal cusps) for individuals of Colobinae, Cercopithecinae and *Victoriapithecus*, as well as least squares linear regressions expressing the relationship between the two variables for each of these groups.

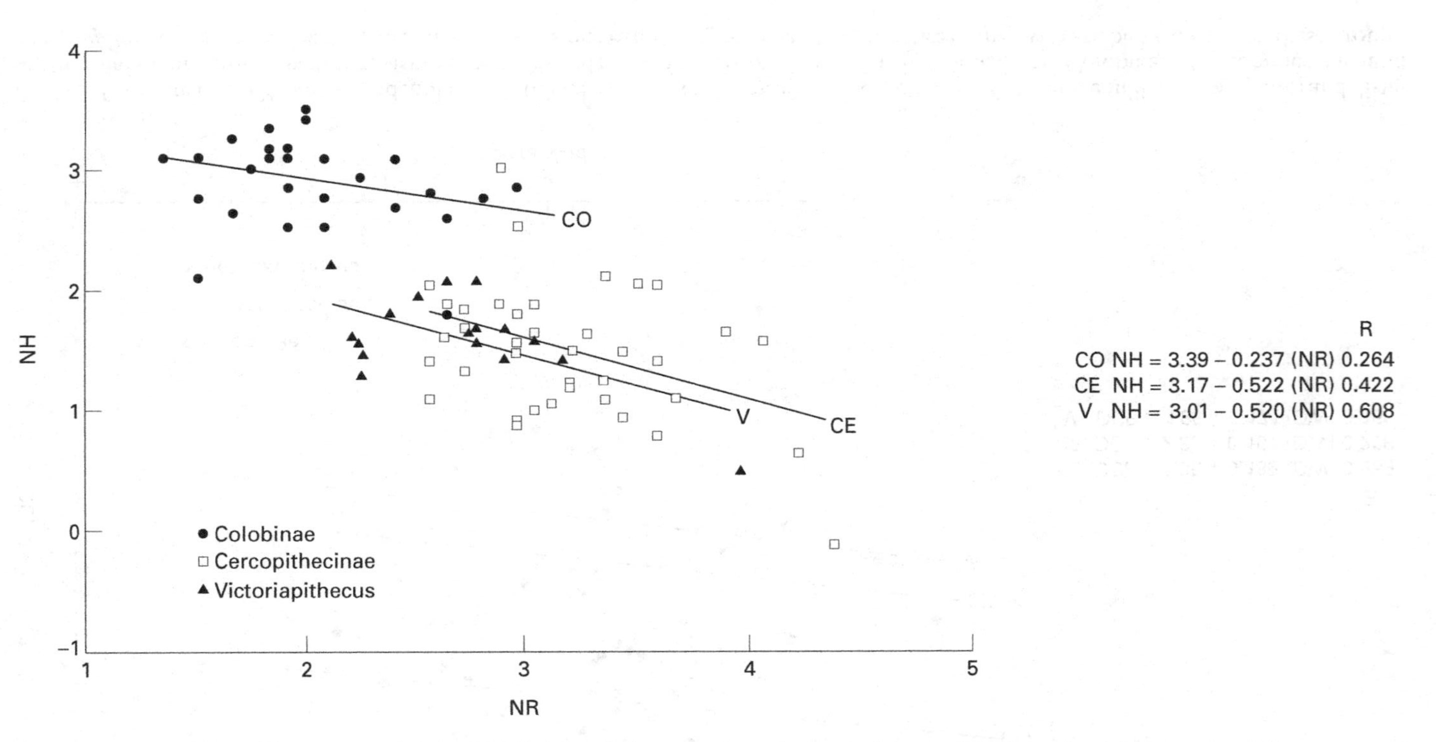

Fig. 6.4. Bivariate plot of lower second molar NH (height of the crown from the base of the median lingual notch to metaconid tip, measured vertically) and NR (Crown height measured from the base of the median lingual notch to the cervix directly below) for individuals of Colobinae, Cercopithecinae and *Victoriapithecus*, as well as least squares linear regressions expressing the relationship between the two variables for each of these groups.

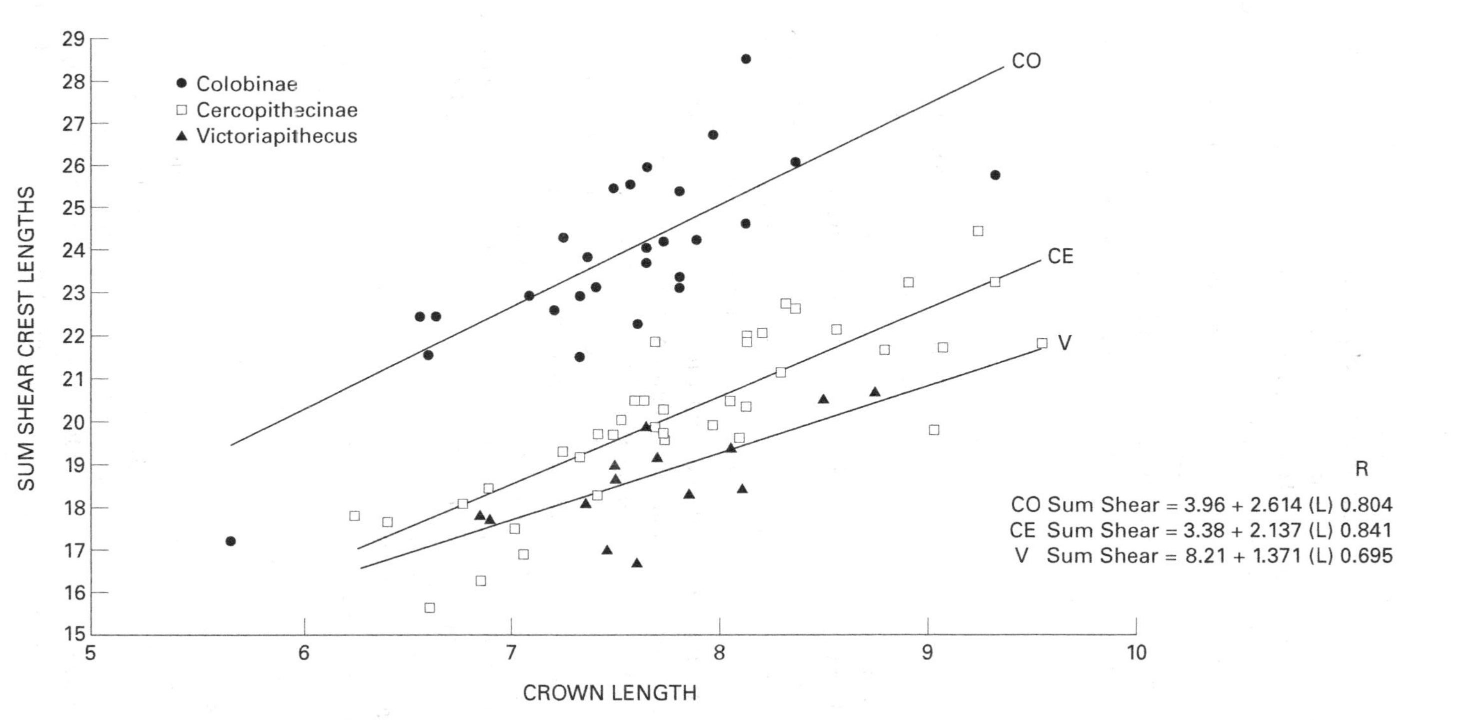

Fig. 6.5. Bivariate plot of lower second molar sum shear crest lengths (sum of eight shear crests – margins of cusps, such as the pre- and post-protocrista) and crown length (maximum mesodistal length of the crown) for individuals of Colobinae, Cercopithecinae and *Victoriapithecus*, as well as least squares linear regressions expressing the relationship between the two variables for each of these groups.

The presence of the pitted microwear on occlusal enamel (Palmer *et al.*, 1998), as well as of microwear on non-occlusal surfaces of *Victoriapithecus* (Ungar and Teaford, 1996), corroborates reconstruction of its diet as consisting of hard fruits or seeds. Heavily pitted wear facets on victoriapithecid cusps result from abrasion between cusps and contact with food, and closely resemble those observed on anubis baboons from the Awash. Examination of enamel microwear on *Victoriapithecus* M_2s reveals that pits make up 46% of its phase II (grinding) facet features (Palmer *et al.*, 1998), comparable to frequencies found among modern hard-object feeders (Teaford *et al.*, 1996).

Analysis of anatomical features

Insight as to why bilophodonty might have evolved is provided by reconstructing the sequence of anatomical changes that led to the emergence of cercopithecoid bilophodonty. Lower molars of propliopithecids are similar to those of cercopithecoids (and unlike the Y-5 pattern found among pliopithecids and hominoids) in having their four main cusps directly oppose each other. If such a condition were primitive for catarrhines, it might have been a precursor to bilophodonty. Otherwise, if the Y-5 pattern were primitive, a distal shift in the position of buccal cusps would have taken place as perhaps the initial step toward the evolution of bilophodonty.

Unlike hominoids and basal catarrhines, modern cercopithecoids have no hypoconulid on dp4, M1 or M2. However, within the *Victoriapithecus* sample from Maboko, hypoconulids occur on 88% of dp4s, 86% of M1s, and 25% of M2s. Position and size of these hypoconulids are variable, with the smallest cusps being more buccally positioned and often fused with the distal end of the posthypocristid (Fig. 6.6). Hence, loss of the dp4/M1/M2 hypoconulid in cercopithecoids probably involved a buccal migration, rather than having lingually migrated and merged with the entoconid as suggested by others (Voruz, 1970; Delson, 1975a,b). It is possible that expansion in the size of the entoconid causes the change in position and eventual loss of the hypoconulid in cercopithecoids. In cercopithecids the entoconid is almost as large as the hypoconid, but in other catarrhines the former is smaller.

All lower molars of *Victoriapithecus* have distal transverse lophids (hypolophids), but 30% of the dp_4s do not and are not bilophodont (Benefit 1994). It was suggested that lower molars of *Prohylobates* from North Africa are incompletely bilophodont (Simons, 1969; Delson, 1979), but known dentitions were too worn to prove this. A mandible with M1–M3 of

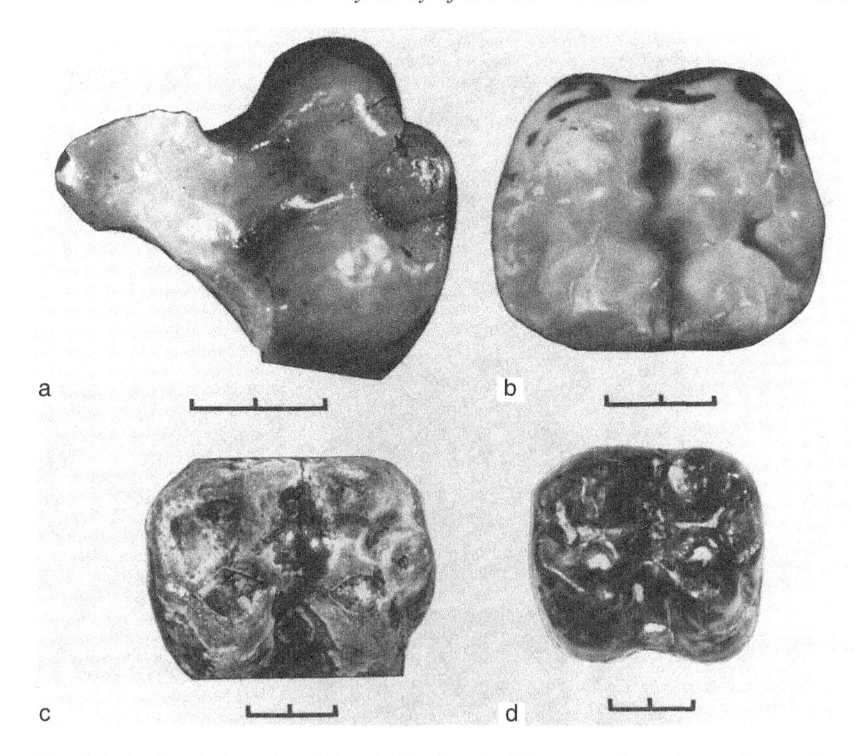

Fig. 6.6. Occlusal views (mesial to left) of some *Victoriapithecus* lower first molars with hypoconulids. Upper row: unworn right M_1s KNM-MB 310 (left) and KNM-MB 324 (right). Lower row: worn left M_1s KNM-Mb 11674 (left) and KNM-MB 20765 (right).

P. tandyi provides support for this hypothesis (Simons, 1994). As for loss of the hypoconulid, formation of the hypolophid is probably related to an increase in the size of the entoconid. The expansion of the occlusal aspect of the entoconid may have led to a meeting of the occlusal faces of the entoconid and hypoconid and concomitant formation of a distal lophid, as well as peripheralization of the hypoconulid.

Unlike upper molars and deciduous premolars of modern monkeys, victoriapithecids exhibit variable development of the primitive catarrhine *crista obliqua*, trigon basin, and isolation of the hypocone (Fig. 6.7). The *crista obliqua* is present on the majority of dp^4s (89%), M^1s (91%) and M^2s (71%), and some M^3s (33%). The crest is composed of a lingual portion extending from the postprotocrista and a buccal portion arising from the side of the metacone, the intersection of which is demarcated by a shallow groove and often a thickening of enamel. On many specimens (especially

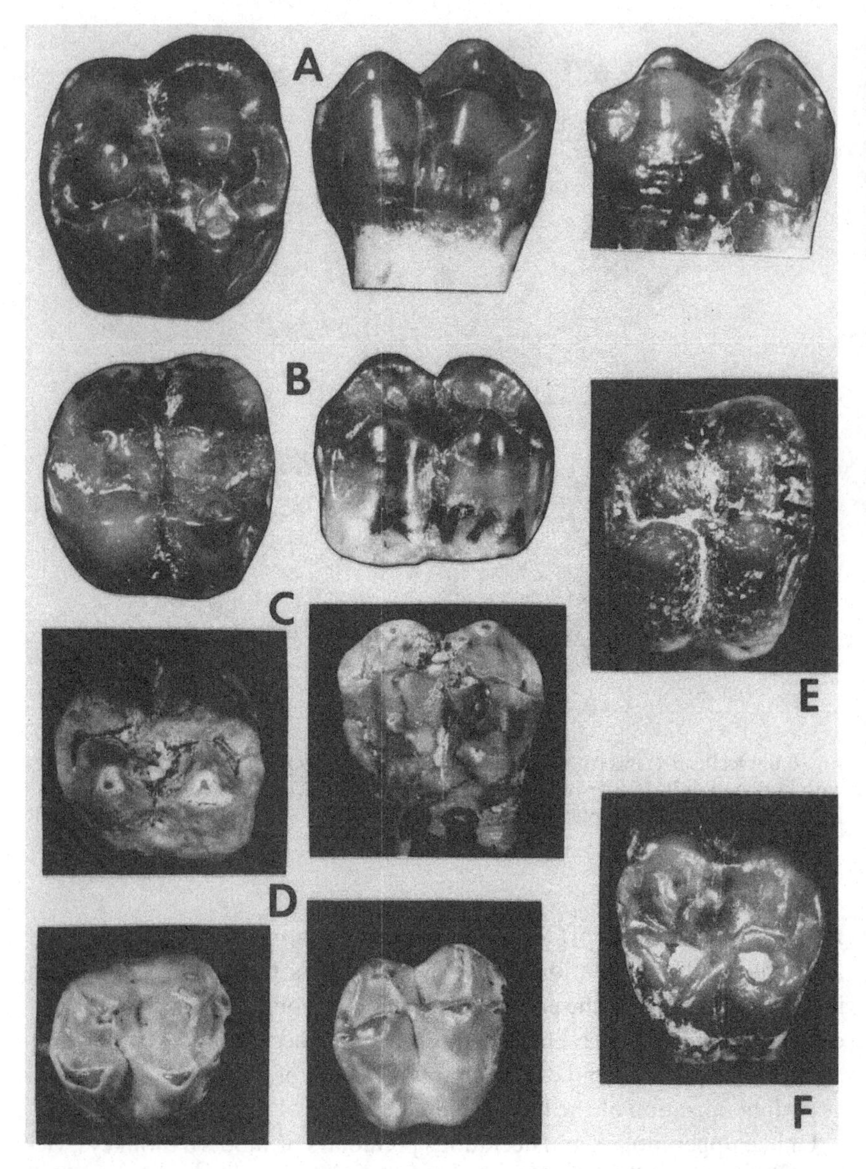

Fig. 6.7. *Victoriapithecus* upper molars with variable expression of the crista obliqua. A: occlusal (mesial to right), lingual (mesial to right) and buccal (mesial to left) views of right M^2 KNM-MB 20798. B: occlusal (mesial to right) and buccal (mesial to left) views of right M^2 KNM-MB 20803. C: occlusal (mesial to right) and lingual (mesial to left) views of left M^2 KNM-MB 19858. D: occlusal (mesial to right) and lingual (mesial to right) views of right M1 KNM-MB 20789. E: occlusal view (mesial to right) of right M^2 KNM-MB 20800. F: occlusal view of left M^2 KNM-MB 20802 (mesial to left).

dp^{4}s and M^{1}s) the crest is well-developed and oblique, but on others the postprotocrista is oriented distally and the metacone crest transversely. In the latter case, the postprotocrista is high and merges with the prehypocrista to form the median lingual notch as in bilophodont cercopithecoids. When this occurs, the metacone crest is weakly-developed and grades into the floor of the trigon basin. The co-occurrence of the weakly-developed metacone crest and transverse distal loph, on a few *Victoriapithecus* upper molars, indicates that the *crista obliqua* has not become the distal loph. Instead the loss of one crest and emergence of another seem to be related to the same phenomenon, i.e. the expansion of occlusal surfaces of the hypocone and metacone and size of the central basin. This model is similar to that of Remane (1951), but differs from that of Butler and Mills (1959) who suggested that the lingual portion of the *crista obliqua* disappears in cercopithecoids, and from Leakey (1985) who interpreted the distal loph as a remnant of the metacone crest.

Lucas and Teaford (1994) suggest that cercopithecoid loph(id)s evolved to produce cusps with a wedge-shape, not seen among other catarrhines. Such wedges are more efficient at producing initial fractures on tough items such as seeds, while the blade-like shear crests on the margins of these wedges are pre-adapted for slicing (Lucas and Teaford, 1994: 192). Increase in the size of the cercopithecoid central basin (trigon) and loss of the *crista obliqua* may be necessary for accomodation of the larger and more wedge-shaped entoconid, which rests in the basin during occlusion. The presence of distal loph(ids) creates a well-demarcated distal shelf which, with the mesial shelf of the following molar, creates a basin-like configuration between cusps. Cercopithecoid bilophodonty permits their teeth to act as a series of smoothly interlocking basins and wedge-shaped cusps (Kay, 1977a, 1978). These changes may have taken place to increase the efficiency with which the earliest cercopithecoids could fracture and masticate hard seeds and fruits with their molar teeth.

Both *Victoriapithecus* upper and lower molars are extremely flared, with the sides of the crown expanding in width below cusp tips. In some instances, exaggerated bulging occurs on *Victoriapithecus* molars below the median cleft on one or both sides of the crown (Fig. 6.7a). Strasser and Delson (1987) interpreted the exaggerated molar flare as resulting from the incorporation of primitive catarrhine cingula onto the side of the crown, with cercopithecids sharing a derived further loss of this feature. However, since the (sometimes) cingulum-like bulging does not originate from the ectoflexid in the median buccal cleft of lower molars, and is not continuous with the mesial lingual cleft, it may not be homologous to true catarrhine cingula. In most instances, enamel on the bulging area has a rough appearance due to the presence of perikymata bands, which are often pitted and possibly hypoplastic, possibly indicating developmental stress. Benefit (1993) suggested that exaggerated bulging in *Victoriapithecus* may be related to a closer approximation of cusp tips than is seen in modern monkeys, rather than to changes below the median cleft. Although regressions of crown width and length show the lower molars of *Victoriapithecus* to be absolutely no wider than those of modern monkeys, significant differences exist in terms of the proximity of the cusps (Fig. 6.4). *Moeripithecus* shares high flare and close cusp approximation with *Victoriapithecus*, and has been suggested as a cercopithecoid ancestor (Szalay and Delson, 1979). If true, a high degree of crown flare in Miocene cercopithecoids may reflect heritage as well as function.

The origin of cercopithecoid bilophodonty appears to be related to increasing grinding surface area by increasing occlusal surfaces of the entoconid and hypocone (Kay, 1977a, 1978; Maier, 1977a,b; this study). However, this explanation does not fully explain the loss of a *crista obliqua*, which in *Victoriapithecus* exhibits heavily pitted microwear features and seems to have acted as an extension of the protocone grinding surface (Fig. 6.5). Lucas and Teaford (1994) argue that, whereas other anthropoid molar cusps are blade-like in design, the transverse lophs of cercopithecoid cusps transform them into wedges. The series of wedges and lophs on bilophodont molars permitted the earliest monkeys to efficiently masticate hard fruits and seeds. It was at the origin of Colobinae that the bilophodont molars were altered to enhance their shearing capacity and reduce their rate of wear.

Paleoecological context of the hominoid/cercopithecoid divergence

Since the earliest monkeys ate hard fruits and seeds, under what conditions would such an adaptation have been advantageous? Most scenarios explaining the divergence of cercopithecoids and hominoids from a common ancestor assume that they occupied the same ecosystems, and responded to the same changes in microclimate throughout their evolutionary history (Napier, 1970; Andrews, 1981; Temerin and Cant, 1983; Andrews and Aiello, 1984). They view Miocene apes as monomorphic in their dentition and dietary preferences, treating all as consumers of ripe fruit (Andrews, 1981; Andrews and Aiello, 1984). Many also assume that primitive cercopithecoids were postcranially conservative arboreal quadrupeds, whereas Miocene apes were more derived toward arboreal suspensory postures if not locomotion (Napier, 1970; Temerin and Cant, 1983). However, examination of patterns of distribution and relative abundance of ape and cercopithecoid fossils at early and middle Miocene localities, of Miocene ape dental diversity, and of Miocene catarhine postcrania provide strong evidence against all of these assumptions.

Locomotor adaptations

It was the origin of Old World monkeys, not apes, that was linked to a shift in locomotor adaptation. Victoriapithecids (as indicated by analysis of *Victoriapithecus* remains from Maboko, since they are unknown for *Prohylobates*) and cercopithecids differ from other primates in having a narrow distal humerus, well-defined median trochlear keel, posteriorly rather than laterally-oriented medial epicondyle, well-defined ischial callosities, and restricted hip and ankle joints (Napier and Davis, 1959; Jolly, 1967; Rose, 1983; Harrison, 1989; McCrossin and Benefit, 1992, 1994). The more limited elbow, hip and ankle joint flexibility of the earliest Old World monkeys provided them with enhanced running abilities, which are particularly useful on the ground. The early monkeys were undoubtedly as terrestrial as modern vervets (Von Koenigswald, 1969; Delson, 1975a; Senut, 1986; Harrison, 1989; McCrossin and Benefit, 1992, 1994). One feature believed to indicate an arboreal habitus for *Victoriapithecus* was an isolated humeral head, thought to be higher than the greater tubercle (Harrison, 1989). However, when this specimen is oriented using a complete *Victoriapithecus* humerus recently found at Maboko to judge its correct anatomic orientation, the greater tubercle is higher than the humeral head as for terrestrial primates (McCrossin, 1995; McCrossin *et al.*, 1998). The same is true for all proximal humeri of *Victoriapithecus*. Hence, the earliest known Old World monkeys were not designed to live an agile life in the trees like their Oligocene ancestors, but to exploit resources both on the ground and in the trees.

Relative to known limb elements of the basal catarrhine *Aegyptopithecus*, Miocene apes were less-derived postcranially than cercopithecoid monkeys. Aside from lacking an entepicondylar foramen on the distal humerus (a catarrhine feature shared with cercopithecoids), small-bodied ape postcrania are so similar to those of generalized arboreal quadrupedal platyrrhines and propliopithecids that their hominoid status has been questioned (Harrison, 1987; Rose *et al.*, 1992). Miocene larger-bodied apes, such as *Proconsul*, are derived in terms of having a broader trochlea than capitulum and a moderately well-developed lateral trochlear keel on the distal humerus (Napier and Davis, 1959; Rose, 1983). The few Miocene ape proximal humeri that are known do not possess the suspensory shoulder joint characteristic of modern hominoids (McCrossin, 1992, 1994a,b, 1995; McCrossin and Benefit, 1994, 1997; Benefit and McCrossin, 1995). Claims to the contrary for a range of fossil hominoids are based on hand remains (Meldrum and Pan, 1988), reconstructed humeral/femoral indices (Moya Sola and Kohler, 1996), humeral shaft fragments (Rose, 1993), derived morphology of other postcrania (Harrison, 1986; Sarmiento, 1987), or fragmentary scapular glenoid fossae (Xiao, 1981; Ciochon and Etler, 1994; Gebo *et al.*, 1997), that do not reveal the actual orientation of the humeral head. With a few exceptions, Miocene hominoids were agile arboreal quadrupeds who used a combination of below- and above-branch postures, like those used by modern platyrrhines (LeGros Clark and Thomas, 1951; Napier and Davis, 1959; Morbeck, 1983; Rose, 1993).

Dental evidence

It is among Miocene hominoids, rather than cercopithecoids, that adaptations for folivory are found (Kay, 1977c; Benefit, 1991, 1993; Gitau and Benefit, 1995; Kay and Ungar, 1997). Of Miocene hominoids that coexisted with monkeys, early Miocene *Simiolus enjiessi* and middle Miocene *Simiolus leakeyorum* and *Nyanzapithecus pickfordi* resemble modern folivores in having molars that are unusually long and narrow, with high occlusal relief, and long shear crests relative to molars of contemporary apes (Table 6.5). Long parallel scratches make up 88% of all microwear features within a 0.2 mm^2 area of *Simiolus leakeyorum* M_2 protoconid and hypoconid grinding facets, and short parallel scratches make up 86% of microwear within the same area of *Nyanzapithecus pickfordi* M_2s (Palmer *et al.*, 1998). These percentages clearly place them among the most specialized of extant folivores (Teaford *et al.*, 1996).

Environmental reconstructions

The earliest monkeys appear to have been highly restricted in the range of environments that they occupied. With the possible exception of Napak V, none of the victoriapithecid localities is reconstructed as rainforest (Bishop, 1968). Whereas apes are diverse and abundant at early Miocene rainforest localities such as Songhor, Koru, and Rusinga, cercopithecoids are conspicuously absent from such deposits. Monkeys are absent from the middle Miocene site of Fort Ternan which is reconstructed as open woodland (Gentry, 1970; Churcher, 1970; Evans *et al.*, 1981; Kappelman, 1991), but also seems to have elements of savanna (Shipman *et al.*, 1981; Retallack *et al.*, 1990; Retallack, 1992) and forest (Pickford, 1983; Cerling *et al.*, 1991). Monkeys are present and highly abundant in lower stratigraphic levels (Beds 3 and 5) of the Maboko formation dated as older than 14.7 million years (Myr), but seem to disappear in more recent levels (Bed 12) which are dated as between 13.8 and 14.7 Myr in age. No obvious environmental differences between upper and lower levels at Maboko have been identified, to date.

The *Prohylobates* sites of Gebel Zelten and Wadi Moghara are reconstructed as deltaic riverine forest with savannah in the hinterland (Savage and Hamilton, 1973). The paleoenvironment of Maboko Beds 3 and 5, at which *Victoriapithecus* is more abundant than any other mammal (33% of all mammal bones found in the deposit are of *V. macinnesi*), has been interpreted as riverine woodland with thick brush cover (Evans *et al.*, 1981), or as similar to the "nyika" semi-arid *Acacia–Commiphora* woodland with grass patches and gallery forest along streams, and some nearby forest patches (Pickford, 1983). Although the anomalurid *Zenkerella* was identified at Maboko, and thought to indicate the presence of forest (Andrews *et al.*, 1981), specimens have been re-identified as dry-adapted naked mole rats (Cifelli *et al.*, 1986). *In situ* remains of bathyergids have been found in the deposits since excavations began there in 1987, but anomalurids have yet to be recovered from Maboko (Winkler, 1994). Analysis of the rich bird and rodent faunas, as well as of the seeds of swamp sedges, recently collected at Maboko indicate that a great diversity of microenvironments, with dry open woodland adjacent to swamps, and seasonally flooding streams, existed at the site during the middle Miocene (Benefit and McCrossin 1992; McCrossin and Benefit, 1994).

Of the diverse range of Miocene ape species, few coexisted with early and middle Miocene monkeys. The only site at which Miocene Old World monkeys may occur with *Proconsul* (specifically *P. major*) is Napak V (Pilbeam and Walker, 1968). At Buluk (17 Ma), *Prohylobates* occurs with *Afropithecus* (Leakey, 1985; Leakey and Leakey, 1986), and middle Miocene *Victoriapithecus* is found with *Kenyapithecus* at Maboko Beds 3 and 5, Kipsaramon, and Nachola (Von Koenigswald, 1969; Benefit, 1993; Pickford, 1987; Hill personal communication). However, cercopithecoids are not present at all *Afropithecus* and *Kenyapithecus* localities, because they are absent from Moroto, Bed 12 at Maboko, and Fort Ternan.

Hard object feeding

The early to middle Miocene monkeys and large-bodied hominoids, with which they occur, apparently shared dietary and locomotor adaptations. *Afropithecus* and *Kenyapithecus* fed upon hard fruits and seeds, using their anterior teeth to prepare tough foods (McCrossin and Benefit, 1993a,b, 1994, 1997; Benefit and McCrossin, 1995; Leakey and Walker, 1997). As for *Victoriapithecus*, microwear on *Kenyapithecus* molars clearly associates with modern hard object feeders in the percentage of pits and pit width. However, *Kenyapithecus* has much wider pits (18.2 μm) than *Victoriapithecus* (13.3 μm), as well as a higher percentage of pits (61% versus 46%), and may have had a more specialized diet (Palmer *et al.*, 1998).The adaptations of these hard object feeding apes differed significantly from those of *Proconsul* and *Limnopithecus*, which are inferred to have eaten softer fruits, and from more folivorous hominoid taxa (Table 6.5; Kay, 1977c, Benefit, 1991, 1993; Gitau and Benefit, 1995; Gitau, 1995). Although *Afropithecus* was apparently an arboreal quadruped (Leakey *et al.*, 1988; Rose, 1993), *Kenyapithecus* was the first, and so far only, Miocene ape recognized to have a semi-terrestrial pattern of locomotion, similar to that of *Mandrillus* and *Cercocebus* (McCrossin, 1994a,b, 1995, 1996; McCrossin and Benefit, 1994; Benefit and McCrossin, 1995). Terrestrial adaptations of the postcrania shared by *Kenyapithecus* and *Victoriapithecus* include: proximal extension of the greater tubercle above the humeral head, postero-medially oriented medial epicondyle of the distal humerus, a long and retroflexed olecranon process, short and straight phalanges, and an adducted big toe (McCrossin, 1994a,b, 1995, 1996; McCrossin and Benefit, 1992, 1994, 1997; Benefit and McCrossin, 1995; McCrossin *et al.*, 1998).

Hard object feeding, in combination with semi-terrestriality, seems to have allowed *Kenyapithecus* and *Victoriapithecus* to cope with the changing vegetation of the eastern African middle Miocene by exploiting a disturbed mosaic of seasonally flooded and wooded environments where food sources were abundant closer to the ground and seasonally unpredictable (Benefit, 1987; McCrossin, 1994a; McCrossin and Benefit, 1997). It would also have permitted them to exploit fallen fruits (McCrossin, 1994a), a resource that is crucial to allowing some semi-terrestrial primates, such as the ring-tailed lemur, to survive the dry season (Sussman, 1977). Whereas apes such as *Kenyapithecus* used their anterior teeth to open hard fruits and seeds, monkeys such as *Victoriapithecus* would have performed this task with their bilophodont molars. Soft fruit-eating arboreal apes such as *Proconsul* disappear after the early Miocene, while monkeys and large-

Table 6.5. *Comparison of propliopithecid, victoriapithecid, and Miocene hominoid mean lower second molar indices thought to be functionally correlated to diet*

	N	L/MW	FL	NHNR	SUMS/L
Propliopithecid					
P. haekeli	2	123	67	41	222
M. markgrafi	1	95	45	59	196
A. zeuxis	2	103	52	67	–
Victoriapithecid					
V. macinnesi	82	111	51	74	243
P. tandyi	3	104	(76)	(21)	–
P. simonsi	1	95	(53)	(20)	(250)
P sp. (Buluk)	3	104	(61)	(50)	(261)
Cercopithecid					
Colobinae	58	125	70	150	311
Cercopithecinae	78	116	56	58	250
Miocene hominoid					
A turkanaensis	WK 2	119	44	–	189
P. africanus	RU 8	117	57	51	206
	CA 1	116	–	–	215
P. major	SO 2	115	47	63	215
	CA 1	105	58	57	219
	LG 2	110	52	56	225
	NP 2	119	42	49	207
P. nyanzae	RU 5	118	55	58	213
K. africanus	MB 6	111	61	71	214
R. gordoni	SO 6	127	63	96	221
N. pickfordi	MB 2	154	78	86	–
D. macinnesi	RU 5	122	47	70	229
	CA 2	117	–	–	–
L. legetet	RU 1	111	70	169	–
	CA 4	118	57	140	–
	KO 2	122	64	114	–
	LG 5	114	52	95	–
L. evansi	SO 6	119	58	93	230
K. songhorensis	SO 2	116	54	76	222
M. clarki	CA 5	118	48	84	–
	LG 1	127	60	47	–
S. enjiessi	WK 2	125	–	94	239
S. leakeyorum	MB 7	131	54	130	239
S. sp.	FT 2	142	70	151	–

Notes:
L/MW: (crown length/mesial crown width) × 100; Fl(flare): (distance between mesial cusps tips/mesial crown width) × 100; NH/NR: (crown height above the base of median notch (lingual on lowers, buccal on uppers)/height of the crown immediately below the median notch) × 100; SUMS/L: (sum of shear crest lengths/crown length) × 100.
For Miocene hominoids A.: *Afropithecus*; P.: *Proconsul*: K: *Kenyapithecus*; R.: *Rangwapithecus*; N.: *Nyanzapithecus*; D.: *Dendropithecus*; L.: *Limnopithecus*; K.: *Kalepithecus*; M: *Micropithecus*; S.: *Simiolus*; CA: Chamtwara; FT: Fort Ternan; KO: Koru; LG: Legetet; MB: Maboko; NP: Napak; RU: Rusinga; SO: Songhor; WK: Kalodir.
Source: (Data from Benefit, 1993.)

bodied apes, as well as the most arboreal and folivorous of the hominoids (*Simiolus* and *Nyanzapithecus*), survived the change toward drier and more seasonal climatic conditions (Benefit, 1991, 1993, 1999; Gitau and Benefit, 1995; Gitau, 1995; McCrossin *et al.*, 1998). Following this first wave of ape extinctions at the end of the early Miocene, the extinction of highly folivorous and harder-object feeding small-bodied apes occurs by the end of the middle Miocene in Africa, and victoriapithecids are replaced by cercopithecids.

Emergence of cercopithecid dentition and dietary adaptations

Colobines

The earliest known cercopithecid, *Microcolobus tugenensis* from 11 Ma deposits at Ngeringerowa in Kenya, is also the earliest known colobine monkey (Benefit and Pickford, 1986). Although both cercopithecid subfamilies are thought to have diverged by this time, the oldest cercopithecine is not found until 7 Ma, at the site of Marceau in North Africa (Szalay and Delson, 1979; Pickford, 1987). Late Miocene and early Pliocene cercopithecids of both subfamilies seem to retain the mainly frugivorous diets, semi-terrestrial mode of locomotion and small- to medium-body size of their victoriapithecid ancestors. The late Miocene Eurasian colobine *Mesopithecus* and cercopithecids of unknown subfamily from the 6 Ma North African site of Sahabi were semi-terrestrial in their adaptations (Delson, 1973; Meikle, 1987). At least one colobine (*Libypithecus*) retained a victoriapithecid-type of cranial morphology (Benefit and McCrossin, 1997), and another (*Microcolobus*) a *Victoriapithecus*/cercopithecine-like mandibular symphysis (Benefit and Pickford, 1986). *Microcolobus* possessed all of the dental features that separate colobine from cercopithecine monkeys (taller shear crests, stronger occlusal cusp relief, and larger central basins), but its fairly conservative molar indices indicate that it ate twice as many fruits (55%) as leaves (28%) (Table 6.6; Benefit, 1987, 1990; Benefit and Pickford, 1986). The lower shear crests and cusp relief of the next oldest colobine, *Mesopithecus*, indicate that it too was more frugivorous than folivorous (Benefit, 1990; Benefit and Pickford, 1986).

It was not until the Plio-Pleistocene that Colobinae and Cercopithecinae experienced a radiation in terms of diversity in size, cranio-facial morphology, diet, locomotor pattern and habitat preference, that was accompanied by a major increase in numbers of species. In sub-Saharan Africa alone, at least 8 colobine and 18 cercopithecine species occur during this time period

Table 6.6. *Predicted diets for fossil cercopithecoids*

				Cusp	
Species	Site	*N*	Shear F/L	Relief F/L	Flare F/L
Fossil Colobinea					
Microcolobus tugenensis	NG	1	54/28		
Colobinae indet sp.	NK	1	75/9	40/43	82/4
Colobinae indet sp.	Omo	15	37/47	23/65	
Colobus sp.	ER	7	49/34	30/53	
Cercopithecoides kimeui	ER	7	47/35	30/53	
	Old	2	40/44	40/43	35/49
Cercopithecoides williamsi	ER	5	23/60	3/82	34/50
	SK	1		1/82	
Paracolobus ado	Lae	12	39/44	12/70	55/29
Paracolobus mutiwa	Omo	38	31/54	29/53	40/45
	ER	2	58/25	19/73	21/62
Rhinocolobus turkanaensis	Omo	59	38/46	18/65	46/38
	ER	9	58/24	19/73	21/62
Fossil Cercopithecinae					
Cercopithecus sp.	LB	5	67/17	54/27	48/36
Cercocebus sp.	ER	30	75/10	62/22	92/0
Cercocebus sp.	Old	3		74/9	
Dinopithecus ingens	SK	24	69/14	63/21	66/19
Gorgopithecus major	KA	15	64/27	39/31	66/19
Papio sp.	Old	5	96/0	48/35	39/45
Papio angusticeps	T	6	46/37	63/20	
	KA	11	60/24	54/29	53/31
Papio robinsoni	SK	15	61/22	55/28	
Parapapio sp.	Lae	25	57/29	67/17	83/0
Parapapio broomi	STS	12	57/25	50/33	60/25
Parapapio jonesi	KP	1	83/0		
	STS	17	58/33	59/24	60/25
	SK	10	67/15	53/30	95/0
Parapapio whitei	STS	9	65/21	47/37	36/47
Parapapio sp.	ER	11	58/24	51/33	
Theropithecus brumpti	ER	23	52/31	28/55	
	Omo	6	35/46		13/70
Theropithecus darti	SK	11	34/49	6/84	95/0
Theropithecus oswaldi	ER	119	61/24	31/52	64/24
	Old	32	64/18		100/0
	Ol	168	67/19	11/80	93/0
Theropithecus quadratirostris	Omo	1	28/60		

Notes:
NG: Ngerngerowa; NK: Nakali; ER: Koobi Fora; Old: Olduvai; SK: Swartkrans; LAE: Laetoli; LB: Leboi; KA: Kromdraai; T: Taung; STS: Sterkfontein; Ol: Olorgesailie; KP: Kanapoi; F/L: fruit/leaves.
Source: (Based on regression equations and data in Benefit, 1987 and Benefit and McCrossin, 1990.)

(Fig. 6.8). All of these species were endemic to Africa, with no cercopithecids migrating into sub-Saharan Africa from Eurasia following their migration to that continent prior to 8 Ma. Fossil colobines *Paracolobus, Rhinocolobus* and *Cercopithecoides* were two to three times larger than extant African colobines and have no modern analogs (Leakey, 1982; Fleagle, 1988). The only small- to medium-sized (4–8 kg) cercopithecoids from this time period are fossils attributed to *Cercopithecus* and *Colobus* (Eck and Howell, 1972; Szalay and Delson, 1979; Leakey, 1988). Of the African monkeys whose postcrania are known, all cercopithecines and two of the four colobine genera with known postcrania (*Cercopithecoides* and Colobinae sp. A) were terrestrial (Leakey, 1982; Birchette, 1981; Ciochon, 1993). *Paracolobus* and *Rhinocolobus* were largely arboreal (Leakey, 1982; Birchette, 1981; Ciochon, 1993). Crania of the long-snouted *Paracolobus* and *Rhinocolobus* resemble *Libypithecus* in retaining a number of primitive cercopithecoid features, but *Cercopithecoides* had a shorter snout and more globular cranium similar to modern colobines (Leakey, 1982; Benefit and McCrossin, 1991, 1993, 1997; Hynes and Benefit, 1995). Plio-Pleistocene cercopithecines were more derived than colobines in their cranial vault shape, and some papionins exhibited extremely long snout lengths.

Like their Miocene predecessors, most African Plio-Pleistocene colobines are reconstructed as consuming fewer leaves than modern African *Colobus guereza* and *Procolobus badius*, the annual diets of which consist respectively of 78% and 46–73% leaves annually, depending on habitat and seasonality (Oates, 1994). This is because fossil colobine molars were conservative in terms of shear crest length and degree of flare (Table 6.6). Yet, Plio-Pleistocene colobine molars were as derived as their modern counterparts in having high occlusal cusp relief. For example, on the basis of lower second molar shear quotients and flare, populations of *Paracolobus* (Laetoli and Omo), *Rhinocolobus* (Koobi Fora), and *Colobus* (Koobi Fora) are estimated to have included more fruits than leaves in their annual diets, but the opposite is true when their diet is reconstructed based on their high cusp relief (Table 6.6). Dental microwear on molars of these fossil monkeys resembles that of *Colobus guereza* (Teaford and Leakey, 1992; Lucas and Teaford, 1994). It is possible that shear crest length lagged behind cusp relief as colobine monkeys adapted to folivory.

All aspects of molar morphology indicate that of the terrestrial fossil colobines, *Cercopithecoides williamsi* (especially from Swartkrans) was a highly derived folivore, whereas *Cercopithecoides kimeui* had the lowest cusp relief of all the Plio-Pleistocene colobines and was the least derived toward folivory (Table 6.6). Both species exhibit an excessively rapid rate of

Fig. 6.8. Geographic distribution of Plio-Pleistocene cercopithecoids in subsaharan Africa.

occlusal wear that is atypical of members of their subfamily. Either they consumed ground level browse covered with gritty dust, or they were grazers with fewer adaptations for resisting tooth wear than the modern gelada. *Cercopithecoides* is the only fossil colobine found at predominantly open savannah localities such as Olduvai, Sterkfontein, Swartkrans and Kromdraai (Leakey, 1982; Boaz, 1977). *Paracolobus* is associated with mainly forest and wooded environments, but also occurs at Laetoli which was dominated by woodland to open savannah with no major rivers or lakes. *Rhinocolobus* seems to occur only where there is evidence of riverine forest or woodland (Leakey, 1982).

In light of the dental morphology and reconstructed diets of Miocene and Pliocene colobines, it was suggested that the earliest members of this subfamily had frugi-folivorous diets similar to those of the Asian *Presbytis* (Benefit and Pickford, 1986; Benefit, 1987, 1990), such as *P. melalophus* and *P. rubicunda* (Bennett and Davies, 1994). As the number and diversity of field projects studying the ecology of modern colobines has expanded, ecologists have concluded that several colobines (especially *Colobus, Procolobus, Semnopithecus, Presbytis*, and *Nasalis*) rely as much on a diet of seeds as they do leaves (Davies and Oates, 1994). Comparative studies of gut morphology have led Chivers (1994: 224) to conclude that seed-eating may have preceded browsing during the evolutionary history of foregut-fermenting colobines. Although all colobines share a specialized foregut, the stomach is largest in the more folivorous species and smallest in the more frugivorous ones (Chivers, 1994: 225). Foregut-fermentation of seed coats by colobines permits the release of rich nutrients from the cotyledon. Hence, enlargement of the stomach for digestion of seeds may have been a preadaptation to the more highly specialized folivory and larger stomachs of species like *Procolobus badius*. Consequently, both fossil and neontological data lead one to the same conclusion, that the primitive colobine diet was one that included as many fruit/seeds as leaves.

Cercopithecines

Based on analysis of their molar morphology, it seems that many Plio-Pleistocene cercopithecines species from eastern Africa were highly frugivorous (Table 6.6; Benefit, 1987: table 66, 1990; Benefit and McCrossin, 1990). *Parapapio* from Kanapoi, *Cercocebus* from Koobi Fora and Olduvai, and *Papio* from Olduvai were found to be the most committed to frugivory, whereas *Parapapio* from Laetoli and Koobi Fora are the least frugivorous of this group (Table 6.6; Benefit, 1987; Benefit and McCrossin, 1990). However, among the southern African species only the diet of *Dinopithecus ingens* was found to be highly frugivorous (Table 6.6). Diets of other southern African *Parapapio* and *Papio* are reconstructed as being composed of 48–58% fruits and 20–30% leaves, in similar proportions to that of *Parapapio* from Laetolil. Similarity between microwear on the teeth of eastern African *Parapapio* and fossil colobines provide further evidence that they included leaves in their diet (Teaford and Leakey, 1992; Lucas and Teaford, 1994). In addition, isotopic analysis confirms that *Papio robinsoni* from Swartkrans had a C_3 based diet which is usually associated with browsing (Lee-Thorp and van der Merwe, 1989).

Molars of fossil and living *Theropithecus* are the most derived of the cercopithecines because of the presence of accessory cuspules along shear crests bordering mesial and distal shelves (Jolly, 1972). As in Plio-Pleistocene African colobines, their molars combine low shear crest lengths and a high degree of flare with high colobine-like occlusal relief. The close proximity of their molar cusps (typical of other cercopithecines) probably contributes to the rapid rate at which their cusps wear. The diet of modern *Theropithecus gelada* is unique in consisting predominantly (90%) of grasses (Dunbar and Dunbar, 1974; Dunbar, 1983), but it is unclear whether the diet of fossil *Theropithecus* was as specialized. Pleistocene *T. oswaldi* from Koobi Fora, Olorgesailie and Olduvai are predicted to have been more frugivorous than other *Theropithecus*, including the extant gelada, because of their reduced shearing crest lengths (Table 6.6). The presence of deep bucco-lingually oriented parallel on the molars of *T. oswaldi* from Koobi Fora may be indicative of a heavy reliance on the transverse component of mastication, such as is associated with grass-eating in modern *T. gelada.* However, it may have been caused by the inclusion of grit in the diet (Benefit, 1987; Benefit and McCrossin, 1990). *T. oswaldi danieli* from Swartkrans exhibits relatively longer shear crests and higher cusp relief than other Pleistocene *Theropithecus*, indicating greater potential for shearing fibrous foods (Table 6.6). Only its high degree of molar flare makes *T. oswaldi danieli* appear to be somewhat less folivorous on average (Table 6.6), although isotopic analysis showed it to be more of a C_4 dependent grazer than a browser (Lee-Thorp and van der Merwe, 1989).

Molars of mid-Pliocene eastern African species *T. quadratirostrus* and *T. brumpti* are generally less massive and elaborate than those of *T. oswaldi*, with fewer accessory cuspules and infoldings of enamel. They are intermediate in morphology between those of more generalized papionins and more specialized *T. oswaldi*. That *T. brumpti* may have included leaves rather than grass in its diet is indicated by its higher shear crests and thinner enamel (enamel thickness/crown length: 1.16 mm/15.6 mm = 7.4%) than that of *T. oswaldi* (enamel thickness/crown length: 1.5 mm/17.3 mm = 8.7%) with which it coexisted at Koobi Fora. The presence of high shear crests similarly indicate that *T. quadratirostris* may have been a browser rather than a grazer (Table 6.6). Molars of *T. brumpti* from Koobi Fora exhibit a higher degree of flare than conspecifics from the Omo, leading to the population's predicted diet as being more frugivorous than that of the latter's (Table 6.6). Microwear studies by Teaford (1993) indicate that *T. brumpti* and *T. quadratirostris* may have consumed some fruits. However, pitted microwear is often caused by attrition and contact between teeth during occlusion, rather than by contact with food items (Ungar and

Teaford, 1996). Hence, the low frequency of pitted microwear that is observed on the teeth of these Pliocene *Theropithecus*, as well as the the rapid rate of cusp deformation observed for the molars of *T. brumpti*, can be attributed to the close proximity of the molar cusps (Benefit, 1987, 1990; Benefit and McCrossin, 1990). Whether their teeth were designed for the consumption of fruits or leaves, a diet of abrasive grass has not been suggested for these animals.

The relative proportions of fruits, leaves, and grasses in the diets of these large-bodied cercopithecines is probably related to whether they occupied forest or woodland habitats versus open or treeless savannas. African grasslands are characterized by low and seasonal fruit productivity. Extant baboons exploiting forested habitats tend to eat higher quantities of fruits than grasses or herbs, while the opposite is true of baboons living in scrub savannah, which tend to supplement their diet with grasses (Dunbar and Dunbar, 1974; Hamilton *et al.*, 1978; Dunbar, 1983). The eastern African cercopithecines found to be highly frugivorous in this study are reconstructed as having occupied forest or woodland habitats (Leakey, 1982). The least frugivorous of the eastern African baboons, *Parapapio* sp. comes from the site of Laetoli which is reconstructed as a dry savannah environment (Leakey, 1982). Indications that the southern African baboons were less frugivorous than those from eastern Africa is similarly consistent with reconstruction of the southern cave deposits as representing drier, more open savannah habitats than was typical of sites such as Koobi Fora and the Omo (Boaz, 1977).

The inclusion of more fruits in the diet of fossil *Theropithecus* indicates that its habitat may have been characterized by a higher degree of tree cover than that occupied by the extant species *T. gelada* (Leakey, 1993). *Theropithecus oswaldi*, which probably consumed grasses rather than leaves, is likely to have occupied open country habitats adjacent to more wooded areas, such as grasslands growing along the margins of shallow lakes where seasonal flooding inhibited the growth of trees, as suggested by Jolly (1972). Postcranial studies indicate that *T. brumpti* may have been less cursorial than *T. oswaldi*, and may have occupied a forest habitat similar to that of the extant mandrill (Ciochon, 1986, 1993; Krentz, 1993).

Dietary and habitat preferences influenced the patterns of species diversity and relative abundance of Plio-Pleistocene *Theropithecus* and other baboons.Throughout this period, the abundance of fossils and diversity of *Parapapio* and *Papio* species is low in eastern Africa but high in southern Africa, whereas the opposite is true for *Theropithecus* (Fig. 6.9). If the origin of the genus *Theropithecus* is linked to the beginnings of leaf- or fruit-eating in a forest dwelling baboon (*T. brumpti*), the paucity of *Theropithecus* in southern Africa may be explained by the absence of forested environments in that region. This hypothesis is consistent with the absence of leaf-eating colobine monkeys other than the terrestrial *Cercopithecoides* in southern Africa. Alternatively, *Theropithecus* in southern Africa may have suffered from competition with grass-eating savannah adapted *Papio* and *Parapapio*. *Parapapio* was present in southern Africa during the late Miocene at Langebaanweg (Grine and Hendey, 1981), but *Theropithecus* did not occur in the area until the middle Pliocene at the site of Makapansgat. If *Theropithecus* were endemic to eastern Africa, as seems likely, they may have arrived in southern Africa after the *Papio* and *Parapapio* species had successfully filled the grass-eating niches available to monkeys, inhibiting *Theropithecus* from "swamping" the southern grasslands with its high population numbers, as *T. oswaldi* did in eastern Africa (Fig. 6.10).

Eastern African *Cercocebus, Papio*, and *Parapapio* would have competed for forest resources with *T. brumpti*, as well as with large-bodied colobine monkeys *Paracolobus* and *Rhinocolobus*, which included almost equal portions of fruits and leaves in their diets (Fig. 6.11; Benefit, 1987, 1990). Since *Papio* baboons did not become the dominant savannah monkey in eastern Africa until after the demise of *T. oswaldi*, it is possible that competition with *T. oswaldi* prevented the baboons from taking advantage of grassland resources at an earlier time. Because of the extinction of the large-bodied terrestrial colobines, *Parapapio*, and several species of *Theropithecus* and *Papio* during the Pleistocene, cercopithecine diversity in modern savannah habitats of eastern and southern Africa are about the same, with the exception of highland areas of Ethiopia where relict populations of *Theropithecus* occur.

Concluding remarks

The fossil record gives us a very different impression of Old World monkey and ape divergence, and the emergence of modern cercopithecid subfamilies, than does the comparison of living forms. The evolution of cercopithecoid bilophodonty was related to an adaptation for the consumption of hard fruits and seeds rather than leaves. Both Miocene apes and cercopithecoids differ from suspensory modern hominoids in being quadrupedal. The earliest Miocene hominoids were generalized arboreal quadrupeds, whereas the earliest cercopithecines had a derived postcranial morphology designed for a semi-terrestrial and cursorial mode of locomotion. Early

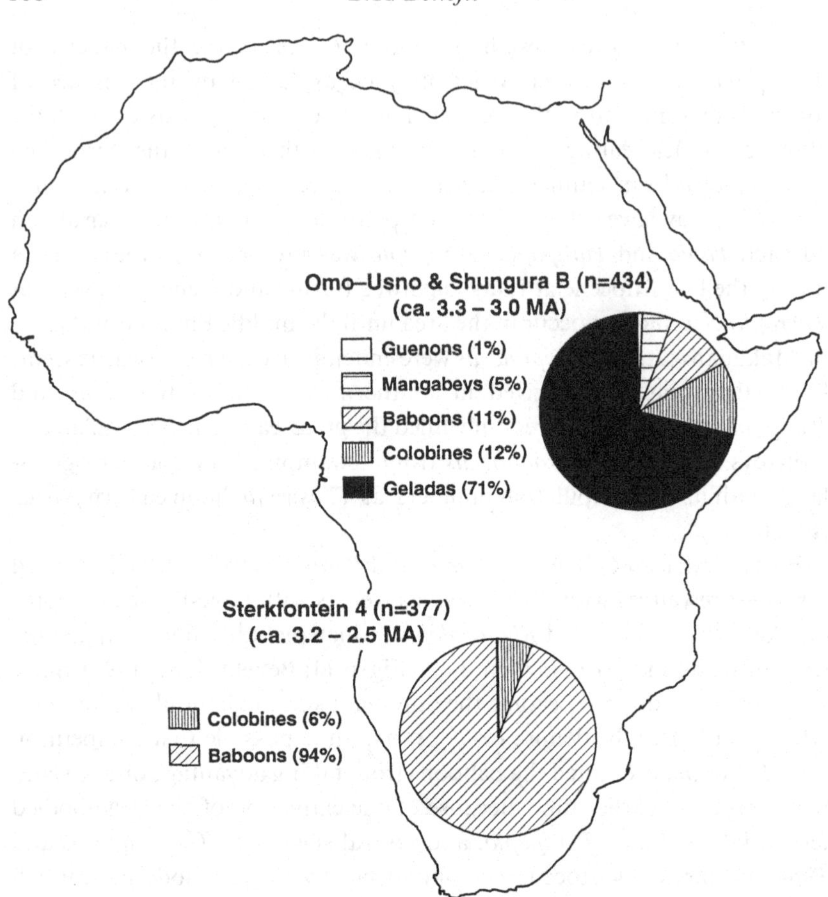

Fig. 6.9. Relative abundance of cercopithecoid taxa in eastern (exemplified by Omo Group Usno Formation and Member B of the Shungura Formation, Eck 1976) and southern Africa (exemplified by Sterkfontein Member 4, Brain, 1981) approximately 3.3–2.5 Ma (Brown and Nash, 1976; Brown *et al.*, 1985; Feibel *et al.*, 1989; Vrba, 1982). The term baboon used in this figure refers to *Papio* and *Parapapio*.

and middle Miocene apes were more diverse, abundant, and broadly-distributed in forest environments than contemporaneous cercopithecoids. Their arboreal habitus and variety of soft fruit-eating and folivorous dental adaptations was well-suited to the more predictable and wetter environments of the early Miocene. Victoriapithecids were restricted to drier, more unpredictable and open environments where their hard fruit/seed eating adaptations and their semi-terrestrial locomotion enabled them to exploit a broad range of resources that were closer to the ground. When the movement of continents contributed to global and regional climatic shifts during

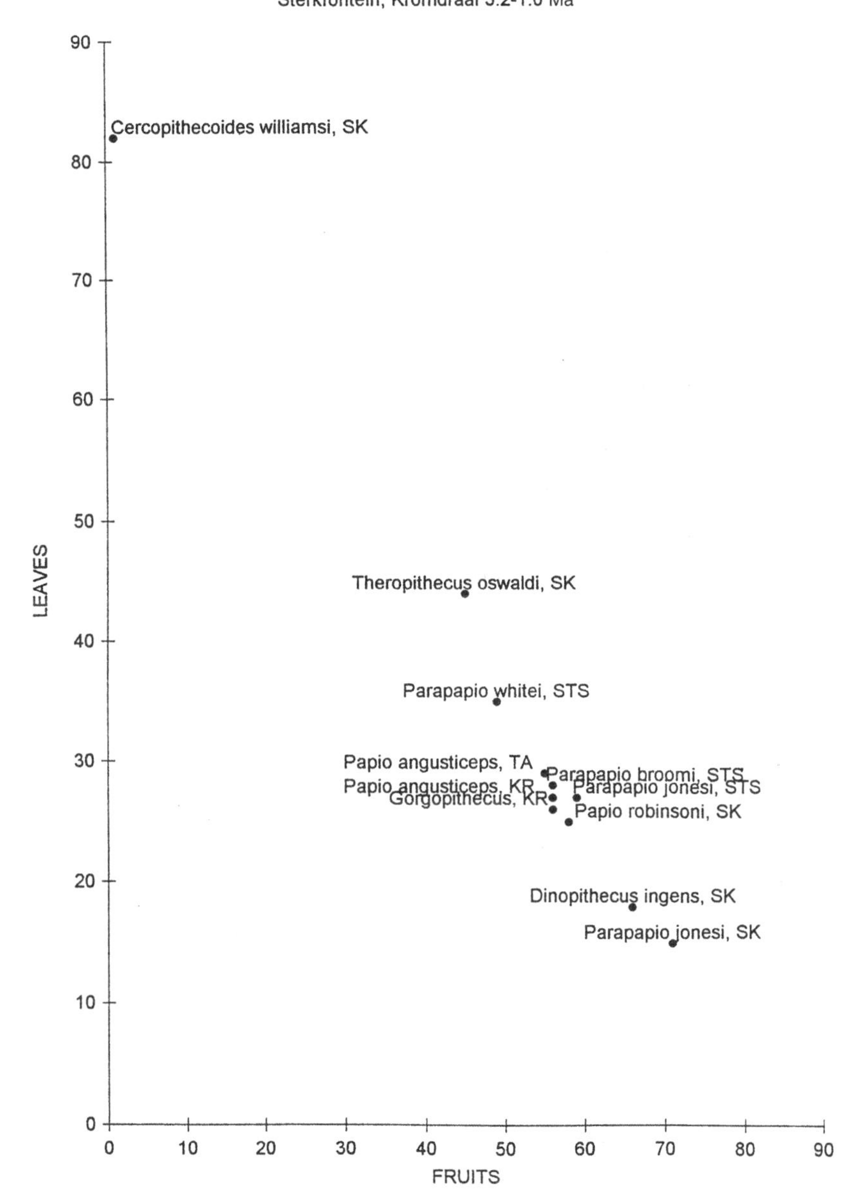

Fig. 6.10. Bivariate plot of reconstructed proportions of leaves and fruits in the annual diets of Plio-Pleistocene cercopithecoids from southern Africa (based on data in Table 6.6).

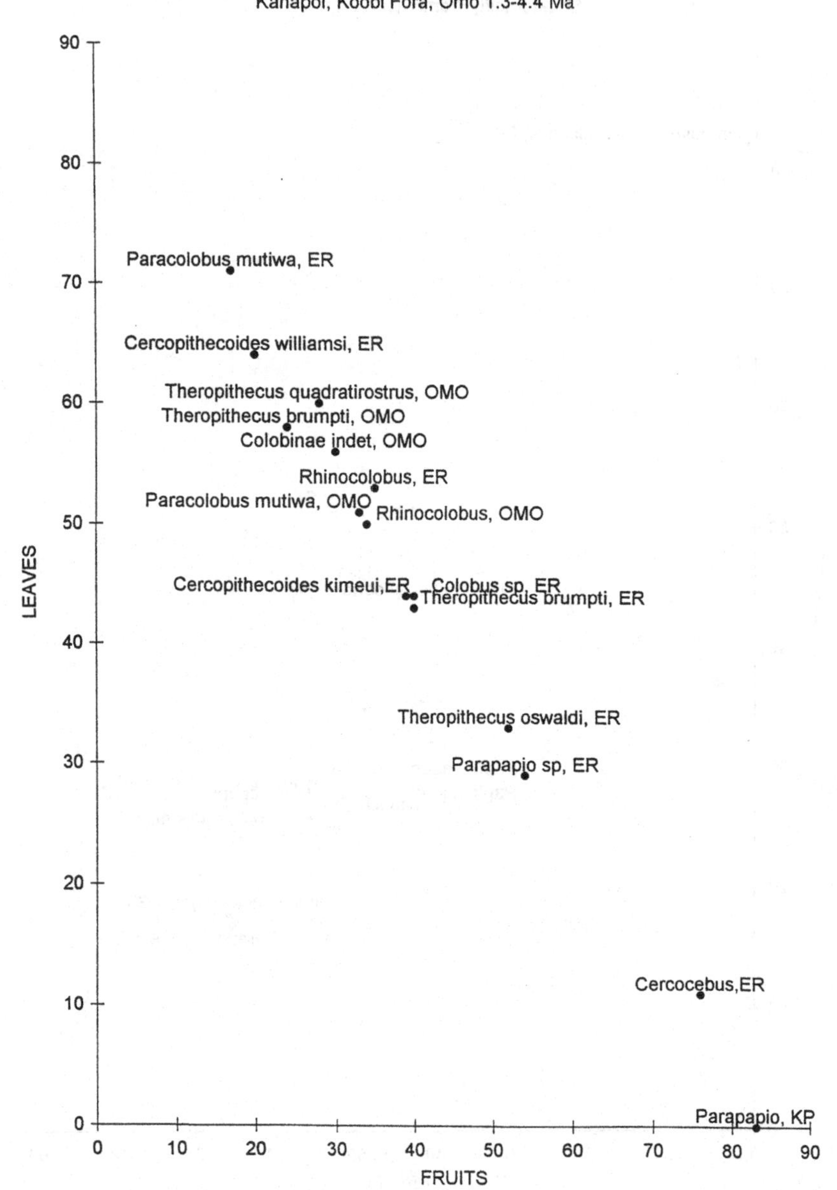

Fig. 6.11. Bivariate plot of reconstructed proportions of leaves and fruits in the annual diets of Plio-Pleistocene cercopithecoids from eastern Africa (based on data in Table 6.6).

the transition from early to middle Miocene, cercopithecoids and hominoids with similar adaptations (e.g. *Victoriapithecus* and *Kenyapithecus*) found themselves in a better position to exploit the woodlands that grew in the seasonally dry and more unpredictable environments.

Dietary reconstructions for the Late Miocene through Pleistocene colobine and cercopithecine monkeys support the interpretation that both subfamilies evolved from a frugivorous ancestor. The earliest colobines and cercopithecines retained many of the victoriapithecid adaptations. Members of both subfamilies were more frugivorous than folivorous, semi-terrestrial in their locomotor adaptations, and medium to small in body size. Dietary adaptations seem to have evolved in a mosaic fashion with cusp relief increasing before shear crest length during the evolution of colobine folivory. During the Plio-Pleistocene a modern type of cercopithecid cranial morphology evolved, colobine monkeys and guenons invaded the trees, large-bodied colobines (*Cercopithecoides*) and even larger-bodied cercopithecines (*Theropithecus*) ate grasses and invaded the southern African savannahs where generalized terrestrial "baboons" flourished. Modern cercopithecid communities of the African savanna are less diverse than those of the Plio-Pleistocene due to the extinction of the large-bodied terrestrial colobines, *Parapapio*, and several species of *Theropithecus* and *Papio.*

Acknowledgements

I want to thank Paul Whitehead and Clifford Jolly for inviting me to make this contribution and making comments on the manuscript, Jolly and Eric Delson for helping to improve this paper, and many people in Kenya (especially Meave Leakey), South Africa and the United States (especially Monte McCrossin) who have helped me with the research.

References

Andrews, P. (1981). Species diversity and diet in monkeys and apes during the Miocene. In *Aspects of Human Evolution*, ed. C.B. Stringer, pp. 25–61. London: Taylor and Francis.

Andrews, P. & Aiello, L. (1984). An evolutionary model for feeding and positional behavior. In *Food Acquisition and Processing in Primates*, ed. D.J. Chivers, B.A. Wood & A. Bilsborough, pp. 422–60. New York: Plenum.

Andrews, P., Meyer, G., Pilbeam, D.R., Van Couvering, J.A. & Van Couvering, J.A.H. (1981). The Miocene fossil beds of Maboko Island, Kenya: Geology, age, taphonomy and paleontology. *J. hum. Evol.* **10**, 35–48.

Benefit, B.R. (1985). Dental remains of *Victoriapithecus* from the Maboko Formation, Kenya. *Society of Vertebrate Paleontology Bulletin* **133**, 21.

Benefit, B.R. (1987). The Molar Morphology, Natural History, and Phylogenetic

Position of the Middle Miocene Monkey *Victoriapithecus*. PhD Dissertation, New York University.

Benefit, B.R. (1990). Fossil evidence for the dietary evolution of Old World monkeys. *Am. J. phys. Anthrop.* **81**, 193.

Benefit, B.R. (1991). The taxonomic status of Maboko small apes. *Am. J. phys. Anthrop.* **12**, 50–1.

Benefit, B.R. (1993). The permanent dentition and phylogenetic position of *Victoriapithecus* from Maboko Island, Kenya. *J. hum. Evol.* **25**, 83–172.

Benefit, B.R. (1994). Phylogenetic, paleodemographic, and taphonomic implications of *Victoriapithecus* deciduous teeth. *Am. J. phys. Anthrop.* **95**, 277–331.

Benefit, B.R. (1995). Earliest Old World monkey skull. *Am. J. phys. Anthrop.* **20**, 64.

Benefit, B.R. (1999). *Victoriapithecus*: the key to Old World monkey and catarrhine origins. *Evol. Anthropol.* **7**, 155–74.

Benefit, B.R. & McCrossin, M.L. (1990). Diet, species diversity and distribution of African fossil baboons. *Kroeber Anthropol. Soc. Papers* **71/72**, 77–93.

Benefit, B.R. & McCrossin, M.L. (1991). Ancestral facial morphology of Old World higher primates. *Proc. Natl. Acad. Sci. USA* **88**, 5267–71.

Benefit, B.R. & McCrossin, M.L. (1992). *Kenyapithecus* from Maboko Island. In *Apes or Ancestors?*, ed. J.A. Van Couvering, pp. 11–12. New York: American Museum of Natural History.

Benefit, B.R. & McCrossin, M.L. (1993). The facial anatomy of *Victoriapithecus* and its relevance to the ancestral cranial morphology of Old World monkeys and apes. *Am. J. phys. Anthrop.* **92**, 329–70.

Benefit, B.R. & McCrossin, M.L. (1995). Miocene hominoids and hominid origins. *Ann. Rev. Anthropol.* **24**, 237–56.

Benefit, B.R. & McCrossin, M.L. (1997). Earliest known Old World monkey skull. *Nature* **388**, 368–71.

Benefit, B.R. & Pickford, M. (1986). Miocene fossil cercopithecoids from Kenya. *Am. J. phys. Anthrop.* **69**, 441–64.

Bennett, E.L. & Davies, A.G. (1994). The ecology of Asian colobines. In *Colobine Monkeys: Their Ecology, Behavior and Evolution*, ed. A.G. Davies & J. F. Oates, pp. 45–74. Cambridge: Cambridge University Press.

Birchette, M.G. (1981). Postcranial remains of *Cercopithecoides*. *Am. J. phys. Anthrop.* **54**, 201.

Bishop, W.W. (1964). More fossil primates and other Miocene mammals from north-east Uganda. *Nature* **203**, 1327–31.

Bishop, W.W. (1968). The evolution of fossil environments in East Africa. *Trans. Leic. Lit. Phil. Soc.* **62**, 22–44.

Boaz, N.T. (1977). Paleoecology of early Hominidae in Africa. *Kroeber Anthropol. Soc. Papers* **50**, 37–62.

Brain, C.K. (1981). *The Hunters or the Hunted? An Introduction to African Cave Taphonomy*. Chicago: University of Chicago Press.

Brown, F.H. & Nash, W.P. (1976). Radiometric dating and tuff mineralogy of Omo Group deposits. In *Earliest Man and Environments in the Lake Rudolf Basin*, ed. Y. Coppens, F.C. Howell, G.Ll. Isaac & R.E.F. Leakey, pp. 50–63. Chicago: University of Chicago Press.

Brown, F.H., McDougall, I., Davies, T. & Maier, R. (1985). An integrated Plio-Pleistocene chronology for the Lake Turkana Basin. In *Ancestors: The Hard Evidence*, ed. E. Delson, pp. 82–90. New York: Alan R. Liss, Inc.

Butler, P.M. & Mills, J.R.E. (1959). A contribution to the odontology of *Oreopithecus*. *Bull. Brit. Mus. Nat. Hist., Geology Series* **4**, 1–30.

Cerling, T.E., Quade, J., Ambrose, S.H. & Sikes, N.E. (1991). Fossil soils from Fort Ternan, Kenya: Grassland or woodland? *J. hum. Evol.* **21**, 295–306.
Chivers, D.J. (1994). Functional anatomy of the gastrointestinal tract. In *Colobine Monkeys: Their Ecology, Behavior and Evolution*, ed. A.G. Davies & J.F. Oates, pp. 205–28. Cambridge: Cambridge University Press.
Churcher, C.S. (1970). Two new Upper Miocene giraffids from Fort Ternan, Kenya, East Africa: *Palaeotragus primaevus* n.sp. and *Samotherium africanum* n. sp. In *Fossil Vertebrates of Africa*, vol. 1, ed. L.S.B. Leakey, **2**, 1–105. London: Academic Press.
Cifelli, R.L., Ibui, A.K., Jacobs, L.L. & Thorington, R.W. (1986). A giant tree squirrel from the late Miocene of Kenya. *J. Mammal.* **67**, 274–83.
Ciochon, R.L. (1986). The Cercopithecoid Forelimb: Anatomical Implications for the Evolution of African Plio-Pleistocene Species. PhD Dissertation, University of California, Berkeley.
Ciochon, R.L. (1993). Evolution of the cercopithecoid forelimb. Phylogenetic and functional implications from morphometric analyses. *Univ. Calif. Publ. Geol. Sci.* ***138***, 1–251.
Ciochon, R.L. & Etler, D.A. (1994). Reinterpreting primate past diversity. In *Integrative Paths to the Past: Paleoanthropological Advances in Honor of F.C. Howell.* ed. R.S. Corruccini & R.L. Ciochon, pp. 95–122. New York: Prentice-Hall.
Davies, A.G. & Oates, J.F. (1994). *Colobine Monkeys: Their Ecology, Behavior and Evolution*. Cambridge: Cambridge University Press.
Delson, E. (1973). Fossil Colobine Monkeys of the Circum-Mediterranean Region and the Evolutionary History of the Cercopithecidae (Primates, Mammalia). PhD Dissertation, Columbia University.
Delson, E. (1975a). Evolutionary history of the Cercopithecidae. In *Approaches to Primate Paleobiology. Contributions to Primatology*, vol. 5, ed. F.S. Szalay, pp. 167–217. Basel: Karger.
Delson, E. (1975b). Toward the origin of the Old World monkeys. In *Evolution des vertebres – Problemes Actuels de Paleontologie Actuel C.N.R.S. Coll. Int.* **18**, 839–50.
Delson, E. (1979). *Prohylobates* (Primates) from the Early Miocene of Libya: a new species and its implications for cercopithecid origins. *Geobios*, 839–48.
Dunbar, R.I.M. (1983). Theropithecines and hominids: contrasting solutions to the same ecological problem. *J. hum. Evol.* **12**, 647–58.
Dunbar, R.I.M. & Dunbar, P. (1974). Ecological relations and niche separation among sympatric terrestrial primates in Ethiopia. *Folia Primatol.* **21**, 36–60.
Eck, G.G. (1976). Cercopithecoidea from Omo Group deposits. In *Earliest Man and Environments in the Lake Rudolf Basin*, ed. Y. Coppens, F.C. Howell, G. Ll. Isaac & R.E.F. Leakey, pp. 332–44. Chicago: University of Chicago Press.
Eck, G.G. & Howell, F.C. (1972). New fossil *Cercopithecus* material from the lower Omo Basin, Ethiopia. *Folia Primatol.* **18**, 325–55.
Evans, E.M.N., Van Couvering, J.A.H. & Andrews, P. (1981). Palaeoecology of Miocene sites in western Kenya. *J. hum. Evol.* **10**, 35–48.
Feibel, C.S., Brown, F.H. & McDougall, I. (1989). Stratigraphic context of fossil hominids from the Omo Group deposits: Northern Turkana Basin and Ethiopia. *Am. J. phys. Anthrop.* **78**, 595–622.
Fleagle, J.G. (1988). *Primate Adaptation and Evolution*. New York: Academic Press.
Fleagle, J.G., Bown, T.M., Harris, J.M., Watkins, R.W. & Leakey, M.G. (1997). Fossil monkeys from Northern Kenya. *Am. J. phys. Anthrop. (Suppl.)* **24**, 111.

Gebo, D.L., MacLatchy, L., Kityo, R., Deino, A., Kingston, J. & Pilbeam, D. (1997). A hominoid genus from the early Miocene of Uganda. *Science* **276**, 401–4.

Gentry, A.W. (1970). The Bovidae (Mammalia) of the Fort Ternan fossil fauna. In *Fossil Vertebrates of Africa*, vol. 1, ed. L.S.B. Leakey, pp. 243–342. London: Academic Press.

Gitau, S.N. (1995). Toward an Understanding of the Small-bodied Ape *Simiolus leakeyorum* from Maboko Island, Kenya. M.A. Thesis, Southern Illinois University.

Gitau, S.N. & Benefit, B.R. (1995). New evidence concerning the facial morphology of *Simiolus leakeyorum* from Maboko Island. *Am. J. phys. Anthrop. (Suppl.)* **20**, 99.

Grine, F.E. & Hendey, Q.B. (1981). Earliest primate remains from South Africa. *S. Afr. J. Sci.* **77**, 374–6.

Groves, C.P. (1989). *A Theory of Human and Primate Evolution*. Oxford: Clarendon Press.

Hamilton, W.J., Buskirk, R.E. & Buskirk, W.H. (1978). Omnivory and utilization of food resources by chacma baboons, *Papio ursinus*. *Am. Nat.* **112**, 110–20.

Harrison, T. (1986). A reassessment of the phylogenetic relationships of *Oreopithecus bambolii* Gervais. *J. hum. Evol.* **15**, 541–84.

Harrison T. (1987). The phylogenetic relationships of the early catarrhine primates: a review of the current evidence. *J. hum. Evol.* **18**, 3–54.

Harrison, T. (1989). New postcranial remains of *Victoriapithecus* from the middle Miocene of Kenya. *J. hum. Evol.* 18, 3–54.

Hynes, T. & Benefit, B.R. (1995). Phylogenetic relationships of the long-snouted fossil colobines. *Am. J. phys. Anthrop., Suppl.* **20**, 115.

Jolly, C.J. (1967). The evolution of baboons. In *The Baboon in Medical Research,* vol. 2, ed. H. Vagtborg, pp. 23–50. Austin: University of Texas Press.

Jolly, C.J.(1970). The large African monkeys as an adaptive array. In *Old World Monkeys: Evolution, Systematics and Behavior*, ed. J.R. Napier & P.H. Napier, pp. 141–74. New York: Academic Press.

Jolly, C.J. (1972). The classification and natural history of *Theropithecus (Simopithecus)* (Andrews, 1916) baboons of the African Plio-Pleistocene. *Bull. Brit. Mus. Nat. Hist., Geology Series* **22**, 1–123.

Kappelman, J. (1991). The paleoenvironment of *Kenyapithecus* at Fort Ternan. *J. Hum. Evol.* **20** 95–129.

Kay, R.F. (1975). The functional adaptations of primate molar teeth. *Am. J. phys. Anthrop.* **43**, 195–215.

Kay, R.F. (1977a). Post-Oligocene evolution of catarrhine diets. *Am. J. phys. Anthrop.* **47**, 141–2.

Kay, R.F. (1977b). The evolution of molar occlusion in the Cercopithecidae and early catarrhines. *Am. J. phys. Anthrop.* **46**, 327–52.

Kay, R.F. (1977c). Diets of early Miocene African hominoids. *Nature* **269**, 628–30.

Kay, R.F. (1978). Molar structure and diet in extant Cercopithecidae In *Studies in the Development, Function, and Evolution of Teeth*. ed. P.M. Butler & K. Joysey, pp. 309–39. Academic Press: London.

Kay, R.F. (1981). The nut-crackers – a new theory of the adaptations of the Ramapithecinae. *Am. J. phys. Anthrop.* **55**, 141–52.

Kay, R.F. (1984). On the use of anatomical features to infer foraging behavior in extinct primates. In *Adaptations for Foraging in Non-human Primates: Contributions to an Organismal Biology of Prosimians, Monkeys, and Apes,*

ed. P.S. Rodman & J.G.H. Cant, pp. 21–53. New York: Columbia University Press.
Kay, R.F. & Covert, H. (1984). Anatomy and behavior of extinct primates. In *Food Acquisition and Processing in Primates*, ed. D.J. Chivers, B.A. Wood & A. Bilsborough, pp. 467–508. New York: Plenum.
Kay, R.F. & Hylander, W.L. (1978). The dental structure of mammalian folivores with special reference to primates and phalangeroids (Marsupialia). In *The Ecology of Arboreal Folivores*, ed. G.G. Montgomery, pp. 173–93. Washington, D.C.: Smithsonian Institution Press.
Kay, R.F. & Ungar, P.S. (1997). Dental evidence for diet in some Miocene catarrhines with comments on the effects of phylogeny on the interpretation of adaptation. In *Function, Phylogeny, and Fossils: Miocene Hominoid Evolution and Adaptations,* ed. D.R.Begun, C.V. Ward & M.D. Rose, pp. 131–52. New York: Plenum.
Krentz, H.B. (1993). Postcranial anatomy of extant and extinct species of *Theropithecus*. In *Theropithecus: The Rise and Fall of a Primate Genus*, ed. N.G. Jablonski, pp. 383–424. Cambridge: Cambridge University Press.
Leakey, M.G. (1982). Extinct large colobines from the Plio-Pleistocene of Africa. *Am. J. phys. Anthrop.* **58**, 153–72.
Leakey, M.G. (1985). Early Miocene cercopithecids from Buluk, Northern Kenya. *Folia Primatol.* **44**, 1–14.
Leakey, M.G. (1988). Fossil evidence for the evolution of guenons. In *A Primate Radiation: Evolutionary Biology of the African Guenons*, ed. A. Gautier-Hion, F. Bourliere & J-P. Gautier, pp. 7–12. Cambridge: Cambridge University Press.
Leakey, M.G. (1993). Evolution of *Theropithecus* in the Turkana Basin. In *Theropithecus: The Rise and Fall of a Primate Genus*, ed. N.G. Jablonski, pp. 85–123. Cambridge: Cambridge University Press.
Leakey, M.G. & Walker, A.C. (1997). *Afropithecus*: Function and phylogeny. In *Function, Phylogeny, and Fossils: Miocene Hominoid Evolution and Adaptations,* ed. D.R.Begun, C.V. Ward & M.D. Rose, pp. 225–40. New York: Plenum.
Leakey, M.G., Leakey, R.E., Richtsmeier, J.T., Simons, E.L. & Walker, A.C. (1991). Similarities in *Aegyptopithecus* and *Afropithecus* facial morphology. *Folia Primatol.* **56**, 65–85.
Leakey, R.E. & Leakey, M.G. (1986). A new Miocene hominoid from Kenya. *Nature* **342**, 143–6.
Leakey, R.E., Leakey, M.G. & Walker, A.C. (1988). Morphology of *Afropithecus turkanensis* from Kenya. *Am. J. phys. Anthrop.* **76**, 289–307.
Le Gros Clark, W.E. & Thomas, D.P. (1951). Associated jaws and limb bones of *Limnopithecus macinnesi. Brit. Mus. Nat. Hist., Fossil Mammals of Africa* **3**, 1–27.
Lee-Thorp, J.A.& Van der Merwe, N.J. (1989). Isotopic evidence for dietary differences between two extinct baboon species from Swartkrans. *J. hum. Evol.* **18**, 183–90.
Lucas, P.W. & Teaford, M.F. (1994). Functional morphology of colobine teeth. In *Colobine Monkeys: Their Ecology, Behavior and Evolution*, ed. Davies, A.G. & Oates, J.F. pp. 173–204. Cambridge: Cambridge University Press.
Maier, W. (1977a). Die Evolution der bilophodonten Molaren der Cercopithecoidea. *Z. Morphol. Anthropol.* **68**, 26–56.
Maier, W. (1977b). Die bilophodonten Molaren der Indriidae (Primates) – Ein evolutionsmorphologischer Modellfall. *Z. Morphol. Anthropol.* **68**, 307–344.

McCrossin, M.L. (1992). An oreopithecid proximal humerus from the middle Miocene of Maboko Island, Kenya. *Int. J. Primatol.* **13**, 659–77.

McCrossin, M.L. (1994a). The Phylogenetic Relationships, Adaptations, and Ecology of *Kenyapithecus*. PhD Dissertation, University of California at Berkeley.

McCrossin, M.L. (1994b). Semi-terrestrial adaptations of *Kenyapithecus*. *Am. J. phys. Anthrop. (Suppl.)* **18**, 142–3.

McCrossin, M.L. (1995). New perspectives on the origins of terrestriality among Old World higher primates. *Am. J. phys. Anthrop. (Suppl.)* **20**, 147.

McCrossin, M.L. (1996). A reassessment of forelimb evidence for the phylogenetic relationships of *Kenyapithecus* and other large-bodied hominoids of the middle-late Miocene. *Am. J. Phys. Anthrop. (Suppl.)* **22**, 161–2.

McCrossin, M.L. & Benefit, B.R. (1992). Comparative assessment of the ischial morphology of *Victoriapithecus macinnesi*. *Am. J. phys. Anthrop.* **87**, 277–90.

McCrossin, M.L. & Benefit, B.R. (1993a). Recently recovered *Kenyapithecus* mandible and its implications for great ape and human origins. *Proc. Natl. Acad. Sci. USA* **90**, 1962–6.

McCrossin, M.L. & Benefit, B.R. (1993b). Clues to the relationships and adaptations of *Kenyapithecus africanus* from its mandibular and incisor morphology. *Am. J. phys. Anthrop. (Suppl.)* **16**, 143.

McCrossin, M.L. & Benefit, B.R. (1994). Maboko Island and the evolutionary history of Old World monkeys and apes. In *Integrative Paths to the Past: Paleoanthropological Advances in Honor of F.C. Howell.* ed. R.S. Corruccini & R.L. Ciochon, pp. 95–122. New York: Prentice-Hall.

McCrossin, M.L.& Benefit, B.R. (1997). On the relationships and adaptations of *Kenyapithecus*, a large-bodied hominoid from the middle Miocene of eastern Africa. In *Function, Phylogeny, and Fossils: Miocene Hominoid Evolution and Adaptations*, ed. D.R. Begun, C.V. Ward & M.D. Rose, pp. 241–67. New York: Plenum.

McCrossin, M.L., Benefit, B.R., Gitau, S., Palmer, A.K. & Blue, K.T. (1998) Fossil evidence for the origins of terrestriality among Old World higher primates. In *Primate Locomotion: Recent Advances*, ed. E.L. Strasser, J. Fleagle, A. Rosenberger & H.M. McHenry, pp. 353–86. New York: Plenum Press.

Meikle, W.E. (1987). Fossil Cercopithecidae from the Sahabi Formation. In *Neogene Paleontology and Geology of Sahabi*, ed. N.T. Boaz, A. el-Arnauti, A.W. Gaziry, J. de Heinzelin & D. Dechant Boaz, pp. 119–27. New York:A.R. Liss.

Meldrum, D.J. & Pan Y. (1988). Manual proximal phalanx of *Laccopithecus robustus* from the latest Miocene site of Lufeng. *J. hum. Evol.* **17**, 719–31.

Morbeck, M.E. (1983). Miocene homonid discoveries from Rudabanya: implications from the postcranial skeleton. In *New Interpretations of Ape and Human Ancestry*, ed. R.L. Ciochon & R.S. Corruccini, pp. 369–404. Plenum Press: New York.

Moya-Sola, S.& Kohler, M. (1993). Recent discoveries of *Dryopithecus* shed new light on evolution of great apes. *Nature* **365**, 543–5.

Moya-Sola, S. & Kohler, M. (1996). A *Dryopithecus* skeleton and the origins of great ape locomotion. *Nature* **379**, 156–9.

Napier, J.R. (1970). Paleoecology and catarrhine evolution. In *Old World Monkeys: Evolution, Systematics and Behavior*, ed. J.R. Napier & P.H. Napier, pp. 53–96. New York: Academic Press.

Napier, J.R. & Davis, P.R. (1959). The forelimb skeleton and associated remains of *Proconsul africanus*. *Fossil Mammals of Africa* **16**, 1–69.
Oates, J.F. (1994). The natural history of African colobines. In *Colobine Monkeys: Their Ecology, Behavior and Evolution*, ed. A.G. Davies & J.F. Oates, pp. 75–128. Cambridge: Cambridge University Press.
Palmer, A.K., Benefit, B.R., McCrossin, M.L. & Gitau S.M. (1998). Paleoecological implications of dental microwear analysis for the middle Miocene primate fauna from Maboko Island, Kenya. *Am J. Phys. Anthropol. (Suppl.)* **26**, 175. (Abstract.)
Pickford, M. (1983). Sequence and environments of the lower and middle Miocene hominoids of western Kenya. In *New Interpretations of Ape and Human Ancestry*, ed. R.L. Ciochon & R.S. Corruccini, pp. 421–39. New York: Plenum Press.
Pickford, M. (1987). The chronology of the Cercopithecoidea of East Africa. *Hum. Evol.* **2**, 1–17.
Pickford, M.H.L., Senut, B., Hadoto, D., Musisi, J. & Kariira, C. (1986). Nouvelles decouvertes dans le Miocene inferieur de Napak, Ouganda Oriental. *C. R. Acad. Sci. (Paris)* **302**, 47–52.
Pilbeam, D.R. (1982). New hominoid skull material from the Miocene of Pakistan. *Nature* **295**, 232–4.
Pilbeam, D. & Walker, A. (1968). Fossil monkeys from the Miocene of Napak, North-east Uganda. *Nature* **220**, 657–60.
Remane, A. (1951). Die entstehung der bilophodontie bei den Cercopithecoidea. *Anat. Anz.* **98**, 161–5.
Retallack, G.J. (1992). Middle Miocene fossil plants from Fort Ternan (Kenya) and evolution of African grasslands. *Paleobiology* **18**, 383–400.
Retallack, G.J., Dugas, D.P. & Bestland, E.A. (1990). Fossil soils and grasses of a middle Miocene East African grassland. *Science* **247**, 1325–8.
Rose, M.D. (1983). Miocene hominoid postcranial morphology: monkey-like, ape-like, or both? In *New Interpretations of Ape and Human Ancestry*, ed. R.L. Ciochon & R.S. Corruccini, pp. 405–17. New York: Plenum Press.
Rose, M.D. (1993). Locomotor anatomy of Miocene hominoids. In *Postcranial Adaptation in Nonhuman Primates*, ed. D.L. Gebo, pp. 252–72. DeKalb: Northern Illinois University Press.
Rose, M.D., Leakey, M.G., Leakey, R.E. & Walker, A.C. (1992). Postcranial specimens of *Simiolus enjiessi* and other primitive catarrhines from the early Miocene of Lake Turkana, Kenya. *J. hum. Evol.* **22**, 171–237.
Sarich, V.M. & Cronin, J. (1976). Molecular systematics of the primates. In *Molecular Anthropology*, ed. M. Goodman & R. Tashian, pp. 141–70. New York: Plenum.
Sarmiento, E.E. (1987) The phyletic position of *Oreopithecus* and its significance in the origin of the Hominoidea. *Am. Mus. Novit.* **2881**, 1–44.
Savage, R.J.G. & Hamilton, R. (1973). Introduction to the Miocene mammal faunas of Gebel Zelten, Libya. *Bull. Brit. Mus. Nat. Hist., Geol. Ser.* **22**, 483–511.
Senut, B. (1986). Nouvelle decouvertes de restes postcraniens de primates (Hominoidea et Cercopithecoidea) sur le site Maboko au Kenya occidental *C. R. Acad. Sci. (Paris)* **303**, 1359–62.
Shipman, P., Walker, A., Van Couvering, J.A.H. & Hooker, P.J. (1981). The Fort Ternan hominoid site, Kenya: Geology, taphonomy and paleoecology. *J. hum. Evol.* **10**, 49–72.
Simons, E.L. (1969). Miocene monkeys (*Prohylobates*) from northern Egypt. *Nature* **223**, 687– 89.

Simons, E.L. (1970). The deployment and history of Old World monkeys (Cercopithecidae, Primates). In *Old World Monkeys: Evolution, Systematics, and Behavior*, ed. J.R. Napier & P.H. Napier, pp. 97–137. New York: Academic Press.

Simons, E.L. (1987). New faces of *Aegyptopithecus* from the Oligocene of Egypt. *J. hum. Evol.* **16**, 273–89.

Simons, E.L. (1994). New monkey (*Prohylobates*) and an ape humerus from the Miocene Moghara formation of northern Egypt. *Proc. XIV Int. Primatol. Cong*. pp. 247–53.

Simons, E.L. & Delson, E. (1978). Cercopithecidae and Parapithecidae. In *Evolution of African Mammals*, ed. V.J. Maglio & H.B.S. Cooke, pp. 100–19. Cambridge, MA: Harvard University Press.

Strasser, E. (1997). Cladistic analysis of the cercopithecoid foot. *Am. J. phys. Anthrop., Suppl.* **24**, 222.

Strasser, E. & Delson, E. (1987). Cladistic analysis of cercopithecid relationships. *J. hum. Evol.* **16**, 81–99.

Susssman, R.W. (1977). Feeding behavior of *Lemur catta* and *Lemur fulvus*. In *Primate Ecology: Studies of Feeding and Ranging Behavior in Lemurs, Monkeys, and Apes*, ed. T.H. Clutton-Brock, pp. 1–37. London: Academic Press.

Szalay, F.S. & Delson, E. (1979). *Evolutionary History of the Primates*. New York: Academic Press.

Teaford, M.F. (1993). Dental microwear and diet in extant and extinct *Theropithecus*: Preliminary analyses. In *Theropithecus: The Rise and Fall of a Primate Genus*, ed. N.G. Jablonski, pp. 331–49. Cambridge: Cambridge University Press.

Teaford, M.F. & Leakey, M.G. (1992). Dental microwear and diet in Plio-Pleistocene cercopithecoids from Kenya. *Am. J. phys. Anthrop., Suppl.* **14**, 160–1.

Teaford, M.F., Maas, M.C. & Simons, E.L. (1996). Dental microwear and microstructure in early Oliogocene primates from the Fayum, Egypt:Implications for diet. *Am. J. phys. Anthrop.* **101**, 527–43.

Temerin, L.A. & Cant, J.G.H. (1983). The evolutionary divergence of Old World monkeys and apes. *Am. Nat.* **122**, 335–51.

Ungar, P.S. & Teaford, M.F. (1996) Preliminary examination of non-occlusal dental microwear in anthropoids:implications for the study of fossil primates. *Am. J. phys. Anthrop.* **100**, 101–14.

Vogel, C. (1966). Morphologische studien am gesichtschadel Catarrhiner primaten. *Bibl. Primatol.* **4**, 1–226.

Vogel, C. (1968). The phylogenetical evaluation of some characters and some morphological trends in the evolution of the skull in catarrhine primates. In *Taxonomy and Phylogeny of Old World Primates with References to the Origin of Man*, ed. B. Chiarelli, pp. 21–55. Turin: Rosenberg and Sellier.

Von Koenigswald, G.H.R. (1969). Miocene Cercopithecoidea and Oreopithecoidea from the Miocene of East Africa. In *Fossil Vertebrates of Africa*, vol. 1, ed. L.S.B. Leakey, pp. 39–51. London: Academic Press.

Voruz, C. (1970). Origine des dents Bilophodontes des Cercopithecoidea. *Mammalia* **34**, 269–93.

Vrba, E. (1982). Biostratigraphy and chronology, based particularly on Bovidae, of southern hominid-associated assemblages: Makapansgat, Sterkfontein, Taung, Kromdraai, Swartkrans; also Elandsfontein, Saldanha, Broken Hill (now Kabwe) and Cave of Hearths. In *Pretirage, Premier Congres*

International de Paleontologie Humaine, II. ed. H. de Lumley & M.-A. de Lumley, pp. 707–52. Nice: Centre National de Recherche Scientifique.

Winkler, A.J. (1994). Middle Miocene rodents from Maboko Island, western Kenya: Contributions to understanding small mammal evolution during the Neogene. *J. Vertebr. Paleontol.* **14**, 53A.

Xiao, M. (1981). Discovery of fossil hominoid scapula at Lufeng, Yunnan. *J. Yunnan Prov. Mus.* **30**, 41–4.

7

Geological context of fossil Cercopithecoidea from eastern Africa

TOM GUNDLING AND ANDREW HILL

Introduction

The major catarrhine radiation documented in the early Miocene of Africa produced a variety of intriguing primate forms, including the earliest known members of the Cercopithecoidea. In terms of taxonomic and ecological diversity, Old World monkeys (OWM) are the most successful group of anthropoid primates, and currently occupy a variety of broad ecological zones and more specific habitat types throughout Asia, Africa, and, until recently, Europe. In examining the fossil record for this group, however, it becomes apparent that although cercopithecoids are known for the last 20 million years (Myr), the modern high level of diversity is a relatively recent phenomenon.

This chapter provides a review of the distribution of east African fossil cercopithecoids in time and space. Except for two early Miocene north African sites, eastern Africa provides all of the evidence for cercopithecoid evolution in the Old World prior to about 8.5 million years ago (Ma), and for the whole of Africa prior to about 5 Ma. Consequently, most of the evolutionary events that we know about in the lineage are documented from sites in this region. Our aim is simply to summarize geological data which form a basis for hypotheses of faunal change and diversification within the superfamily. We provide a catalog of the major east African Neogene sites at which monkeys are found along with a brief account of the geological context, particularly the relative stratigraphic position and absolute dating of fossil sites when available.

Fossil cercopithecoids are found throughout most of the Neogene in east Africa, and as with their ape contemporaries, unresolved issues of systematics and ecology persist. From an ecological perspective, for example, evidence about the nature of the original OWM niche, and also about the timing and environmental circumstances surrounding the evolution of extant cercopithecoid specializations, such as the colobine adaptation to a folivorous diet, continues to be elusive. Concerning systematics, while extinct monkeys from the early and middle Miocene of Africa are widely accepted as true cercopithecoids, there remains some discussion as to their precise phylogenetic position. Benefit (1993; Chapter 6) has pointed out that the two modern subfamilies share derived features, particularly in the dentition, indicating that they are more closely related to each other than either is to the more primitive Miocene monkeys. Consequently, she places the Miocene forms within their own family, the Victoriapithecidae. We adopt this position, and view victoriapithecids as the sister clade of modern monkeys, realizing that the former may not be directly ancestral to the latter.

These unresolved issues and others can only be approached through careful systematic and environmental analyses firmly grounded in an accurate geochronologic sequence. We need to establish more precisely the timing of events within the evolution of OWM in order to help address problems concerning the phylogeny and ecological diversification of this important group of primates. In the record as we know it at present, victoriapithecids are found in the early and middle Miocene, while cercopithecids are found from the later Miocene to the present. This is how our account is divided. It is further organized around the geological formations in which relevant sites occur, since formations serve as rough proxies of time and space. The formations are ordered chronologically, although in some instances there is much temporal overlap. We consider occurrences from the earliest appearance of cercopithecoids in the record until just after 1 Ma. Some early radiometric dates have been corrected using more recent decay and abundance constants according to the formula of Ness *et al.* (1980). They are indicated in the text by (nc). Finally, equal attention is given to formations from the earlier part of the record, despite the numerical paucity of cercopithecoid specimens from most of these localities. Some of these formations provide only one specimen, compared, for example, to over 6000 specimens from the Shungura Formation. However, *because* of the general paucity of specimens from the earlier part of the record these instances are in some ways more crucial to the overall picture of cercopithecoid evolution and diversification than later parts of the record, after the modern pattern of the cercopithecoid diversity has been established.

Early to middle Miocene

The oldest known cercopithecoid specimens are attributed to the genera *Victoriapithecus* and *Prohylobates*. However, the two genera are morphologically similar, differing mainly in overall size, and may eventually be united, with *Prohylobates* having priority (Leakey, 1985; but see Benefit, 1993). Victoriapithecids are known from Napak in Uganda, probably older than 18.3 Ma (nc) and maybe as old as 19.5 Ma, to 12.5 Ma in the Ngorora Formation, Tugen Hills, Kenya. In sub-Saharan Africa just two time successive species of *Victoriapithecus* are recognized, *V. macinnesi* from most sites, and an unnamed species of *Victoriapithecus* from the Ngorora Formation. Apart from a cranium (Benefit, 1995; Chapter 6) and a variety of postcrania (Senut, 1987; Harrison, 1989), most victoriapithecid fossils are isolated teeth and mandibular fragments, and are known from only a handful of sites in east Africa (Leakey, 1985; Benefit, 1994; Hill *et al.*, 1999b) and two sites in north Africa (Fourtau, 1918; Simons, 1969, 1994; Delson, 1979). Cercopithecoids from the north African sites may be older than most of the earliest occurrences in east Africa, though they are poorly dated. Wadi Moghara in northern Egypt has produced five fragmentary mandibles of *Prohylobates tandyi*, and is estimated to be approximately 18–17 Ma based on faunal comparison with dated sites in east Africa (Simons, 1994; Miller and Simons, 1996). The second north African site, Gebel Zelten in Libya, from which the mandible of *Prohylobates simonsi* was recovered, is estimated to 17–15 Ma (Delson, 1979; Miller and Simons, 1996).

Napak, Uganda

Probably the earliest known monkeys in eastern Africa, and possibly anywhere in the world, occur at the site of Napak in northern Uganda (Bishop, 1958a,b). A single upper first or second molar from Napak V and a partial frontal bone from Napak IX were originally attributed to the Cercopithecinae and Colobinae, respectively (Pilbeam and Walker, 1968). It now appears that the frontal belongs to the primitive catarrhine genus *Micropithecus* (Fleagle and Simons, 1978) but the molar is cercopithecoid, and is designated *Victoriapithecus* sp. by Szalay and Delson (1979). Additionally, Pickford and Senut (1988) report a cercopithecoid canine and a proximal left radius and ulna from the site.

The regional geology of Napak was studied by King (1949), and subsequent work on the fossil sites, including details of local geology, fauna, and dating was reported by Bishop (1958a,b, 1963a,b, 1964, 1968, 1971; Bishop and Whyte, 1962). The primates come from sites formed under subaerial conditions. Bishop (1964) records a date on biotite from the top of the sequence containing the site Napak I of 19.5 ± 2 Ma (nc). A more extensive series of dates was established by Bishop *et al.* (1969) from various horizons within the Napak volcanic center giving dates ranging from 32.1 to 6.9 Ma (nc). However, the only ones of relevance to the fossiliferous horizons are from the same biotites related to site Napak I, which are 18.3 ± 0.4 Ma (nc), but these dates may be discrepantly low.

Bakate Formation, Buluk, Kenya

Besides Napak, victoriapithecid remains are found at only a few other east African sites, all in Kenya. At the northern Kenya site of Buluk (Harris and Watkins, 1974), 16 specimens attributed to *Prohylobates* sp. were recovered from channel fill deposits within the Miocene Bakate Formation (Leakey, 1985). Leakey uses the senior genus *Prohylobates* to reflect her reservations regarding a distinction between the two genera, rather than to signify the presence of monkeys at Buluk that are different than *Victoriapithecus* from other east African sites.

The monkeys, and other vertebrate and plant fossils, come from deposits within the Buluk Member of the Bakate Formation, which are capped by a basalt flow dating to 17.2 ± 0.2 Ma using the K/Ar method (McDougall and Watkins, 1985). Although no radiometric dates from the underlying Irile Member are available, the lack of unconformities within the sequence suggests that the fossiliferous sediments were laid down no more than about 18 Ma. These ages are concordant with other K/Ar dates from feldspars found higher up in the sequence. The environment of the fossils' accumulation is interpreted as a shallow interdistributary or behind-shore lagoon (Harris and Watkins, 1974).

Loperot, Kenya

Undescribed cercopithecoid specimens, including a molar and part of a mandible collected by Patterson, and attributed to *Victoriapithecus*, are known from a long sedimentary sequence at Loperot, west of Lake Turkana in northern Kenya (Leakey, 1985). Baker *et al.* (1971) give a range from 18.0 to 16.2 Ma (nc) on a basalt flow overlying the fossiliferous beds and Mead (1975) cites a date of 17.1 ± 1.0 Ma (nc) but provides few details. Boschetto *et al.* (1992) report age determinations of 13.9 ± 0.2 Ma and

15.0 ± 0.2 Ma, and attribute the discrepancy to argon loss. They conclude that 15.0 Ma is a minimum age for the overlying basalt, but there is no maximum age for the fauna, some of which appears to be significantly older, perhaps approaching that of Losodok and Meswa Bridge. The *Victoriapithecus* fossils are not closely related to the basalt (M.G. Leakey, pers. comm.). These specimens may eventually prove to be some of the oldest known monkeys.

Muruyur Formation, Tugen Hills, Kenya

Kipsaramon is a site complex in the Muruyur Formation near the base of the Tugen Hills sequence west of Lake Baringo, Kenya (Bishop *et al.*, 1971; Pickford, 1988; Hill *et al.*, 1991), studied by Hill and colleagues as part of the Baringo Paleontological Research Project (BPRP). In the Muruyur Formation, fossils of *Victoriapithecus* are found at several different localities. One set of sites is associated with outcrops of a widespread bone bed (BPRP#89) (Hill *et al.*, 1991). A wide range of fauna has come from these sites, which are fairly closely constrained to 15.5 Ma by bracketing K–Ar dates. The *Victoriapithecus* material consists of a mandible, a number of isolated teeth and a proximal femur, none yet described in detail. The associated fauna, particularly the rodents (Winkler, 1992), suggests a predominantly forested habitat.

Another occurrence is slightly higher in the sequence from a site (BPRP#122) dated to younger than 15.4 Ma. Here, isolated *Victoriapithecus* teeth are associated with many teeth and a partial skeleton of a hominoid attributable to *Equatorius africanus.* Behrensmeyer *et al.* (1999) give further details of the Kipsaramon site complex, including a set of more finely resolved radiometric determinations.

Maboko Formation, Kenya

Maboko Island is the most prolific site for early cercopithecoid fossil evidence (von Koenigswald, 1969; Harrison, 1989; Benefit, 1993) and is the type site for *Victoriapithecus*. A history of exploration is given by Pickford (1986a), and other relevant information is provided by McCrossin and Benefit (1994). As the subject of several expeditions since the 1930s, the Maboko deposits have produced at least 1485 permanent and deciduous dental remains assignable to *Victoriapithecus*, along with many postcranial elements. Benefit's (1993, 1994) analyses of the dentition show that one variable species, *Victoriapithecus macinnesi*, is present in large numbers, in addition to a number of other non-cercopithecoid catarrhine species. Based on a study of the postcrania, Harrison (1989) also concludes that one dimorphic species is represented. However, in an examination of post-crania from Maboko and also Nyakach and Napak, Senut (1987) states that two distinct morphologies exist, which might indicate two species, one more terrestrially adapted than the other. She does not, however, suggest that this represents the appearance of the two modern subfamilies, as had Delson (1975) and Szalay and Delson (1979). Delson (1994) now accepts a single species at Maboko.

Recent Ar–Ar dating from two levels in the sequence confirms the middle Miocene age of the deposits originally established on faunal grounds (Andrews *et al.*, 1981). Feibel and Brown's (1991) analysis produced a date for a phonolite capping the sequence of 13.80 ± 0.04 Ma and a second date of 14.71 ± 0.16 Ma for a tuff within Bed 8. The cercopithecoid fossils with well established provenience all come from sediments further down in the sequence, Beds 3 and 5 in particular, and are older than 14.71 Ma. No dates are given for the bottom of the sequence but there is no reason to suspect that the monkeys are significantly older than 15 Ma.

Majiwa and Ombo are sites in western Kenya placed by Pickford (1986a) in the Maboko Formation, and are of broadly similar age. Early collections were made by Owen and by Louis Leakey. Andrews *et al.* (1981) give a date of 11.8 Ma and 12.1 Ma for phonolites overlying the Majiwa sediments. Although Pickford (1986a,b) records *Victoriapithecus* from Majiwa, other authors have asserted that the specimen is not cercopithecoid, but probably a carnivore (Senut, 1987; Leakey, 1988; Harrison, 1989). Pickford (1986a) does not record any cercopithecoids as coming from Ombo, yet Szalay and Delson (1979) discuss a broken *Victoriapithecus* molar from that location, which is housed in the Natural History Museum in London (Delson, pers. comm.).

Aka Aiteputh Formation, Samburu Hills, Kenya

The Aka Aiteputh Formation crops out in the Samburu Hills on the eastern side of the northern Kenya Rift, and has been studied by Ishida and colleagues (Ishida, 1984; Matsuda *et al.*, 1986; Nakaya, 1994). Nakaya (1994) documents *Victoriapithecus* sp. as part of an overall faunal list from the formation, and cites an age range of 15 to 11.5 Ma based on fauna and K–Ar dating. Notably, the fauna suggests an age older than the radiometric ages indicate.

Nachola Formation, Kenya

Ishida and colleagues have also described fossils found in the Nachola area, to the east of the Samburu Hills (Ishida, 1984; Matsuda *et al.*, 1986). Pickford *et al.* (1984) list possible cercopithecoids from the Emuruilem Member of the Nachola Formation, and Pickford and Senut (1988) mention cercopithecoid material although no other details are given. Matsuda *et al.* (1984, 1986) cite radiometric dates of 11.8 to 10.1 Ma for Nachola, but more recent estimates suggest an age nearer to 15 Ma (Ishida, pers. comm.).

Nyakach Formation, Kenya

The geology and history of exploration of Nyakach in western Kenya are discussed in Pickford (1986a), and he records *Victoriapithecus* from three and possibly four sites, further suggesting that the sediments accumulated in a floodplain setting. In addition, a cercopithecoid distal humerus is described by Pickford and Senut (1988). Pickford (1986b), while recording single teeth of this genus from only two sites in the Nyakach Formation, gives a fission track date of 13.4 ± 1.3 Ma for a tuff in the middle of the sequence, according with the fauna which closely matches that of Maboko.

Ngorora Formation, Tugen Hills, Kenya

The Ngorora Formation (Bishop and Chapman, 1970; Bishop *et al.*, 1971; Bishop and Pickford, 1975, 1978a; Hill *et al.*, 1985; Hill, 1995) spans almost 5 Myr, from 13.2 Ma to about 8.5 Ma (Deino *et al.*, 1990) and contains many sites which document faunal change through the Miocene. One site (BPRP#38) in the Kabasero section of the formation has produced a number of cercopithecoid teeth and is dated to 12.5 Ma. The first specimen, discovered by the Aguirre–Leakey expedition, though not recorded by them (Aguirre and Leakey, 1974), was referred to the Cercopithecoidea by Benefit and Pickford (1986). They suggest that it might belong to *Microcolobus tugenensis*, a colobine from higher in the sequence, and in later publications allude to the specimen as colobine (Pickford, 1987; also see Delson, 1994). Subsequent work (Hill *et al.*, 1999b) retrieved nine additional teeth, which show this taxon to be a new species of *Victoriapithecus*. The species has not yet been named, in the hope that better material will be found to serve as a type specimen. The primate is accompanied by a diverse fauna (Hill *et al.*, 1999b), and there is evidence that the immediate area was occupied by lowland rainforest a few hundred thousand years earlier (Jacobs and Kabuye, 1987; Jacobs and Winkler, 1992; Jacobs and Deino, 1996).

Late Miocene to the present

To date, victoriapithecids are unknown beyond the middle Miocene. After about 10 Ma, all identifiable east African cercopithecoid fossils can be placed within one of the two extant subfamilies, although many of the extinct genera are so different from modern monkeys (especially among the Colobinae) that it is difficult to link them directly to modern representatives (Leakey, 1982). Modern genera do not appear until the early Pliocene, with the possible exception of *Theropithecus*, and a fully modern monkey fauna does not appear until well into the Pliocene.

Ngorora Formation, Tugen Hills, Kenya

The first appearance in Africa, and possibly in the world, of monkeys attributable to either modern subfamily occurs in the youngest part of the Ngorora Formation. An almost complete mandible of a diminutive colobine comes from the Ngeringerowa exposures at the top of the formation (BPRP#25), and is the type specimen of *Microcolobus tugenensis* (Benefit and Pickford, 1986). The site has been well-dated to between 9.5 and 9 Ma by BPRP (Deino, pers. comm.), thus predating the earliest appearance of the colobine *Mesopithecus* from the late Miocene of Europe (MN 11/12), with one possible exception (Delson, 1994; Andrews *et al.*, 1996). An upper premolar, tentatively referred to *Mesopithecus* based on its size, from the site of Wissburg in Germany may be of MN9 age and therefore 11–10 Ma. However, some fossils from this area are probably younger inclusions, so more secure dating is desirable for this specimen (Andrews *et al.*, 1996).

Nakali, Kenya

Two cercopithecoid fragments from Nakali are recorded as Cercopithecidae indet. by Aguirre and Leakey (1974; also see Aguirre and Guerin, 1974). Additionally, a lower molar (?M1) in a fragment of mandible, discovered by Meave Leakey in 1978, is described as an indeterminate genus of colobine by Benefit and Pickford (1986). Although interesting faunally (Hill, 1987), Nakali is not well dated. A plausible estimate is between 9 and 7 Ma.

Mpesida Beds, Tugen Hills, Kenya

The Mpesida Beds constitute a formation in the Tugen Hills succession (Bishop *et al.*, 1971; Hill *et al.*, 1985; Hill *et al.*, 1986; Hill, 1995; Hill *et al.*, 1999a). Recent work by BPRP shows them to be geographically more extensive than thought previously, but they are relatively short in duration, bracketed by flows of the Kabarnet Trachytes.These give dates ranging between 7.5 and 7 Ma for lower flows, and 6.3 to 6.2 Ma for overlying flows. Apart from an indeterminate primate astragalus, a single M3 of a colobine was recently found at site BPRP#85 (Hill *et al.*, 1999a). It is one of the earliest colobines larger than *Microcolobus* known from Africa, along with specimens from the Nkondo Formation (see below) and from Menacer, Algeria which are roughly the same age. There is evidence of a forest in the formation including a number of trunks in growth position, buried under a volcanic ash.

Nawata/Nachukui Formations, Lothagam, Kenya

Lothagam is a fossiliferous succession to the southwest of Lake Turkana originally investigated in 1967 by a Harvard expedition (Patterson *et al.*, 1970). Smart (1976) provides an early summary of the fauna, and Hill *et al.*, (1992) give an updated faunal list. More recent work (Leakey *et al.*, 1996) has redefined the local stratigraphy and will undoubtedly provide better information in the near future. Two formations are recognized at Lothagam, the Nawata and the Nachukui. The lower member of the Nawata Formation ranges from older than 7.9 Ma to 6.57 Ma, while the upper Nawata member falls between 6.24 and 5.5 Ma. The lower members of the Nachukui Formation crop out at Lothagam (but have different names to their equivalents at West Turkana), and are probably 5.0 to younger than 4.0 Ma. Most of the Lothagam fauna originates from the upper Nawata and lowermost Nachukui Formations (Leakey *et al.*, 1996), and since no clear distinction is made within the literature, the two formations are here taken together. Lothagam monkeys include cf. *Parapapio* and cf. *Cercocebus* which are listed as coming from Lothagam members 1B and 1C (Smart, 1976; Hill *et al.*, 1992). Leakey *et al.* (1996) also report two unnamed colobine species from this location.

The younger sediments within the Lothagam sequence, which are contemporaneous with the older levels of the Nachukui Formation, have produced a single cercopithecoid lower molar. Patterson *et al.* (1970) refer this specimen to *Simopithecus* sp. from Member 3, Leakey (1993) refers it to

Theropithecus, and it is illustrated and described by Delson (1993). Within the new stratigraphic arrangement (Leakey *et al.*, 1996), the tooth comes from the Kaiyumung Member which is considered a temporal equivalent to the Lonyumun and Lomekwi Members of West Turkana, and is just younger than 4.0 Ma. Therefore, this is the oldest known specimen of the genus.

Nkondo Formation, Uganda

Within the western rift valley, Miocene, Pliocene, and Pleistocene sediments straddling the Uganda–Zaire border have produced a few monkey fossils. The oldest are from the Nkondo Formation in Uganda, where two colobine M3s have been recovered (Senut, 1994). They have been assigned to cf. *Paracolobus*, and are dated between 6.5 and 6.2 Ma based on comparison with fauna from the Tugen Hills succession (Pickford *et al.*, 1993).

Ongoliba Beds, Democratic Republic of Congo (former Zaire)

This formation in the Western Rift Valley has produced only a single cercopithecoid M3, but it is possibly the earliest cercopithecine recorded in eastern Africa, and perhaps in the world. However, isolated teeth from Menacer (formerly Marceau), Algeria which have been estimated at 7.5–7.0 Ma, may be as old (Thomas and Petter, 1986). Hooijer (1963) designates Ongoliba as late Miocene in age, based mainly on fossil fauna, but it could be significantly older than that. Pickford *et al.* (1993) compare the Ongoliba sites faunally to the Nkondo Formation in Uganda, and on this basis suggest an age of about 6 Ma. Based on cusp height and the relative position of the hypoconulid, Hoojier (1963) assigned the monkey specimen to the subfamily Cercopithecinae without further discussion. Szalay and Delson (1979) tentatively referred it to the same ?*Macaca* species as at Menacer, and Delson (1980) attributes the specimen to the Papionini.

Lukeino Formation, Tugen Hills, Kenya

The Lukeino Formation overlies the Kabarnet Trachytes in the Tugen Hills sequence, and for most of its outcrop is overlain by the Kaparaina Basalts. Bracketing ages on these units are 6.3 to 6.2 Ma beneath, and 5.6 Ma above (Hill *et al.*, 1986). There are many sites within the formation, but despite a good fauna (Bishop *et al.*, 1971; Pickford, 1975, 1978b; Hill *et al.*, 1985, 1986; Hill, 1995) only a few monkeys have been collected. There are a few postcranial specimens, some of them dubious, fragmentary canine teeth, and a molar attributable to Cercopithecidae.

Adu-Asa Formation, Middle Awash, Ethiopia

Both colobines and cercopithecines have been recovered from members of the Adu–Asa Formation, which is latest Miocene to early Pliocene in age, and in any case older than 4.4 Ma (Kalb *et al.*, 1982a,b; Kalb, 1993). Isolated teeth of a small papionin, along with a mandible of an indeterminate colobine have been collected, in addition to a femur and a few teeth of a large colobine that the authors attribute to cf. *Paracolobus chemeroni*.

Chemeron Formation, Tugen Hills, Kenya

The Chemeron Formation of the Tugen Hills sequence overlies the Kaparaina Basalts dated at around 5.6 Ma. It is composed of sediments from this age onwards to near 1.6 Ma at the top, where it is overlain over much of its outcrop by the Ndau Trachymugearite (Hill *et al.*, 1986). There are many sites through the whole 4 Myr time span of the formation, a number of which contain monkeys. There are some isolated cercopithecine teeth from sites fairly low in the section, such as Kibingor (BPRP#1) and Sagatya (BPRP#78) which are probably around 4.5–4 Ma, and from Moisionin (BPRP#79) which is between 4 and 3.5 Ma.

The best specimen of the unusual, large-bodied colobines that begin to appear in the Pliocene of east Africa (Leakey, 1982) comes from higher in the Chemeron Formation. The fossil is a nearly complete skeleton, the type specimen of *Paracolobus chemeroni* Leakey, 1969 (Leakey, 1969; Birchette, 1982). It was found at a site in the Kapthurin River (BPRP#97; EAGRU JM90/91), dated by BPRP to about 3 Ma (Deino, pers. comm.). It is accompanied at that site by the type specimen, a skull, of *Papio baringensis* Leakey, 1969, and another mandible of *P. baringensis* comes from the same site (Leakey and Leakey, 1976). Eck and Jablonski (1984, 1987) believe these should be *Theropithecus baringensis*. Delson and Dean (1993) tentatively agree with this attribution, and Leakey (1993) accepts this specimen as an early representative in the *T. brumpti* lineage, placing it in a subspecies, *T. brumpti baringensis*. Jolly (pers. comm.) does not agree. Other specimens of *P. chemeroni* associated with *T. baringensis* have been collected recently from other Chemeron sites around this age or up to half a million years older (e.g. BPRP#134, BPRP#154). Additional material attributable to *T.*

baringensis is found at other sites (e.g. BPRP#136; BPRP#155; BPRP#160) ranging in age from about 3 to 2 Ma.

Sagantole Formation, Middle Awash, Ethiopia

This series of early Pliocene fossiliferous exposures overlying the Adu-Asa Formation (>4.4 Ma) has an overlying datable layer which is 3.65 Ma (Kalb *et al.*, 1982a; Kalb, 1993). At Aramis, the type locality of the hominid genus *Ardipithecus* (White *et al.*, 1994), the faunal list names three cercopithecoid taxa. One colobine species, cf. sp. "A" accounts for over 30% of all vertebrate fossils collected. In addition, cf. *Parapapio* sp. is recorded and one or two specimens of cf. *Paracolobus* sp. may be present (Delson, pers. comm.). The Aramis locality has been estimated at 4.387 ± 0.031 Ma, based on Ar–Ar dating of a tuff underlying the fossiliferous sediments (WoldeGabriel *et al.*, 1994). Aramis has been reconstructed as at least partially forested based on the abundance of colobine monkeys, the recovery of certain fossil flora and the presence of antelopes which today inhabit wooded environments. However, colobine cf. sp. "A" may be a semi-terrestrial monkey, perhaps indicating a more open environment (Delson, pers.comm). Other sites within this formation have also produced monkeys, including a crushed cranium of *Parapapio* sp., a mandible of *Theropithecus oswaldi* cf. *darti*, and some isolated teeth of a small papionin (Kalb *et al.*, 1982b).

Kanapoi, Kenya

Kanapoi, which has produced part of the hypodigm for the hominid species *Australopithecus anamensis*, is a site to the southwest of Lake Turkana, which was first surveyed in the 1960s (Patterson *et al.*, 1970). Patterson (1968) reports *Parapapio jonesi* from Kanapoi, and the site has also yielded specimens of *Parapapio* aff. *ado* and a new, as yet unnamed, colobine genus (Leakey *et al.*, 1995). A radiometric date of 4.17 ± 0.02 Ma is given for a tuff underlying the fossils, and a second date of 3.41 ± 0.04 is given for a capping basalt layer.

Koobi Fora Formation, Kenya

Comprised of sediments exposed along the Omo River in southern Ethiopia, together with those to the east and west of Lake Turkana in northern Kenya, the Omo Group provides a continuous record of faunal change over the last four million years in a variety of ecological settings (Coppens *et al.*, 1976; Feibel *et al.*, 1991).

Abundant fossil sites to the east of Lake Turkana in Kenya, occurring in the Koobi Fora Formation provide some of the earliest evidence for modern OWM genera that still live in the area today (Feibel *et al.*, 1991), as well as three genera of extinct large colobine monkeys (Leakey, 1982). The oldest specimens of *Theropithecus brumpti* (along with those from Usno) were recovered within the Lokochot member, dated to about 3.5 Ma, and *T. oswaldi* appears around 2.4 to 2.0 Ma (Leakey, 1993). This species is the most common throughout the sequence, becoming extinct only 500 to 700 Ka. The oldest *Papio* fossils are younger than the most recent *Theropithecus* specimens. This supports the theory that *Papio* locally supplanted *Theropithecus* during the middle to late Pleistocene, the latter being restricted today to the Ethiopian highlands. *Parapapio* is found within the upper Burgi Member. A small *Cercopithecus* mandible with M2,3 was recovered from deposits dated about 2.6 Ma and *Cercocebus* does not appear at Koobi Fora until 1.6 Ma (Leakey, 1988).

Colobines are also well represented at Koobi Fora. Leakey (1982) reports *Rhinocolobus turkanaensis*, *Paracolobus mutiwa* and *Cercopithecoides williamsi* from sediments below the KBS tuff (1.9 Ma), and also present in this interval is *Cercopithecoides kimeui*, which persists until about 1.5 Ma. The earliest evidence for the genus *Colobus* is about 1.6 Ma at Koobi Fora (Leakey, 1988).

At Allia Bay, monkey fossils have been derived from within or just beneath the equivalent of the Moiti tuff in the Koobi Fora sequence. This tuff has been dated to 3.89 ± 0.02 Ma (Coffing *et al.*, 1994). The monkeys listed as part of the Allia Bay fauna are *Parapapio* sp. and Colobinae indet.

Nachukui Formation, Kenya

Since the early 1980s, survey and excavation to the west of Lake Turkana have produced many new fossil sites and filled in some of the gaps in the sequence at Koobi Fora, particularly the time range between 2.5 and 2.0 Ma. Harris *et al.* (1988) provide descriptions of the cercopithecoids from this formation. *Theropithecus brumpti* is found within levels dating between 3.4 and 2.5 Ma, and 10 specimens of *T. oswaldi* have been recovered from younger parts of the sequence. A partial mandible of *Parapapio ado* was found within the Lomekwi Member, as were two specimens of another taxon that may be *P.* cf. *whitei*. The Colobinae are represented by a partial skeleton of *Paracolobus mutiwa* from the upper Lomekwi member (about 3.0 Ma).

Usno Formation, Ethiopia

Sediments of the Usno Formation crop out along the Omo River to the northwest of the main Shungura Formation exposures, and span the same time range as from the Basal Member to Member C of the latter (Brown and Feibel, 1991). Leakey (1988) reports two upper molars of *Cercopithecus* sp. dated to 2.9 Ma, and Leakey (1987) describes specimens attributed to *Rhinocolobus turkanaensis* and *Paracolobus mutiwa* from the formation. Both Eck (1976) and Leakey (1993) report *Theropithecus brumpti*, including the type cranium of *T. quadratirostris*, which may, like *T. baringensis*, represent an early form in the *T. brumpti* lineage. Delson and Dean (1993), however, place *"T." quadratirostris* from Shungura and Usno within *Dinopithecus*, a subgenus of *Papio*.

Shungura Formation, Ethiopia

In Ethiopia, fossil assemblages recovered to the west of the Omo river drainage derive from a mostly fluvial context. Approximately 6500 cercopithecoid specimens, mostly isolated teeth, have been recovered from the Shungura and Usno Formations, although none are known from the older Mursi Formation (greater than 4 Ma). Cercopithecoid specimens are concentrated at the low end of the sequence, between about 3 and 2.4 Ma, after which the numbers drop dramatically. Eck (1977) suspects that this is due to the encroachment of a more arid environment, and uses faunal, sedimentologic, and palynologic evidence for support.

Leakey (1987) discusses four colobine species from the Shungura Formation. *Rhinocolobus turkanaensis*, *Paracolobus mutiwa* and the smaller Colobinae indet. (sp. "A") are known from members A through G, while four specimens of *Colobus* sp. have been collected in members K and L (1.54 Ma). Eck and Jablonski (1987) describe 32 specimens of *Theropithecus brumpti* from sites dated between about 2.9 and 2.0 Ma, and Eck (1987a) describes 25 of the numerous specimens of *T. oswaldi* and two fossils of what may be *T. darti* from deposits dated between 2.4 and 2.0 Ma. The genus *Cercopithecus*, of which very few definitive fossils are known, has been recovered from member B (2.95 Ma), member G (2.32 Ma) and from deposits dated at 1.45 Ma (Eck 1977, 1987b; Leakey, 1988). Finally, Eck (1977) mentions the occurrence of *Parapapio* while Delson (1984) gives a range for this genus of upper B to member K and perhaps L. Eck (1977) also mentions *Papio* in the Shungura Formation, and Delson and Dean (1993) illustrate *Papio (Dinopithecus) quadratirostris* from upper B through lower G.

Laetolil Beds, Laetoli, Tanzania

Laetoli is best known for the preservation of fossil footprints within a volcanic ash, including those of hominids and cercopithecoids, but the fauna includes a large sample of cercopithecine and colobine specimens, mostly teeth and jaws. These are mainly attributed to *Parapapio ado* and cf. *Paracolobus* sp. (Leakey and Delson, 1987), although two much rarer species are also known. One is a smaller colobine represented by four teeth and possibly a partial femur, and the other is a large papionin, of which a dP4 and a distal humerus were collected. The latter is listed as cf. *Papio* sp.

The majority of the fossils from Laetoli are derived from the upper Laetolil Beds, which have been radiometrically dated between 3.76 and 3.46 Ma (Drake and Curtis, 1987). To date, cercopithecoid fossils have not been found in the lower beds. No large standing body of water seems to have been present during the period in which the fossils were accumulating, and there is a complete lack of water-dwelling forms such as hippos and crocodiles (Harris, 1987). This has implications for the type of vegetation supported there, and hence the vertebrate fauna. The Pliocene environment has been likened to that of the modern Serengeti, although Andrews (1989) concludes that significant tree cover existed based on overall community structure of the fauna.

Warwire Formation, Uganda

Overlying the Nkondo Formation in the Western Rift Valley of Uganda is the Warwire Formation, dated to about 3.6 Ma (Pickford *et al.*, 1993). A colobine right distal humerus and two, perhaps three, parts of a papionin humerus are described by Senut (1994), who assigns the former to cf. *Paracolobus* and the latter to cf. *Parapapio*.

Hadar Formation, Ethiopia

Noted for its abundant remains of *Australopithecus afarensis*, the Hadar region within the Afar depression of Ethiopia has also produced a great number of cercopithecoid fossils. Many of these specimens are classified as *Theropithecus* cf. *darti*, making Hadar and the Middle Awash the only sites that record this species outside of South Africa (Eck, 1993, but see the Shungura Formation). Delson (1984) mentions partial crania of *Parapapio* cf. *jonesi* from the Kada Hadar member. There are some colobine fossils as well, which White (1984) lists as *Rhinocolobus turkanaensis*, and Delson

(1994) mentions four jaws and a partial humerus of this species from the Sidi Hakoma and Denen Dora members. He also notes a mandible and humerus of colobine sp. "A" from the Sidi Hakoma member and a cranium and partial skeleton of the same species from nearby Leadu, which is poorly dated.

Dating has been somewhat problematic at Hadar, but recent 40Ar–39Ar dates and tephra-correlation with the Turkana Basin have resolved some of the major issues. It now appears that the main fossiliferous sediments are bracketed by the Sidi Hakoma tuff (SHT) at 3.4 ± 0.03 Ma and the BKT-2 tuff, which is 2.9 ± 0.08Ma (Walter and Aronson, 1993; Brown, 1995). The Hadar region is usually described as swamp-like, perhaps with more arid regions in the distance.

Ndolanya Beds, Laetoli, Tanzania

The Ndolanya Beds overlie the Laetolil Beds at Laetoli, and they are bracketed radiometrically between 3.46 and 2.41 Ma (Drake and Curtis, 1987). A few isolated cercopithecoid teeth derive from locality 7E in the upper part of the Ndolanya Beds, which are perhaps 2.6 or 2.5 Ma in age, based on associated fauna (Hay, 1987). A worn M3 could be from *Parapapio ado*, and an I1 is cf. *Paracolobus* sp. (Leakey and Delson, 1987).

Chiwondo Beds, Malawi

In Malawi, towards the southernmost end of the Great Rift Valley, a series of fossiliferous exposures were first described in the 1920s. These are broadly grouped as the Chiwondo Beds of Pliocene age, and the overlying Chitimwe Beds, which are of Pleistocene age. Later, collections made by J.D. Clark's expeditions in the 1960s included a cercopithecoid molar, incisor, and a few postcranial fragments, which were assigned to *Papio*. More recent collections include three additional monkeys, assigned to *Theropithecus* sp. and *Parapapio* sp. (Bromage and Schrenk, 1986; Bromage *et al.*, 1995).

Although no radiometric dates are given for these exposures, they are important in providing a geographic link between east African and southern African faunal assemblages. Based on biostratigraphic comparisons, Kaufulu *et al.* (1981) identify two time levels: one is early to middle Pliocene; and the other is a late Pliocene age for more northern localities. Further work by Bromage *et al.* (1995) suggests three moderately distinct age groups. One is early Pliocene, greater than 4.0 Ma, a second is between

3.76 and 2.0 Ma, while the youngest specimens are listed as less than 1.6 Ma. Unfortunately, no exact stratigraphic position of the monkeys is given, and there are additional general problems with some of the biostratigraphy (Hill, 1995).

Reconstruction of community structure or paleoenvironments is hampered by the fragmentary and reworked nature of the assemblages (Schrenk *et al.*, 1995). Bovids, which occupy a variety of habitats are found throughout the exposures, suggesting a wide range of available biomes that the monkeys could have utilized.

Fejej Formation, Ethiopia

To the east of the Omo River, the Fejej Formation in Ethiopia, although only recently surveyed, has produced fossils dated from the early Pliocene to the early Pleistocene (Asfaw *et al.*, 1991, Kappelman *et al.*, 1996). Like the Omo Shungura deposits, these deposits are largely fluviatile in nature, and a few cercopithecine remains have been collected from two sites (Asfaw *et al.*, 1991). At the FJ5 locality, a "partial papionin skeleton" has been recovered. A tuff overlying the sediments at this locality is chemically very similar to the Orange Tuff found elsewhere in the Omo group and which dates between 1.87 and 1.65 Ma. Other fauna from this site corroborates this suggestion. Sediments from locality FJ1 underlie a tuff corresponding to the KBS tuff and the Shungura H-2 tuff which are well-dated at 1.88 Ma. The only monkey known from this site is *Theropithecus oswaldi*.

Kanam, Kenya

On the southern shores of the Winam Gulf of Lake Victoria, there are numerous outcrops of Plio-Pleistocene sediments. The Kanam exposures lie to the west of the somewhat younger Kanjera sediments (see below). Harrison and Harris (1996) have recently described specimens originally collected by Louis Leakey in the 1930s from the site of Kanam East. They attribute some of the material to the genera *Colobus*, *Lophocebus*, and *Cercopithecus*, but the remains are considered too fragmentary to be allocated to species. Since the provenience is unknown for these fossils, it is impossible to give reliable ages for them. The overall diversity of the assemblage suggests to the authors a forested or woodland habitat.

Matabaietu Formation, Middle Awash, Ethiopia

Four monkey taxa have been recovered from late Pliocene sediments of the Matabaietu Formation, which contain a fauna similar to the upper Hadar Formation (Kalb *et al.*, 1982a,b; Kalb, 1993). There are two cercopithecine taxa represented; the maxillary dentition of a female *Dinopithecus* cf. *ingens* and many crania, teeth, and postcrania of *Theropithecus oswaldi oswaldi*. The latter is very similar to the form from Olduvai Bed I, the lower units at Koobi Fora and especially Kanjera. Two colobines are also listed, a mandible of *Paracolobus chemeroni*, and a humerus of an indeterminate species.

Kaiso Formation, Kaiso Village, Uganda

A right maxillary fragment containing M2,3 attributed to *Theropithecus oswaldi* is listed by Cooke and Coryndon (1970) from the Kaiso Formation in Uganda. The specimen was collected at Kaiso Village, which is considered from the Later Kaiso Formation and dated to about 2 Ma.

Lusso Beds, Upper Semliki, Democratic Republic of Congo (former Zaire)

On the D.R. Congo side of the western rift valley, two monkeys have been collected from the Lusso Beds (Boaz, 1990) that are approximately contemporary with the Later Kaiso Formation in Uganda (Pickford *et al.*, 1993). One species is *Theropithecus* sp., which is illustrated and described in Delson (1993), the other is Colobinae gen. et sp. indet.

Olduvai Beds, Tanzania

Olduvai Gorge in Tanzania is a conspicuous geological feature cut into the northern Serengeti Plains which is a result of both fluvial processes and local tectonic faulting (Hay, 1990). At the base of Bed I, the Naabi Ignimbrites are dated to approximately 2 Ma, overlain by a fairly continuous sequence up until at least the Brunhes/Matuyama boundary, around 780 Ka (Walter *et al.*, 1991; Tamrat *et al.*, 1995). During the early to middle Pleistocene, a lake existed in the Olduvai region that fluctuated with the prevailing regional climate and occasionally shifted its location and drainage pattern as a result of local faulting (Hay, 1990). Reconstruction of the paleoclimate throughout this period suggests that, compared with present conditions, more humid and more arid periods occurred at various times (Hay, 1990).

Few colobines are known from Olduvai, although it is the type locality of *Cercopithecoides kimeui* from Beds II and III (Leakey, 1982). Delson (1984) also refers some specimens from Beds I–III to cf. *Colobus*. Papionins are well represented by many specimens of *Theropithecus oswaldi*, as well as *Cercocebus* sp. (Leakey and Leakey, 1973, 1976). Delson (1984) lists *Parapapio* from Bed I and possibly younger sediments, and possibly Papio from Bed III and above.

Kanjera Formation, Kenya

Monkeys were first discovered at Kanjera in 1911, including what was subsequently named the type specimen of *Simopithecus oswaldi* Andrews, 1916, later assigned to *Theropithecus*. Recently, more specimens of *Theropithecus oswaldi* have been recovered (Plummer and Potts, 1989), bringing the total to 85 identifiable parts from at least 13 individuals (Behrensmeyer *et al.*, 1995). No other species have been recorded from this location. Field notes from previous expeditions, along with chemical proveniencing using the color and matrix adhering to the fossils, indicate that all of the monkeys have come from the early Pleistocene KN-2a level. (Plummer *et al.*, 1994).

These more recent excavations at Kanjera have also led to a refined understanding of the age and depositional environments in that region (Behrensmeyer *et al.*, 1995). Using local paleomagnetic stratigraphy, and constrained by the presence of certain marker fossils (three pig species and one deinothere), the Kanjera Formation is estimated to span 1.76 to about 0.5 Ma. KN-2a deposits range from 1.76 to 1.1 Ma, although a maximum of about 1.3 Ma is more likely.

Chemoigut Formation, Chesowanja, Kenya

Several early Pleistocene sites crop out in the Chesowanja area, on the eastern side of Lake Baringo, Kenya. A rich fauna was collected in the region, which includes remains of *Theropithecus* sp.(Carney *et al.*, 1971; Bishop *et al.*, 1975, 1978). The Chemoigut Formation is probably a little older than 1.4 Ma.

Olorgesailie Formation, Kenya

Dense concentrations of ancient stone tools first led to excavations at Olorgesailie carried out by the Leakeys in the 1940s. Excavations by them and others led to the recovery of abundant remains of a large form of *Theropithecus*, *T. oswaldi leakeyi*, the only known cercopithecoid from this site (Jolly, 1972). The high number of specimens, age profile and modification to the bones have suggested to some researchers that early humans may have hunted the monkeys (Shipman *et al.*, 1981).

New radiometric dates indicate that the deposits cover a greater temporal span than originally thought. ^{40}Ar–^{39}Ar dates throughout the 14 member sequence range from 0.99 to 0.49 Ma (Deino and Potts, 1990). A lake existed in the region for much of its depositional history, as revealed by thick diatomaceous layers. Sediments more characteristic of fluvial activity are also present within several of the members.

Wehaietu Formation, Middle Awash, Ethiopia

The youngest sediments in the Middle Awash region, which are dated from the middle Pleistocene to the present preserve an essentially modern monkey fauna, with the exception of the continued presence of *Theropithecus oswaldi* cf. *leakeyi* (Kalb *et al.*, 1982a,b; Kalb, 1993). A skull of this species was recovered along with evidence for the extant species *Papio* cf. *hamadryas*, *Cercopithecus* sp. and *Colobus* cf. *guereza*.

Discussion and conclusion

Ultimately, resolution of questions concerning evolutionary patterns depend upon the nature of the fossil record for the group of organisms under investigation. Ideally, abundant, well-preserved fossils are recovered in contexts that lend themselves to accurate dating and whose taphonomic history does not preclude attempts at reconstructing the local environment at the time of deposition. Unfortunately, terrestrial depositional settings rarely provide such ideal circumstances. Typically, isolated skeletal elements are found scattered within water or wind-worked sediments, having been transported sometimes great distances from their original habitat.

In east Africa we are fortunate to be provided with many fossil-bearing exposures as a result of persistent tectonic activity throughout the Neogene. This activity resulted in the exposure of ancient sediments and allows radiometric analysis of the many volcanic tuffs that intercalate the fossil bearing strata. The importance of this cannot be overstated, as accurate temporal placement of fossils is an essential first step to discussing broader evolutionary issues such as reconstructing phylogeny or correlating appearances or extinctions with global and local climatic events.

This review of the cercopithecoid fossil record in eastern Africa emphasizes a number of features about their distribution and diversity. It is convenient to break up the cercopithecoid fossil record into three broad time spans, starting with the first appearance and dispersal of primitive victoriapithecids during the early and middle Miocene. This is followed by the apparent extinction of these forms and the appearance of the modern subfamilies during the late Miocene and earliest Pliocene, a time period for which we have very little fossil evidence from east Africa. Finally, from the early Pliocene until the present, the fossil record is drastically improved, and OWM appear to radiate taxonomically and ecologically (Table 7.1).

During the early Miocene, OWM represent a minor component of a very diverse catarrhine fauna. Latest Oligocene and earliest Miocene sites in Kenya, such as Meswa Bridge, Lothidok and Bukwa, have failed to produce any monkey fossils although several catarrhine species have been recovered. Due to poor sample sizes at these locations, however, it is difficult to assert with any real confidence that monkeys had not yet evolved. The oldest fossil monkey in east Africa, and perhaps the world, appears at Napak, Uganda, dated possibly as old as 19.5 Ma. A few fragmentary mandibles are known from north Africa that may be a bit younger. Unlike in the earliest Miocene, this paucity seems to have real biological meaning since some contemporary well-sampled locations, such as Rusinga Island and Songhor in Kenya, have produced many fossil catarrhines, but not a single monkey.

Throughout the early and middle Miocene, cercopithecoids are not diverse, being confined to one, possibly two genera. Although present at several late early Miocene and middle Miocene sites, only Maboko Island, and to a lesser degree Buluk, have yielded a large number of specimens. At younger middle Miocene sites, monkeys remain very sparse in the record. After about 15 Ma there is an almost complete gap in the record lasting until about 9.5 to 9 Ma when the first colobine is known from the Ngorora Formation. When identifiable to species, all victoriapithecids belong to *Victoriapithecus macinnesi*, with the exception of the new species at site BPRP #38 in the Tugen Hills succession. Victoriapithecids last appear in the fossil record at this site, dated to 12.5 Ma.

Almost nothing is known about catarrhine evolution during the late Miocene in east Africa. What we can say is that the colobines and, less certainly, the cercopithecines had evolved by, or during, this time. There are no victoriapithecids known from the late Miocene, and they may well have become extinct. Colobines appear before cercopithecines, in keeping with the general pattern which is found in the late Miocene and early Pliocene

Table 7.1. *Chronological summary of locations and cercopithecoid taxa discussed*

Geographical location	Formation	Age range (Ma)	Victoriapithecidae	Colobinae	Cercopithecinae	Other
Middle Awash, Ethiopia	Wehaietu	Mid Pleist–Present		*Colobus*	*Theropithecus*	
					Papio	
					Cercopithecus	
Olorgesailie, Kenya	Olorgesailie	0.99–0.49			*Theropithecus*	
Chesowanja, Kenya	Chemoigut	>1.4			*Theropithecus*	
Kanjera, Kenya	Kanjera	1.76–0.5			*Theropithecus*	
Olduvai Gorge, Tanzania	N/A	1.88–present		*Cercopithecoides*	*Theropithecus*	
				cf. *Colobus*	*Cercocebus*	
					?*Papio* or *Parapapio*	
Lusso, Zaire	N/A	2		Colobinae indet.	*Theropithecus*	
Kaiso Village, Uganda	Kaiso	2			*Theropithecus*	
Middle Awash, Ethiopia	Matabaietu	Late Pliocene		*Paracolobus*	*Theropithecus*	
				Colobinae indet.	*Dinopithecus*	
Kanam, Kenya	N/A	Pliocene		*Colobus*,	*Cercopithecus*	
					Lophocebus	
Fejej, Ethiopia	Fejej	early Pliocene–Pleistocene			Papionini indet.	
					Theropithecus	
Chiwondo, Malawi	N/A	>4.0–<1.6			*Theropithecus*	
					Parapapio	
					Papio	
Laetoli, Tanzania	Ndolanya	3.46–2.41		cf. *Paracolobus*	?*Parapapio*	
Hadar, Ethiopia	Hadar	3.4–2.9		*Rhinocolobus*	*Theropithecus*	
				Colobinae sp. “A”	*Parapapio*	
Warwire, Uganda	Warwire	3.6		cf. *Paracolobus*	cf. *Parapapio*	
Laetoli, Tanzania	N/A	3.76–3.46		Colobinae indet.	*Parapapio*	
				cf. *Paracolobus*	cf. *Papio*	

Table 7.1. (*cont.*)

Geographical location	Formation	Age range (Ma)	Victoriapithecidae	Colobinae	Cercopithecinae	Other
Omo, Ethiopia	Shungura	3.89–1.0		*Rhinocolobus* *Paracolobus* Colobinae sp. "A" *Colobus*	*Theropithecus* *Parapapio* *Papio* (*Dinopithecus*) *Cercopithecus*	
Omo, Ethiopia	Usno	>3.89–3.0		*Rhinocolobus* *Paracolobus*	*Theropithecus* *Papio* (*Dinopithecus*) *Cercopithecus*	
West Turkana, Kenya	Nachukui	>3.89–1.0		*Paracolobus*	*Theropithecus* *Parapapio*	
East Turkana, Kenya	Koobi Fora	>3.89–0.4		*Rhinocolobus* *Paracolobus* *Cercopithecoides* *Colobus* Colobinae indet.	*Theropithecus* *Parapapio* *Papio* *Cercopithecus* *Cercocebus*	
Kanapoi, Kenya	Kanapoi	4.17–3.41		Colobinae gen. nov.	*Parapapio*	
Middle Awash, Ethiopia	Sagantole	4.4–3.65		Colobinae sp. "A" cf. *Paracolobus*	Papionini indet. *Theropithecus* cf. *Parapapio*	
Tugen Hills, Kenya	Chemeron	5.6–1.6		*Paracolobus*	*Theropithecus* Cercopithecinae indet.	
Middle Awash, Ethiopia	Adu Asa	>4.4		Colobinae indet. cf. *Paracolobus*	*Papionini* indet.	
Tugen Hills, Kenya	Lukeino	6.3/6.2–5.6			Cercopithecidae indet.	

Ongoliba, Zaire	N/A	6			Cercopithecinae indet.	
Nkondo-Kaiso, Uganda	Nkondo	6.5–6.2			cf. *Paracolobus*	
Lothagam, Kenya	Nachukui Nawata	7.9–4.0		Colobinae indet. (2 spp.)	cf. *Parapapio* cf. *Cercocebus*	
Tugen Hills, Kenya	Mpesida	7.5/7.0–6.3/6.2		Colobinae indet.		
Nakali, Kenya	N/A	9.0–7.0				Cercopithecidae indet.
Tugen Hills, Kenya	Ngorora	12.5–9.0	*Victoriapithecus*	*Microcolobus*		
Nyakach, Kenya	Nyakach	13.4	*Victoriapithecus*			
Nachola, Kenya	Nachola	15; 11.8–10.1				Cercopithecoidea indet.
Samburu Hills, Kenya	Aka Aiteputh	15–11.5	*Victoriapithecus*			
Maboko Island, Kenya	Maboko	>14.71	*Victoriapithecus*			
Tugen Hills, Kenya	Muruyur	15.5–15.4	*Victoriapithecus*			
Loperot, Kenya	N/A	18–16.2;17.1;>15	*Victoriapithecus*			
Gebel Zelten, Libya	N/A	17–15	*Prohylobates*			
Buluk, Kenya	Bakate	>17.2	*Prohylobates*			
Wadi Moghara, Egypt	Moghara	18–17	*Prohylobates*			
Napak, Uganda	N/A	19.5; 18.3	*Victoriapithecus*			

Notes:
For details of stratigraphic placement, taxonomy and dating method(s) see text and references therein.

throughout Europe and Asia. *Microcolobus* is probably the earliest cercopithecid in the world, and the Nakali monkeys are of about the same age. Other species of colobine are known by 6.5 Ma from the Mpesida Beds of the Tugen Hills, at Lothagam, and specimens from the Nkondo Formation in Uganda may be as old. The first recorded cercopithecines in east Africa come from Ongoliba, D.R. Congo, with an estimated faunal date of about 6 Ma, and also from the upper Nawata Formation at Lothagam. These fossils, although sparse, signal an increase in diversity by the latest Miocene, with both colobines and cercopithecines present in the east African record. The earliest levels at which more than one species is present at the same site occur near the Pliocene boundary, in the Chemeron Formation of the Tugen Hills, and the Adu–Asa Formation in the Middle Awash of Ethiopia.

With the benefit of a much improved fossil record in the later Pliocene and Pleistocene, it is clear that monkeys had established themselves as the most common and diverse catarrhine group throughout the entire Old World. Many modern genera only appear in the latest Pliocene and Pleistocene, unlike the platyrrhine monkey pattern where specimens clearly related to modern forms appear in the Miocene. Monkeys are abundant, and display significant diversity, in sharp contrast to the complete absence from the record of non-bipedal apes. This transition presumably took place during the late Miocene, but the paucity of fossil evidence precludes any real attempts at pinpointing the timing or environmental backdrop within which it occurred.

Nonetheless, various attempts to explain the seeming disparate evolutionary patterns between monkeys and non-cercopithecoid catarrhines throughout the Neogene have been made (Napier, 1970; Simons, 1972; Andrews, 1982; Temerin and Cant, 1983; Pickford, 1987; Delson, 1994; McCrossin and Benefit, 1994). These explanations have often relied almost solely on data concerning modern catarrhine ecology, yet modern monkeys are so diverse and apes so rare that trying to reconstruct primitive diets, habitats, or other behaviors is problematic at best. We cannot rely on modern species' behavior and adaptations entirely, especially when extinct species display clear morphological differences from living relatives, particularly in the case of the Colobinae. Also, many of these hypotheses invoke widespread, often global climate change as a causal mechanism. However, the effects of such large scale climatic events on local environments are not well known.

The fossil record affords us a reasonably clear picture of catarrhine evolution during the early Miocene and from the Pliocene to the present. The very poorly sampled segment of the record from about 15 Ma to around 5 Ma makes it very difficult to assess the truth or likelihood of the above hypotheses (Hill and Gundling, unpub. data). The transition from a primate fauna dominated by numerous, taxonomically diverse "ape" species to one in which monkeys abound was undoubtedly complex. Models providing simple causal relationships involving presumed differences between these two groups within the context of a changing physical environment require further scrutiny.

Acknowledgements

The authors thank Paul Whitehead and Cliff Jolly for the invitation to contribute to this volume, and for providing helpful comments on an early draft of the manuscript. Thanks also go to Meave Leakey for overall comments and especially for unpublished information on the dating at Loperot. Rick Potts gave useful information on Kanjera, Kanam and Olorgesailie and Eric Delson provided suggestions on the final draft.

References

Aguirre, E. & Leakey, P. (1974). Nakali: nueva fauna de Hipparion del Rift Valley de Kenya. *Estudios geol. Inst. Invest. geol. Lucas Mallada* **30**, 219–27.

Aguirre, E. & Guerin, C. (1974). Première découverte d'un Iranotheriinae (Mammalia, Perissodactyla, Rhinocerotidae) en Afrique: *Kenyatherium bishopi* nov. gen. nov. sp. de la formation vallésienne (Miocène supérieur) de Nakali (Kenya). *Estudios geol. Inst. Invest. geol. Lucas Mallada* **30**, 229–33.

Andrews, C.W. (1916). Note on a new baboon (*Simopithecus oswaldi*, gen. et sp. n.) from the (?) Pliocene of British East Africa. *Ann. Mag. Nat. Hist.* **18**, 410–19.

Andrews, P. (1982). Ecological polarity in primate evolution. *Zool. J. Linn. Soc.* **74**, 233–44.

Andrews, P. (1989). Palaeoecology of Laetoli. *J. hum. Evol.* **18**, 173–81.

Andrews, P., Meyer, G., Pilbeam, D., Van Couvering, J.A. & Van Couvering, J.A.H. (1981). The Miocene fossil beds of Maboko Island, Kenya: geology, age, taphonomy and palaeontology. *J. hum. Evol.* **10**, 123–8.

Andrews, P., Harrison, T., Delson, E., Bernor, R.L. & Martin, L. (1996). Distribution and biochronology of European and Southwest Asian Miocene catarrhines. In *The Evolution of Western Eurasian Neogene Mammal Faunas*, ed. R.L. Bernor, V. Fahlbusch & H.W. Mittmann, pp. 168–206. New York: Columbia University Press.

Asfaw, B., Beyene, Y., Semaw, S., Suwa, G., White, T. & WoldeGabriel, G. (1991). Fejej: a new paleoanthropological research area in Ethiopia. *J. hum. Evol.* **21**, 137–43.

Baker, B.H., Williams, L.A.J., Miller, J.A. & Fitch, F.J. (1971). Sequence and geochronology of the Kenya Rift volcanics. *Tectonophysics* **11**, 191–215.

Behrensmeyer, A.K., Deino, A., Hill, A., Kingston, J. & Saunders, J. (1999). Geology of the Muruyur Formation at Kipsaramon, Tugen Hills, Kenya. (In press.)

Behrensmeyer, A.K., Potts, R., Plummer, T., Tauxe, L., Opdyke, N. & Jorstad, T. (1995). The Pleistocene locality of Kanjera, Western Kenya: stratigraphy, chronology and paleoenvironments. *J. hum. Evol.* **29**, 247–74.

Benefit, B.R. (1993).The permanent dentition and phylogenetic position of *Victoriapithecus* from Maboko Island, Kenya. *J. hum. Evol.* **25**, 83–172.

Benefit, B.R. (1994). Phylogenetic, paleodemographic, and taphonomic implications of *Victoriapithecus* deciduous teeth from Maboko, Kenya. *Am. J. phys. Anthrop.* **95**, 277–331.

Benefit, B.R. (1995). Earliest Old World Monkey skull. *Am. J. phys. Anthrop.* (Suppl.) **20**, 64. (Abstract.)

Benefit, B.R. & Pickford, M. (1986). Miocene fossil cercopithecoids from Kenya. *Am. J. phys. Anthrop.* **69**, 441–64.

Birchette, M.G. (1982). *The Postcranial Skeleton of* Paracolobus chemeroni. PhD Thesis, Harvard University.

Bishop, W.W. (1958a). The mammalian fauna and geomorphological relations of the Napak volcanics, Karamoja. *Geological Survey of Uganda Records*, 1957–1958, pp. 1–18.

Bishop, W.W. (1958b). Miocene mammalia from the Napak volcanics, Karamoja, Uganda. *Nature* **182**, 1480–2.

Bishop, W.W. (1963a). Uganda's animal ancestors. *Uganda Wildl. Sport* **3**(3).

Bishop, W.W. (1963b). The later Tertiary and Pleistocene in eastern equatorial Africa. In *African Ecology and Human Evolution*. ed. F.C. Howell & F. Bourlière, pp. 246–75. London: Methuen.

Bishop, W.W. (1964). More fossil primates and other Miocene mammals from north-east Uganda. *Nature* **203**, 1327–31.

Bishop, W.W. (1968). The evolution of fossil environments in east Africa. *Trans. Leicester lit. phil. Soc.* **62**, 22–44.

Bishop, W.W. (1971). The late Cenozoic history of east Africa in relation to hominoid evolution. In *The Late Cenozoic Glacial Ages*, ed. K. Turekian, pp. 493–527. New Haven: Yale University Press.

Bishop, W.W. & Chapman, G.R. (1970). Early Pliocene sediments and fossils from the northern Kenya Rift Valley. *Nature* **226**, 914–18.

Bishop, W.W., Chapman, G.R., Hill, A. & Miller, J.A. (1971). Succession of Cainozoic vertebrate assemblages from the northern Kenya Rift Valley. *Nature* **233**, 389–94.

Bishop, W.W., Hill, A. & Pickford, M.H.L. (1978). Chesowanja: a revised geological interpretation. In *Geological Background to Fossil Man*, ed. W.W. Bishop, pp. 309–27. Geological Society of London: Scottish Academic Press.

Bishop, W.W., Miller J.A. & Fitch F.J. (1969). New Potassium–Argon age determinations relevant to the Miocene fossil mammal sequence in East Africa. *Am. J. Sci.* **267**, 669–99.

Bishop, W.W. & Pickford, M.H.L. (1975). Geology, fauna and palaeoenvironments of the Ngorora Formation, Kenya Rift Valley. *Nature* **254**, 185–92.

Bishop, W.W., Pickford, M.H.L. & Hill, A. (1975). New evidence regarding the Quaternary geology, archaeology and hominids of Chesowanja, Kenya. *Nature* **58**, 204–8.

Bishop, W.W. & Whyte, F. (1962). Tertiary mammalian faunas and sediments in Karamoja and Kavirondo, east Africa. *Nature* **196**, 1283–7.

Boaz, N. (1990). The Semliki Research Expedition: history of investigation, results, and background to interpretation. In *Evolution of Environments and Hominidae in the African Western Rift Valley. Virginia Museum of Natural*

History Memoir No. 1., ed. N.T. Boaz, pp. 3–14. Martinsville: Virginia Museum of Natural History.

Boschetto, H.B., Brown, F.H. & McDougall, I. (1992). Stratigraphy of the Lothidok Range, northern Kenya, and K/Ar ages of its Miocene primates. *J. hum. Evol.* **22**, 47–71.

Bromage, T.G. & Schrenk, F. (1986). A cercopithecoid tooth from the Pliocene of Malawi. *J. hum. Evol.* **15**, 497–500.

Bromage, T.G., Schrenk F. & Juwayeyi Y.M. (1995). Paleobiogeography of the Malawi Rift: age and vertebrate paleontology of the Chiwondo Beds, northern Malawi. *J. hum. Evol.* **28**, 37–57.

Brown, F.H. (1995). The potential of the Turkana basin for paleoclimatic reconstruction in East Africa. In *Paleoclimate and Evolution with Emphasis on Human Origins*, ed. E.S.Vrba, G.H. Denton, T.C. Partridge & L.H. Burckle, pp. 319–30. New Haven: Yale University Press.

Brown, F.H. & Feibel, C.S. (1991). Stratigraphy, depositional environments and palaeogeography of the Koobi Fora Formation. In *Koobi Fora Research Project*, vol. 3, *The Fossil Ungulates: Geology, Fossil Artiodactyls, and Palaeoenvironments*, ed. J.M. Harris, pp. 1–30. Oxford: Clarendon Press.

Carney, J., Hill, A., Miller, J. & Walker, A. (1971). Late australopithecine from Baringo District, Kenya. *Nature* **230**, 509–14.

Coffing, K., Feibel, C., Leakey, M. & Walker, A. (1994). Four-million-year-old hominids from East Lake Turkana, Kenya. *Am. J. phys. Anthrop.* **93**, 55–65.

Cooke, B. & Coryndon, S. (1970). Pleistocene mammals from the Kaiso Formation and other related deposits in Uganda. In *Fossil Vertebrates of Africa*, vol. II. ed. L.S.B. Leakey & R.J.G. Savage, pp. 107–224. London: Academic Press.

Coppens, Y., Howell, F.C., Isaac, G.L. & Leakey, R.E.F. (1976). *Earliest Man and Environments in the Lake Rudolf Basin*. Chicago: University of Chicago Press.

Deino, A. & Potts, R. (1990). Single-crystal 40Ar/39Ar dating of the Olorgesailie Formation, southern Kenya Rift. *J. Geophysical Res.* **95**, 8453–70.

Deino, A., Tauxe, L., Monaghan, M. & Drake, R. (1990). 40Ar/39Ar age calibration of the litho- and paleomagnetic stratigraphies of the Ngorora Formation, Kenya. *J. Geol.* **98**, 567–87.

Delson, E. (1975). Evolutionary history of the Cercopithecidae. In *Approaches to Primate Paleobiology. Contrib. Primatol.* **5**, 167–217. Basel: Karger.

Delson, E. (1979). *Prohylobates* (Primates) from the Early Miocene of Libya: A new species and its implications for cercopithecoid origins. *Geobios* **12**, 725–33.

Delson, E. (1980). Fossil macaques, phyletic relationships and a scenario of deployment. In *The Macaques: Studies in Ecology, Behavior and Evolution*, ed D.G. Lindburg, pp. 10–30. New York: von Nostrand Reinhold.

Delson, E. (1984). Cercopithecid biochronology of the African Plio-Pleistocene: Correlation among eastern and southern hominid-bearing localities. *Cour. Forsch. Inst. Senckenberg* **69**, 199–218.

Delson, E. (1993). *Theropithecus* fossils from Africa and India and the taxonomy of the genus. In *Theropithecus: The Rise and Fall of a Primate Genus*, ed. N.G. Jablonski, pp. 157–89. Cambridge: Cambridge University Press.

Delson, E. (1994). Evolutionary history of the colobine monkeys in paleoenvironmental perspective. In *Colobine Monkeys: Their Ecology, Behavior and Evolution*. ed. A.G. Davies & J.F Oates, J.F., pp. 11–43. Cambridge: Cambridge University Press.

Delson, E. & Dean, D. (1993). Are *Papio baringensis* R. Leakey (1969), and *P.*

quadratirostris Iwamoto (1982), species of *Papio* or *Theropithecus*? In *Theropithecus: The Rise and Fall of a Primate Genus*, ed. N.G. Jablonski, pp. 125–56. Cambridge: Cambridge University Press.

Drake, R. & Curtis, G.H. (1987). K-Ar Geochronology of the Laetoli fossil localities. In *Laetoli: A Pliocene Site in Northern Tanzania*, ed. M.G. Leakey & J.M. Harris, pp. 48–52. Oxford: Clarendon Press.

Eck, G.G. (1976). Cercopithecoidea from Omo Group deposits. In *Earliest Man and Environments in the Lake Rudolf Basin*, ed. Y. Coppens, F.C. Howell, G.Ll. Isaac & R.E.F. Leakey, pp. 332–44. Chicago: University of Chicago Press.

Eck, G. (1977). Diversity and frequency distribution of Omo Group Cercopithecoidea. *J. hum. Evol.* **6**, 55–63.

Eck, G. (1987a). *Theropithecus oswaldi* from the Shungura Formation, Lower Omo Basin, Southwestern Ethiopia. In *Les Faunes Plio-Pléistocènes de la Vallée de L'Omo (Ethiopie)*, Tome 3 *Cercopithecidae de la formation de Shungura*, ed. Y. Coppens & F.C. Howell, pp. 123–41. Paris: Editions du Centre National de la Recherche Scientifique.

Eck, G. (1987b). Plio-Pleistocene specimens of *Cercopithecus* from the Shungura Formation, Southwestern Ethiopia. In *Les Faunes Plio-Pléistocènes de la Vallée de L'Omo (Ethiopie)*, Tome 3 *Cercopithecidae de la formation de Shungura*, ed. Y. Coppens and F.C. Howell, pp. 142–7. Paris: Editions du Centre National de la Recherche Scientifique.

Eck, G. (1993). *Theropithecus darti* from the Hadar Foramtion, Ethiopia. In *Theropithecus: The Rise and Fall of a Primate Genus*, ed. N.G. Jablonski, pp. 15–83. Cambridge: Cambridge University Press.

Eck, G. & Jablonski, N. (1984). A reassessment of the taxonomic status and phyletic relationships of *Papio baringensis* and *Papio quadratirostris* (Primates, Cercopithecidae). *Am. J. phys. Anthrop.* **65**, 109–34.

Eck, G. & Jablonski, N. (1987). The skull of *Theropithecus brumpti* compared with those of other species of the genus *Theropithecus*. In *Les Faunes Plio-Pléistocènes de la Vallée de L'Omo (Ethiopie)*, Tome 3 *Cercopithecidae de la formation de Shungura*, ed. Y. Coppens & F.C. Howell, pp. 12–122. Paris: Editions du Centre National de la Recherche Scientifique.

Feibel, C.S. & Brown, F.H. (1991). Age of the primate-bearing deposits on Maboko Island, Kenya. *J. hum. Evol.* **21**, 221–5.

Feibel, C.S., Harris, J.M. & Brown, F.H. (1991). Palaeoenvironmental context for the Late Neogene of the Turkana Basin. In *Koobi Fora Research Project*, vol. 3, *The Fossil Ungulates: Geology, Fossil Artiodactyls, and Palaeoenvironments*, ed. J.M. Harris, pp. 321–46. Oxford: Clarendon Press.

Fleagle, J.G. & Simons, E.L. (1978). *Micropithecus clarki*, a small ape from the Miocene of Uganda. *Am. J. phys. Anthrop.* **49**, 427–40.

Fourtau, R. (1918). Contribution a l'étude des vertèbrés miocènes de l'Egypte. Survey Department: Ministry of Finance (Government Press Cairo), pp. 1–121.

Harris, J. M. (1987). Summary. In *Laetoli: A Pliocene Site in Northern Tanzania*. ed. M.G. Leakey and J.M. Harris, pp. 524–31. Oxford: Clarendon Press.

Harris, J.M., Brown, F.H. & Leakey, M.G. (1988). Stratigraphy and paleontology of Pliocene and Pleistocene Localities West of Lake Turkana, Kenya. *Contr. Sci.* **399**, 1–128.

Harris, J.M. & Watkins, R. (1974). New early Miocene vertebrate locality near Lake Rudolf, Kenya. *Nature* **252**, 576–7.

Harrison, T. (1989). New postcranial remains of *Victoriapithecus* from the middle Miocene of Kenya. *J. hum. Evol.* **18**, 3–54.

Harrison, T. & Harris, E.E. (1996). Plio-Pleistocene cercopithecids from Kanam East, western Kenya. *J. hum. Evol.* **30**, 539–61.
Hay, R.L. (1987). Geology of the Laetoli Area. In *Laetoli: A Pliocene Site in Northern Tanzania*. ed. M.G. Leakey & J.M. Harris, pp. 23–47. Oxford: Clarendon Press.
Hay, R.L. (1990). Olduvai Gorge: a case history in the interpretation of hominid paleoenvironments in East Africa. In *Establishment of a Geologic Framework for Paleoanthropology*, ed. L.F. Laporte, pp. 23–37. *Geol. Soc. Am. Spec. Pap.* No. 242. Boulder, CO: Geological Society of America.
Hill, A. (1987). Causes of perceived faunal change in the Neogene of east Africa. *J. hum. Evol.* **16**, 583–96.
Hill, A. (1995). Faunal and environmental change in the Neogene of east Africa: evidence from the Tugen Hills sequence, Baringo District, Kenya. In *Paleoclimate and Evolution, with Emphasis on Human Origins*, ed. E.S. Vrba, G.H. Denton, T.C. Partridge & L.H. Burkle, pp. 178–93. New Haven: Yale University Press.
Hill, A., Behrensmeyer, A.K., Brown, B., Deino, A., Rose, M., Saunders, J., Ward, S. & Winkler, A. (1991). Kipsaramon: a lower Miocene hominoid site in the Tugen Hills, Baringo District, Kenya. *J. hum. Evol.* **20**, 67–75.
Hill, A., Curtis, G. & Drake, R. (1986). Sedimentary stratigraphy of the Tugen Hills, Baringo District, Kenya. In *Sedimentation in the African Rifts*, ed. L. Frostick, R.W. Renaut, I. Reid and J.J. Tiercelin, pp. 285–95. Geological Society of London Special Publications No. 25. Oxford: Blackwell.
Hill, A., Drake, R., Tauxe, L., Monaghan, M., Barry, J.C., Behrensmeyer, A.K., Curtis, G., Fine Jacobs, B., Jacobs, L., Johnson, J. & Pilbeam, D. (1985). Neogene palaeontology and geochronology of the Baringo Basin, Kenya. *J. hum. Evol.* **14**, 759–73.
Hill, A., Kingson, J.& Deino, A. (1999a). A colobine and new dates from the Mpesida Beds, Tugen Hills, Kenya. *J. hum. Evol.* (In press.)
Hill, A., Leakey, M., Kingston, J. & Ward, S. (1999b). New cercopithecoids and a hominoid from 12.5 Ma in the Tugen Hills succession, Kenya. *J. hum. Evol.* (In press.)
Hill, A., Ward, S. & Brown, B. (1992). Anatomy and age of the Lothagam mandible. *J. hum. Evol.* **22**, 439–51.
Hoojier, D.A. (1963). Miocene Mammalia of the Congo. *Ann. Mus. Roy. Afr. Cent., Ser. 8, Sci. Geol.* **46**, 1–71
Ishida, H. (1984). Outline of the 1982 survey in Samburu Hills and Nachola area, northern Kenya. *Afr. Stud. Monogr. (Suppl.)* **2**, 1–14.
Jacobs, B. & Deino, A. (1996). Test of climate-leaf physiognomy regression models, their application to two Miocene floras from Kenya, and 40Ar/39Ar dating of the Late Miocene Kapturo site. *Palaeogeogr. Palaeoclimatol. Palaeoecol.* **123**, 259–71.
Jacobs, B. & Kabuye, C. (1987). A middle Miocene (12.2 my old) forest in the East African Rift Valley, Kenya. *J. hum. Evol.* **16**, 147–55.
Jacobs, B. & Winkler, A. (1992). Taphonomy of a middle Miocene autochthonous forest assemblage, Ngorora Formation, central Kenya. *Palaeogeogr. Palaeoclimatol. Palaeoecol.* **99**, 31–40.
Jolly, C.J. (1972). The classification and natural history of *Theropithecus* (*Simopithecus*) (Andrews, 1916), baboons of the African Plio-Pleistocene. *Bull. Brit. Mus. (Nat. Hist.), Geol.* **22**, 1–122.
Kalb, J.E. (1993). Refined stratigraphy of the hominid-bearing Awash Group, Middle Awash Valley, Afar Depression, Ethiopia. *Newsl. Stratigr.* **29**, 21–62
Kalb, J.E., Oswald, E.B., Tebedge, S., Mebrate, A., Tola, E. & Peak, D. (1982a).

Geology and stratigraphy of Neogene deposits, Middle Awash Valley, Ethiopia. *Nature* **298**, 17–25.
Kalb, J.E., Jolly, C.J., Tebedge, S., Mebrate, A., Smart, C., Oswald, E.B., Whitehead, P.F., Wood, C.B., Adefris, T. & Rawn-Schatzinger, V. (1982b). Vertebrate faunas from the Awash Group, Middle Awash Valley, Afar, Ethiopia. *J. Vertebr. Paleontol.* **2**, 237–58.
Kappelman, J., Swisher, C.C., Fleagle, J.G., Yirga, S., Bown, T.M. & Feseha, M. (1996). Age of *Australopithecus afarensis* from Fejej, Ethiopia. *J. hum. Evol.* **30**, 139–46.
Kaufulu, Z., Vrba, E. & White, T.D. (1981). Age of the Chiwondo Beds, Northern Malawi. *Ann. Trans. Mus.* **33**, 1–8.
King, B.C. (1949). The Napak area of southern Karamoja. *Geol. Surv. Uganda Mem.* **5**.
Leakey, M.G. (1982). Extinct large colobines from the Plio-Pleistocene of Africa. *Am. J. phys. Anthropol.* **58**, 153–72.
Leakey, M.G. (1985). Early Miocene Cercopithecoids from Buluk, Northern Kenya. *Folia primatol.* **44**, 1–14.
Leakey, M.G. (1987). Colobinae (Mammalia, Primates) from the Omo Valley, Ethiopia. In *Les Faunes Plio-Pléistocènes de la Vallée de L'Omo (Ethiopie)*, Tome 3 *Cercopithecidae de la formation de Shungura*, ed. Y. Coppens & F.C. Howell, pp. 148–69. Paris: Editions du Centre National de la Recherche Scientifique.
Leakey, M.G. (1988). Fossil evidence for the evolution of the guenons. In *A Primate Radiation: Evolutionary Biology of the African Guenons*, ed. A. Gautier-Hion, F. Bourlière, J.-P. Gautier & J. Kingdon, pp. 7–12. Cambridge: Cambridge University Press.
Leakey, M.G. (1993). Evolution of *Theropithecus* in the Turkana Basin. In *Theropithecus: The Rise and Fall of a Primate Genus*. ed. N.G. Jablonski, pp. 85–123. Cambridge: Cambridge University Press.
Leakey, M.G. & Delson, E. (1987). Fossil Cercopithecidae from the Laetolil Beds. In *Laetoli: A Pliocene Site in Northern Tanzania*, ed. M.G. Leakey & J.M. Harris, pp. 91–107. Oxford: Clarendon Press.
Leakey, M.G. & Leakey, R.E.F. (1973). Further evidence of *Simopithecus* (Mammalia, Primates) from Olduvai and Olorgesailie. In *Fossil Vertebrates of Africa*, vol. 3, ed. L.S.B. Leakey, R.J.G. Savage & S.C. Coryndon, pp. 101–20. New York: Academic Press.
Leakey, M.G. & Leakey, R.E.F. (1976). Further Cercopithecinae (Mammalia, Primates) from the Plio/Pleistocene of East Africa. In *Fossil Vertebrates of Africa*, vol. 4, ed. R.J.G. Savage and S.C. Coryndon, pp. 121–46. New York: Academic Press.
Leakey, M.G., Feibel, C.S., McDougall, I. & Walker, A. (1995). New four-million-year-old hominid species from Kanapoi and Allia Bay, Kenya. *Nature* **376**, 565–71.
Leakey, M.G., Feibel, C.S., Bernor, R.L., Harris, J.M., Cerling, T.E., Stewart, K.M., Storrs, G.W., Walker, A., Werdelin, L. & Winkler, A.J. (1996). Lothagam: a record of faunal change in the Late Miocene of East Africa. *J. Vertebr. Paleontol.* **16**, 556–70.
Leakey, R.E.F. (1969). New Cercopithecidae from the Chemeron Beds of Lake Baringo, Kenya. In *Fossil Vertebrates of Africa*, vol. 1, ed. L.S.B. Leakey, pp. 53–73. New York: Academic Press.
Matsuda, T., Torii, M., Koyaguchi, T. Makinouchi, T., Mitsushio, H. & Ishida, S. (1984). Fission track, K-Ar age determinations and palaeomagnetic

measurements of Miocene volcanic rocks in the western area of Baragoi, northern Kenya: ages of hominoids. *Afr. Stud. Monogr. (Suppl.)* **2**, 57–66.

Matsuda, T., Torii, M., Koyaguchi, T. Makinouchi, T., Mitsushio, H. & Ishida, S. (1986). Geochronology of Miocene hominoids east of the Kenya Rift Valley. In *Primate Evolution*, ed. J.G. Else & P.C. Lee, pp. 35–45. Cambridge: Cambridge University Press.

McCrossin, M. & Benefit, B.R. (1994). Maboko Island and the evolutionary history of Old World monkeys and apes. In *Integrative Paths to the Past*, ed. R.S. Corruccini & R.L. Ciochon, pp. 95–122. Englewood Cliffs: Prentice Hall.

McDougall, I. & Watkins R.T. (1985). Age of hominoid-bearing sequence at Buluk, northern Kenya. *Nature* **318**, 175–8.

Mead, J.G. (1975). A fossil beaked whale (Cetacea: Ziphiidae) from the Miocene of Kenya. *J. Paleont.* 49, 745–51.

Miller, E.R. & Simons E.L. (1996). Age of the first cercopithecoid, *Prohylobates tandyi*, Wadi Moghara, Egypt. *Am. J. phys. Anthropol. (Suppl.)* **22**, 168–9. (Abstract.)

Nakaya, H. (1994). Faunal change of Late Miocene Africa and Eurasia: Mammalian fauna from the Namurungule Formation, Samburu Hills, Northern Kenya. *Afr. Stud. Monogr. (Suppl.)* **20**, 1–112.

Napier, J.R. (1970). Paleoecology and catarrhine evolution. In *Old World Monkeys: Evolution, Systematics, and Behavior*, ed. J.R. Napier & P.H. Napier, pp. 53–95. New York: Academic Press.

Ness, G., Levi, S. & Crouch, R. (1980). Marine magnetic anomaly time scales for the Cenozoic and late Cretaceous: a precis, critique, and synthesis. *Rev. Geophys. Space Phys.* **18**, 753–70.

Patterson, B. (1968). *Parapapio jonesi* from Kanapoi, Kenya. *Breviora* **282**, 1–4.

Patterson, B., Behrensmeyer, A.K. & Sill, W.D. (1970). Geology and fauna of a new Pliocene locality in north-western Kenya. *Nature* **226**, 918–21.

Pickford, M.H.L. (1975). Late Miocene sediments and fossils from the northern Kenya Rift Valley. *Nature* **256**, 279–84.

Pickford, M.H.L. (1978a). Geology, palaeoenvironments and vertebrate faunas of the mid-Miocene Ngorora Formation, Kenya. In *Geological Background to Fossil Man*, ed. W.W. Bishop, pp. 237–62. Geological Society of London: Scottish Academic Press.

Pickford, M.H.L. (1978b). Stratigraphy and mammalian palaeontology of the Late Miocene Lukeino Formation, Kenya. In *Geological Background to Fossil Man*, ed. W.W. Bishop, pp. 263–78. Geological Society of London: Scottish Academic Press.

Pickford, M.H.L. (1986a). Cainozoic paleontological sites of western Kenya. *Munch. Geowiss. Abh.* **8**, 1–151.

Pickford, M.H.L. (1986b). The geochronology of Miocene higher primate faunas of East Africa. In *Primate Evolution*, ed. J.G. Else & P.C. Lee, pp. 19–33. Cambridge: Cambridge University Press.

Pickford, M.H.L. (1987). The chronology of the Cercopithecoidea of east Africa. *Human Evol.* **2**, 1–17.

Pickford, M.H.L. (1988). Geology and fauna of the middle Miocene hominoid site at Muruyur, Baringo District, Kenya. *Human Evol.* **3**, 381–90.

Pickford, M.H.L., Nakaya, H., Ishida, H. & Nakano, Y. (1984). The biostratigraphic analyses of the faunas of the Nachola area and Samburu Hills, northern Kenya. *Afr. Stud. Monogr. (Suppl.)* **2**, 67–72.

Pickford, M.H.L. & Senut, B. (1988). Habitat and locomotion in Miocene cercopithecoids. In *A Primate Radiation: Evolutionary Biology of the African*

Guenons, ed. A. Gautier-Hion, F. Bourlière, J.-P. Gautier & J. Kingdon, pp. 35–53. Cambridge: Cambridge University Press.

Pickford, M.H.L., Senut, B. & Hadoto, D. (1993). *Geology and Palaeobiology of the Albertine Rift Valley Uganda-Zaire*, Vol. I: *Geology*, pp. 195–205. Orleans: CIFEG Occasional Publication, (1993/24).

Pilbeam, D. & Walker, A. (1968). Fossil monkeys from the Miocene of Napak, north-east Uganda. *Nature* **220**, 657–60.

Plummer, T.W., Kinyua, A.M. & Potts, R. (1994). Provenancing of hominid and mammalian fossils from Kanjera, Kenya, using EDXRF. *J. Archaeol. Sci.* **21**, 553–63.

Plummer, T.W. & Potts, R. (1989). Excavations and new findings at Kanjera, Kenya. *J. hum. Evol.* **18**, 269–76.

Schrenk, F., Bromage, T.G., Gorthner, A. & Sandrock, O. (1995). Paleoecology of the Malawi Rift: Vertebrate and invertebrate faunal contexts of the Chiwondo Beds, northern Malawi. *J. hum. Evol.* **28**, 59–70.

Senut, B. (1987). Upperlimb skeletal elements of Miocene cercopithecoids from East Africa: implications for function and taxonomy. *Hum. Evol.* **2**, 97–106.

Senut, B. (1994). Cercopithecoidea Néogenes et Quaternaires du Rift Occidental (Ouganda). In *Geology and Palaeobiology of the Albertine Rift Valley, Uganda-Zaire*, vol. II: *Palaeobiology*, ed. B. Senut & M. Pickford, pp.195–205. Orleans: CIFEG Occasional Publication (1994/29).

Shipman, P., Bosler, W. & Davis, K.L. (1981). Butchering of giant geladas at an Acheulian site. *Curr. Anthrop.* **22**, 257–68.

Simons, E.L. (1969). Miocene monkeys (*Prohylobates*) from northern Egypt. *Nature* **223**, 687–9.

Simons, E.L. (1972). *Primate Evolution*. New York: Macmillan.

Simons, E.L. (1994). New monkeys (*Prohylobates*) and an ape humerus from the Miocene Moghara Formation of northern Egypt. In *Proceedings of the XIVth International Primatological Conference. 1992*, pp. 247–53. Strasbourg, France.

Smart, C. (1976). The Lothagam 1 fauna: Its phylogenetic, ecological, and biostratigraphic significance. In *Earliest Man and Environments in the Lake Rudolf Basin*, ed. Y. Coppens, F. C. Howell, G.Ll. Isaac & R.E.F. Leakey, pp. 361–9. Chicago: University of Chicago Press.

Szalay, F.S. & Delson, E. (1979). *Evolutionary History of the Primates*. New York: Academic Press.

Tamrat, E., Thouveny, N., Taieb, M. & Opdyke, N.D. (1995). Revised magnetostratigraphy of the Plio-Pleistocene sedimentary sequence of the Olduvai Formation (Tanzania). *Palaeogeogr. Palaeoclimatol. Palaeoecol.* **114**, 273–83.

Temerin, L.A. & Cant, J.G.H. (1983). The evolutionary divergence of Old World monkeys and apes. *Am. Nat.* **122**, 335–51.

Thomas, H. & Petter, G. (1986). Revision de la faune de mammifères du Miocène supérieur de Menacer (Ex-Marceau), Algerie: Discussion sur l'âge du gisement. *Geobios* **19**, 357–73.

von Koenigswald, G.H.R. (1969). Miocene cercopithecoidea and oreopithecoidea from the Miocene of East Africa. In *Fossil Vertebrates of Africa*, vol. 1, ed. L.S.B. Leakey, pp. 39–52. New York: Academic Press.

Walter, R.C. & Aronson, J.L. (1993). Age and source of the Sidi Hakoma Tuff, Hadar Formation, Ethiopia. *J. hum. Evol.* **25**, 229–40.

Walter, R.C., Manega, P.C., Hay, R.L., Drake, R.E. & Curtis, G.H. (1991) Laser-

fusion 40Ar-39Ar dating of Bed I, Olduvai Gorge, Tanzania. *Nature* **354**, 145–9.
White, T.D (1984) Hadar biostratigraphy and hominid evolution. *J. Vertebr. Paleontol.* **4**, 575–83.
White, T.D., Suwa, G. & Asfaw, B. (1994). *Australopithecus ramidus*, a new species of early hominid from Aramis, Ethiopia. *Nature* **371**, 306–12.
Winkler, A. (1992). Systematics and biogeography of middle Miocene rodents from the Muruyur Beds, Baringo District, Kenya. *J. Vertebr. Paleontol.* **12**, 236–49.
WoldeGabriel, G., White, T.D., Suwa, G., Renne, P., de Heinzelin, J., Hart, W.K. & Helken, G. (1994). Ecological and temporal placement of early Pliocene hominids at Aramis, Ethiopia. *Nature* **371**, 330–3.

8

The oro-facial complex in macaques: tongue and jaw movements in feeding

KAREN HIIEMAE

Introduction

No mammal survives to reproductive age without first suckling and then adapting to a diet of semi-solid or solid foods during weaning. That process is associated with significant ontogenetic change in the anatomy of the oro-facial complex, and in the mechanisms by which food is moved through the mouth to the oropharynx and into the digestive tract. Much attention has been paid to the morphology of cercopithecoid teeth and jaws (Warwick James, 1960; Swindler, 1976; Kay, 1978; Lucas and Teaford, 1994) and there is even more information on chewing in humans. However, the anatomy and role of the soft tissues, especially the tongue-hyoid complex, have largely been ignored. It is now clear that the feeding process depends as much on the soft tissues of the mouth and pharynx as on the teeth and jaws.

The events involved in moving food from the external environment into the mouth and through the pharynx into the gastrointestinal (GI) tract constitute the *feeding sequence* (Fig. 8.1). This is conventionally described (based on non-primate studies) as having three sequential elements, usually occurring *seriatim:* ingestion; mastication (chewing); and deglutition (swallowing). In all mammals so far studied, these processes depend on patterned behavior of the hard and soft tissue components of the system. Two fundamental mechanisms are involved: food transport, which is a tongue dependent function; and food breakdown, which is largely achieved by the relative movement of upper and lower teeth as a function of jaw movement (Fig. 8.1). Food transport involves two distinct processes: *Stage I Transport*, in which food is moved from the lips/incisal area of the mouth to the post-canines; and *Stage II Transport*, in which material is moved through the pillars of the fauces into the oropharynx, and is either immediately swallowed or retained for the addition of further aliquots and then swallowed.

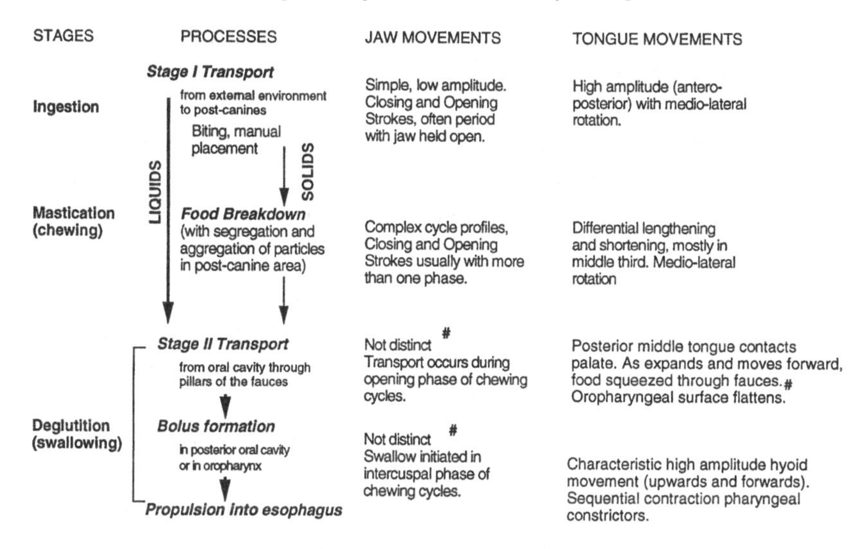

Fig. 8.1. The stages in the feeding sequence with their constituent patterns of activity and movements. # indicates behavior different from that seen in non-primate mammals.

Feeding has been analysed in some mammalian species using cinefluorographic/videofluorographic techniques: i.e. opossum (Hiiemae, 1976; Hiiemae *et al.*, 1978; Crompton, 1989); cat (Thexton *et al.*, 1980; Thexton and McGarrick, 1988, 1989); hyrax (Franks *et al.*, 1985; German and Franks, 1991); rabbit (Ardran *et al.*, 1958; Schwarz *et al.*, 1989; Cortopassi and Muhl, 1990); bat (De Gueldre and De Vree, 1984), and miniature pig (Herring and Scapino, 1973). These studies show that the mechanisms of intra-oral food processing and transport are similar in mammals with orofacial morphologies adapted for different primary diets (Turnbull, 1970). The movements of the jaw and tongue are rhythmic and cyclical. In 1923, based on neurophysiological studies in rabbits and cats, Bremer described a *"centre de correlation"* in the hindbrain as the source of the motor output generating these movements. His results were confirmed and expanded by Dellow and Lund (1971). There is a "central pattern generator" (CPG) in the hindbrain that regulates the rhythmic jaw movements of feeding. Whether the rhythmic movements of the tongue are also controlled by the same CPG is an open question. The process of swallowing is controlled by a "swallowing center" in the hindbrain (Doty and Bosma, 1956; Doty, 1968; Miller, 1982). It is increasingly accepted that the mechanisms by which feeding is centrally regulated may be similar in all mammals, including anthropoids and humans, albeit that the actual process is highly dependent on the anatomy of the system and is continually modulated by feedback to the CPG from system sensory receptors (Thexton, 1992; Hiiemae, 1993).

This chapter has two functions: (1) to report what is now known about the mechanisms of feeding in macaques; and (2) to attempt to set that information in its functional and evolutionary context. It will address the questions:

- How is food processed from ingestion to deglutition in *Macaca fascicularis* as an exemplar of cercopithecoid feeding behavioral mechanisms?
- How do the processes in *M. fascicularis* compare with those documented in non-primate mammals and in humans?
- What are the feeding mechanisms (transport and swallowing) in neonatal macaques?
- How does the behavior of the oro-facial system change during infancy?
- Can studies in anthropoids, such as macaques, illuminate the evolutionary process which, in humans, has reconciled the biomechanical demands of feeding with those required for speech?

Since much of the recent experimental work is attributable to a small group of investigators (including the author and her colleagues), references to that work are limited, after the first citation, for ease of reading.

Experimental methods

Earlier studies of feeding in macaques (Luschei and Goodwin, 1974; Byrd *et al.*, 1978) focused on patterns of jaw movement and associated jaw muscle activity. Hylander and Crompton (1986) and Hylander *et al.*, (1987) used cinefluorographic, strain gauge, and electromyographic (EMG) techniques to examine patterns of bone strain on the lower jaw and face.

To examine the integration of tongue, soft palate, hyoid complex and jaw movements in mammals and to correlate those with the position of the food in the mouth, high speed (100 f.p.s.) cineradiography is essential. To make the food radiopaque, it is usually mixed (semi-solids, liquids) or coated (solids) with barium sulphate. Recordings are normally made in the lateral projection. Given the radiolucency of the soft tissues, and the requirement for the analysis and quantification of their movements over time and relative to each other, densely radiopaque markers are used as reference points (Hiiemae *et al.*, 1995; Thexton and Crompton, 1998). Markers are surgically inserted below the gustatatory epithelium on the dorsal surface of the tongue just behind its tip, in its middle, and just anterior to the sulcus terminalis. Reference points are digitized and their Cartesian coordinates trigonometrically manipulated to show: (1) the movements of one structure, e.g. the tongue (Fig. 8.2); or (2) the tongue relative to other components, e.g. the hard palate or the upper occlusal plane, in the vertical and/or horizontal planes over time (Fig. 8.3) or in space (Fig. 8.4). Marker positions relative to upper or lower jaws are plotted to track movement through a single jaw movement cycle ("loops") or plotted for complete sequences to show (as seen in two dimensions) the total range of marker position in complete sequences (Fig. 8.4). The area/volume occupied by each marker in each sequence is its 'envelope of motion'. This differs between food types (Fig. 8.4). The Euclidean distance between individual tongue markers is used to measure intrinsic patterns of anteroposterior expansion (lengthening) and contraction (shortening) in the *anteroposterior* axis (German *et al.*, 1989; Hiiemae *et al.*, 1995). In all of the adult macaque studies reported here, behavior was recorded in frontal view using black and white cinephotography (50/ 100 f.p.s.) electronically time-linked to the X-ray cine, so that the movements of the cheeks, lips, and tongue could be analyzed. Detailed descriptions of the methods used are given in Franks *et al.* (1984), Hylander and Crompton (1986), German *et al.* (1989), Hiiemae *et al.* (1995), and Thexton and Crompton (1998).

While the anatomical correlations between tooth form and diet in mammals have been exhaustively described, and the mechanics of food breakdown and post-canine dental morphology examined (Lucas and Luke, 1984; Hiiemae and Crompton 1985; Lucas and Teaford, 1994), the physical properties of most natural foods are poorly quantified. Lucas and co-workers (Lucas, 1989; Lucas and Corlett, 1991; Corlett and Lucas, 1990) examined masticatory behavior in adult macaques in relation to natural dietary items. However, since such tropical forest products are not readily available to experimentalists, the studies reported here used bananas, apples, and hard solids (monkey chow) as "test foods".

The oro-facial complex

The primary components of the mammalian oro-facial complex (Hiiemae and Crompton, 1985; Fig. 8.1) are: (1) the upper jaw, the upper dentition, and the included hard palate, with the lower jaw, mandibular teeth and the adductor muscles; (2) the tongue and hyo-laryngeal complex – i.e. the tongue itself, the hyoid, and the muscles suspending the hyoid from the mandibular symphysis, the cranial base, and anchoring it to the sternum;

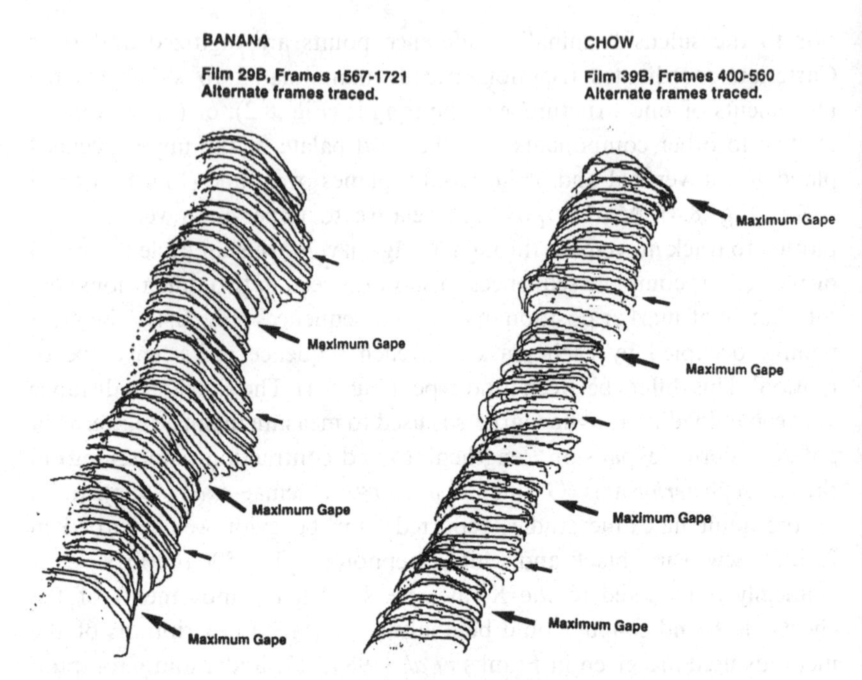

Fig. 8.2. Changes in tongue surface profile over time in the same macaque feeding on two foods of different consistency. Alternate frames of film taken at 100 f.p.s. were projected, the tongue profile and tongue marker positions (dots) traced using the hard palate shadow and the (fixed) markers in the upper canine and first molar as reference points. The data were digitized and printed using Jandel 3D PC software. The frame count begins at the top of each montage. The small arrows show the time at which the jaw begins to open at the end of IP. The jaw movement profiles for the first two cycles on banana correspond to that shown for 'Stage II Transport' in Fig. 8.5, the third to that for a rapid-sequence chewing cycle in the same figure. In contrast, the profiles for chow were as for early (first two) and middle (third) chewing cycles in Fig. 8.5. (From Hiiemae *et al.*, 1995. Reproduced with permission from Elsevier Science.)

and (3) the soft palate and the constrictor muscles of the pharynx. Many of the muscles of the hyo-laryngeal complex also function as depressors of the mandible.

The hyoid is the posterior skeletal element in the tongue base and anchors the larynx. Its position determines the anteroposterior dimension of the pharynx just above the entry to the esophagus. In describing the movements of the tongue and their correlation with jaw movements in feeding (Figs. 8.2, 8.5), it is important to note that three mechanisms can operate concurrently or separately as a function of its muscular anatomy: (1) change in the gross position of the tongue in space, produced largely by changes in the length of its base (hyoid-symphysis); (2) changes in the

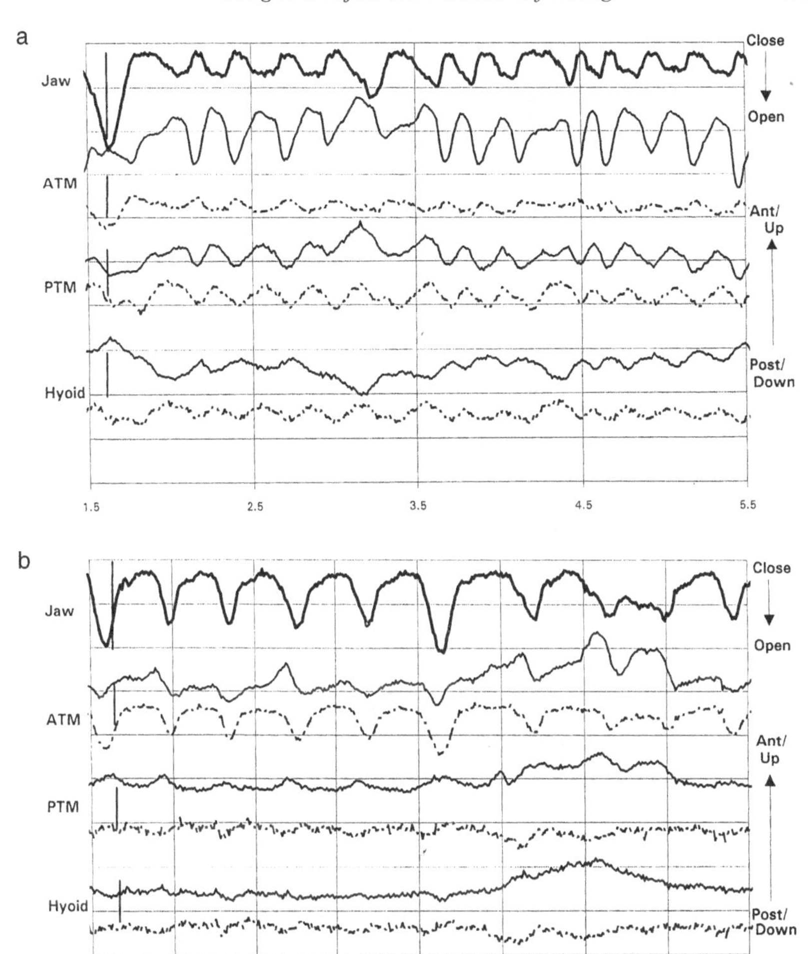

Fig. 8.3. Movement of the jaw (Y axis) and two tongue and the hyoid marker (X axis, antero-posterior movement, solid line, Y axis, vertical movement, hatched line) over time (4 sec) relative to the upper occlusal plane (represented by a line connecting radio-opaque markers inserted into the upper canine and first molar). The vertical bars represent one cm.
a. Feeding on soft food (banana). High amplitude antero-posterior movements of the anterior part of the tongue are associated with low amplitude jaw movements. Note that the tongue markers reach their most anterior position in synchrony, and just before maximum gape (see Fig. 8.5). Vertical movement of the anterior tongue marker (ATM) is highly correlated with that of the jaws (see also b), but that for the posterior marker (PTM) and the hyoid reflects their independent movement.
b. Feeding on hard food (chow). The amplitude of jaw movements is greater than in 'a', and cycle durations longer. Antero-posterior tongue movements are attentuated (except in the last 3 cycles) with minimal PTM and Hyoid movement.

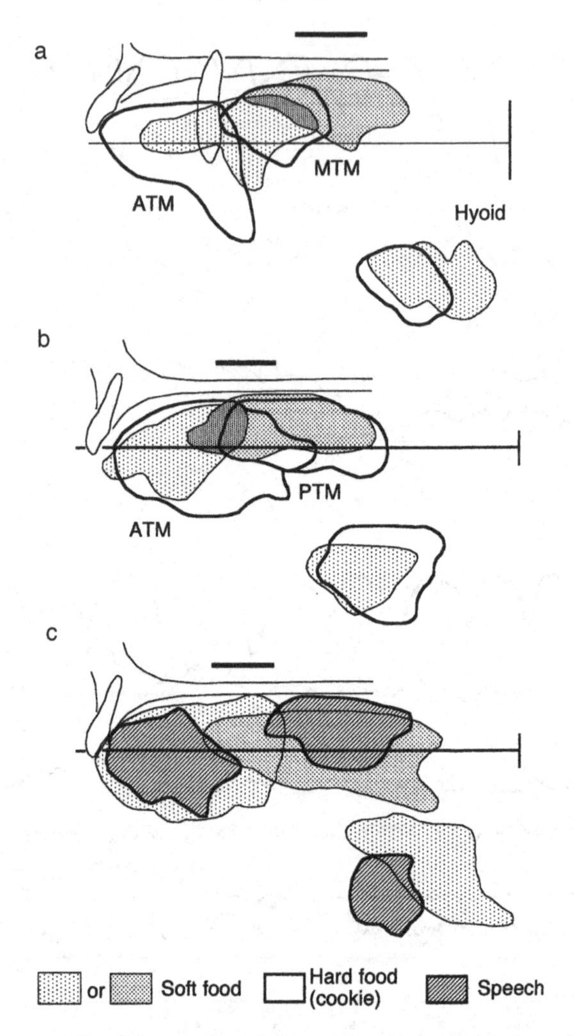

Fig. 8.4 The spatial distribution of marker position (domains) in sample complete sequences.
a. Female macaque eating banana (solid shading) and chow (black outlines).
b. Female human subject eating chicken spread (soft food, analogous to banana, solid shading) and hard biscuit (analogous to chow, black outline).
c. Male human subject eating chicken spread and speaking (dark shading).
The horizontal heavy line is the upper occlusal reference plane. The palate, upper incisor and canine (macaque) are drawn to illustrate their relative positions. Much tongue movement occurs above the upper occlusal plane in direct relation to the palate (shown here slightly separated for clarity). The horizontal scale is held constant but note that the vertical scale (at end upper occlusal plane) is halved for *H. sapiens,* thus diminishing the actual vertical separation of the hyoid and tongue marker domains.
The records in 'a' and 'b' include swallows as does 'c'. The important finding is the position of the 'hyoid domain' for speech ('Grandfather passage', see text).

position of its body and surface relative to its base; and (3) changes within its body. Contraction of the geniohyoid, the largest suprahyoid muscle, shortens the distance between the hyoid and symphysis, pulling the hyoid forward, or depressing the mandible or both. The infrahyoid muscles can depress the hyoid, or, acting with the posterior suprahyoids (particularly digastric), pull the hyoid backwards or maintain its position despite contraction in geniohyoid. The body of the tongue can be pulled forward by contraction in genioglossus. Conversely, the hyoglossus and styloglossus can pull it backwards and downwards or upwards respectively. Contraction of any of the muscles entering the tongue, but with an external attachment (genioglossus, hyoglossus, palatoglossus, styloglossus), produces shape change within the tongue. The intrinsic muscles (longitudinal, transverse, vertical) all play a role in changing overall tongue shape. Given this complex anatomy (essentially the same in macaques and humans, although the longer and shallower oral cavity in the former affects the angulation of their fibers), it is not surprising that the tongue can be considered a "muscular hydrostat" (Kier and Smith, 1985).

The anatomy of the muscular cheeks and the lips in cercopithecoids and *Homo* differs from that in the other mammals studied experimentally. The postcanine area of the mouth is covered by buccinator which has its modiolus far forward just behind the angle of the lips, with its fibers blending with those of orbicularis oris. The cheeks are lined internally with oral mucosa, which folds back onto the alveoli of the jaws well above the occlusal surfaces of the teeth to form the vestibules of the mouth. The highly mobile lips can be apposed to form a complete anterior oral seal. These muscular soft tissues allow for more complex behaviors in the management of the food in the mouth (Lucas and Luke, 1984; Lucas *et al.*, 1986). Macaques have highly extensible ventral evaginations of the cheeks in the molar region, which serve as cheek pouches. Food cached in the pouches is forced into the oral cavity proper for chewing by powerful contractions of the facial muscles, platysma, and buccinator, with the occasional helping hand.

The spatial relationships of the oro-facial and oro-pharyngeal complexes in macaque differ from those in most other mammals as the cranium is placed more vertically on the vertebral column (Campbell, 1967), so the long axes of the mouth and the pharynx are more acutely angled. However, no cercopithecoid has achieved the verticality of the cervical vertebral column found in humans.

The anatomy of the oro-facial complex changes during infancy (Crelin, 1965). The eruption of the deciduous and permanent dentitions, change facial height and length. This is conspicuously so in male cercopithecoids, given the canine sexual dimorphism characteristic of these monkeys.

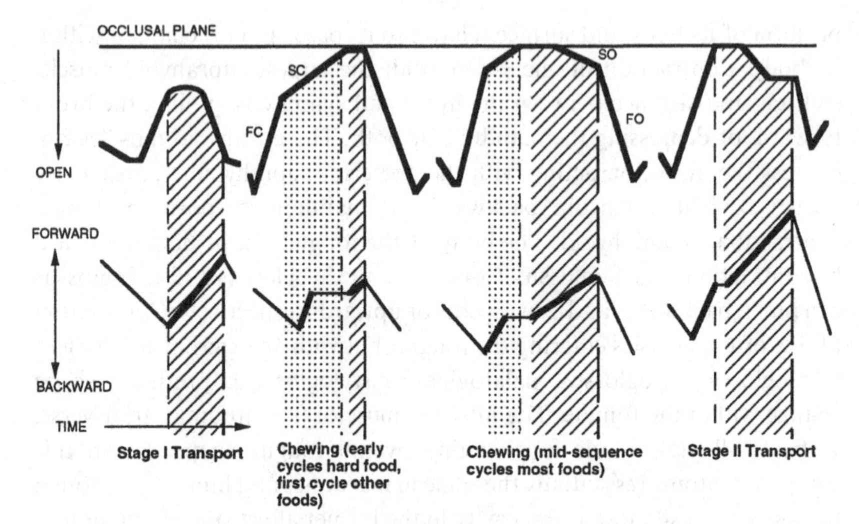

Fig. 8.5. Diagrammatic representation of jaw and tongue movements (anterior tongue marker) at various stages in the feeding sequence (see Fig.8.1). The vertical shorter dashed line indicates the start of IP, the longer dashed lines the last rate change in Opening before maximum gape. The correlation between the amplitude of tongue forward movement and the last rate change in Opening is shown (hatched columns). The longer this distance, the greater the degree of food transport (Stage I and especially Stage II). In contrast, movement of triturated material into the oropharynx in non-primate mammals studied normally occcurs as the tongue is travelling backwards (see text). No rate change occurs during closing in Stage I or the Stage II transport cycle as shown, that does not mean that no triturition of food is occurring in the latter rather that it cannot be detected in the films. Tooth–food–tooth contact is occuring in IP rather than earlier during Closing (dotted columns show the time spent in food breakdown).

Ontogenetic changes in the cercopithecoid skull have been described (Moore and Lavelle, 1974; Maier, Chapter 5; Ravosa and Profant, Chapter 9). However, when the mechanisms of feeding are considered, the most significant changes are those associated with the hyo-laryngeal complex (Laitman and Reidenberg, 1993). In many mammals, including neonate cercopithecoids and hominids, the larynx is described as "intra-narial". While an unfortunate term, since the larynx is not actually *in* the nares, it is positioned high in the nasopharynx allowing food to be collected for bolus formation in the valleculae, i.e. "round" rather than "above" the laryngeal aditus (Thexton and Crompton, 1998). With facial maturation, the hyo-laryngeal complex descends in the neck, so that the hyoid reaches an "average" position at the level of the lower border of the mandible. This descent creates a vertical separation between the posterior oral seal (soft palate–posterior tongue) and the epiglottis–laryngeal aditus. The greater the separation, the greater the risk of aspiration of food into the lungs with its pathological consequences.

The feeding sequence in macaques

The movement of food from the lips to the esophagus depends on two fundamental processes: transport and food breakdown (Fig. 8.1) The former is a tongue dependent function. The latter depends on tooth–food–tooth contact driven by jaw movement, although the tongue and cheeks play a major role in managing the food and insuring that particles needing breakdown are positioned on the occlusal table for each chewing stroke. The relative importance of these two processes in any feeding sequence in macaques is highly correlated with the initial consistency of the food. When solid food is rendered "swallowable" is highly correlated with its initial condition (Hiiemae *et al.*, 1978, 1995, 1996; Thexton *et al.*, 1980; Thexton and Hiiemae, 1997; Palmer *et al.*, 1995, and unpubl. data). Clearly, as in non-primate mammals, liquids can be transported and immediately swallowed, as can some semi-solid, thixotropic materials such as soft fruits (e.g. banana). Solids must be reduced, mixed with saliva, and the resultant smaller particles aggregated into a bolus (Hector and Linden, 1987).

In mastication, solid food is primarily reduced by the interaction of upper and lower postcanine teeth as the jaws move into occlusion. It is possible that further reduction occurs as the teeth leave occlusion in the earliest stage of opening (Hiiemae and Kay, 1973; Kay and Hiiemae, 1974). The tongue may also have an important function in "mushing" food by compressing it against the hard palate.

An important difference between macaques (and humans; Hiiemae *et al.*, 1996) and the other mammals studied to date, is the ability of macaques, when feeding on solid food, to swallow boli of chewed food at intervals during the feeding sequence. All mammals swallow at regular intervals when lapping liquids or consuming semisolids (Hiiemae *et al.*, 1978; Thexton and McGarrick, 1988). Infant macaques (German *et al.*, 1992), in common with other non-primate mammals feeding on liquids, merge the mechanisms of Stage I and Stage II Transport into an integrated transport process such that new aliquots of liquid enter the mouth (or are expressed from the nipple) as previous aliquots are swallowed. *Prima facie*, the ability of adult macaques to swallow fully triturated food, while still chewing particles of the original solid ingestate, is attributable to the segregation and aggregration functions of the tongue coupled with the additional element of food position control afforded by the muscular cheeks. These "mid-sequence" swallows occur as "intercalated events" in chewing cycles. It follows that, in macaques and *Homo*, a tongue-based function (transport) can occur synchronously with a jaw-based function (chewing).

The number of chewing cycles before and between swallows is also correlated with the initial consistency of the food – the harder the food, the greater the number of intervening chewing cycles (Thexton and Hiiemae, 1997). Tooth form and jaw movement pattern have traditionally dominated evaluation of the oro-facial complex in primates. While it can be argued that this focus may be inappropriate if we are interested in the evolution of function rather than form, description of the complex mechanisms involved in feeding is simplified if jaw movements are first described.

Jaw movements in feeding

Both jaws move in feeding. Movement of the upper (a function of cranio-cervical movement) is normatively small, but occurs in ingestion when large food items are held between the incisors before a portion is finally "bitten off". All jaw movements are cyclical, but cycles vary in amplitude and complexity (Figs. 8.1 and 8.3). Cycles have two basic movements: closing and opening separated by a period, the intercuspal phase (IP), in which no vertical movement can be seen in fluorographs although some transverse movement is occurring (frontal view films). The duration of IP varies but is always longer in "swallow cycles" and shortest in simple transport (Stage I) cycles (Fig. 8.5). The closing movement may have two phases: fast close (FC) and slow close (SC), the latter commencing at tooth–food–tooth contact. The amplitude of the FC–SC transition is a function of the size of the food particle between the teeth (Van der Bilt *et al.*, 1991). If the food (e.g. banana) offers no or minimal resistance to the jaw closing movement, a distinct SC phase may not be present. Opening may be a simple movement but usually consists of one or two, sometimes three, phases (Fig. 8.5). In macaques, there is no "distinct" Stage II Transport or swallow cycle jaw movement profile, although the more complex opening strokes are associated with extensive tongue movement (see below).

Tongue movements in feeding

Patterns of tongue movement are complex. The tongue can change shape from cycle to cycle (Fig. 8.2) depending on the instantaneous condition of the food and the stage in sequence, given that both transport and chewing functions can occur simultaneously. However, the tongue movements of

Stage I and Stage II Transport are the same regardless of original food type.

The aggregation of triturated food particles into a midline position for bolus formation requires their medial movement onto the dorsum of the tongue from the occlusal surfaces of the postcanines. There is, at present, no information on the details of that transport mechanism although tongue rotation is known to be involved (Abd el Malek, 1955; Tomura *et al.*, 1981; Franks *et al.*, 1984; Hiiemae *et al.*, 1995, 1996). All transport mechanisms involve anteroposterior movements of the tongue, whether produced by movement of the tongue as a whole, or intrinsic tongue expansion and contraction (lengthening/shortening in its anteroposterior axis) coupled with varying degrees of rotation in the coronal plane about its longitudinal anteroposterior axis. There is evidence from human studies (Hamlett, 1989; Stone, 1991) that the tongue is capable of complex transitional shape changes producing furrows and depressions on its surface. While there is no equivalent data for macaques, the envelopes of motion and spatial domains of tongue markers (Fig. 8.4) recorded during normal feeding in macaque and in human subjects (Hiiemae *et al.*, 1995; Palmer *et al.* 1997) suggest that the macaque tongue may exhibit comparably complex intrinsic shape changes.

Mechanisms of Stage I Transport

Unless manually placed into the postcanine region of the mouth, a "bite" of solid food is moved posteriorly from the incisors to the postcanines by a series of low amplitude, somewhat irregular, jaw movements associated with rhythmic tongue movements (German *et al.*, 1989). The bite is moved both backwards and forwards as the tongue moves anteroposteriorly and is rotated. Backwards movement dominates, since forward movement of the bite is prevented by the incisors. Where well-developed, as in opossum and cat, the palatal rugae serve the same function in non-primate mammals. The actual mechanism of transport involves rhythmic expansion and contraction of the tongue. In Stage I cycles, the tongue expands and moves forward below the bite as the jaws open. As the jaws approach maximum gape (widest opening), the tongue begins to move backwards and upwards, continuing that movement during jaw closing so pressing the food against the palate as it is carried backwards and laterally towards the postcanines of one side until the jaws reach maximum closure (minimum gape). As the food is "trapped" behind the incisors and against the palate, the tongue again expands below and round the bite, as the jaw opens. The cycle is then repeated. Stage I Transport depends on a simple integration of low amplitude jaw movement with anteroposterior and rotational movements of the tongue to move the bite posterolaterally without affecting its physical dimensions/characteristics. The number of cycles depends on the anteroposterior dimensions of the dental arcade – generally the greater the distance of the postcanines from the incisors/lips, the more cycles. This process is attenuated in humans, given the parabolic dental arch, and can occur in one jaw movement cycle or in a very fast series of low amplitude "jaw oscillations" (Hiiemae *et al.*, 1996). Stage I Transport of liquids has not been studied in adult macaques, but almost certainly depends on the same tongue mechanism to carry the fluid posteriorly in the mid-line.

Mechanisms of Stage II Transport

Franks *et al.* (1984) describe the process, by which food is moved through the fauces into the oropharynx for deglutition in macaques, as a "squeeze back" mechanism with the tongue working against the hard palate. A bolus of food is collected on the dorsum of the tongue; the tongue surface anterior to the bolus rises, contacting the palate; and the tongue surface–palate contact is maintained and moves backwards as the tongue itself moves forward – i.e. successively more posterior points on the middle tongue surface maintain the tongue–palate contact. At the same time as the tongue and palate are (to use an analogy) acting like fingers compressing toothpaste up a tube, the posterior (oropharyngeal) surface of the tongue behind the contact is flattening, creating a channel between the elevated posterior part of the soft palate and the oropharyngeal surface of the tongue through which the food can be pushed into the oropharynx.

This propulsive mechanism depends on a forward movement of the tongue against the hard palate, in contrast to the "carry back on the tongue" mechanism normative in the non-primate manimals studied. The latter depends on the posterior movement of the tongue, which occurs during the late opening and closing movements of the jaws (see Fig. 8.5, i.e. it occurs during the tongue movements shown by the finer lines). However, a squeeze-back mechanism has been reported in opossums drinking liquids (Crompton pers. comm.) and cats (Hiiemae *et al.*, 1978). The mechanism used to move triturated material from the posterior intra-oral surface of the tongue into the oropharynx in macaques is not a uniquely anthropoid phenomenon, but rather an example of functional selection between two (so far identified) methods. The dominance of the squeeze-back mechanism in adult anthropoids may be linked to changes in the anatomy of the oropharynx and so the capacity of the valleculae. These differences are discussed under "Swallowing".

Swallowing

While Stage II Transport is a clearly defined and documented process in all mammals, it has (until recently; Palmer *et al.*, 1992) been difficult to equate behaviors in experimental animals with those described in the literature on human swallowing (Miller, 1982, 1986). The problem results from the use of the (artificially) isolated or "command" swallow as the experimental paradigm in human subjects. When the totality of the feeding process is examined, there is strong evidence for equating the human "oral phase of swallowing" with Stage II Transport (Palmer *et al.*, 1992; Hiiemae and Palmer, 1999). This interpretation becomes even more significant when the process in non-human mammals is considered, since triturated food is routinely transported into the oropharynx (a "pre-bolus") in one Stage II Transport cycle, to be followed by one or more additional pre-boli. The definitive bolus is formed in the oropharynx. Contrary to previously reported results (based on the "command swallow"), bolus formation normatively occurs in the oropharynx when subjects eat natural bite sizes of solid foods. Bolus accumulation can take several seconds (Hiiemae and Palmer, 1999).

Mechanisms of swallowing

The "pharyngeal stage" of swallowing in the adult involves moving the bolus from the pharyngeal surface of the tongue across the pharynx and into the esophagus. The process involves the synchronous movements of the hyoid, epiglottis, vocal folds, the tongue, epiglottis, and the soft palate. The hyoid moves upwards and forwards, effectively shortening the vertical height of the oropharynx and assisting in opening the esophagus; the epiglottis tilts over the laryngeal aditus; the true and false vocal folds close, sealing the larynx, as the soft palate elevates to contact the posterior pharyngeal wall, so closing off the nasopharynx. As the posterior surface of the tongue pushes backwards, the bolus leaves the oropharyngeal surface of the tongue moving towards the esophagus. Its movement is facilitated by a wave of contraction (from above downwards) of the pharyngeal constrictors. However, it must be emphasized that in normal feeding, Stage II Transport and the pharyngeal phase of swallowing form a continuum. A bolus is assembled on the posterior surface of the tongue, moved through the pillars of the fauces, and propelled across the pharynx. Even in those frequent cases where some portion of the triturated food reaches the pharynx and is held in the valleculae in one cycle, and a second larger volume is moved in the next Stage II cycle, the two become confluent as the second aliquot enters the pharynx.

There are significant differences between the mechanism of swallowing in adult macaques and non-primate mammals (Thexton and Crompton, 1998). Food can be stored in the valleculae and flow round the larynx through the pyriform fossae, without prejudice to the airway, in mammals with an "intranarial" larynx. Given the relatively high position of the hyo-laryngeal complex in infant cercopithecoids, questions have been raised as to whether they use the same swallowing mechanisms as non-primates. In a recent cinefluorographic study on infant macaques (German *et al.*, 1992), a change in swallowing behaviour between early and late infancy was observed. In the very young infant (24 days), the epiglottis did not bend, and the milk flowed laterally through the pyriform recesses although the soft palate was elevated and in contact with the posterior pharyngeal wall. In contrast, in the older infant (37 days), evidence of epiglottic bending was present. The authors hypothesize that an ontogenetic change in the mechanism of swallowing occurs in macaques associated with weaning. This finding suggests that changes in the mechanics of swallowing precede the grosser anatomical changes in the oro-facial complex associated with eruption of the deciduous and permanent dentitions.

Tongue–jaw linkages in feeding

In non-primate mammals, the movements of the tongue and jaws are highly correlated. The tongue begins its forward movement at the end of closing, and reverses direction as the jaws approach maximum opening, reversing again when the teeth reach minimum gape or full occlusion (Thexton, 1984; German and Franks, 1991).The only predictable temporal linkage between tongue and jaw movement cycles in macaques occurs during jaw opening (Fig. 8.5) – the reversal from forward to backward movement of the anterior part of the tongue occurs within 30 milliseconds of the final rate change before maximum gape (Hiiemae *et al.*, 1995). This pattern of forward tongue movement during opening, with a reversal at or before maximum gape, followed by backwards movement during late opening and through closing, is found in other mammals. Given the basic similarities, the differences between macaque and (for example) hyrax are more interesting.

Allowing for the capacity of parts of the macaque tongue, specifically the middle section (as measured from radiopaque markers in lateral projection) to change shape independently of the anterior or posterior (largely oro-pharyngeal) segments during chewing, the available data suggest the following. In Stage I Transport, the tongue is moving as a single unit although the antero-posterior movement (expansion-contraction) is greatest in the anterior third (Fig. 8.2). In chewing, forward tongue movement is minimal and expansion/ contraction of the anterior third is limited. However, substantial volumetric change occurs in the middle third, related to the postcanine teeth. Backward movement ceases before or at the the FC–SC transition (Fig. 8.5), when forward and upward movement begins. As the teeth approach full occlusion, forward movement stops. It resumes during IP, i.e. after all closing movement has *ceased.* In the few cases for which records are available for "middle segment tongue markers" positioned laterally but in the postcanine area, they show vertical movements during SC and IP, which suggest that the tongue is "hollowing out" to receive the products of the associated chewing stroke. It is reasonable to suggest that that part of the tongue related to the postcanine teeth is changing shape in response to the presence of both adequately triturated and inadequately reduced food.

The plasticity, as compared with other mammals, of the temporal relationships between tongue and jaw movements in the macaque, is paralleled in humans (Palmer *et al.*, 1997). It appears that these higher primates retain the core mammalian pattern (i.e. "active" tongue movement occurs during jaw opening and early closing) but that the timing of transitional events is not stereotypic cycle to cycle (Hiiemae *et al.*, 1995). This is not surprising, given the the role of the tongue in feeding and the differences between jaw movement profiles both within and between sequences, when macaques feed on foods of different initial consistency.

Tongue–hyoid movement and the capacity for speech

Homo sapiens is the only mammal in which an "accessory" oro-facial activity – speech – has become *the* dominant behavior. It has long been argued that the anatomy of the oro-facial complex in hominids evolved to meet the biomechanical demands of complex speech (Lieberman, 1975, 1984). Lieberman and others have hypothesized that the elongation of the pharynx, produced by (ontogenetic/phylogenetic) hyo-laryngeal descent coupled with the retention of a short (anteroposterior) oral cavity, was the anatomical prerequisite for the intra-oral modulation of the laryngeal sounds intrinsic to modern speech. The trend to facial shortening in primates began long before the advent of bipedalism, and accelerated during the evolution of the hominids. The arguments as to precisely when speech evolved apart, it remains the case that without the ability to drink and feed, hominids would not have evolved to the point where modulated sound production (vocalization) could become the vowel-consonant combinations of nuanced interpersonal communication.

Do the experimental studies shed any light on this question? When the envelopes of motion for tongue and hyoid markers in macaques and humans feeding on soft and hard food are compared, the patterns are much the same although the vertical amplitude of the marker "domain" is greater in the latter. This is to be expected, given the greater vertical dimension of the human oral cavity (Fig. 8.4). There are distinct differences in the position of the tongue marker envelopes, relative to the upper incisors and palate, between the two types of food in both species. The hyoid domain also differs between food types. More interesting, the spatial domain in which swallowing occurs differs from that for chewing cycles. It follows that the *mechanisms* of ingestion, transport, and chewing in the macaque are essentially the same as in modern *H. sapiens.* As part of an ongoing videofluorographic study of tongue-jaw linkages in humans, using radiopaque tongue markers, subjects have been asked to read the "Grandfather passage", a short text which includes almost all the vowel-consonant combinations in English. Preliminary results show the rates of tongue marker movement in speech fall within the range for feeding, i.e. they are not "faster" as has been thought. Importantly, given the accoustic models for speech (e.g. Lieberman *et al.*, 1992), the hyoid domain in speech is very different from that used in feeding. When the XY positions of the hyoid, measured relative to the upper, or the lower, occlusal planes for all completed feeding sequences (solid foods and liquids) are plotted, and the XY positions for hyoid in speech are also plotted for each subject, then using the statistical test (Hald, 1965) for the differences between the means of two bivariate (two-dimensional) populations where the number of data points for feeding are in the range 2200–2700 and for speech 3000–3500, then we find that the hyoid domain used in speech is very highly significantly different ($p > 0.0000001$, n 26 data pairs). In speech the hyoid moves in a constrained domain, anterior to that used in feeding, such that the anteroposterior dimension of the oropharynx is increased (Hiiemae *et al.*, 1999).

Discussion

The anatomy of the orofacial complex in cercopithecoids, as exemplified by *M. fascicularis*, differs from that in other non-anthropoid primates, but shares common features with *H. sapiens* and (therefore by inference) with earlier hominids. Equally, when the general trends in the evolution of the primate skull are considered (Campbell, 1967, fig. 4.12), the major changes in the proportions and position of the jaws relative to the cranium and orbits result in a shortening and deepening of the lower third of the face and an increasing verticality of the cervical vertebral column. These skeletal changes are paralleled by changes in soft tissue anatomy. Chewing is only one component of the feeding process, albeit that the hard tissues are the only components of the complex preserved in the fossil record and so available as specimens.

Although jaw movements in *Tupaia* have been reported (Hiiemae and Kay, 1973), there is more information available on feeding mechanisms in the American opossum *(Didelphis virginiana)*. It is reasonable to assume that the latter represents the system in early primates. There is a wealth of data on feeding in modern humans, albeit few studies have focused on the process in its entirety. *Macaca fascicularis* is the only anthropoid primate in which both tongue and jaw movement patterns have been studied. While in no way suggesting a simple evolutionary lineage, "*Tupaia–Macaca–Homo*", the available *experimental* data suggest how the basic pattern seen in mammals such as *Didelphis* has been modified with the evolution of mobile cheeks and lips coupled with the forward migration of the foramen magnum.

These changes can be summarized as follows:

- In all mammals studied, and by extension, all mammals (save anteaters and their ilk which swallow their ingested food whole) the feeding process involves the same three stages (Fig. 8.1). Macaques and humans have developed the capacity routinely to chew, transport, and swallow solid food concurrently. In other mammals, for which experimental data is available, those processes occur *seriatim* for the single bite (pellet) of solid food supplied.
- The temporal relationship between tongue and jaw movement cycles is looser in macaques. While the basic cycle of tongue forward movement in jaw opening and backwards in jaw closing is retained, the expression of that cycle and its exact linkage to jaw movement events is variable and is correlated with the state of the food in the mouth cycle to cycle.
- There is evidence that, in tongue-dependent behaviors (Stage I and

Stage II Transport), jaw movements are secondary and may be constrained by the biomechanics of the tongue.

- A mechanism for Stage II Transport occasionally seen in non-primate mammals ("squeeze back") is normative in anthropoids and humans.
- Infant macaques swallow liquids in the same way as non-primate mammals, but begin to show a somewhat different adult pattern within the second month *post partum.*

The studies completed to date, albeit on a small range of mammals with different evolutionary lineages, show a common underlying cyclical pattern that integrates jaw and tongue movements in feeding. The process has the same stages in all, with cycle profiles that change as the sequence proceeds. However, as compared with non-primates, macaques show greater variability cycle to cycle during chewing. Lund (1991) argues that the masticatory rhythm is generated in the brainstem by a CPG common to all mammals and that the expression of that rhythm is modulated by sensory input from the periphery. Based on the evidence reported here, it can be argued that the tongue and palate must make an important contribution to that input given their role in the intra-oral management of food. While some differential expansion and contraction within the tongue has been reported in cats (Thexton and McGarrick, 1989), it is tempting to speculate that the much more complex differential movements observed in macaques, and seen in human feeding, were present in early hominids and provided the substrate upon which the complex shape changes used in speech could develop.

Acknowledgements

On behalf of my colleagues and friends (especially R.Z. German, W.L.Hylander, R.F. Kay, J.B. Palmer and A.J.Thexton), I wish to acknowledge the debt we all owe Professor A.W. Crompton, whose appreciation of the potential of high speed cinefluorography for the examination of orofacial behavior made all our studies possible.

References

Abd el Malek, (1955). The part played by the tongue in mastication and deglutition. *J. Anat.* **89**, 250–4.

Ardran, G.H., Kemp F.H. & Ride, W.D.L. (1958). A radiographic analysis of mastication and swallowing in the domestic rabbit *(Oryctolagus cuniculus. L.). Proc. Zool. Soc. Lond.* **130**, 257–74.

Bremer, F. (1923). Physiologie nerveuse de la mastication chez le chat et le lapin. *Arch. Int. Physiol. Biochim.* **21**, 667–81

Byrd, K.E., Milberg, D.J. & Luschei, E.S. (1978). Human and macaque mastication: a quantitative study. *J. Dent. Res.* **57**, 834–43.

Campbell, B. (1967). *Human Evolution*. London: Heinemann Educational Books.

Corlett, R.T. & Lucas P.W. (1990). Alternative seed-handling strategies in primates: seed-spitting by long-tailed macaques *(Macaca fascicularis)*. *Oecologia* **82**, 166–71.

Cortopassi, D. & Muhl, Z. (1990). Videofluorographic analysis of tongue movement in the rabbit *(Oryctolagus cuniculus)*. *J. Morphol.* **204**, 139–46

Crelin, J.F. (1965). *The Human Vocal Tract: Anatomy, Function, Development and Evolution.* New York: Vantage.

Crompton, A.W. (1989). The evolution of mammalian mastication. In *Complex Organismal Functions: Integration and Evolution in Vertebrates*, ed. D.B. Wake & G. Roth, pp. 23–39. New York: Wiley.

De Gueldre, G. & de Vree, F. (1984). Movements of the mandible and tongue during mastication and swallowing in *Pteropus giganteus* (Megachiroptera). A cineradiographical study. *J. Morphol.* **179**, 95–114.

Dellow, P.G. & Lund, J.P. (1971). Evidence for the central timing of rhythmical mastication. *J. Physiol. (Lond.)* **215**, 1–13.

Doty, R.W. (1968). Neural organisation of deglutition. In *Handbook of Physiology, Alimentary Canal.* ed. C.F. Code, pp. 1861–902. Washington, DC: American Physiological Society.

Doty, R.W. & Bosma, J.F. (1956). An electromyographic study of reflex deglutition. *J. Neurophysiol.* **19**, 44–60.

Franks, H.A., German, R.Z. & Crompton, A.W. (1984). Intra-oral food transport in the macaque. *Am. J. phys. Anthrop.* **65**, 275–82.

Franks, H.A., German, R.Z., Crompton, A.W. & Hiiemae, K.M. (1985). Mechanisms of intra-oral transport in an herbivore, the hyrax. *Arch. Oral Biol.* **30**, 539–44.

German, R.Z., Crompton, A.W., Levitch L.C. & Thexton A.J. (1992). The mechanism of suckling in two species of infant mammal: Miniature pigs and long-tailed macaques. *J. Exp. Zool.* **261**, 322–30.

German, R.Z. & Franks, H.A. (1991). Timing in the movements of the jaws, tongue and hyoid during feeding in the hyrax (*Procavia syriacus*). *J. Exp. Zool.* **257**, 34–42.

German, R.Z., Saxe, S., Crompton, A.W. & Hiiemae, K.M. (1989). Mechanism of food transport through the anterior oral cavity in anthropoid primates. *Am. J. phys. Anthrop.* **80**, 369–77.

Hamlett, S.L. (1989). Dynamic aspects of lingual propulsive activity in swallowing. *Dysphagia* **4**, 136–45.

Hald, A. (1965). Statistical Theory with Engineering Applications. New York: John Wiley and Sons.

Hector, M.P. & Linden, R.W.A. (1987). The possible role of periodontal mechanoreceptors in the control of parotid secretion in man. *Q. J. Exp. Physiol.* **72**, 285–301.

Herring, S.W. & Scapino, R.P. (1973). Physiology of feeding in miniature pigs. *J. Morphol.* **141**, 427–60.

Hiiemae, K.M. (1976). Masticatory movements in primitive mammals. In *Mastication*, ed. D.J. Anderson & B. Matthews, pp. 105–18. Bristol: Wright and Sons.

Hiiemae, K.M. (1993). Process and mechanism: mechanoreceptors in the mouth as the primary modulators of rhythmic behavior in feeding? In *Sensory*

Research: Multimodal Perspectives, ed. R.T. Verrillo, pp. 263–84. Hillsdale NJ: Lawrence Erlbaum.
Hiiemae, K.M. & Crompton, A.W. (1985). Mastication, food transport and swallowing. In *Functional Vertebrate Morphology*, ed. M. Hildebrand, D.M. Bramble, K.F. Liem and D.B. Wake, pp. 262–90. Cambridge MA: The Belknap Press of Harvard University Press.
Hiiemae K.M. & Kay, R.F. (1973). Evolutionary trends in the dynamics of primate mastication. In *Craniofacial Biology of Primates*, ed. M.R. Zingeser, pp. 28–64. Symposium, 4th International Congress of Primatology **3**. Basel: Karger.
Hiiemae, K.M., Hayenga, S.M. & Reese A. (1995). Patterns of jaw and tongue movement in a cinefluorographic study of feeding in the macacque. *Arch. Oral Biol.* **40**, 229–46.
Hiiemae, K.M., Heath, M.R., Heath, G,. Kazazoglu, E., Murray J., Sapper D. & Hamblett, K. (1996). Natural bites, food consistency and feeding behaviour in man. *Arch. Oral Biol.* **41**, 175–89.
Hiiemae, K.M. & Palmer, J.B. (1999). Food transport and bolus formation during complete sequences on foods of different initial consistency. *Dysphagia* **14**, 31–42.
Hiiemae, K.M. Palmer, J.B. Medicis, S.W., Jackson, B.S. & Hegener, J. (1999). Talking and eating: how the hyoid complex accomodates both activities. Paper presented at IADR Vancouver, March 12. *J. Dent. Res.* (Special issue). (Abstract.)
Hiiemae, K.M., Thexton, A.J. & Crompton, A.W. (1978). Intra-oral transport: a fundamental mechanism of feeding? In *Muscle Adaptation in the Craniofacial Region*, ed. D.Carlson & J. McNamara, pp. 181–208. Monograph No. 8. University of Michigan.
Hylander, W.L. & Crompton A.W. (1986). Jaw movements and patterns of mandibular bone strain during mastication in the monkey *Macaca fascicularis. Arch. Oral Biol.* **31**, 841–8.
Hylander, W.L., Johnson, K.R. & Crompton, A.W. (1987). Loading patterns and jaw movements during mastication in *Macaca fascicularis:* a bone-strain, electromyogrphic and cinefluorographic analysis. *Am. J. phys. Anthrop.* **72**, 287–314.
Kay, R.F. (1978). Molar structure and diet in extant Cercopithecidae. In *Development, Function and Evolution of Teeth*, ed. P.M. Butler & K.A. Joysey, pp. 309–34. London: Academic Press.
Kay, R.F. & Hiiemae, K.M. (1974). Jaw movement and tooth use in recent fossil primates. *Am. J. phys. Anthrop.* **40**, 227–56.
Kier, W. & Smith, K.K. (1985). Tongues, tentacles and trunks: the biomechanics of movement in muscular hyrostats. *Zool. J. Linn. Soc.* **83**, 307–24.
Laitman, J.T. & Reidenberg, J.S. (1993). Specialisations of the human upper respiratory and upper digestive systems as seen through comparative and developmental anatomy. *Dysphagia* **8**, 318–25.
Lieberman, P. (1975). *On the Origins of Language: An Introduction to the Evolution of Speech.* New York:Macmillan.
Lieberman, P. (1984). *The Biology and Evolution of Language.* Cambridge, MA: Harvard University Press.
Lieberman, P., Laitman, J.T., Reidenberg, J.S. & Gannon, P.J. (1992). The anatomy, physiology, accoustics and perception of speech: essential elements in analysis of the evolution of human speech. *J. hum. Evol.* **23**, 447–67.
Lucas, P.W. (1989). A new theory relating seed processing by primates to their

relative tooth sizes. In *The Growing Scope of Human Biology. Proceedings of the Australasian Society for Human Biology*, vol. 2, ed. L. H. Schmitt, L. Freedman & N.W. Bruce, pp 37–49. Perth: Centre for Human Biology, University of Western Australia.
Lucas P.W. & Corlett, R.T. (1991). Relationship between the diet of *Macaca fascicularis* and forest phenology. *Folia Primatol.* **57**, 201–15.
Lucas, P.W. & Luke, D.A. (1984). Basic principles of food breakown. In *Food Acquisition and Processing in Primates*, ed. D.J. Chivers, B.A. Wood & A. Bilborough, pp. 283–301. New York: Plenum Press.
Lucas, P.W., Ow, R.K.K., Ritchie, G.M., Chew, C.L. & Keng, S.B. (1986). Relationship between jaw movement and food breakdown in human mastication. *J. Dental Res.* **65**, 400–4.
Lucas, P.W. & Teaford, M.F. (1994). Functional morphology of colobine teeth. In *Colobine Monkeys: Their Ecology, Behaviour and Evolution*, ed. A.G. Davies & J.F. Oates, pp. 173–204. Cambridge: Cambridge University Press.
Lund, J.P. (1991). Mastication and its control by the brainstem. *Crit. Rev. Oral Biol. Med.* **2**, 33–64.
Luschei, E.S. & Goodwin, G.M. (1974). Patterns of mandibular movement and jaw muscle activitity in the monkey. *J. Neurophysiol.* **37**, 954–66
Miller, A.J. (1982). Deglutition. *Physiol. Rev.* **62**, 129–84.
Miller, A.J. (1986). Neurophysiological basis of swallowing. *Dysphagia* **1**, 91–100.
Moore, W.J. & Lavelle, C.L.B. (1974). *Growth of the Facial Skeleton in the Hominoidea.* London: Academic Press.
Palmer, J.B., Hiiemae, K.M. & Liu, J. (1995). Motions of the tongue and jaw in feeding. *Arch. Phys. Med. Rehabil.* **76**, 1026.
Palmer, J.B. Hiiemae, K.M. & Lui, J. (1997). Tongue-jaw linkages in feeding. *Arch. Oral Biol.* **42**, 429–41.
Palmer, J.B., Rudin, N.J., Lara, G. & Crompton, A.W. (1992). Coordination of mastication and swallowing. *Dysphagia* **7**, 187–200.
Schwarz, G., Enomoto, S., Valiquette, C. & Lund, J.P. (1989). Mastication in the rabbit: a description of movement and muscle activity. *J. Neurophysiol.* **62**, 273–87.
Stone, M. (1991). Toward a model of three dimensional tongue movement. *J. Phonetics* **19**, 309–20.
Swindler, D.R. (1976). *Dentition of Living Primates.* London: Academic Press.
Thexton, A.J. (1992). Mastication and swallowing: an overview. *Brit. Dent. J.* **173**, 197–206.
Thexton, A.J. (1984). Jaw, tongue and hyoid movement: a question of synchrony? *J. Royal Soc. Med.* **77**, 1010–19.
Thexton, A.J. & Crompton, A.W. (1998). The control of swallowing. In *The Scientific Basis of Eating*, ed. R.W.A. Linden, *Frontiers of Oral Biology*, vol 9, pp. 168–222. Basel: Karger.
Thexton, A.J. & Hiiemae, K.M. (1997). The effect of food consistency on jaw movement in the macaque: a cineradiographic study. *J. Dent. Res.* **76**, 552–60.
Thexton, A.J., Hiiemae, K.M. & Crompton A.W. (1980). Food consistency and particle size as regulators of masticatory behavior in the cat. *J. Neurophysiol.* **44**, 456–74.
Thexton, A.J. & McGarrick, J.D. (1988). Tongue movement of the cat during lapping. *Arch. Oral Biol.*. **33**, 331–9.
Thexton, A.J. & McGarrick, J.D. (1989). Tongue movement in the cat during the intake of solid food. *Arch. Oral Biol.* **34**, 239–48.

Tomura Y., Ide, Y. & Kamijo, Y. (1981). Studies on the morphological changes of the tongue movements during mastication by X-ray TV cinematography. In *Orofacial Sensory and Motor Functions*, eds Y. Kawamura & R. Dubner, pp. 45–52. Tokyo: Quintessence Publishing.

Turnbull, W.D. (1970). Mammalian masticatory apparatus. *Fieldiana Geolog.* ***18***, 153–356.

Warwick James, W. (1960). *The Jaws and Teeth of Primates.* London: Pitman Medical Publishing.

Van der Bilt, A., Van der Glas, H.W., Olthoff, L.W. & Bosman, F. (1991). The effect of particle size reduction on the jaw gape in human mastication. *J. Dent. Res.* **70**, 931–7.

9

Evolutionary morphology of the skull in Old World monkeys

MATTHEW J. RAVOSA AND LORNA P. PROFANT

Introduction

In an early review of Old World monkeys, Schultz (1970: 41) comments that "In most of their basic morphological characters . . . Old World monkeys are much more uniform than the other major groups of primates". On the other hand, cercopithecid subfamilies clearly evince divergent functional specializations of the skull. "Colobines apparently have optimized bite-force magnitudes at the expense of a reduction in jaw gape in order to masticate leaves more efficiently. An increase in jaw gape is . . . advantageous to more frugivorous and/or terrestrial primates since they eat large food objects, which require extensive incisal preparation, and/or because of canine displays or canine slashing" (Hylander, 1979b:229).

Experimental studies have been instrumental in characterizing dynamic functional determinants of skull form in Old World monkeys and other primates (Luschei and Goodwin, 1974; McNamara, 1974; Hylander, 1979a–c, 1984, 1985; Bouvier and Hylander, 1981; Hylander *et al.*, 1987, 1991a,b, 1992, 1998; Dechow and Carlson, 1990). In turn, morphological studies of cercopithecid subfamilies have greatly enhanced our knowledge of the functional bases of such craniodental variation (Hylander, 1975; Walker and Murray, 1975; Kay, 1978; Kay and Hylander, 1978; Bouvier, 1986a,b; Ravosa, 1988, 1990, 199la–c, 1996; Lucas and Teaford, 1994).

Some research on the cercopithecid skull emphases the role of phylogeny in channeling morphological variation at the inter- and intra-specific level (Freedman, 1962; Fooden, 1975, 1988, 1990; Cochard, 1985; Cheverud and Richtsmeier, 1986; Leigh and Cheverud, 1991; Ravosa, 1991a,c; Shea, 1992; Richtsmeier *et al.*, 1993; Profant and Shea, 1994; Ravosa and Shea, 1994; Profant, 1995). Cranial ontogenetic scaling data (Shah and Leigh, 1995) have been useful in testing phylogenetic hypotheses regarding papionin affinities based on molecular systematic analyses (Disotell *et al.*, 1992;

Disotell, 1994; van der Kuyl *et al.*, 1995). Work on *Victoriapithecus* (Benefit and McCrossin, 1991, 1993a; Benefit, 1995) sheds further light on earlier discussions of craniofacial form in ancestral cercopithecoids and catarrhines (Vogel, 1968; Delson, 1975; Delson and Andrews, 1975; Szalay and Delson, 1979; Strasser and Delson, 1987).

This chapter aims to summarize and highlight research on the cercopithecid skull since 1970. We focus on functional morphology and ontogeny, assessing the current state of knowledge and providing additional analyses of function, ontogeny, and evolution of the craniofacial skull.

Functional analyses of the Old World monkey skull

Experimental and morphological studies of mandibular form and function

Parasagittal bending of the corpus and dorsoventral shear of the symphysis

Experiments on *Macaca fascicularis* indicate that the balancing-side mandibular corpus experiences parasagittal bending during mastication (Hylander, 1979a,b, 1981, 1984; Hylander *et al*,. 1998). This bending causes dorsoventral shear stress at the mandibular symphysis proportional to the force transmitted across the symphysis from the balancing to the working side of the jaw (Hylander, 1977, 1979a,b, 1984).

Research on *Macaca* and *Papio* indicates that a tough diet requires greater balancing-side jaw-muscle force recruitment (Luschei and Goodwin, 1974; Hylander, 1979a–c; Hylander and Johnson, 1985, 1994; Hylander *et al.*, 1992, 1998). This increases the likelihood of fatigue failure of the corpus and symphysis because of greater repetitive or cyclical loading (Hylander, 1979b; Bouvier and Hylander, 1981).

Jaw-scaling analyses in living and fossil cercopithecids indicate that colobines have relatively deeper corpora in the molar region than cercopithecines (Fig. 9.1). This difference appears designed to counter greater parasagittal bending of the balancing-side corpus during the mastication of a more obdurate diet of leaves and seeds (Hylander, 1979b; Bouvier, 1986a; Ravosa, 1996). Colobines also possess relatively larger symphyses to resist increased dorsoventral shear of the symphysis during chewing and biting of a tougher diet (Figs. 9.2 and 9.3; Ravosa 1996). Ontogenetic comparisons of mandibular cross-sectional measures between *Nasalis larvatus* and *M. fascicularis* demonstrate similar subfamily differences in jaw scaling, with the more folivorous proboscis monkey having a more robust mandible than the macaque (Ravosa, 1991c).

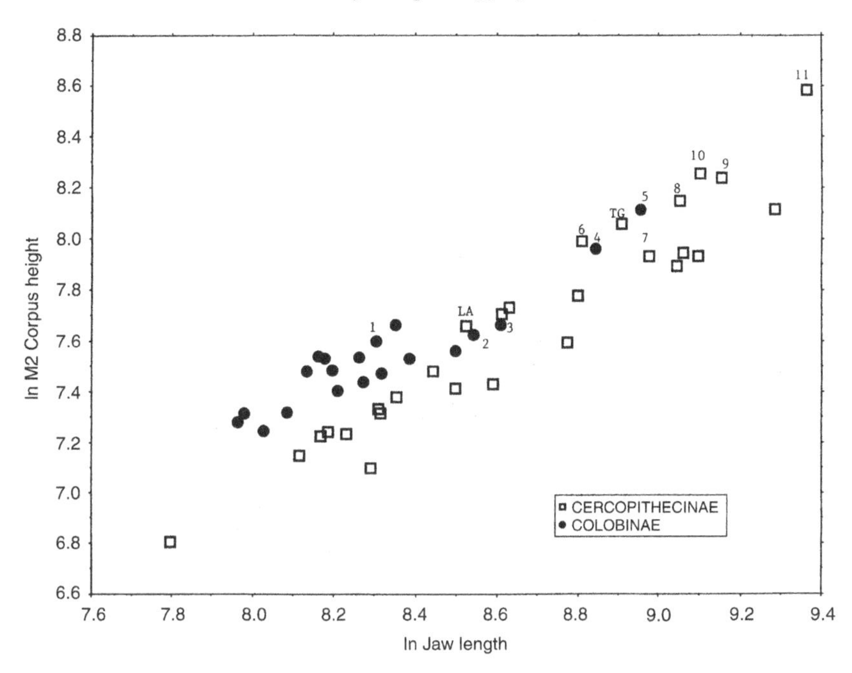

Fig. 9.1. A plot of ln corpus height versus ln jaw length in a pooled-sex sample of 29 cercopithecine species (squares) and 22 colobine species (circles). Analysis of covariance (ANCOVA, $p<0.01$) indicates that the colobine regression line for corpus height is significantly transposed above that for cercopithecines (cf., Bouvier, 1986a; Ravosa, 1996). This supports the hypothesis that leaf-, seed- and grass-eating primates often have more robust jaws than do more frugivorous and omnivorous taxa due to greater parasagittal bending of the balancing-side mandibular corpus during mastication. Note that living and fossil *Theropithecus*, all of which have or are inferred to have had tougher diets, also plot at the upper end of the cercopithecine scatter much like the colobines. *Lophocebus albigena* also plots high, due to greater parasagittal bending of both corpora during habitual, powerful incisal biting (cf., Bouvier 1986a). (Fossil key: 1 = *Mesopithecus pentelici*; 2 = colobine species " A "; 3 = *Cercopithecoides williamsi*; 4 = *Dolichopithecus ruscinensis*; 5 = *Rhinocolobus turkanensis*; 6 = *Paracolobus chemeroni*; 7 = *Parapapio jonesi*; 8 = *Dinopithecus ingens*; 9 = *Paradolichopithecus arvernensis*; 10 = *Papio baringensis*; 11 = *Theropithecus oswaldi*; 12 = *T. brumpti*; TG = *T. gelada*; LA = *Lophocebus albigena*.)

Theropithecus gelàda, a large "grazer" (Dunbar, 1977), also has more robust corpora than other cercopithecines (Hylander, 1979b; Bouvier, 1986a). Most large-bodied extinct papionins are similar to geladas, and living and fossil colobines, in possessing relatively deep jaws (Fig. 9.1; Ravosa, 1996). As many of these taxa are inferred to have had tough diets (Jolly, 1970a,b; Benefit and McCrossin, 1990; Delson and Dean, 1993; Jablonski, 1993a; Teaford, 1993; Lucas and Teaford, 1994), this morphological pattern

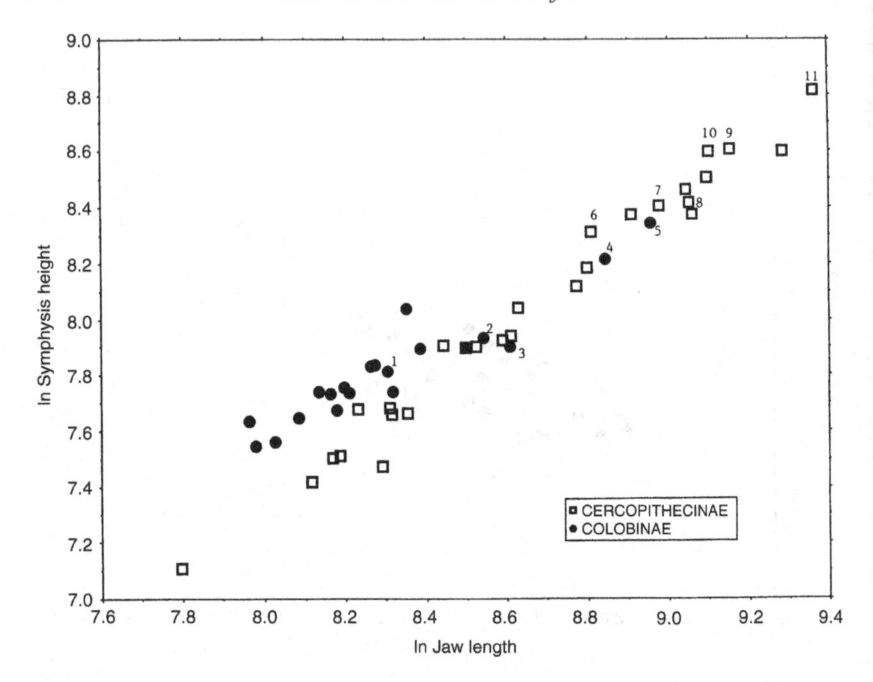

Fig. 9.2. A plot of ln symphysis height versus ln jaw length in 29 cercopithecine (squares) and 22 colobine (circles) species. Note that colobine symphysis dimensions are typically transposed above those for cercopithecines, thus supporting the hypothesis that primates like colobines with tougher diets experience greater dorsoventral shear than other taxa (cf., Ravosa 1996). (Fossil key follows Fig. 9.1)

is due to greater balancing-side muscle-force recruitment during powerful chewing and biting (Luschei and Goodwin, 1974; Hylander *et al.*, 1992). Therefore, perhaps the reason cercopithecine jaw-scaling patterns converge on the colobine scatter at larger sizes is due to the effects of obdurate diets in the larger extinct papionins (Ravosa, 1996).

Experimental and morphological analyses of incision in macaques indicate that both mandibular corpora experience parasagittal bending (Hylander, 1979a,b, 1981). This explains the presence of greater mandibular depth in the postcanine region of species like *Lophocebus albigena* (Fig. 9.1), known to engage in habitual, forceful incisal biting (Hylander, 1979b; Bouvier, 1986a,b).

Lateral transverse bending of the symphysis and corpus ("wishboning")

In macaques, the mandibular symphysis and corpus experience lateral transverse bending or "wishboning" during mastication, due to a laterally-directed component of jaw-muscle force from the balancing-side deep-

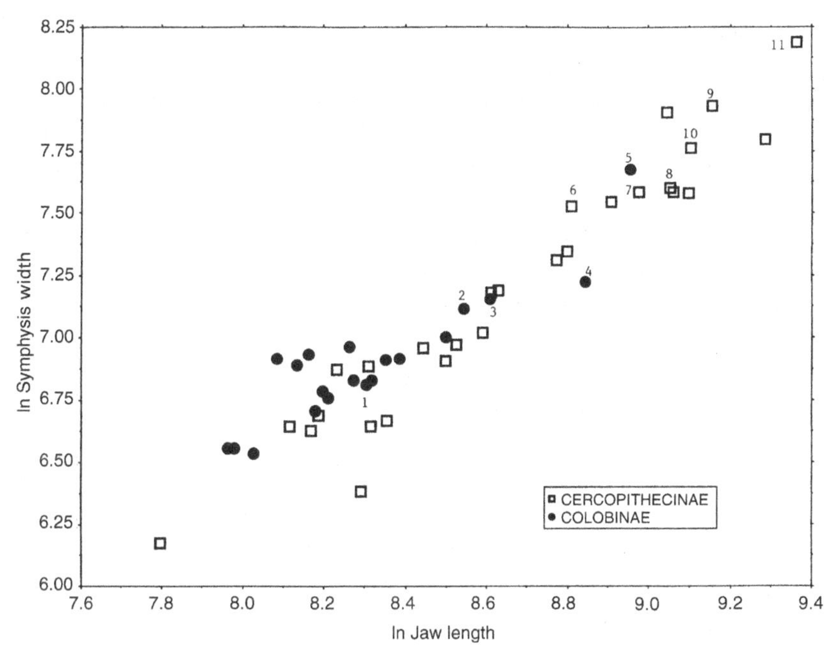

Fig. 9.3. A plot of ln symphysis width versus ln jaw length in 29 cercopithecine (squares) and 22 colobine (circles) taxa. Colobine symphysis means again tend to plot above those for cercopithecines, thus suggesting that, as a group, they likely experience greater wishboning. This pattem is surprising given the presence of subfamily differences in the scaling of symphyseal curvature (Figure 9.4) (cf., Ravosa, 1996). (Fossil key follows Figure 9.1.)

masseter muscle at the end of the masticatory power stroke, and a laterally-directed component of bite force (and perhaps muscle force) on the working side (Hylander, 1984, 1985; Hylander *et al.*, 1987, 1998; Hylander and Johnson, 1994).

In cercopithecines, size-related changes in the relation between mandibular length and breadth (=arch width) affect the degree of symphyseal curvature and thus the amount and distribution of wishboning stress at the symphysis (Hylander, 1984, 1985). Since wishboning results in high stress concentrations and high strain magnitudes at the inner surface of the symphysis, increased wishboning stress is best resisted by increasing the labiolingual width of the symphysis. In cercopithecines, as jaw length scales with positive allometry versus jaw breadth, and all taxa presumably experience wishboning, symphysis width scales positively to counter allometric increases in wishboning due to greater symphyseal curvature in larger forms with more elongate jaws (Hylander, 1984, 1985; Vinyard and Ravosa, 1998).

Colobines exhibit similar positive allometry of jaw length versus jaw breadth (least-squares slope of 1.794 vs. 1.622 for cercopithecines), but do not show the cercopithecine pattem of positive allometry of symphyseal width (slope of 1.224 vs. 0.875 for colobines) (Ravosa, 1996). Thus, although colobines exhibit allometric increases in symphyseal curvature, wishboning levels do not appear to increase proportionally. At a common jaw length, colobines have greater jaw breadths, resulting in a lesser degree of symphyseal curvature than a cercopithecine of similar size (Fig. 9.4). Nonetheless, especially at smaller sizes, colobines possess more robust symphyses than cercopithecines (Figs. 9.2, 9.3). The presence of subfamily scaling differences means that colobines may, on average, experience elevated wishboning levels because of a tougher diet requiring greater balancing-side muscle force recruitment (Ravosa, 1996). If this loading regime exists in all catarrhines (Hylander, 1985), the jaw-muscle activity pattern underlying colobine wishboning becomes less pronounced with size, which is opposite the cercopithecine pattern. Alternatively, perhaps there are subfamily differences in wishboning and dorsoventral shear levels, both of which significantly influence symphyseal form.

Axial torsion of the corpus and vertical bending of the symphysis

During both incision and unilateral mastication, the mandibular corpora of macaques are twisted about their long axes (Hylander, 1979a, 1981). During mastication and incision, axial torsion of both corpora is due to the overall jaw-muscle force resultants, both of which lie lateral to the mandible on each side, as well as the working-side bite force during mastication. Axial twisting everts the lower borders and inverts the upper borders of both corpora, such that the symphysis is bent vertically (Hylander, 1979b, 1981). This symphyseal bending causes compression along the alveolar surface of the symphysis and tension along its base.

Analysis of mandibular allometry in Old World monkeys demonstrates that, in addition to having deeper corpora, colobines also possess relatively thick corpora (Fig. 9.5). This is an adaptation to greater axial torsion produced during the mastication of tough leaves and seeds (Bouvier, 1986a; Ravosa,1996). Comparisons of New and Old World monkey jaw-scaling patterns indicate differences in the relative thickness of the corpus, with cercopithecoids having thicker corpora (Bouvier, 1986b), probably due to greater masseter recruitment and thus greater torsion of the corpus (Bouvier, 1986b; Bouvier and Tsang, 1990).

Among Old World monkeys, fossil taxa tend to have relatively wider corpora relative to extant species (Fig. 9.5; Ravosa, 1996). Macaques,

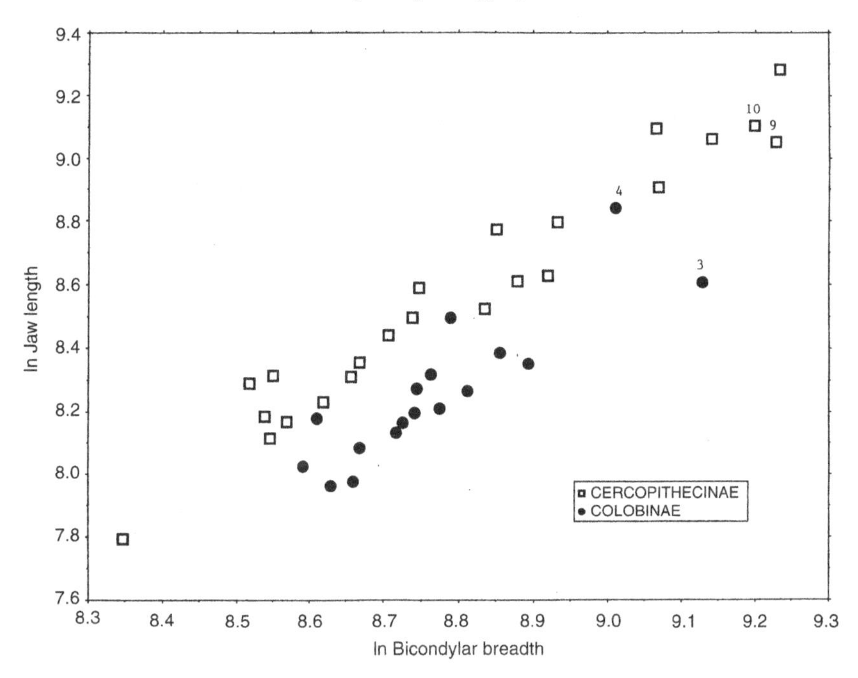

Fig. 9.4. A plot of ln jaw length versus ln jaw breadth in 29 cercopithecines (squares) and 22 colobines (circles). This indicates that, at a given jaw length, cercopithecines exhibit increased symphyseal curvature relative to colobines. In each group, the degree of curvature also increases with size, such that larger taxa have more curved symphyses than smaller species (cf., Ravosa, 1996). (Fossil key follows Figure 9.1.)

which ingest tougher foods than their congeners – *M. thibetana* (Takahashi and Pan, 1994) and *M. fuscata* (Antón, 1996) – also have relatively thicker corpora. Antón (1996) notes parallels in the way *M. fuscata* differs in skull form from other macaques and in the way colobines and cercopithecines differ (Ravosa, 1990). In fossil and extant *Theropithecus*, and perhaps in other monkeys with obdurate diets, a wider corpus may reflect a more laterally displaced masseter (Jablonski, 1993a; Antón, 1996) and greater emphasis during chewing by the masseter and medial pterygoid.

Temporomandibular joint loading and function

Theoretical debate has focused on the load-bearing nature of the mammalian temporomandibular joint (TMJ), with morphological evidence marshalled both for and against the influence of joint-reaction forces on TMJ form (Hylander, 1992). *In vivo* data from macaque species indicate that the TMJ is loaded primarily in compression during incision and mastication (Hylander, 1979c, 1992; Hylander and Bays, 1979; Brehnan *et al.*, 1981).

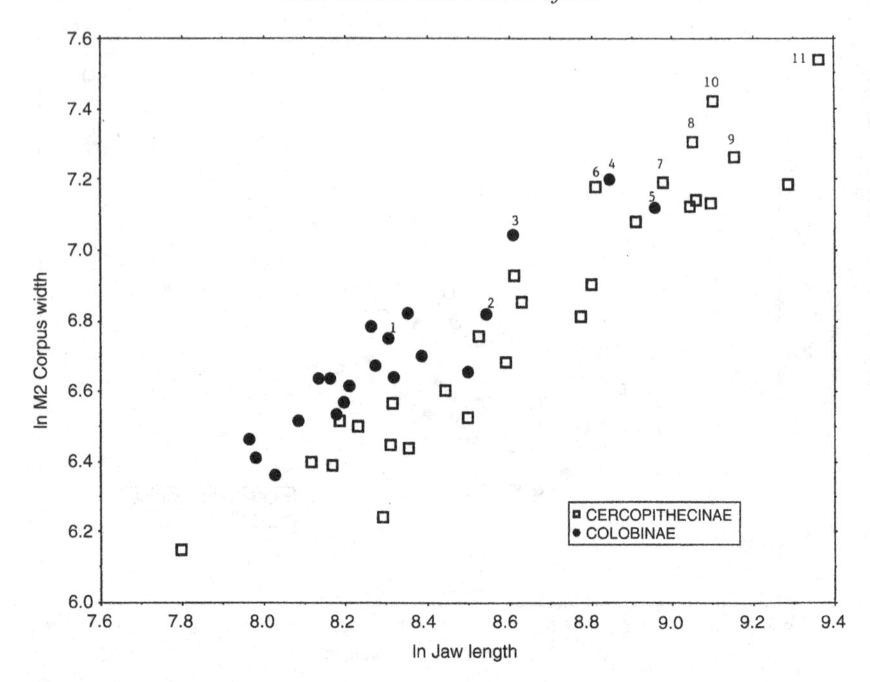

Fig. 9.5. A plot of ln corpus width versus ln jaw length in 29 cercopithecines (squares) and 22 colobines (circles). Colobine corpus measurements are transposed above those for cercopithecines (cf, Bouvier 1986a; Ravosa 1996). Once more, this supports claims that primate leaf- and seed-eaters exhibit relatively larger mandibles. Note that some extinct monkeys have wider corpora than other subfamily members, which likely indicates the presence of elevated axial torsion. (Fossil key follows Figure 9.1.)

On occasion, the working-side TMJ is loaded in tension during biting along the most posterior molars. During mastication, relative levels of TMJ reaction forces vary between working and balancing sides. However, during vigorous molar biting and chewing, the balancing-side TMJ is loaded more than is the working side. Finally, strain levels at the TMJ are higher during incision than during unilateral mastication (Hylander, 1979c, 1992; Hylander and Bays, 1979; Brehnan *et al.*, 1981).

Colobines have mediolaterally wider condyles than cercopithecines, suggesting greater transverse and rotational (about the anteroposterior axis) movements of the condyle during the mastication of a fibrous diet of leaves and/or seeds (Bouvier, 1986a). Cercopithecines have anteroposteriorly longer condyles, apparently due to greater incisal preparation of fruits and nuts, and perhaps more frequent agonistic canine displays (Bouvier 1986a).

Pan *et al.* (1995) found that *Rhinopithecus roxellana* has a longer condyle, thicker corpus and deeper symphysis than macaques and leaf monkeys, suggesting a greater emphasis on incision. Like colobines, leaf-, seed- and grass-eating cercopithecines (such as *M. thibetana* and *Theropithecus*) possess wider condylar heads (Bouvier, 1986a; Jablonski, 1993a; Takahashi and Pan, 1994). This is due to an expansion of the lateral aspect of the condyle, thus suggesting that the lateral portion of the working-side TMJ is loaded more than the medial portion during mastication (Hylander, 1992).

Jablonski's (1993a) recent functional analysis of TMJ morphology in extinct species of *Theropithecus* shows that, compared to its sister taxa, *T. oswaldi* possesses TMJ modifications related primarily to powerful chewing. In particular, *T. oswaldi* lacks an extension of the articular surface onto the posteromedial aspect of the condylar head combined with an anteroposteriorly short lateral portion of the glenoid cavity; both of which suggest a reduced emphasis on canine displays (Jablonski, 1993a). Jablonski argues that the medial elongation of the glenoid of *T. oswaldi* facilitates greater anteroposterior translation of the working-side condyle during unilateral mastication. While this explanation is plausible, Jablonski (1993a) finds that *T. gelada* and *T. oswaldi* exhibit similar size-adjusted measures of condylar width, which brings into question the exact nature of the purported functional association between condylar and glenoid form.

Jaw function and relative canine size: T-complex and gape

Jolly (1970a,b) identifies a suite of craniodental characters shared by early hominids and large, extinct forms of *Theropithecus* (but not *T. brumpti*): a robust mandibular corpus in the molar region; deep posterior maxilla; vertical ascending ramus; anteriorly set, vertically-oriented temporalis; marked postorbital constriction; and relatively small incisors and canines. Jolly argued that this "T-complex" is linked to a dietary shift emphasizing small, tough objects, which was fundamental to the evolutionary differentiation of basal australopithecines from a chimplike ancestor.

Martin (1993) claims that, as living *Theropithecus* have relatively small brains and average-sized canines, the functional argument linking the "T-complex" requires further scrutiny. In fact, Jolly's (1970b) original model focuses solely on morphological patterns in fossil papionins, so that Martin's conclusions based on extant *Theropithecus* are irrelevant. In addition, predictions about relative brain size were never incorporated into the "T-complex" model. As for canine size, *T. baringensis* and more recent members of the *T. oswaldi* lineage have relatively smaller canines (Jablonski, 1993a), much as Jolly (1970b) suggests. Several recent analyses of dental function also confirm Jolly's inferences about the dietary proclivities of extinct papionins (Benefit and McCrossin, 1990; Delson and Dean, 1993; Jablonski, 1993a; Teaford, 1993). Fossil *Theropithecus* possess somewhat thicker corpora than other larger-bodied baboons (Fig. 9.5; Ravosa, 1996). It appears that Jolly (1970b) is correct in identifying several functional parallels in the evolution of craniomandibular form in fossil *Theropithecus* and basal hominids.

Following from prior work on the functional and developmental relationship between jaw size and canine size in Old World monkeys (Siegel, 1972; Lucas, 1981, 1982) and biomechanical work on gape in mammals (Herring, 1972, 1975; Herring and Herring, 1974; Emerson and Radinsky, 1980; Lucas *et al.*, 1986), Ravosa (1990) used an allometric analysis to investigate the functional significance of differences in facial form between cercopithecines and colobines. Cercopithecines were examined for specializations related to the possession of enlarged canines used in agonistic displays, while colobines were examined for the presence of features increasing the efficiency of processing a tougher, obdurate diet (Hylander, 1979b).

Results indicate that cercopithecines have a facial skull that increases with stronger positive allometry than that of colobines. Given a similar angle between the maxilla and the mandible during maximum jaw opening, a more elongate facial skull facilitates increased gape and the presence of larger canines. Cercopithecines also have a larger gonial angle, which enhances gape by increasing the distance between upper and lower canines for a given amount of mandibular retraction (Ravosa, 1990). On the other hand, at a given basicranial length, colobines possess a relatively longer masseter lever arm, a shorter facial skull, and a larger masseter/medial-pterygoid complex (Ravosa, 1990). Colobines can recruit a lower masseter force to produce similar molar bite forces as cercopithecines (Hylander, 1979b; Ravosa, 1990, 1996), which appears important when ingesting greater amounts of food and greater percentages of tough foodstuffs over the course of a day. Apart from having a somewhat shorter face, species of *Theropithecus* variably exhibit relatively longer masseter lever arms, higher jaw joints, and other craniodental specializations similar to leaf- and seed-eating colobines (Jolly, 1970a,b; Jablonski, 1993a; Ravosa, 1996).

Experimental and morphological studies of circumorbital form and function

Facial torsion during unilateral mastication

Greaves (1985, 1995) suggests that forces experienced during molar biting and chewing produce a net twisting of the facial skull about the cranium's long axis. Owing to to its position at the interface of the upper face and cranial vault, the circumorbital region is purportedly designed to resist facial torsion (see also Rosenberger, 1986). Under a torsional loading regime, principal strains at the circumorbit are predicted to lie 45 degrees relative to the long axis of the cranium. It follows from the facial-torsion model that the directions of tensile and compressive strains should exhibit a reversal pattern when molar biting and chewing shifts from one side of the face to the other side.

Experimental work in *Papio anubis* and *M. fascicularis* provides mixed support for the facial-torsion model. Controlling for diet, the cercopithecine circumorbit experiences much lower strain levels than along the mandibular corpus (Hylander *et al.*, 1991a,b). The presence of this significant strain gradient along the facial skull suggests that the circumorbital loading regime in question has a secondary influence on circumorbital form, regardless of whether principal-strain directions correspond to that specific model (Hylander *et al.*, 1991a,b; Hylander and Johnson, 1992; Hylander and Ravosa, 1992). A study of circumorbital strains, produced during tetanic stimulation of the jaw-adductor muscles in *Cercopithecus aethiops*, also indicates very low strain magnitudes (Oyen and Tsay, 1991). Strain magnitudes from both experimental analyses demonstrate that cercopithecid circumorbital structures are overbuilt to counter masticatory stresses.

As this is not the case for the cercopithecine mandible, it is unlikely that the primary factors underlying supraorbital torus formation are masticatory in nature. Although *Macaca* and *Papio* evince a characteristic reversal pattern, when mastication shifts from one side of the face to the other side, circumorbital strain directions in both papionins are *opposite* predictions of the facial-torsion model, such that the working-side browridge experiences compression during molar biting, not tension (Hylander *et al.*, 1991a,b; Hylander and Johnson, 1992; Hylander and Ravosa, 1992). Thus, in the case of the primate supraorbital torus, "there has been evolutionary selection for an increase in the size of a structure composed of tissue whose sensitivity to load bearing is now superfluous" (Lanyon and Rubin, 1985: 1).

Scaling comparisons indicate that cercopithecines and colobines share similar allometric trajectories for browridge height and projection (Ravosa, 1988, 1991b; Hylander and Ravosa, 1992). Given the presence of significant subfamily differences in mandibular corpus and symphysis scaling patterns (Bouvier, 1986a; Ravosa, 1996), this suggests a secondary role for facial torsion, or any masticatory loading regime, as a determinant of circumorbital form (Ravosa, 1988, 1991b; Hylander and Ravosa, 1992). Since colobines have significantly larger jaws because of the greater stresses encountered during mastication of a tougher diet, then colobine supraorbital size should also be relatively larger if similarly affected by subfamily differences in stress levels. Ontogenetic data from *M. fascicularis* also provide only limited support for the effect of facial torsion on circumorbital morphology (Ravosa, 1991a; Hylander and Ravosa, 1992).

Frontal bending during anterior dental loading

Investigators have argued that supraorbital torus formation is an adaptive response to anterior dental loading (Hooton, 1936; Endo, 1966, 1970; Oyen *et al.*, 1979; Wolpoff, 1980, 1996; Oyen and Russell, 1982; Russell, 1982, 1985). This model predicts that incisor-bite forces transmitted superiorly deform the glabellar region of the browridge, and that bilateral contraction of the jaw adductors results in bending of the interorbit and dorsal orbit in the frontal plane. The model also predicts that greater supraorbital stress is produced during incision than during mastication.

In a study of olive baboons, Oyen *et al.* (1979) posit that ontogenetic and interspecific changes in the elongation of the incisor load arm (relative to the masseter lever arm) require compensatory increases in jaw-adductor force levels to maintain similar incisor-bite forces. Enlarged browridges are thought to resist such increased muscle forces (see also Russell, 1982). Russell (1985) suggests that in order to counter anterior dental loads transmitted superiorly, supraorbital torus formation occurs in primates lacking a more vertically-oriented frontal bone.

Macaques and baboons are well-suited for testing the anterior dental loading model, since they possess relatively large incisors (Hylander, 1975), combined with significant ontogenetic increases in browridge development (Oyen *et al.*, 1979; Ravosa, 1991a,c). *In vivo* results offer mixed support for the anterior dental loading model. Circumorbital strain directions during incision are as predicted by this model, but the presence of a significant strain gradient indicates that the browridges are overbuilt to counter routine masticatory loads (Hylander *et al.*, 1991a,b). Circumorbital strain levels during incision are not higher than those produced during mastication, which does not support the anterior dental loading model (Hylander *et al.*, 1991a,b; Hylander and Johnson, 1992; Hylander and Ravosa, 1992).

Interspecific scaling studies suggest a secondary role for incisor loading on the development of circumorbital form (Ravosa, 1988, 1991b; Hylander and Ravosa, 1992). Cercopithecines and colobines show similar browridge scaling patterns, which counters predictions of the anterior dental loading model. Moreover, contrary to the predictions of Oyen *et al.*, (1979), colobines exhibit size-related increases in browridge dimensions despite isometry of the incisor load arm relative to the masseter lever arm (Ravosa, 1988, 1990, 1991b). Macaque ontogenetic analyses also point to a minor role for anterior dental loading on browridge formation (Ravosa, 1991a; Hylander and Ravosa, 1992).

Browridge development, allometry and spatial factors

Analyses of interspecific primate samples, and an ontogenetic sample of *M. fascicularis*, demonstrate that both supraorbital torus height and projection are correlated with skull size (Ravosa, 1988, 199la–c). These studies also indicate that browridge elongation is influenced by variation in the position of the orbits relative to the frontal lobes (Moss and Young, 1960; DuBrul, 1965; Smith and Ranyard, 1980; Shea, 1986; Ravosa, 1988, 1991a,b; Hylander and Ravosa, 1992; Vinyard, 1994). This spatial variation is described by an angular measure describing the orientation of the orbital axis relative to the frontal lobes, and a linear measure of the anteroposterior disjunction between the orbital aperture and anterior neurocranium (Ravosa, 1988, 1991a,b).

Interspecific and ontogenetic studies of Old World monkeys show that the purported association between forehead development, facial orientation, and incisor loading (Russell, 1985) actually tracks the spatial relationship between browridge projection and orbit orientation relative to the anterior cranial base (Ravosa, 1988, 1991a,b). Males and females of *M. fascicularis* exhibit ontogenetic scaling of browridge and postorbital-bar dimensions, such that sexual differences in adult circumorbital morphology are due simply to the differential extension of common growth patterns (Ravosa 1991a,c). The putative link between browridge formation and the elongation of dental load arms relative to masticatory muscle lever arms (Oyen *et al.*, 1979; Russell, 1982) is shown to be a spurious correlation reflecting increases in overall skull size during growth (Ravosa, 1991a).

Interorbital function and variation

Experimental analyses also have implications for interpreting the significance of interorbital form and function between colobines and cercopithecines (Verheyen, 1962; Vogel, 1968; Delson, 1975; Szalay and Delson, 1979; Strasser and Delson, 1987). Based on a morphological study of the Eskimo skull, Hylander (1977) posits that a wider interorbit may be designed to resist greater parasagittal/sagittal shear forces during mastication and incision. He also notes a link between folivory and interorbital width in primates such as colobines.

Strain data from the rostral interorbital region in papionins (Hylander *et al.*, 1991a,b) permit a direct investigation of the functional significance of differences in interorbital form between colobines and cercopithecines. During mastication, strain magnitudes at the papionin interorbit are quite low relative to mandibular levels. As noted above, craniofacial strain gradients suggest a secondary role of masticatory stress on circumorbital form. Moreover, strain-direction data do not indicate the presence of interorbital shearing. Taken a step further, *in vivo* data suggest that subfamily variation in interorbital width is unrelated to masticatory stress levels and, thus, to differences in dietary consistency. Experimental data are crucial for testing the validity of a functional model, as comparative criteria alone would have supported a biomechanical explanation for the relatively wide interorbits in colobines, i.e. a tougher diet resulting in greater cyclical loading of the interorbit.

In a study of Old World monkeys, and 40 cercopithecine species in particular, Benefit and McCrossin (1993b: 143) conclude that "the exclusively lacrimal fossa does not seem to be solely correlated to facial elongation . . . but instead seems also to be associated with possession of an extremely narrow interorbital septum and occupation of terrestrial, open-country habitats." Unfortunately, they offer little quantitative data to support the putative link between lacrimal fossa position, i.e. whether the lacrimal fossa is formed exclusively or only in part by the lacrimal bone, interorbital form, and terrestriality (greater protection from sunlight and/or dust). Using Benefit and McCrossin's (1993b) data on intraspecific variation in lacrimal fossa configuration combined with our data on interorbital scaling in 20 cercopithecine species, we re-examined their argument. To facilitate this analysis, six categories were chosen to designate the percentage of adults in each species possessing a lacrimal fossa formed solely by the lacrimal bone.

Kruskal-Wallis analysis of variance ($p<0.01$) in 20 cercopithecine species indicates that larger taxa do exhibit a significantly greater percentage of adults with an exclusively lacrimal fossa (Vogel, 1968; Delson, 1975; Strasser and Delson, 1987; Fig. 9.6). Controlling for face length, taxa with relatively narrow interorbits do not exhibit a higher incidence of the exclusively lacrimal condition (contra Benefit and McCrossin, 1993b). Of the four species plotting below the regression line, only one has an exclusively lacrimal fossa (a "rank" value of six), while the remaining three taxa have lower rank scores of one, two, and three (Fig. 9.6). Thus, although these three monkeys have quite narrow interorbital septa, less than half of the adults in each species have a lacrimal fossa formed solely by the lacrimal. Moreover, terrestrial open-country forms do not evince relatively narrow interorbits (only *Erythrocebus* falls below the line; Fig. 9.6) nor, when holding face length constant, do such taxa evince a greater percentage of the exclusively lacrimal condition relative to less terrestrial monkeys. Interorbital width scales negatively (slope $=0.659$, $r=0.796$, $p=0.001$), such that the two sets of factors posited to be independent by Benefit and McCrossin (1993b), in fact, covary with size in cercopithecines – larger species with more elongate faces and a more exclusively lacrimal fossa *also* exhibit relatively narrow interorbits and tend to be terrestrial (but not necessarily open country).

Ontogenetic analyses of the Old World monkey skull

Our present understanding of the evolutionary implications of patterns of postnatal skull development in Old World monkeys is derived primarily from comparative investigations of cross-sectional growth patterns. This section will focus on those studies of cranial evolution using data on ontogenetic allometry, thereby foregoing an exhaustive review of numerous longitudinal analyses of cranial size and shape.

Comparative studies of craniofacial ontogeny

Intersexual comparisons

The comparative growth literature is dominated by analyses of intersexual variation in craniofacial size and shape (*Macaca*: Fooden, 1975, 1988, 1990; Byrd and Swindler, 1980; Cochard, 1985; Cheverud and Richtsmeier, 1986; Ravosa, 1991a,c; Richtsmeier *et al.*, 1993. *Papio*: Freedman, 1962; Leigh and Cheverud, 1991. *Nasalis*: Ravosa, 1991c). These studies show that patterns of ontogenetic allometry are largely concordant in males and females, such that dramatic intersexual differences in adult craniofacial proportions are often simple allometric correlates of larger skull/body size in males (Gould, 1975; Shea, 1985a). Thus, morphological differences between adults of each sex are linked to the degree of allometry in a system, such that structures growing with strong positive allometry (e.g. browridges, Ravosa, 1991a,c) will be more dimorphic. That some regions of the skull grow with more allometry may be the result of sexual selection. Not all structures, however, are ontogenetically scaled. In macaques, the upper

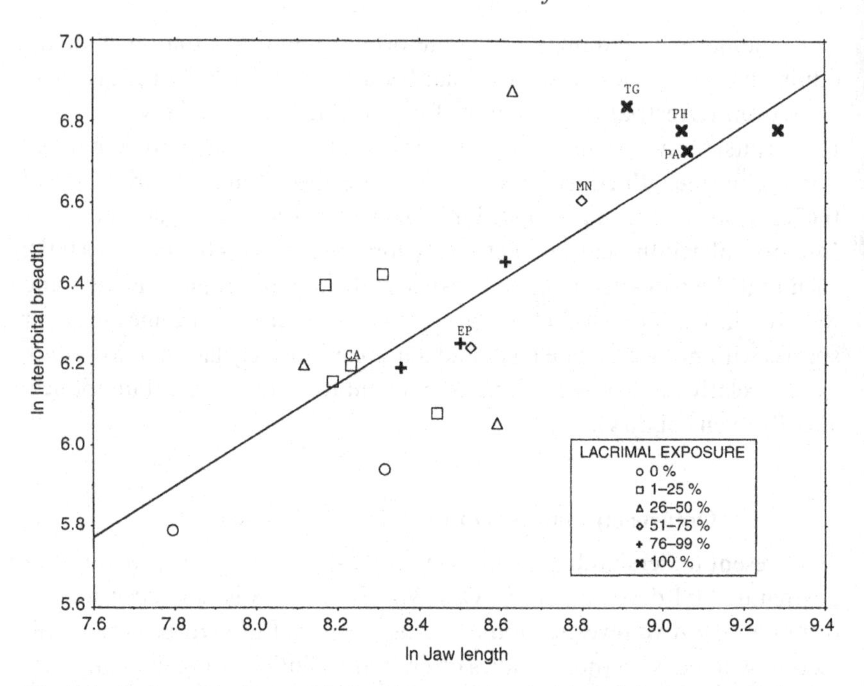

Fig. 9.6. A plot of ln interorbital width versus ln jaw length in 20 cercopithecine taxa. The percentage of individuals within each species having a lacrimal fossa formed exclusively by the lacrimal are from Benefit and McCrossin (1993b) and classed into six groups. Larger taxa exhibit a greater percentage of individuals with an exclusively lacrimal configuration. However, monkeys with relatively narrow interorbits, i.e., those that plot below the regression line, do not exhibit a greater percentage of adults with an exclusively lacrimal fossa. In addition, highly terrestrial cercopithecines (see Key) do not have more narrow interorbits, nor do they have higher lacrimal-exposure percentages as compared to other taxa. (Key: CA = *Cercopithecus aethiops*; EP = *Erythrocebus patas*; MN = *Macaca nemestrina*; TG = *Theropithecus gelada*; PH = *Papio hamadryas*; PA = *P. anubis*.)

canines are relatively larger in males than in females of similar size due to the differential importance of canines in male agonistic displays (Cochard, 1985; Ravosa, 1991c).

Various workers have used adult data to analyze patterns of sexual dimorphism from both allometric (Albrecht, 1978, 1980; Siebert *et al.*, 1984; Kieser and Groeneveld, 1987a,b) and non-allometric (Jablonski, 1986; Pan *et al.*, 1993; Jablonski and Pan, 1995) perspectives. Without knowing how proportions vary during cranial ontogeny, however, it is often impossible to discern how much evolutionary importance to ascribe a sexual difference in the absolute or relative size of a feature. This problem is exemplified by an adult intraspecific study of *Papio ursinus*, in which facial length was found to scale differently in males and females (Kieser and Groeneveld, 1987b). In contrast, an ontogenetic analysis of this taxon indicates that differences in adult proportions "must be ascribed to the similar rate of relative growth proceeding for different lengths of time or/and at different speeds in the two sexes" (Freedman, 1962: 127). The disparate results of these studies can probably be traced to the broader size range and correspondingly higher correlation values encompassed by ontogenetic versus adult data. Though desirable in theory, because of the paucity of subadults of rare (e.g. *Rhinopithecus*, Jablonski and Pan, 1995) or extinct taxa, growth analyses are admittedly not always feasible in practice.

Interspecific comparisons

Despite the relative prevalence of studies dealing with intersexual variation in cranial ontogeny, the allometric underpinnings of interspecific differences in skull morphology have been little explored. The earliest comparative study of growth allometry in the Old World monkeys was carried out by Freedman (1962), who concludes that variation in facial length among living and fossil baboons is linked primarily to the differential extension of a common ontogenetic trajectory. Fooden (1975) arrives at a similar conclusion in an ontogenetic comparison of liontail and pigtail macaques. Shea (1992) demonstrates that pattems of postnatal allometry in the guenon skull (and postcranium) are largely concordant in talapoin and mustached monkeys, thus supporting Verheyen's (1962) argument that the unique adult proportions exhibited by the former taxon are simply allometric correlates of smaller body size. Indeed, Shea (1992) presents compelling evidence that variation in adult morphology among all species in the *Cercopithecus* group, including patas and swamp monkeys, is largely due to the effects of size divergence operating on shared networks of relative growth. Thus, diversification in cranial form among the Cercopithecini may have been channeled by a common developmental program terminating at different adult body sizes.

Profant (1995) compares patterns of ontogenetic allometry in the cranium among three papionins (*M. fascicularis*, *M. nemestrina*, *Papio cynocephalus*). For most dimensions, patterns of relative growth are indistinguishable between the two macaques, and adult means for other *Macaca* are positioned on or very near the ontogenetic regression line (Figs. 9.7 and 9.8). This corresponds to Albrecht's (1978) observation that adult craniofacial variation in non-Sulawesi macaques is linked to a latitudinal size gradient operating within and between taxa (but see Fooden, 1988, and below). A principal components analysis, combining macaques and baboons

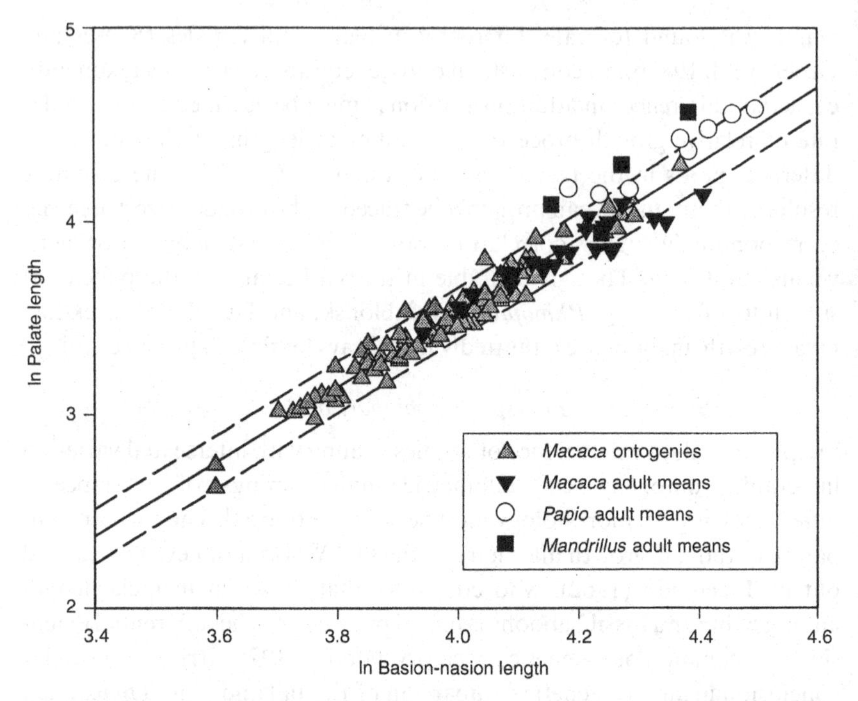

Fig. 9.7. A plot showing a pervasive pattem of ontogenetic scaling for relative palate length in macaques. ANCOVAs ($p<0.01$) yield no slope orY-intercept differences between ontogenetic sequences of *Macaca fascicularis* and *M. nemestrina* (gray triangles). In most cases, adult male and female means for 11 additional macaque species (solid triangles) lie inside 95 % confidence limits on the predicted Y-values derived from the ontogenetic regression, indicating that a significant component of the adult variation in relative facial length is size related. The slope of the ontogenetic regression line for *Papio cynocephalus* is significantly greater than in macaques. Moreover, adult means for 5 baboon taxa (open circles) are positioned mostly above the upper limit of the macaque prediction interval, which shows that baboons have longer faces than macaques at similar skull lengths. The position of drills and mandrills (solid squares), well above the baboon adult cluster, suggests that these taxa have undergone selection for an even longer palate, and provides some preliminary support for the claim that elongate facial skulls were independently derived in *Papio* and *Mandrillus*.

(Fig. 9.8), reveals that the first component, which can be treated as a vector of ontogenetic scaling (Shea, 1985b), accounts for 98% of the variance in papionin craniofacial proportions, with only 2% of the variance unexplained by developmental allometric factors. Thus, multivariate allometric growth vectors in papionins are concordant. Profant's (1995) ontogenetic study helps to clarify the evolutionary implications of analyses by Cheverud (1989), who compared adult phenotypic variation and covaria-

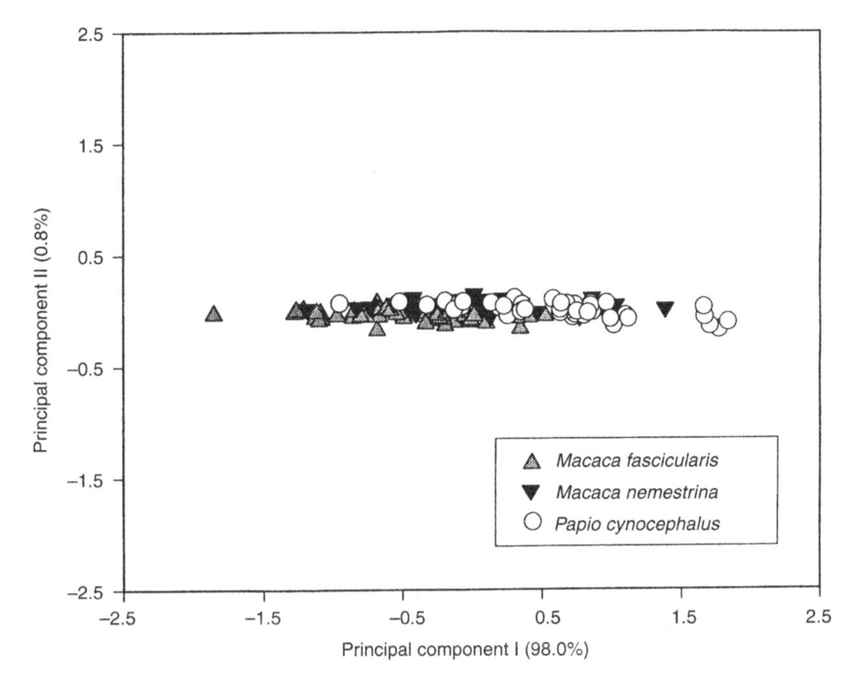

Fig. 9.8. A plot of individual scores on the first and second principal components from a PCA on the covariance matrix of log-transformed craniofacial dimensions for a sample combining ontogenies of *Macaca fascicularis* (gray triangles), *M. nemestrina* (solid triangles) and *Papio cynocephalus* (open circles). Variable loadings on the first principal component are all positive; this component, which summarizes 98 % of the variance in papionin craniofacial proportions, can be interpreted as a multivariate vector of ontogenetic scaling. ANCOVAs and ANOVAs ($p<0.01$) indicate that the slight separation between macaque species along the second component axis is statistically significant. Surprisingly, *P. cynocephalus* and *M. nemestrina* (but not *M. fascicularis*) are indistinguishable on this component, providing support for the hypothesis that inherited allometric factors have been important in patterning interspecific differences in adult craniofacial proportions among macaques and baboons.

tion patterns in the skull of seven papionins, and concluded that these patterns are strikingly similar across this tribe.

As might be expected in generic- and higher-level scaling comparisons, there are several cases in which various macaque species show divergent cranial growth allometries (Fooden, 1975, 1988). Bivariate scaling comparisons between macaques and baboons yield certain significant differences, perhaps reflecting divergent adaptive histories of these clades. For instance, based on qualitative observations (Szalay and Delson, 1979; Benefit and McCrossin, 1993a,b), it is commonly assumed that baboons have relatively longer faces than macaques (Fig. 9.7). Profant's (1995) analysis provides support for this claim, using an ontogenetic criterion of subtraction (also Swindler *et al.*, 1973). In fact, due to the high correlation between face length and canine length, perhaps the evolution of baboon facial elongation is linked to selection for larger canines (as weapons) in a terrestrial habitat (Ravosa, 1990; Jablonski, 1993b). As the best case for adaptation can be made for adult morphologies whose allometric growth trajectories deviate from those of their sister taxa, presumably in response to selection for novel form-function complexes, ontogenetic data provide a biologically meaningful baseline for characterizing such variation (Gould, 1966, 1975; Lauder, 1982; Shea, 1985a, 1992). On the other hand, without prior theoretical and/or experimental studies of factors influencing the allometric coefficient of a specific growth trajectory, it is difficult to detect size-correlated or size-required functional changes within and among species.

Elsewhere in the Papionini, historical allometric factors have been frequently invoked to explain craniofacial evolution in extant and extinct *Theropithecus*. Jolly (1972) hypothesizes that many of the morphological trends characterizing the evolutionary history of this genus, including anterior dental reduction, increased complexity of the cheek teeth and greater ascending ramus, and posterior maxillary height, may reflect "allometric and ontogenetic trends" (Jolly, 1972: 87) related to phyletic size increase. Variations on this argument have since been elaborated by Jablonski (1986, 1993a; Eck and Jablonski, 1984), who makes an explicit appeal to "phylogenetic factors" as an explanation for the evolutionary conservatism of patterns of sexual dimorphism across species of *Theropithecus*. As investigations thus far are based solely on adult data, it would be of great interest to examine the position of the fossil taxa relative to an ontogenetic sequence for the living gelada baboon.

Profant and Shea (1994) carried out the first subfamily-wide investigation of allometric influences on diversity in cercopithecine craniofacial proportions. They find that papionins show significantly greater variability in relative facial length than do cercopithecins, largely as the result of size divergence operating on stronger underlying scaling trajectories (reflected by significantly higher bivariate slopes) in the former tribe. The Papionini evince a broader range of morphological differentiation, despite less intratribal divergence in skull and body size than in the Cercopithecini (the range of mean male body sizes in cercopithecins exceeds that in papionins by nearly a factor of two – Fleagle 1988). This study offers compelling evidence that inherited allometric factors have played a far greater role in patterning craniofacial diversity in all Cercopithecinae than is generally appreciated.

Within colobines, the ontogenetic underpinnings of interspecific variation in craniofacial proportions remain wholly unexplored. This deficit may be attributable to the impression that the Colobinae are relatively homogeneous in both skull form and body size (e.g., Napier, 1985) and are therefore of less inherent interest from an allometric standpoint. While the colobine size range is more restricted than in cercopithecines, the existence of large-bodied extant (*Nasalis*, *Rhinopithecus*) and fossil (*Paracolobus*, *Rhinocolobus*, Figs. 9.1–9.5) colobines suggests several potential avenues of allometric inquiry. Analyses in this regard might include an ontogenetic study of the presence in some large colobines of such cercopithecine-like characters as elongate faces, long, thin nasals, narrow interorbits and posteriorly-oriented mandibular rami (Szalay and Delson, 1979; Benefit and McCrossin, 1993a) as well as lower level studies of ontogenetic scaling in sister taxa differing in size (*Pygathrix–Rhinopithecus*: Jablonski and Peng, 1993; *Simias–Nasalis*: Groves, 1970, 1989).

That adult craniofacial proportions, as well as angular relationships, tend to be more isometric in colobines than in cercopithecines may be a response to the biomechanical demands of the colobine diet of greater percentages of leaves and/or seeds (cf. Verheyen, 1962; Vogel, 1968; Ravosa, 1988, 1990, 1996; Ravosa and Shea, 1994). However, the developmental underpinnings of this interspecific pattern have not been examined. It would be of interest to know if geometric similarity in adult skull form is produced as a consequence of size divergence operating on inherited growth trajectories that are (for functional reasons noted above) mostly isometric (ontogenetic scaling), or if it reflects the outcome of selection to maintain constancy of shape at different adult sizes (biomechanical scaling; cf. Gould, 1966; 1975). The former possibility would be supported if ancestral growth allometries are largely concordant across a clade, whereas a pattern of randomly distributed allometric dissociations would constitute indirect evidence for active selection.

Ravosa (1991c) was the first to systematically compare patterns of cranial growth allometry between a cercopithecine (*M. fasciculatis*) and a colobine (*Nasalis larvatus*) monkey. As might be predicted for these dietarily divergent taxa, scaling relationships are largely discordant. Nevertheless, nearly one-third of the 28 bivariate comparisons indicate similar patterns of relative growth between the two monkeys. This finding is intriguing in view of Ravosa and Shea's (1994: 816) observation that adults of both subfamilies exhibit striking similarities in angular relationships, especially in the upper face, which are suggestive of "some potentially basic pattern features underlying the diversity in facial form among cercopithecid monkeys." This highlights the need for a more comprehensive inter-subfamily analysis of relative growth patterns in the cranium.

Relative facial length in Old World monkey evolution

Relative facial length has played a more prominent role than any other character in discussions of the evolutionary morphology of Old World monkeys at all taxonomic levels. Interest in the dramatic proportional changes that accompany facial growth in baboons can be traced back to Zuckerman (1926), whose work was popularized by Huxley (1932) in his seminal study of relative growth. Elongate faces are widely distributed among extant members of the clade (*Nasalis*, *Papio*, *Mandrillus*, *Theropithecus* and certain macaques like *Macaca nigra*) and are also a salient feature of both the cercopithecine and colobine fossil records (*Libypithecus*, *Dolichopithecus*, *Dinopithecus*, *Gorgopithecus*, *Theropithecus*). The most long-faced taxa are also among the largest cercopithecids ever, suggesting that there is an allometric component to the recurrence of facial lengthening. Indeed, it may be that the potential for moderate to extreme facial elongation is resident in the developmental program of every family member (though only fully realized at large body sizes). There has never been a rigorous, comparative ontogenetic analysis of allometric influences on relative facial length in Old World monkeys.

The polarity of long faces and other cranial morphoclines has recently come under scrutiny as part of a phylogenetic assessment of newly recovered fossils of the earliest known Old World monkey, *Victoriapithecus* (Benefit and McCrossin, 1991, 1993a; Benefit, 1995). The ancestral facial morphotype of cercopithecoids (and basal catarrhines) was originally reconstructed as colobine-like, with a relatively broad interorbit and a short, wide face (Vogel, 1968; Delson, 1975; Delson and Andrews, 1975). This reconstruction rested on the assumption that the phenetic resemblance between colobines and gibbons was the result of common descent from a relatively short-faced ancestor. Thus, the long, deep faces characteristic of cercopithecines were viewed as autapomorphic.

As the *Victoriapithecus* remains preserve features of the facial skull which were previously unknown, Benefit and McCrossin (1991, 1993a) present a revised reconstruction of the basal Old World monkey facial morphotype as moderately long-faced, with long, tapered nasals, a narrow interorbit, and a moderately tall facial height. This scenario suggests that shorter-faced colobines (and gibbons) exhibit the derived condition, whereas the moderately prognathic proboscis monkey may represent a relict member of an early radiation of longer-faced colobines. In this regard, it is important to stress that it has never been adequately demonstrated that *Nasalis* facial length exceeds that predicted by extrapolating various colobine growth trajectories into the range occupied by larger adult proboscis monkeys. A better knowledge of the allometric basis (and potential evolutionary lability) of relative face length would lend an invaluable developmental perspective to Benefit and McCrossin's scenario and would also aid in identifying patterns of covariation among cranial characters deemed important in cercopithecoid evolution (cf. Cheverud, 1989). If, for example, canine length (Siegel, 1972; Lucas, 1981; Ravosa, 1990; Jablonski, 1993b), interorbital narrowing, and elongate nasals (Delson, 1975; Szalay and Delson, 1979; Strasser and Delson, 1987; Ravosa and Profant this study) are all correlated with facial elongation, then it is unnecessary to invoke separate evolutionary events to explain the presence of these traits in larger-bodied, longer-faced Old World monkeys.

The controversy surrounding baboon affinities is an excellent example of a problem that could be clarified through the examination of ontogenetic data. Extreme facial elongation has traditionally been considered a synapomorphy linking *Papio* and *Mandrillus* as sister taxa, to the exclusion of *Theropithecus* (Delson, 1975; Strasser and Delson, 1987). Disotell *et al.* (1992) and van der Kuyl *et al.* (1995) proffer mitochondrial DNA sequence data supporting a *Papio–Theropithecus* clade and a more distantly related *Mandrillus–Cercocebus* clade. This resurrects the notion that the prognathic faces of *Theropithecus*, *Mandrillus* and *Papio* are "of no phyletic significance, since facial elongation is a feature accompanying increased body-size in most if not all Cercopithecidae" (Jolly, 1970a: 164; also Cronin and Meikle, 1979; Groves, 1989).

Because a growth trajectory can provide a biologically interpretable baseline from which to characterize adaptation, developmental covariance structures can be a powerful tool for identifying subtle divergences where adult morphology might suggest homology (Gould, 1966, 1975; Lauder, 1982; Shea, 1985a, 1992). The hypothesis that *Papio* and *Mandrillus* evolved their longer faces via alternative developmental pathways would be supported by the finding that adult craniofacial proportions in these two genera are underlain by significantly different allometric covariance patterns, especially if *Theropithecus* and *Papio* share one suite of patterns while *Mandrillus* and *Cercocebus* share another. Profant's (1995) results indicate that drill and mandrill adult means are positioned above the baboon ontogenetic regression for palate length, implying a departure in underlying allometries (Fig. 9.7). Shah and Leigh (1995), however, observe that *Mandrillus* shows greater affinities with *Papio* than with *Cercocebus* for most craniofacial growth trajectories. Further work is clearly needed to ascertain if ontogenetic data can aid in the resolution of morphological and molecular phylogenies of the papionins. Unraveling the significance of relative facial length continues to be one of the challenges facing Old World monkey systematists (Groves, 1978; Szalay and Delson, 1979; Strasser and Delson, 1987; Benefit and McCrossin, 1993a; Disotell, 1994).

Conclusions

The past decades have witnessed a considerable increase in the extent and types of research on the evolution of the Old World monkey skull. This chapter has reviewed a series of functional and ontogenetic analyses of the craniofacial complex, largely to identify outstanding issues in morphological, paleontological, and experimental research. Several conclusions follow.

Inasmuchas Old World monkeys exhibit marked diversity in adult cranial proportions and body size, they are an excellent group in which to investigate the role of ontogenetic and allometric factors in patterning morphological evolution within and among extant and extinct claries. Apart from higher-level taxonomic comparisons, given the presence of marked subspecific and latitudinal variation in *Macaca* (Fooden, 1975; Albrecht, 1980; Fooden and Albrecht, 1993) and *Colobus* (Hull, 1979), future work might be directed at elucidating the ontogenetic bases of geographic variation in craniofacial form in cercopithecids. Finally, comparative ontogenetic analyses may reveal that it is inappropriate to impute evolutionary significance to a given deviation from an empirical regression line. This point was cogently argued by Shea (1992), who showed that interspecific residuals for various guenon craniodental measures used by Martin and MacLarnon (1988) to construct a novel phylogeny for *Cercopithecus*, were equalled or exceeded in magnitude by the data scatter about growth trajectories for mustached and talapoin monkeys.

To date, there has been considerable comparative and experimental research on skull form and function. That said, more study could be directed at incorporating fossil monkeys, many of which were larger than living cercopithecids (e.g. Ravosa, 1996). There is also a dearth of work integrating data on both skull function and ontogeny; analyses in this regard can offer a new perspective on prior interspecific research (Oyen *et al.*, 1979; Dechow and Carlson, 1990; Ravosa, 1991a,c; Emel and Swindler, 1992; Vinyard and Ravosa, 1998). In papionins, for instance, a study of mandibular ontogeny suggests that interspecific patterns of symphyseal allometry largely result from the ontogenetic process of maintaining functional equivalence at the symphysis *vis-à-vis* increased symphyseal curvature during postnatal development (Vinyard, 1996; Vinyard and Ravosa, 1998). This highlights the need for comparative ontogenetic and interspecific information on the scaling of bone-strain and jaw-muscle activity, both of which are presently unavailable.

Acknowledgements

Curators and staff at the following museums kindly offered access to cercopithecid crania: Field Museum of Natural History, American Museum of Natural History, Museum of Comparative Zoology, National Museum of Natural History, and the Natural History Museum, London. K. Hiiemae, W. Hylander, C. Jolly, B. Shea, C. Vinyard, P. Whitehead and an anonymous reviewer provided helpful comments.

References

Albrecht, G.H. (1978). The craniofacial morphology of the Sulawesi macaques. *Contrib. Primatol.* **13**, 1–151.

Albrecht, G.H. (1980). Longitudinal, taxonomic, sexual, and insular determinants of size variation in pigtail macaques, *Macaca nemestrina*. *Int. J. Primatol.* **1**, 141–52.

Antón, S.C. (1996). Cranial adaptation to a high attrition diet in Japanese macaques. *Int. J. Primatol.* **17**, 401–27.

Benefit, B.R. (1995). Earliest Old World monkey skull. *Am. J. phys. Anthrop.* Suppl. **20**, 64.

Benefit, B.R. & McCrossin, M.L. (1990). Diet, species diversity and distribution of African fossil baboons. *Kroeber Anthropol. Soc. Pap.* **71/72**, 77–93.

Benefit, B.R. & McCrossin, M.L. (1991). Ancestral facial morphology of Old World higher primates. *Proc. Nat. Acad. Sci. USA* **88**, 5267–71.

Benefit, B.R. & McCrossin, M.L. (1993a). Facial anatomy of *Victoriapithecus* and its relevance to the ancestral cranial morphology of Old World monkeys and apes. *Am. J. phys. Anthrop.* **92**, 329–70.

Benefit, B.R. & McCrossin, M.L. (1993b). The lacrimal fossa of Cercopithecoidea, with special reference to cladistic analysis of Old World monkey relationships. *Folia Primatol.* **60**, 133–45.

Bouvier, M. (1986a). A biomechanical analysis of mandibular scaling in Old World monkeys. *Am. J. phys. Anthrop.* **69**, 473–82.

Bouvier, M. (1986b). Biomechanical scaling of mandibular dimensions in New World monkeys. *Int. J. Primatol.* **7**, 551–67.

Bouvier, M. & Hylander, W.L. (1981). Effect of bone strain on cortical bone structure in macaques (*Macaca mulatta*). *J. Morphol.* **167**, 1–12.

Bouvier, M. & Tsang, S.M. (1990). Comparison of muscle weight and force ratios in New and Old World monkeys. *Am. J. phys. Anthrop.* **82**, 509–15.

Brehnan, K., Boyd, R.L., Laskin, J.L. Gibbs, C.H. & Mahan, P.E. (1981). Direct measurement of loads at the temporomandibular joint in *Macaca arctoides. J. Dental Res.* **60**, 1820–4.

Byrd, K.E. & Swindler, D.R. (1980). Palatal growth in *Macaca nemestrina. Primates* **21**, 253–61.

Cheverud, J.M. (1989). A comparative analysis of morphological variation patterns in the papionins. *Evolution* **43**, 1737–47.

Cheverud, J.M. & Richtsmeier, J.T. (1986). Finite-element scaling applied to sexual dimorphism in rhesus macaque (*Macaca mulatta*) facial growth. *Syst. Zool.* **35**, 381–99.

Cochard, L.R. (1985). Ontogenetic allometry of the skull and dentition of the rhesus monkey (*Macaca mulatta*). In *Size and Scaling in Primate Biology*, ed. W.L. Jungers, pp. 231–56. New York: Plenum.

Cronin, J.E. & Meikle, W.E. (1979). The phyletic position of *Theropithecus*: congruence among molecular, morphological, and paleontological evidence. *Syst. Zool.* **28**, 259–69.

Dechow, P.C. & Carlson, D.S. (1990). Occlusal force and craniofacial biomechanics during growth in rhesus monkeys. *Am. J. phys. Anthrop.* **83**, 219–38.

Delson, E. (1975). Evolutionary history of the Cercopithecidae. In *Approaches to Primate Paleobiology. Contributions to Primatology*, Number 5, ed. F.S. Szalay, pp. 167–217. Basel: Karger.

Delson, E. & Andrews, P. (1975). Evolution and interrelationships of the catarrhine primates. In *Phylogeny of the Primates: A Multidisciplinary Approach*, ed. W.P. Luckett & F.S. Szalay, pp. 405–46, New York: Plenum.

Delson, E. & Dean, D. (1993). Are *Papio baringensis* R. Leakey, 1969, and *P. quadratirostris* lwamoto, 1982, species of *Papio* or *Theropithecus*? In *Theropithecus: Rise and Fall of a Primate Genus*, ed. N. G. Jablonski, pp. 125–56. Cambridge: Cambridge University Press.

Disotell, T.R. (1994). Generic level relationships of the Papionini (Cercopithecoidea). *Am. J. phys. Anthrop.* **94**, 47–57.

Disotell, T.R., Honeycutt, R.L. & Ruvolo, M. (1992). Mitochondrial DNA phylogeny of the Old World monkey tribe Papionini. *Mol. Biol. Evol.* **9**, 1–13.

DuBrul, E.L. (1965). The skull of the lion marmoset *Leontideus rosalia* Linneaus: a study in biomechanical adaptation. *Am. J. phys. Anthrop.* **23**, 261–76.

Dunbar, R.I.M. (1977). Feeding ecology of gelada baboons: a preliminary report. In *Primate Ecology: Studies of Feeding and Ranging in Lemurs, Monkeys and Apes*, ed. T.H. Clutton-Brock, pp. 251–73. London: Academic Press.

Eck, G.G. & Jablonski, N.G. (1984). A reassessment of the taxonomic status and phyletic relationships of *Papio baringensis* and *Papio quadratirostris* (Primates: Cercopithecidae). *Am. J. phys. Anthrop.* **65**, 109–34.

Emel, L.M. & Swindler, D.R. (1992). Underbite and the scaling of facial dimensions in colobine monkeys. *Folia Primatol.* **58**, 177–89.

Emerson, S.B. & Radinsky, L. (1980). Functional analysis of sabertooth cranial morphology. *Paleobiology* **6**, 295–312.

Endo, B. (1966). Experimental studies of the mechanical significance of the form of the human facial skeleton. *J. Fac. Sci. Univ. Tokyo, Section V, Anthropol.* **3**, 1–106.

Endo, B. (1970). Analysis of stress around the orbit due to masseter and temporalis muscles. *J. Anthropol. Soc. Nippon* **78**, 251–66.

Fleagle, J. G. (1988). *Primate Adaptation and Evolution*. New York: Academic Press.
Fooden, J. (1975). Taxonomy and evolution of liontail and pigtail macaques. (Primates: Cercopithecidae). *Field. Zool. New Ser*. **67**, 1–169.
Fooden, J. (1988). Taxonomy and evolution of the *sinica* group of macaques. 6. Interspecific comparisons and synthesis. *Field. Zool. New Ser*. **45**, 1–44.
Fooden, J. (1990). The bear macaque, *Macaca arctoides*: a systematic review. *J. Human Evol*. **19**, 607–86.
Fooden, J. & Albrecht, G.H. (1993). Latitudinal and insular variation of skull size in crab-eating macaques (Primates, Cercopithecidae: *Macaca fascicularis*). *Am. J. phys. Anthrop*. **92**, 521–38.
Freedman, L. (1962). Growth of muzzle length relative to calvaria length in *Papio*. *Growth* **162**, 117–28.
Gould, S.J. (1966). Allometry and size in ontogeny and phylogeny. *Biol. Rev.* **41**, 587–640.
Gould, S.J. (1975). Allometry in primates, with emphasis on scaling and the evolution of the brain. In *Approaches to Primate Paleobiology. Contributions to Primatology*, No. 5, ed. F. S. Szalay, pp. 244–92. Basel: Karger.
Greaves, W.S. (1985). The mammalian postorbital bar as a torsion-resisting helical strut. *J. Zool*. **207**, 125–36.
Greaves, W.S. (1995). Functional predictions from theoretical models of the skull and jaws in reptiles and mammals. In *Functional Morphology in Vertebrate Paleontology*, ed. J.J. Thomason, pp. 99–115. Cambridge: Cambridge University Press.
Groves, C.P. (1970). The forgotten leaf-eaters, and the phylogeny of the Colobinae. In *Old World Monkeys: Evolution, Systematics and Behavior*, ed. J.R. Napier & P.H. Napier, pp. 555–87. New York: Academic Press.
Groves, C.P. (1978). Phylogenetic and population systematics of the mangabeys (Primates: Cercopithecoidea). *Primates* **19**, 1–34.
Groves, C.P. (1989). *A Theory of Human and Primate Evolution*. Oxford: Clarendon.
Herring, S.W. (1972). The role of canine morphology in the evolutionary divergence of pigs and peccaries. *J. Mammal*. **53**, 500–12.
Herring, S.W. (1975). Adaptations for gape in the hippopotamus and its relatives. *Forma Functio* **8**, 85–100.
Herring, S.W. & Herring, S.E. (1974). The superficial masseter and gape in mammals. *Am. Nat*. **108**, 561–76.
Hooton, E.A. (1936). *Up From the Ape*. New York: Macmillan.
Hull, D.B. (1979). A craniometric study of the black-and-white *Colobus* Eliger 1811 (Primates: Cercopithecoidea). *Am. J. phys. Anthrop.* **51**, 163–82.
Huxley, J. S. (1932). *Problems of Relative Growth*. London: MacVeagh.
Hylander, W.L. (1975). Incisor size and diet in anthropoids with special reference to Cercopithecidae. *Science* **189**, 1095–8.
Hylander, W.L. (1977). The adaptive significance of Eskimo craniofacial morphology. In *Orofacial Growth and Development*, ed. A.A. Dahlberg & T.M. Graber, pp. 129–69. Paris: Mouton.
Hylander, W.L. (1979a). Mandibular function in *Galago crassicaudatus* and *Macaca fascicularis*: an *in vivo* approach to stress analysis of the mandible. *J. Morphol*. **159**, 253–96.
Hylander, W.L. (1979b). The functional significance of primate mandibular form. *J. Morphol*. **160**, 223–40.
Hylander, W.L. (1979c). An experimental analysis of temporomandibular joint reaction forces in macaques. *Am. J. phys. Anthrop*. **51**, 433–56.

Hylander, W.L. (1981). Patterns of stress and strain in the macaque mandible. In *Craniofacial Biology. Monograph 10, Craniofacial Growth Series*, ed. D.S. Carlson, pp. 1–37. Ann Arbor: University of Michigan.
Hylander, W.L. (1984). Stress and strain in the mandibular symphysis of primates: a test of competing hypotheses. *Am. J. phys. Anthrop.* **64**, 1–46.
Hylander, W.L. (1985). Mandibular function and biomechanical stress and scaling. *Am. Zool.* **25**, 315–30.
Hylander, W.L. (1992). Functional anatomy. In *The Temporomandibular Joint. A Biological Basis for Clinical Practice*, ed. B.G. Sarnat & D.M. Laskin, pp. 60–92. Philadelphia: Saunders.
Hylander, W.L. & Bays, R. (1979). An *in vivo* strain-gauge analysis of the squamosal-dentary joint reaction force during mastication and incisal biting in *Macaca mulatta* and *Macaca fascicularis*. *Arch. Oral Biol.* **24**, 689–97.
Hylander, W.L. & Johnson, K.R. (1985). Temporalis and masseter function during incision in humans and macaques. *Int. J. Primatol.* **6**, 289–322.
Hylander, W.L. & Johnson, K.R. (1992). Strain gradients in the craniofacial region of primates. In *The Biological Mechanisms of Tooth Movement and Craniofacial Adaptation*, ed. Z. Davidovitch, pp. 559–69. Columbus: Ohio State University.
Hylander, W.L. & Johnson, K.R. (1994). Jaw muscle function and wishboning of the mandible during mastication in macaques and baboons. *Am. J. phys. Anthrop.* **94**, 523–47.
Hylander, W.L. & Ravosa, M.J. (1992). An analysis of the supraorbital region of primates: a morphometric and experimental approach. In *Structure, Function and Evolution of Teeth*, ed. P. Smith and E. Chernov, pp. 223–55. Tel Aviv: Freund.
Hylander, W.L., Johnson, K.R. & Crompton, A.W. (1987). Loading patterns and jaw movements during mastication in *Macaca fascicularis*: a bone-strain, electromyographic and cineradiographic analysis. *Am. J. phys. Anthrop.* **72**, 287–314.
Hylander, W.L., Johnson, K.R. & Crompton, A.W. (1992). Muscle force recruitment and biomechanical modeling: an analysis of masseter muscle function during mastication in *Macaca fascicularis*. *Am. J. phys. Anthrop.* **88**, 65–387.
Hylander, W.L., Picq, P.G. & Johnson, K.R. (1991a). Function of the supraorbital region of primates. *Arch. Oral Biol.* **36**, 273–81.
Hylander, W.L., Picq, P.G. & Johnson, K.R. (1991b). Masticatory-stress hypotheses and the supraorbital region of primates. *Am. J. phys. Anthrop.* **86**, 1–36.
Hylander, W.L., Ravosa, M.J., Ross, C.F. & Johnson, K.R. (1998). Mandibular corpus strain in primates: further evidence for a functional link between symphyseal fusion and jaw-adductor muscle force. *Am. J. phys. Anthrop.* **107**, 257–71.
Jablonski, N.G. (1986). Patterns of sexual dimorphism in *Theropithecus*. In *Sexual Dimorphism in Living and Fossil Primates*, ed. M. Pickford & A.B. Chiarelli, pp. 171–82. Firenze: Il Sedicesimo.
Jablonski, N.G. (1993a). Evolution of the masticatory apparatus in *Theropithecus*. In *Theropithecus: Rise and Fall of a Primate Genus*, ed. N. G. Jablonski, pp. 299–329. Cambridge: Cambridge University.
Jablonski, N.G. (1993b). Muzzle length and heat loss. *Nature* **366**, 216–17.
Jablonski, N.G. & Pan, R. (1995). Sexual dimorphism in the snub-nosed langurs (Colobinae: *Rhinopithecus*). *Am. J. phys. Anthrop.* **96**, 251–72.

Jablonski, N.G. & Peng, R. (1993). The phylogenetic relationships and classification of the doucs and snub-nosed langurs of China and Vietnam. *Folia Primatol.* **60**, 36–55.

Jolly, C.J. (1970a). The large African monkeys as an adaptive array. In *Old World Monkeys: Evolution, Systematics, and Behavior*, ed. J.R. Napier & P.H. Napier, pp. 141–74. New York: Academic Press.

Jolly, C.J. (1970b). The seed-eaters: a new model of hominid differentiation based on a baboon analogy. *Man* **5**, 5–28.

Jolly, C.J. (1972). The classification and natural history of *Theropithecus (Simopithecus)* (Andrews, 1916), baboons of the African Plio-Pleistocene. *Bull. Brit. Mus. (Nat. Hist.) Geol.* **22**, 1–123.

Kay, R.F. (1978). Molar structure and diet in extant Cercopithecidae. In *Development, Function and Evolution of Teeth*, ed. P.M. Butler & K.A. Joysey, pp. 309–39. New York: Academic Press.

Kay, R.F. & Hylander, W.L. (1978). The dental structure of mammalian folivores with special reference to Primates and Phalangeroidea (Marsupialia). In *The Ecology of Arboreal Folivores*, ed. G. G. Montgomery, pp. 173–91. Washington, DC: Smithsonian Institution.

Kieser, J.A. & Groeneveld, H.T. (1987a). Static intraspecific allometry of jaws and teeth in *Cercopithecus aethiops*. *J. Zool.* **212**, 499–510.

Kieser, J.A. & Groeneveld, H.T. (1987b). Static intraspecific maxillofacial allometry in the Chacma baboon. *Folia Primatol.* **48**, 151–63.

Lanyon, L.E. & Rubin, C.T. (1985). Functional adaptation in skeletal structures. In *Functional Vertebrate Morphology*, ed. M. Hildebrand, D.M. Bramble, K.F. Liem & D.B. Wake, pp. 1–25. Cambridge, MA: Harvard University Press.

Lauder, G.V. (1982). Historical biology and the problem of design. *J. Theor. Biol.* **97**, 57–67.

Leigh, S.R. & Cheverud, J.M. (1991). Sexual dimorphism in the baboon facial skeleton. *Am. J. phys. Anthrop.* **84**, 193–208.

Lucas, P.W. (1981). An analysis of canine size and jaw shape in some Old and New World non-human primates. *J. Zool.* **195**, 437–48.

Lucas, P.W. (1982). An analysis of the canine tooth size of Old World higher primates in relation to mandibular length and body weight. *Arch. Oral Biol.* **27**, 493–6.

Lucas, P.W. & Teaford, M.F. (1994). Functional morphology of colobine teeth. In *Colobine Monkeys: Their Ecology, Behaviour and Evolution*, ed. A.G. Davies & J.F. Oates, pp. 173–203. Cambridge: Cambridge University Press.

Lucas, P.W., Corlett, R.T. & Luke, D.A. (1986). Sexual dimorphism of tooth size in anthropoids. In *Sexual Dimorphism in Living and Fossil Primates*, ed. M. Pickford & A.B. Chiarelli, pp.23–39. Firenze: Il Sedicesimo.

Luschei, E.S. & Goodwin, G.M. (1974). Patterns of mandibular movement and jaw muscle activity during mastication in monkeys. *J. Neurophysiol.* **37**, 954–66.

Martin, R.D. (1993). Allometric aspects of skull morphology in *Theropithecus*. In *Theropithecus: Rise and Fall of a Primate Genus*, ed. N.G. Jablonski, pp. 273–98. Cambridge: Cambridge University Press.

Martin, R.D. & MacLarnon, A.M. (1988). Quantitative comparisons of the skull and teeth in guenons. In *A Primate Radiation: Evolutionary Biology of the African Guenons*, ed. A. Gautier-Hion, F. Bourliere & J.P. Gautier, pp. 160–83. Cambridge: Cambridge University Press.

McNamara, J.A. (1974). An electromyographic study of mastication in the rhesus monkey (*Macaca mulatta*). *Arch. Oral Biol.* **19**, 821–3.

Moss, M.L. & Young, R.W. (1960). A functional approach to craniology. *Am. J. phys. Anthrop.* **18**, 281–92.

Napier, P.H. (1985). *Catalogue of Primates in the British Museum (Natural History) and Elsewhere in the British Isles. Part III. Family Cercopithecidae, Subfamily Colobinae*. London: British Museum (Natural History).

Oyen, O.J. & Russell, M.D. (1982). Histogenesis of the craniofacial skeleton and models of facial growth. In *The Effect of Surgical Intervention on Craniofacial Growth*. Monograph 12, Craniofacial Growth Series, ed. J. A. McNamara, D. S. Carlson & K.A. Ribbens, pp. 361–72. Ann Arbor: University of Michigan.

Oyen, O.J. & Tsay, T.P. (1991). A biomechanical analysis of craniofacial form and bite force. *Am. J. Orthod. Dentofac. Orthop*. **99**, 298–309.

Oyen, O.J., Walker, A.C. & Rice, R.W. (1979). Craniofacial growth in olive baboons (*Papio cynocephalus anubis*): browridge formation. *Growth* **43**, 174–87.

Pan, R.L., Peng, Y.Z., Ye, Z.Z. & Yu, F.H. (1993). Sexual dimorphism of skull and dentition in Phayre's leaf monkey *(Presbytis phayrei)*. *Folia Primatol.* **60**, 230–6.

Pan, R., Peng, Y., Ye, Z., Wang, H. & Yu, F. (1995). Comparison of masticatory morphology between *Rhinopithecus bieti* and *R. roxellana*. *Am. J. Primatol.* **35**, 271–81.

Profant, L. (1995). Historical allometric inputs to interspecific pattems of craniofacial diversity in the cercopithecine tribe Papionini. *Am. J. phys. Anthrop.* (*Suppl.*) **20**, 175.

Profant, L.P. & Shea, B.T. (1994). Allometric basis of morphological diversity in the Cercopithecini vs. Papionini tribes of Cercopithecine monkeys. *Am. J. phys. Anthrop. (Suppl.)* **18**, 162–3.

Ravosa, M.J. (1988). Browridge development in Cercopithecidae: a test of two models. *Am. J. phys. Anthrop*. **76**, 535–55.

Ravosa, M.J. (1990). A functional assessment of subfamily variation in maxillomandibular morphology among Old World monkeys. *Am. J. phys. Anthrop*. **82**, 199–212.

Ravosa, M.J. (1991a). Ontogenetic perspective on mechanical and nonmechanical models of primate circumorbital morphology. *Am. J. phys. Anthrop*. **85**, 95–112.

Ravosa, M.J. (1991b). Interspecific perspective on mechanical and nonmechanical models of primate circumorbital morphology. *Am. J. phys. Anthrop.* **86**, 363–96.

Ravosa, M.J. (1991c). The ontogeny of cranial sexual dimorphism in two Old World monkeys: *Macaca fascicularis* (Cercopithecinae) and *Nasalis larvatus* (Colobinae). *Int. J. Primatol.* **12**, 403–26.

Ravosa, M.J. (1996). Jaw morphology and function in living and fossil Old World monkeys. *Int. J. Primatol.* **17**, 909–32.

Ravosa, M.J. & Shea, B.T. (1994). Pattern in craniofacial biology: evidence from the Old World monkeys (Cercopithecidae). *Int. J. Primatol.* **15**, 801–22.

Richtsmeier, J.T., Cheverud, J.M., Danahey, S.E., Corner, B.M. & Lele, S. (1993). Sexual dimorphism of ontogeny in the crab-eating macaque (*Macaca fascicularis*). *J. Human Evol.* **25**, 1–30.

Rosenberger, A.L. (1986). Platyrrhines, catarrhines and the anthropoid transition. In *Major Topics in Primate and Human Evolution*, ed. B.A. Wood, L. Martin & P. Andrews, pp. 66–88. Cambridge: Cambridge University Press.

Russell, M.D. (1982). Tooth eruption and browridge formation. *Am. J. phys. Anthrop*. **58**, 59–65.

Russell, M.D. (1985). The supraorbital torus: " a most remarkable peculiarity." *Curr. Anthrop.* **26**, 337–60.
Schultz, A.H. (1970). The comparative uniformity of the Cercopithecoidea. In *Old World Monkeys: Evolution, Systematics, and Behavior*, ed. J.R. Napier & P.H. Napier, pp. 39–51. New York: Academic Press.
Shah, N.F. & Leigh, S.R. (1995). Cranial ontogeny in three Papionin genera. *Am. J. phys. Anthrop. (Suppl.)* **20**, 194.
Shea, B.T. (1985a). Ontogenetic allometry and scaling: a discussion based on the growth and form of the skull in African apes. In *Size and Scaling in Primate Biology*, ed. W.L. Jungers, pp. 175–205. New York: Plenum.
Shea, B.T. (1985b). Bivariate and multivariate growth allometry: statistical and biological considerations. *J. Zool.* **206**, 367–90.
Shea, B.T. (1986). On skull form and the supraorbital torus in primates. *Curr. Anthrop.* **27**, 257–59.
Shea, B.T. (1992). Ontogenetic scaling of skeletal proportions in the talapoin monkey. *J. Human Evol.* **23**, 283–307.
Siebert, J.R., Swindler, D.R. & Lloyd, J.D. (1984). Dental arch form in the Cercopithecidae. *Primates* **25**, 507–18.
Siegel, M.I. (1972). The relationship between facial protrusion and root length in the dentition of baboons. *Acta Anat.* **83**, 17–29.
Smith, F.H. & Ranyard, G.C. (1980). Evolution of the supraorbital region in Upper Pleistocene fossil hominids from South-Central Europe. *Am. J. phys. Anthrop.* **53**, 589–610.
Strasser, E. & Delson, E. (1987). Cladistic analysis of cercopithecid relationships. *J. hum. Evol.* **16**, 81–99.
Swindler, D.R., Sirianni, J.E. & Tarrant, L.H. (1973). A longitudinal study of cephalofacial growth in *Papio cynocephalus* and *Macaca nemestrina* from three months to three years. In *Symposium of IVth International Congress of Primatology*, vol. 3: *Craniofacial Biology of Primates*, ed. M.R. Zingeser, pp. 227–40. Basel: Karger.
Szalay, F.S. & Delson, E. (1979). *Evolutionary History of the Primates*. New York: Academic Press.
Takahashi, L.K. & Pan, R. (1994). Mandibular morphometrics among macaques:the case of *Macaca thibetana*. *Int. J. Primatol.* **15**, 597–621.
Teaford, M.F. (1993). Dental microwear and diet in extant and extinct *Theropithecus*: preliminary analyses. In *Theropithecus: Rise and Fall of a Primate Genus*, ed. N.G.Jablonski, pp. 331–49. Cambridge: Cambridge University Press.
van der Kuyl, A.C., Kuiken, C.L., Dekker, J.T. & Goudsmit, J. (1995). Phylogeny of African monkeys based upon mitochondrial 12s rRNA sequences. *J. Mol. Evol.* **40**, 173–80.
Verheyen, W.N. (1962). Contribution à la craniologie comparée des primates. *Ann. Mus. Roy. Afr. Cent.* **105**, 1–247.
Vinyard, C.J. (1994). *A Quantitative Assessment of the Supraorbital Region in Modern Melanesians*. MA Thesis, Dept. of Anthropology, Northern Illinois University.
Vinyard, C.J. & Ravosa, M.J. (1998). Ontogeny, function, and scaling of the mandibular symphysis in papionin primates. *J. Morphol.* **235**, 157–75.
Vogel, C. (1968). The phylogenetical evaluation of some characters and some morphological trends in the evolution of the skull in catarrhine primates. In *Taxonomy and Phylogeny of Old World Primates with References to the Origin of Man*, ed. A.B. Chiarelli, pp. 21–55. Torino: Rosenberg and Sellier.

Walker, P. & Murray, P. (1975). An assessment of masticatory efficiency in a series of anthropoid primates with special reference to the Colobinae and Cercopithecinae. In *Primate Functional Morphology and Evolution*, ed. R.H. Tuttle, pp. 135–50. Paris: Mouton.
Wolpoff, M.H. (1980). *Paleoanthropology*. New York: Knopf.
Wolpoff, M.H. (1996). *Human Evolution*. New York: McGraw-Hill.
Zuckerman, S. (1926). Growth-changes in the skull of the baboon, *Papio porcarius*. *Proc. Zool. Soc. London* **55**, 843–73.

10

Evolutionary endocrinology of the cercopithecoids

PATRICIA L. WHITTEN

Introduction

Thirty years ago, three disciplines – ethology, endocrinology, and ecology – undertook the explanation of primate social behavior. Ethological methods have since become universal in primatology, but endocrine and ecological investigations have maintained greater distance. Despite remarkable similarities in the research plans presented in influential papers in behavioral endocrinology (Beach, 1975) and socioecology (Crook *et al.*, 1976) (see Fig. 10.1), differing methods and priorities (see Table 10.1) set these two research areas on divergent trajectories. The experimental methods of early behavioral endocrinology in which hormones were detected and characterized by their action focused attention on the evidence and mechanisms for hormonal influences on individual behavior with less attention to social context and contingencies (Worthman, 1990). Early socioecology, relying on correlations between gross categories of social system and environment, sidestepped the issue of process (Richard, 1981) while focusing attention on the group as the locus of behavioral evolution. Although these different emphases hampered the integration of endocrine and ecological perspectives, the research areas have independently converged as each has broadened its methodologies and perspectives. The causal focus and experimental approaches of behavioral endocrinology have expanded to include a more synergistic framework and observational approach, termed "socioendocrinology," reflecting an emerging view of the individual as a social organism and new attention to the role of social processes in the regulation of hormone function (Bercovitch and Ziegler, 1990). At the same time, socioecological inquiry has been transformed by sociobiology from a largely inferential systems approach into the deductive study of the adaptive function and genetic bases of sociality, now termed "behavioral ecology," reflecting the rejection

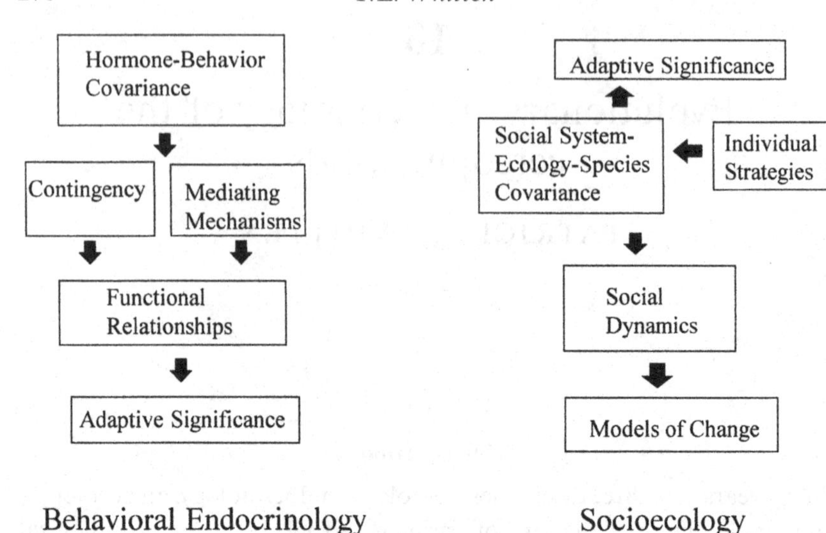

Fig. 10.1. Proposed models for investigations in behavioral endocrinology (Beach, 1975) and socioecology (Crook *et al.*, 1976).

of group selection as a significant force in social evolution and emphasis on the individual, rather than the group, as the unit of evolutionary selection, and a more deterministic view of the role of biology in social behavior (Richard, 1981; Crook, 1989). This chapter will review the implications of these new investigations for the evolution of cercopithecoid behavior and social organization and suggest some directions for future research.

Behavioral and endocrine systems

Primate behavioral endocrinology began with studies of gonadal steroids and reproductive behavior and simple linear models of hormone–behavior causation. That these approaches produced conflicting results is not surprising given the nature of steroid and hormone action. The steroid hormones are of intermediate latency in the temporal spectrum of physiological responses to the environment (Wilson, 1975; Worthman, 1990). The classic actions of steroid hormones occur through the regulation of nuclear transcription, genomic actions with latencies of one hour or more (Wehling, 1994). Because behavioral outcomes often have even longer latencies, occurring 48 hours or more after the steroid has disappeared from the site of action (McEwen *et al.*, 1979), transitory changes in hormone levels could not have immediate effects on behavioral states or acts. Thus genomic steroid actions would not be expected to directly elicit behavior,

Table 10.1. *Methods in behavioral endocrinology and socioecology*

Behavioral endocrinology	Socioecology
Hormone–behavior correlations	Group-habitat correlations
Experimental method	Comparative method
Replication	Rare events
Controlled environments	Unpredictable environments
Captive setting	Natural setting
Pairs or group	Group
Proximate and phylogenetic explanations	Adaptive explanations

but are more likely to provide priming for subsequent behavioral responses, altering the stimulus thresholds required for elicitation of behavior or the sensitivity of sensory systems. These temporal differences in hormonal action and reaction may make it easier to demonstrate the influences of behavioral interactions on steroid hormones than the influences of steroid hormones on behavior. Moderate latency actions, however, are not the only actions of steroids. Recent research has shown that steroids also can exert a wide variety of rapid, nongenomic actions with latencies of only a few seconds or minutes (Wehling, 1994). These short latency actions have been particularly well documented in the brain (McEwen, 1991; Wehling, 1994) where steroids function as ligands for membrane-bound receptors (Wehling, 1994) and as neuromodulators or neurotransmitters (Naftolin, 1994; Garcia-Segura *et al.*, 1994), and contribute to the more immediate modulation of behavior and hormonal feedback (Andrew, 1991; Wehling, 1994).

There is a tendency to see physiology as more immutable, unified, and fundamental than behavior, but in reality both are interactive systems that influence and respond to one another in a context-dependent manner (Gunnar and Mangelsdorf, 1989; Worthman, 1990). On the physiological level, specific hormones are always embedded in more complex feedback systems in which hormone secretion both regulates and is regulated by other hormones. In a parallel fashion, specific behaviors can be seen as responses that regulate and are regulated by physiological states. Thus it is not surprising that there is as much evidence for behavioral influences on hormones as for hormonal influences on behavior, as perceptively noted by Wilson (1975: 154). By the same token, hormonal responsiveness to the social environment should not be taken as evidence against hormonal influences on behavior; for example, the negative feedback effects of estrogen on gonadotropin secretion do not preclude the induction of estrogen biosynthesis by gonadotropins. Similarly, the context-dependent nature of hormone–behavior relationships does not distinguish behavior from other hormonally-mediated processes. Hormone action is by nature context-dependent and cannot be predicted simply by circulating hormone levels. The local action of hormones is contingent upon the availability of hormone receptors and other tissue-specific factors that transduce or inhibit hormonal signals, factors that are themselves products of developmental organization, past history, and current physiological state.

A functional approach provides organizing principles for interpreting hormone–behavior interactions (Gunnar and Mangelsdorf, 1989). From a proximate perspective, the goal of behavioral and physiological response is the modulation of internal and external states to maintain homeostasis (Worthman, 1990). Behavioral processes are particularly effective in mediating responses to changes in the external environment, either altering the environment or changing position within it, and in group-living primates, the social environment is a central focus of these regulatory actions. From an ultimate perspective, of course, homeostatic mechanisms must be subservient to processes that promote reproduction and genetic propogation. Recent reviews have argued that the wide range of targets and actions of hormones and the specificity of localized hormone actions makes them well-suited for the generation of alternative reproductive tactics, mating strategies, and life history trade-offs (Moore, 1991; Ketterson and Nolan, 1992; Clark and Galef, 1995). Although endocrinology traditionally has looked to phylogeny to explain interspecific differences in behavior, many interspecific differences in behavior–endocrine systems may be better explained by adapations to species-specific ecology and mating systems (Moore, 1991). This approach, termed "evolutionary endocrinology," complements the socioendocrine study of individual differences by focusing on the evolution of neuroendocrine regulatory systems as part of adaptation to species-specific contexts.

Cercopithecoid social organization

The most salient adaptive context for cercopithecoids is the social group. The overwhelming majority of Old World monkeys live in heterosexual groups composed of two or more females and one or more males. Most baboon and macaque species live in multimale groups whereas most colobine and guenon species live in unimale groups. In most cercopithecoid species, females generally remain in their natal group and males migrate (Packer and Pusey, 1987). These species are "female-bonded" with marked dominance hierarchies and strong kin-based affiliative relationships among females (Wrangham, 1980). However, the strength of both affiliative bonds and dominance relations vary across species. Intertroop transfers of females appear to be more common in colobines than in cercopithecines (Moore, 1984) and are characteristic in red colobus and olive colobus (Oates, 1994). The hamadryas baboon and DeBrazza's monkey are the only cercopithecine species characterized by female transfer. Kinship bonds and dominance hierarchies are strong in most baboon and macaque species (Wrangham, 1980) but kin-based alliances are weak in mangabeys (Ehardt, 1988; Gust and Gordon, 1994) and colobines (Hrdy, 1977) and dominance relations are muted and subtle in colobines and guenons with an emphasis on social avoidance rather than threat (Rowell, 1988). Intermale tolerance and the degree of monopolization of sexual access by the highest ranking (alpha) male vary considerably (Shively *et al.*, 1982; Caldecott, 1986), as do male-female bonds (Smuts, 1985).

Explaining these interspecific differences has been a major concern of socioecology and behavioral ecology. Socioecology began with associations between broad features of the environment and social structure under the assumption that social organization arose primarily from the constraints of habitat and foraging pressures (Crook and Gartlan, 1966), a focus that was expanded to include morphological and behavior attributes of the individual such as body size, sexual dimorphism, food choice, and intermale tolerance in subsequent investigations (Goss-Custard *et al.*, 1972; Eisenberg *et al.*, 1972). Paired correlations of behavioral and ecological variables (Clutton-Brock and Harvey, 1977) and analyses of within-species variation (Altmann, 1974) provided more rigorous tests of these associations, but only a few studies (eg. Crook *et al.*, 1976; Altmann and Altmann, 1979) attempted to model mechanisms for evolution or dynamics of social systems. With the advent of sociobiology, evolutionary models gained center stage, ushering in a deductive, individual-oriented approach to social evolution (Wilson, 1975). Subsequent research in primate behavioral ecology has focused on the competing demands of interindividual and intergroup competition and on resource and predator defense as the primary factors guiding the evolution of group structure (Wrangham, 1980; van Schaik, 1989; Rodman, 1988). Although the focus has shifted from the optimal group to the strategies of individuals, theoretical models still tend to treat philopatry, group composition, and dominance structure as if they were genetic traits. However, it is clear that these social characteristics must be emergent properties of individual and population attributes (Lott, 1984; Richard, 1985; Rowell, 1994), and therefore interspecific differences in social structure should be associated with interspecific differences in the regulation of individual behavior. Although evolutionary analyses traditionally ignored proximate mechanisms, it has become increasingly clear the study of mechanisms can provide important insights into the function and evolution of behavior (Dawkins, 1989; Krebs and Davies, 1991). Endocrine data can make important contributions to the study of behavioral mechanisms.

Expanded sexuality and primate social strategies

Sir Solly Zuckerman (1932) was the first to posit a causal relationship between expansion of the duration of female sexual activity and primate sociality. He proposed that the permanent social groups of primates were a product of their uninterrupted reproductive life. Although evidence for seasonal (Lancaster and Lee, 1965; Lindburgh, 1987) and cyclic patterns of sexual activity (Carpenter, 1942; Wallen, 1990) in many species has shown that sexuality is not uninterrupted, the capacity to engage in sexual activity outside of the periovulatory period is a characteristic feature of cercopithecoid biology. Whereas the period of behavioral estrus is brief and strictly delimited in most mammals and in many prosimian species, sexual behavior can occur throughout the ovarian cycle in the catarrhine primates and may occur in prolonged "runs" or even during pregnancy in many species (Butler, 1974; Rowell, 1970; Hrdy and Whitten, 1987). Conspicuous coloration and swellings of the perineum also extend and enhance sexual activity beyond the periovulatory period in many cercopithecoid species (Dixson, 1983; Hrdy and Whitten, 1987).

Expanded sexuality has continued to be an important concept in models of primate social behavior, which propose that females use sexuality as a means of procuring benefits from males, such as investment in offspring (Hrdy, 1981; 1988; Hamilton, 1984; Whitten, 1987), aid in competition (Smuts, 1987; Hooks and Green, 1993), or protection from sexual aggression (Smuts and Smuts, 1993) or as a means of reducing the mating success of other females (Small, 1993; Pagel, 1994). These hypotheses have proven difficult to test because the physiological basis and benefits have not been well defined. For example, hypothesized male benefits depend upon the development of later affiliative behaviors as a result of prior mating frequency but little is known about the occurrence of these affiliative relationships or their ontogeny in primate species other than baboons (Smuts, 1985). Although this outcome may appear to demand considerable cognitive ability, it could be produced through neuroendocrine mechanisms similar to those linked to the formation of social bonds in monogamous voles in which two hypothalamic neurochemicals, oxytocin and vasopressin, are released in response to copulation and sexual arousal and facilitate partner preference, mating guarding, and later associations with infants (Insel *et al.*, 1993; Insel and Carter, 1995). Similarly, ability of sexual swellings to enhance paternity certainty or flag ovulation also is untested. Although it generally has been assumed that ovulation is concealed in the absence of sexual swellings (Alexander and Noonan, 1979; Hrdy, 1981; Strassman, 1981; Andelman, 1987; Sillén-Tullberg and Møller, 1993), there is good reason to believe that sexual swellings conceal rather than reveal ovulation (Whitten and Russell, 1996).

Whether expanded sexuality has been fostered by the evolution of supporting physiological or neurological traits or is entirely a byproduct of other adaptations is unclear. One hypothesis holds that expanded sexuality derives from the enlarged cerebral cortex of primates, which widens the range of stimuli capable of releasing copulatory reflexes, reducing dependence on hormonal controls (Beach, 1947; Wrangham, 1993). However, experimental studies have shown that in some circumstances sexual activity can be tightly linked to the ovarian cycle (Wallen and Goy, 1977), depending on the social and environmental setting (Wallen, 1982; Wallen *et al.*, 1984; Zumpe and Michael, 1996). Although hormone independence has been presumed in situation-dependent receptivity, that conclusion may be unwarranted. For example, postconception estrus has a clear endocrine basis associated with elevated estrogen secretion during the shift from luteal to placental control of steroid hormone secretion phase (Bielert *et al.*, 1976; Wilson *et al.*, 1982; Gouzoules and Goy, 1983; Gordon *et al.*, 1991). Alternatively, the expanded sexuality may be a side-effect of primate anatomy. Wallen (1990) has proposed that because primate females need not assume a specific posture in order to copulate, mating is physically possible at any point in the ovarian cycle, whereas copulation of rodent females requires lordosis, a hormonally modulated behavior. Therefore, the primary hormonal influence on sexual activity in primates may be through modulation of the motivation to mate, resulting in a flexible behavioral response that maps most closely onto ovarian hormones when sexual interaction is costly or difficult, so that it is achieved only when motivation, and estrogen levels, are high. For example, presenting and other proceptive behaviors closely parallel rising estrogen levels in the first half of the ovarian cycle in low-ranking rhesus macaque females but are poorly linked to the ovarian cycle in high-ranking females (Wallen *et al.*, 1984, 1990). Similarly, hormonal influences are more evident in one-male groups where female competition is enhanced than in groups with multiple males (Wallen, 1990). Finally, an additional factor may be changes in the ovarian cycle itself. The anthropoid ovarian cycle is distinguished from that of prosimians by the more prolonged duration of the follicular phase, which is about 14 days in the cercopithecoids and hominoids but as short as five days in some strepsirhine primates and by a much shorter luteal phase, which is about 14 days in anthropoids but as long as 24–28 days in strepsirhine primates (Brockman *et al.*, 1995). Because the duration of estrous behavior is dependent upon the duration of estrogen stimulus in other vertebrates (Fabre-Nys *et al.*, 1993), the more prolonged estrogen stimulus resulting from the lengthened follicular phase in cercopithecoids may contribute to the expansion of sexual activity in cercopithecoids (Shideler and Lasley, 1982). Enhanced estrogen exposure also may result from extra-ovarian sources. Primates appear to be exceptional in their reliance on non-gonadal sources for estrogen biosynthesis. Adipose tissue contributes a substantial proportion of circulating estrogen levels in primates (Longcope *et al.*, 1983), providing a persistent source of estrogen stimulus regardless of ovarian function. The contribution of this extra-ovarian estrogen to the expanded sexual capacity of cercopithecoid females has not been explored although there is evidence that fatting may be related to male sexual activity (Bercovitch, 1992; see below).

Androgens, aggression and dominance in males

Primatology also has a long standing interest in dominance hierarchies as organizing factors in social systems, beginning with Zuckerman's (1932) proposal that hierarchies served to control male aggression in the sexually charged primate group. Endocrine studies in male primates were undertaken originally in order to identify the physiological attributes that confer dominance. This goal proved elusive, in part because it suffered from several misperceptions. A consistent endocrine profile of dominance would not be expected because hormones are responsive to social interactions and because dominance is a social relationship rather than a property of the individual (Bernstein, 1981), although there may be some intrinsic qualities of the individual that increase success in competitive encounters (Popp and DeVore, 1979).

Although some early studies suggested that high testosterone levels in males were associated with high rank (Rose *et al.*, 1971), most subsequent studies have found conflicting results. Difficulties in establishing links between testosterone and social rank are not surprising given episodic and diurnal rhythms of testosterone secretion and responsiveness to external events. Although testosterone varies as much as tenfold across individual males, intra-individual variation in testosterone concentrations can be as high as two- to fourfold over the day (Martensz *et al.*, 1987), tenfold over the year in seasonally breeding rhesus macaques (Gordon *et al.*, 1976) and five- to tenfold from day-to-day during the breeding season in vervet monkeys (Steklis *et al.*, 1985). Testosterone concentrations also are highly responsive to social interactions: serum levels fall following defeat or fall in rank (Eberhardt *et al.*, 1980), and during drought (Sapolsky, 1986) or social instability (Sapolsky, 1983) and rise following exposure to estrous females (Bernstein *et al.*, 1977), or winning an aggressive encounter (Rose *et al.*, 1975; Bernstein *et al.*, 1979). Thus the variable relationship between testosterone and rank may be a consequence of the variable association between rank and aggression. Circulating testosterone levels are not correlated with social rank in stable groups of rhesus macaques (Gordon *et al.*, 1976), Japanese macaques (Eaton and Resko, 1974), baboons (Sapolsky, 1983), or vervet monkeys (Steklis *et al.*, 1985; McGuire *et al.*, 1986) where high-ranking males are not characterized by high rates of aggression, but high status is associated with high testosterone concentrations during group formation or reorganization when dominant males are engaged in high rates of aggression (Rose *et al.*, 1971; Keverne *et al.*, 1982; Sapolsky, 1983). These data show that testosterone cannot be related to high rank in a simple causal manner since there is no single androgen profile of dominance (Sapolsky, 1991) and may more closely reflect patterns of aggression and sexual behavior.

Testosterone appears to be more consistently related to rank, however, in species in which the highest-ranking, or alpha male, monopolizes sexual access to females. A tendency for testosterone levels to be highest in the alpha male of the group has been reported for captive groups of vervet monkeys (Steklis *et al.*, 1985) and talapoin monkeys (Eberhardt *et al.*, 1980), and semifree-ranging mandrills (Wickings and Dixson, 1992). The most extreme rank-related differences in serum testosterone levels are seen in mandrills where they are associated with distinct morphological and behavioral variants (Wickings and Dixson, 1992). In this species, serum testosterone levels are five- to tenfold higher in socially-living males than in solitary or socially peripheral males. Group-associated males have a "fatted" appearance, with a broad, flattened rump, intense coloration of the sexual skin on the face and rump, and more marked development of secondary sexual characteristics such as nuchal crest, beard, and shoulder mane. Solitary and socially peripheral males are "nonfatted," with slimmer rumps, less intense coloration of the sexual skin, smaller testicles, and less marked development of secondary sexual characteristics. Fatting occurs seasonally in rhesus macaque (Bercovitch, 1992) and squirrel monkey males (Mendoza *et al.*, 1978; McCamant *et al.*, 1987), but the fatted appearance appears to be permanent in mandrills (Wickings and Dixson, 1992). These features are not simply a consequence of their more peripheral lifestyle, as differences in physical development are apparent well before puberty (Wickings and Dixson, 1992). Moreover, the morphological features are associated with differences in behavior and reproductive success – fatted males form permanent associations with females and their offspring and guard and mate with females during peak sexual swellings, and the alpha male fathers the majority of offspring (Dixson *et al.*, 1993; Wickings *et al.*, 1993). Nonfatted males are solitary or socially peripheral, mate opportunistically and do not guard females, and appear to father no offspring (Dixson *et al.*, 1993). In other species, fatting appears to be responsible for breeding season elevations in circulating estrogen (Bercovitch, 1992; Mendoza *et al.*, 1978; McCamant *et al.*, 1987) and may contribute to seasonal sexual activity (Mendoza *et al.*, 1978), female choice (Boinski, 1987), or reduced feeding during mating competition (Bercovitch, 1992), factors which also might be relevant to the mandrill. The rank-related patterning of serum testosterone in mandrills is of interest in light of the extreme sexual dimorphism of this species. Within-sex polymorphisms reflect alternative reproductive tactics within many vertebrate species (Caro and Bateson, 1986; Moore, 1991), and these mandrill phenotypes likewise may reflect the role of secondary sexual characteristics in intermale competition or female choice.

Some more subtle distinctions in testosterone response also characterize high-ranking males in baboons. In response to stress, luteinizing hormone (LH) and testosterone levels fall in all males, but in high-ranking males, testosterone increases transiently before falling (Sapolsky, 1986). The transient increase in testosterone is a result of reduced testicular responsiveness to the suppressive effects of glucocorticoids and greater activation of the sympathetic nervous system during stress in high-ranking males (Sapolsky, 1991). These differences disappear during periods when the social hierarchy is unstable (Sapolsky, 1991), however, suggesting that they may be a consequence of rank rather than a reflection of intrinsic competitive ability.

These findings show that differences in serum testosterone are associated with behavior patterns and, in some species, with social rank but leave unanswered the actual role of testosterone in male behavior. Transient testosterone responses are most likely to be related to the short latency actions of testosterone since the longer latency, genomic actions of testosterone would be too far removed in time to influence ongoing behavioral interactions and the responses would be too brief to have much influence on later behavioral motiviation. Short latency metabolic actions might provide advantage in muscular response in fights or chases (Sapolsky, 1993). Other short latency actions help to sustain behavior against obstacles and distractions by focusing and stabilizing attention and facilitating repetition of ongoing motor sequences that have previously been successful (Andrew, 1991). These actions would be highly advantageous in intensely aggressive situations, where escalating threat and bluff would demand intense concentration and repeated display.

More persistent differences in testosterone secretion may be related to mating competition. The "challenge hypothesis" argues that high testosterone levels serve primarily to motivate mate defense and other aspects of reproductive competition (Wingfield *et al.*, 1990). In birds, maximal physiological levels of testosterone are well above those necessary for spermatogenesis, maintenance of secondary sexual characteristics, or expression of sexual behavior (Wingfield *et al.*, 1990). In primates, as well, testosterone facilitates spermatogenesis and sexual behavior but appears to have little effect above minimal threshold levels (Marshall *et al.*, 1986; Weinbauer *et al.*, 1988; Michael and Zumpe, 1993). Aggression tracks fluctuations in serum testosterone more closely, and correlations between individual differences in testosterone concentrations and individual differences in rates of aggression have been reported in studies of rhesus macaques (Gordon *et al.*, 1976), stumptailed macaques (Kling and Dunne, 1976), baboons (Sapolsky, 1983), and vervet monkeys (Steklis *et al.*, 1985) although other studies have reported no correlation (Eaton and Resko, 1974; Gordon *et al.*, 1978). Testing the relationship between aggression and physiology may be confounded by the problems of maintaining highly aggressive animals in captivity, resulting in their exclusion from primate colonies (Higley *et al.*, 1992). Moreover, the natural setting may provide more opportunities for elicitation of intense aggression, through social disruptions induced by male mortality, maturation, and immigration. The challenge hypothesis suggests that there should be a correlation between testosterone and mate guarding behavior primarily at the high end of serum concentrations. The relation of testosterone to different aspects of aggression has not been carefully examined in primates, but there is evidence that testosterone can have marked effects on aggressive behavior at maximal concentrations: treatments that produce ten- to twentyfold increases in serum testosterone have been observed to induce marked elevations in aggression (Gordon *et al.*, 1979; Rejeski *et al.*, 1988).

The challenge hypothesis also suggests that testosterone set-points and responsiveness to challenge should vary with the regularity of mating competition. This connection is supported by evidence in birds that testosterone levels vary with the mating system; testosterone is constitutively maintained closer to the physiological maxima in polygynous birds where more prolonged mating effort is required whereas average testosterone concentrations are lower but more responsive to challenges in monogamous birds (Wingfield *et al.*, 1990; Wingfield, 1994).

A similar pattern may distinguish multimale from unimale mating systems in the cercopithecoids. Figure 10.2 shows that the mean of species breeding season averages for male testosterone is significantly higher in species living in multimale groups, where male–male competition should be more constant, than in species with unimale groups where challenges occur only at irregular intervals. Both values are higher than averages for monogamous gibbons (1.6 ng/ml; Coe *et al.*, 1992). This pattern is consistent with the challenge hypothesis but will require further study to determine its implications. The observed differences could reflect interspecific differences in the regulation of testosterone secretion, in social stimuli, or in relative testes size. A role for social stimuli is suggested by evidence that average testosterone levels vary even among baboon populations. For example, testosterone values reported for anubis baboons in the Masai Mara (Sapolsky, 1993) and Gilgil (Sapolsky, 1993) are much higher than values reported for anubis baboons at Awash (Phillips-Conroy *et al.*, 1992). Moreover, although hamadryas baboon males living in their normal unimale groups have lower testosterone levels than anubis males, hamadryas males who have immigrated into multimale anubis groups exhibit the testosterone profiles of anubis males (Phillips-Conroy *et al.*, 1992).

Alternatively, these testosterone patterns may be merely a byproduct of sperm competition. Testes are larger relative to body size in vertebrate species with promiscuous mating systems than in monogamous species, a pattern that has been attributed to sperm competition (Harcourt *et al.*, 1981; Short, 1981; Harvey and Harcourt, 1984; Kenagy and Tombulak, 1986). Therefore, species in multimale groups might have high serum testosterone merely because they have large testes in order to produce large quantities of sperm. Serum testosterone levels are higher in cercopithecoids with testes that are larger than expected for their body size (see Fig. 10.3). However, this correlation is more likely to reflect parallel selection pressures than the constraints of gametogenic capacity, since serum testostosterone

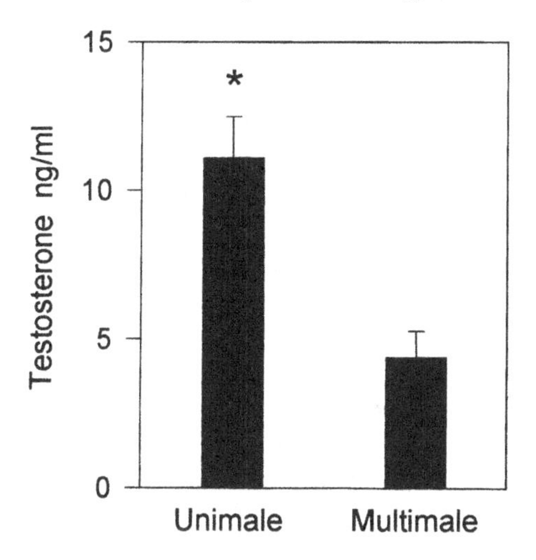

Fig. 10.2. Relation of male serum testosterone concentrations to mating systems in Old World monkeys. Means calculated from data of Coe *et al.*, 1992 and Phillips-Conroy *et al.*, 1992. * indicates that $p<0.05$ in Student's t-test.

levels are not correlated with ejaculate sperm number ($n=9$; $p=0.423$) or concentration ($n=9$; $p=0.272$; data from Møller, 1988). Moreover, testicular steroidogenesis is not directly tied to spermatogenesis or testes size in other mammals. Intratesticular androgen concentrations are considerably in excess of those required for normal spermatogenesis (Weinbauer and Nieschlag, 1993), and the number of androgen-secreting Leydig cells markedly exceeds the number required for normal intratesticular androgen levels (Jégou and Sharpe, 1993), making up 1–60% of the volume of the testes (Fawcett *et al.*, 1973). Although absolutely larger testes do produce more testosterone per day (T/day$=0.05$ testes wt$^{0.6}$; $r=0.79$, $n=9$, $p=0.011$), mammalian testes that are large for body size do not produce more testosterone per unit body weight ($p=0.356$; see Fig.10. 4). In fact, the steroidogenic capacity of Leydig cells may compensate for testes size, since cellular testosterone production rates appear to be elevated in species with testes that are small for body size (see Fig.10.4). Although further investigations will be required, these comparisons suggest that the patterns of testosterone synthesis and secretion can provide insights into mate competition in primates.

Stress responses

Studies of stress responses provide evidence of the costs of group life. Stress can have a variety of adverse effects on individual survival and

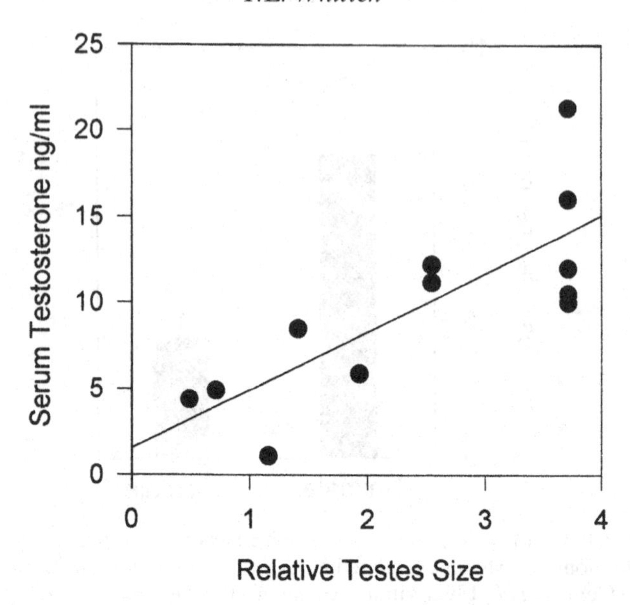

Fig. 10.3. Relation of male serum testosterone concentrations to relative testes size in Old World monkeys. Relative testes size is observed/expected testes size for body weight. Data from Coe *et al.*, 1992; Harvey & Harcourt, 1984.

reproduction including elevated heart rates, lower HDL-cholesterol, altered immune function, atherosclerosis, suppression of ovarian function, and miscarriages (Kaplan, 1986; Kaplan *et al.*, 1990; Adams *et al.*, 1985; Martensz *et al.*, 1987; Abbott, 1987; Harcourt, 1987; Sapolsky, 1993). Although rank-related differences in reproduction often have been attributed to the stress of low (Wasser, 1983; Dunbar, 1984) or high (Packer *et al.*, 1995; Dunbar 1995) rank, little is known about the physiological stress of rank or its influence on ovarian cycles or gestation in wild cercopithecine populations (Harcourt, 1987) and in fact the only available endocrine data on stress in wild females demonstrates an effect of intermale rather than interfemale aggression (Alberts *et al.*, 1992).

Both low rank and social disruptions can stimulate adrenocortical activity (Kaplan, 1986), but these stress responses are context dependent. Elevated cortisol levels have been found in subordinate animals in stable groups of male baboons (Sapolsky, 1983) and captive talapoin monkeys (Eberhardt *et al.*, 1980) and in captive male (Sassenrath, 1970) and female rhesus macaques (Gust *et al.*, 1993). In unstable groups, physiological stress may be positively (Chamove and Bowman, 1976; Scallet *et al.*, 1981) or negatively (Kaplan, 1986; Steklis *et al.*, 1985; McGuire *et al.*, 1986) related to

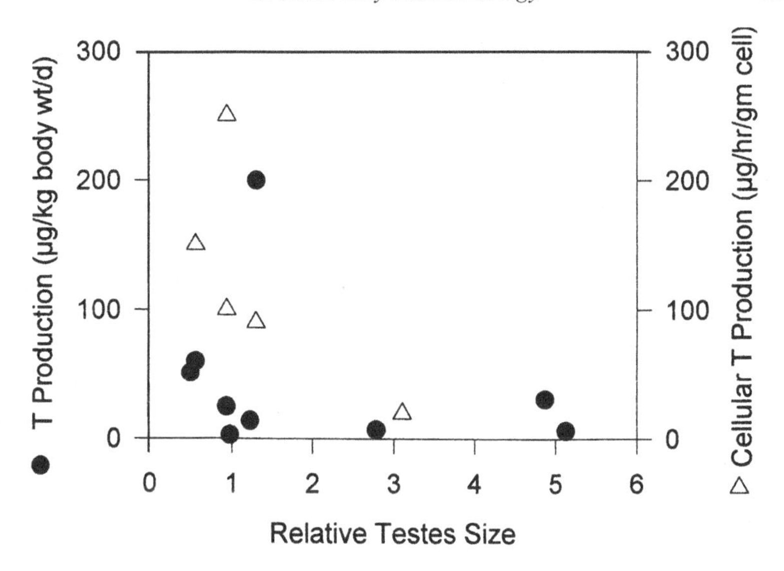

Fig. 10.4. Relation of relative testes size to testes (•) and Leydig cell testosterone production rates (Δ) in domestic mammals. Data from Waites and Setchell, 1990; Kenagy and Tombulak, 1986.

rank or unrelated (Sapolsky, 1983; Gust *et al.*, 1993). During periods when dominance relations are contested in baboons, cortisol increases in males whose social status is threatened but not in males who are actively contesting the rank of dominants (Sapolsky, 1993). An exception, however, is the male who pursues a highly aggressive stance during intertroop transfer, a tactic that results in markedly elevated cortisol in the aggressive male as well as in the rest of the troop (Alberts *et al.*, 1992). Once the hierarchy has stabilized, high-ranking male baboons and talapoin monkeys experience reductions in serum cortisol (Sapolsky, 1993; Eberhardt *et al.*, 1983). In contrast, in vervet monkeys, cortisol levels are unrelated to rank during periods of stability but are elevated during unstable periods in males who eventually achieve high rank (Steklis *et al.*, 1985; McGuirc *et al.*, 1986). Thus it appears that strivers are more stressed in vervets whereas males whose rank is threatened are more stressed in baboons. These differences may represent differences in male social relations and dominance acquisition in baboons and vervet monkeys. In baboons, dominance is constantly contested and maintained by male threat, harassment and intimidation (Sapolsky, 1991), and in fact consequent elevations in cortisol rise with the frequency of these events (Alberts *et al.*, 1992; Sapolsky, 1993). In vervet monkeys, affiliations with high-ranking females are a precondition for elevations in male rank (Raleigh *et al.*, 1991).Males exhibit high arousal and vigilance prior to a rise in rank (McGuire and Raleigh, 1985) but are more relaxed once alpha status is achieved when it is maintained primarily by the behavior of subordinates (Rowell, 1988).

Social disruptions also increase adrenocortical activity and aggression (Kaplan, 1986; Alberts *et al.*, 1992), suggesting one reason why subordinate animals often do not contest their low rank. However, the extent and distribution of these responses varies with interspecific differences in the intensity and expression of interindividual competition. In baboons and macaques, social disruption results in transient elevations in aggression that are followed by social integration and a return to normal levels of aggression and affiliation; guenons, in contrast, exhibit persistent and sometimes lethal aggression following social disruption (Kaplan, 1986) which may be related to their reliance on approach-avoidance rather than stylized gestures to express submission (Rowell, 1988). Similarly, the duration of responses to new group formation varies with intermale tolerance in macaques. In captive lion-tail macaques, a species which may have a unimale mating system in the wild (Green, 1981; Johnson, 1985), males exhibit behavioral tension, aggression, and a lack of affiliative behavior when housed with other males, and cortisol levels remain elevated throughout the period of male–male contact (Clarke *et al.*, 1995). In contrast, in multimale crabeating macaques, males housed together readily develop dominance relationships and thereafter exhibit more affiliative behaviors and less aggression, and cortisol levels decline.

Not surprisingly, social interactions and behavioral profiles are better predictors of physiological stress responses than rank or social group (Sapolsky and Ray, 1989; Kaplan *et al.*, 1991; Ray and Sapolsky, 1992; Gust *et al.*, 1993). The recipient of intense aggression elevates stress responses (Dillon *et al.*, 1992; Gust *et al.*, 1993) whereas affiliative interactions and relationships (Kaplan *et al.*, 1991; Ray and Sapolsky, 1992; Gust *et al.*, 1993) appear to be protective. Social skillfulness, particularly in distinguishing the threatening from nonthreatening presence of a rival, also appears to be a buffer against stress (Sapolsky and Ray, 1989) and may aid in maintaining or establishing high rank (Sapolsky and Ray, 1989; McGuire *et al.*, 1994). Emotional reactivity, or responsiveness to novel stimuli, has been associated with a suite of behavioral and physiological responses in rhesus monkeys such as avoidance of novel stimuli, elevated heartrate and adrenocortical activity, and hyper-responsiveness to stress (Higley and Suomi, 1989). These response styles appear to represent individual differences in personality or temperament since they are apparent early in infancy and show considerable stability across developmental stages and significant heritability although they also can be influenced by early rearing experience (Higley and Suomi, 1989).

Neurochemical bases of temperament

Recent evidence suggests that constitutional differences in neurotransmitter function may underlie individual differences in temperament. Concentrations of neurotransmitters and their metabolites, measured in cerebrospinal fluid (CSF), are stabile over time and are influenced by parental heritage as well as individual experience (Higley *et al.*, 1991, 1993). Individual differences in CSF monoamine metabolite concentrations are associated with individual differences in behavioral responsiveness. For example, in captive vervet monkeys, individual differences in measures of serotonergic function such as the serotonin metabolite 5-hydroxyindoleacetic acid (5-HIAA), whole blood serotonin, and serum tryptophan concentrations are associated with individual differences in affilitative behavior (Raleigh *et al.*, 1981). Treatments that enhance serotonergic function increase grooming and other affiliative behaviors whereas treatments that suppress serotonergic function reduce affiliation and increase aggression (Raleigh *et al.*, 1986). In this species, where alliances with high-ranking females appear to be crucial for male attainment of dominant status (Raleigh and McGuire, 1989), serotonergic mechanisms that mediate affiliative behaviors can significantly influence a male's ability to rise in rank and to achieve alpha status during periods of social upheaval (Raleigh *et al.*, 1991). Pharmacological manipulations that alter serotonergic function determine which of the remaining males will eventually become dominant when the dominant male is removed from the group (Raleigh *et al.*, 1991). The behavioral consequence of treatments that enhance serotonergic function is an increase in affiliation with females, followed by increased female support in coalitions, and only later by an increase in success in intermale competition. In contrast, suppression of serotonergic activity markedly enhanced male aggression to males and to females and reduced affiliation, resulting in reduced female support and increased often severe counter-attacks from females in response to male–female aggression, lower success in intermale competition and spatial and social isolation (Raleigh *et al.*, 1991). In free-ranging rhesus monkeys, low concentrations of CSF 5-HIAA also are associated with excessive aggression (Higley *et al.*, 1992), but falling values at adolescence (Higley *et al.*, 1991) also may help to mediate the normal migration of males from their natal group and concommitant aggression (Berard, 1989) since males who do not disperse from their natal group fail to exhibit the typical adolescent decline in 5–HIAA, retaining the higher levels of juveniles (Kaplan *et al.*, 1995).

Interspecific differences in temperament and behavioral style

These findings suggest that individual differences in endocrine and behavioral responses are linked to individual differences in social predispositions or temperament. Moreover the links between behavior–endocrine responses and social contexts vary with species, reflecting species-specific styles of competitive and affiliative behavior. Thus it seems not unreasonable to postulate that interspecific differences in temperament also may underlie interspecific differences in behavioral styles (Mason, 1990). This view accords with the notion of coevolution of social characteristics (de Waal and Luttrell, 1987; Thierry, 1990) and individual social predispositions as the focus of selection on social systems (Lott, 1984). A number of studies have documented differences among macaque species in dominance style (Shiveley *et al.*, 1982; Thierry, 1990; de Waal and Luttrell, 1987; Chaffin *et al.*, 1995), patterns of sexual behavior and mating strategies (Shively *et al.*, 1982; Caldecott, 1986), and mother–infant interactions (Rosenblum and Kaufmann, 1967) that appear to be related to differences in behavioral reactivity and adrenocortical and cardiovascular response (Clarke *et al.*, 1988, 1994). Rhesus and Japanese macaque societies are nepotistic, with high degrees of contest asymmetry and social intolerance whereas stumptail and bonnet macaque societies are more tolerant, egalitarian, and affiliative and crabeating macaques are intermediate in this spectrum (Shiveley *et al.*, 1982; Thierry, 1990; de Waal and Luttrell, 1987; Chaffin *et al.*, 1995). Nepotistic species like rhesus macaques respond more actively and aggressively to stressful situations with measured increases in cortisol and heartrate (Clarke *et al.*, 1988, 1994), responses that may support their more despotic style. In contrast, less nepotistic species like crabeating macaques respond fearfully to stressful situations with large elevations in cortisol and heartrate (Clarke *et al.*, 1988, 1994), responses that may be more adaptive for maintaining their more contested and individualized relationships. However, highly egalitarian species like bonnet macaques are not the most reactive but are intermediate in physiological responses, which accords with their more passive and affiliative response to stress, perhaps reflecting a de-emphasis on social competition in this species.

Conclusions

Although differences in the emphasis given the individual versus the group hampered the integration of early investigations in behavioral endocrinology and socioecology, socioendocrinology and behavioral ecology have converged in their interest in the socially-situated individual. Early endocrine research seemed to indicate that the behavior of primate males was hormone dependent while that of females was not, a notion fostered by Zuckerman's model of primate sociality based on continual receptivity and the resultant necessity for control of male aggression. Subsequent research demonstrated that hormonal actions are highly context-dependent in both male and female primates, as are hormonal concentrations. Although the organizing role of the dominance hierarchy has given way to individual mating and foraging tactics in behavioral ecology, dominance status has continued to be an important concept in endocrine studies, but here dominance has shifted from a dependent to an independent variable. There is no consistent endocrine profile of rank, but rank can have important influences on hormonal concentrations. The latter findings demonstrate the costs of subordinate status but do not inform behavioral ecology's search for individual difference and the traits responsible for variance in reproductive success. For those issues, new studies of behavioral styles and their linkages with physiology may be more informative. At least some individual differences in temperament appear to have a genetic basis and are associated with individual differences in health or reproductive success and thus could be a focus of selection. Focusing on behavioral style and temperament could provide a useful inroad for delineating the selective consequences of individual variation. More information is needed on hormonal responses and their relation to key behavioral events in field settings, an approach facilitated by the development of methods of fecal steroid analysis (Clarke *et al.*, 1991; Wasser *et al.*, 1991; Strier and Ziegler, 1994; Shideler *et al.*, 1993; Brockman *et al.*, 1995; Stavisky *et al.*, 1995) and the application of sampling and experimental techniques previously confined to the laboratory (Sapolsky, 1993; Kaplan *et al.*, 1995). These techniques will facilitate the comparisons of hormone–behavior interactions in different populations of the same species, to investigate the role of social and ecological context in the development and expression of individual differences.

There has been less progress on the evolution of primate social systems. Neither behavioral ecology nor socioendocrinology has yet produced a model of how individual biology, local ecology, and social processes interact to generate species-specific patterns of social organization, alternative strategies, or contextual response. However, research in both areas has provided the basis for the development of such models, and there have been promising beginnings at several levels of analysis (Wilson, 1975; Hinde, 1979, 1982; Altmann and Altmann, 1979; Richard, 1985; Bronson, 1989; Fentress, 1991; Lott, 1991). Behavioral ecology has helped to define the role of individual strategies in the evolution of social structure but has not attempted to specify the individual attributes that might be responsible for these strategies. Socioendocrinology has not specifically addressed the evolution of social systems, but research reviewed here has shown that the direction and intensity of the influence of social context on hormone actions varies with species differences in mating systems and dominance style, suggesting that interspecific comparisons of behavior–endocrine regulation may be a promising area for future research. Studies of interspecific differences in behavior style in macaques suggest some interesting links with temperament. Of particular interest may be the role of neurochemicals that mediate affiliation and aggesssion such as serotonin, oxytocin, and vasopressin and interspecific differences in hormonal set-points and behavioral thresholds. A focus on social predispositions provides a framework for bridging evolutionary concepts of the individual as the unit of selection and ecological concepts of social systems as adaptive responses to the environment. By enabling more precise predictions, these investigations should enable more rigorous tests of models of the evolution of primate social systems (Dunbar, 1989).

Acknowledgements

Thanks to E.O. Smith and two anonymous reviewers for comments and helpful suggestions.

References

Abbott, D.H. (1987). Behaviourally mediated suppression of reproduction in female primates *J. Zool.* **213**, 455–70.

Adams, M.R., Kaplan, J.R. & Koritnik, D.R. (1985). Psychosocial influences on ovarian endocrine and ovulatory function in *Macaca fascicularis*. *Physiol. Behav.* **35**, 935–40.

Alberts, S.C., Sapolsky, R.M. & Altmann, J. (1992). Behavioral, endocrine, and immunological correlates of immigration by an aggressive male into a natural primate group. *Hormones Behav.* **26**, 167–78.

Alexander, R.D. & Noonan, K.M. (1979). Concealment of ovulation, paternal care, and human social evolution. In *Evolutionary Biology and Human Behavior*, ed. N.A. Chagnon & W. Irons, pp. 436–53. N. Scituate, MA: Duxbury Press.

Altmann, S.A. (1974). Baboons, space, time, and energy. *Am Zool.* **14**, 221–48.
Altmann, S.A. & Altmann, J. (1979). Demographic constraints on behavior and social organization. In *Recent Advances in Primatology*, vol. 1, *Behaviour*, ed. D.J. Chivers & J. Herbert, pp. 407–14. New York: Academic Press.
Andelman, S.J. (1987). Evolution of concealed ovulation in vervet monkeys (*Cercopithecus aethiops*). *Am. Nat.* **129**, 785–99.
Andrew, R.J. (1991). Testosterone, attention and memory. In *The Development and Integration of Behavior. Essays in Honour of Robert Hinde*, ed. P. Bateson, pp. 171–90. New York: Cambridge University Press.
Beach, F.A. (1947). Evolutionary changes in the physiological control of mating behavior in mammals. *Psychol. Revs.* **54**, 297–315.
Beach, F.A. (1975). Behavioral endocrinology: an emerging discipline. *Am Sci.* **63**, 178–87.
Berard, J.D. (1989). Life histories of male Cayo Santiago macaques. *Puerto Rico Health Sci. J.* **8**, 61–4.
Bercovitch, F.B. (1992). Estradiol concentrations, fat deposits, and reproductive strategies in male rhesus macaques. *Horm. Behav.* **26**, 272–82.
Bercovitch, F.B. & Ziegler, T. (1990) *Socioendocrinology of Primate Reproduction*. New York: Wiley-Liss.
Bernstein, I. S. (1981). Dominance: The baby and the bathwater. *Behav. Brain Sci.* **4**, 419–58.
Bernstein, I.S., Rose, R.M. & Gordon, T.P. (1977). Behavioural and hormonal responses of male rhesus monkeys introduced to females in the breeding and nonbreeding seasons. *Anim. Behav.* **25**, 609–14.
Bernstein, I.S., Rose, R.M., Gordon, T.P. & Grady, C.L. (1979). Agonistic rank, aggression, social context, and testosterone in male pigtail monkeys. *Agress. Behav.* **5**, 329–39.
Bielert, C., Czaja, J.A., Eisele, S., Scheffler, G., Robinson, J.A. & Goy, R.W. (1976). Mating in the rhesus monkey (*Macaca mulatta*) after conception and its relationship to oestradiol and progesterone levels throughout pregnancy. *J. Reprod. Fertil.* **46**, 179–87.
Boinski, S. (1987). Mating patterns in squirrel monkeys (*Saimiri oerstedi*). *Behav. Ecol. Sociobiol.* **21**, 13–21.
Brockman, D.K., Whitten P.L., Russell E., Richard, A.F. & Izard, M.K. (1995). Application of fecal steroid techniques to the reproductive endocrinology of Verreauxi's sifaka, *Propithecus verreauxi. Am. J. Primatol.* **36**, 313–25.
Bronson, F.H. (1989). *Mammalian Reproductive Biology*. Chicago: University of Chicago Press.
Butler, H. (1974). Evolutionary trends in primate sex cycles. *Contrib. Primatol.*, vol. 3., pp. 2–35. Basel: S. Karger.
Caldecott, J.O. (1986). Mating patterns, societies and the ecogeography of macaques. *Anim. Behav.* **34**, 208–20.
Carpenter, C.R. (1942). Sexual behavior of free-ranging rhesus monkeys (*Macaca mulatta*). II. Periodicity of estrus, homosexual, autoerotic, and nonconformist behavior. *J. Comp. Psychol.* **33**, 143–62.
Caro, T.M. & Bateson, P. (1986). Organization and ontogeny of alternative tactics. *Anim. Behav.* **34**, 1483–99.
Chaffin, C.L., Friedlen, K. & de Waal, F.B.M. (1995). Dominance style of Japanese macaques compared with rhesus and stumptails. *Am. J. Primatol.* **35**, 103–16.
Chamove, A.S. & Bowman, R.E. (1976). Rank, rhesus social behavior and stress. *Folia primatol.* **26**, 57–66.

Clark, M.M., & Galef, B.G., Jr. (1995). Prenatal influences on reproductive life history strategies. *Tr. Ecol. Evol.* **10**, 151–3.
Clarke, A.S., Mason, W.A. & Mendoza, S.P. (1994). Heart rate patterns in three species of macaques. *Am. J. Primatol. 33*, 133–48.
Clarke, A.S., Mason, W.A. & Moberg, G.P. (1988). Differential behavioral and adrenocortical responses to stress among three macaque species. *Am. J. Primatol.* **14**, 37–52.
Clarke, A.S., Czekala, N.M. & Lindburg, D.G. (1995). Behavioral and adrenocortical responses of male cynomologous and lion-tailed macaques to social stimulation and group formation *Primates* **36**, 41–56.
Clarke, M.R., Zucker, E.L. & Harrison, R.M. (1991). Fecal estradiol, sexual swelling, and sociosexual behavior in free-ranging female howling monkeys in Costa Rica. *Am. J. Primatol.* **24**, 93.
Clutton-Brock, T.H. & Harvey, P.H. (1977). Primate ecology and social organization. *J. Zool. (Lond.)***183**, 1–39.
Coe, C.L., Savage, A. & Bromley, L.J. (1992). Phylogenetic influences on hormone levels across the primate order. *Am. J. Primatol.* **28**, 81–100.
Crook, J.H. (1989). Introduction: socioecological paradigms, evolution and historical perspectives for the 1990s. In *Comparative Socioecology: The Behavioural Ecology Of Humans And Other Mammals*, ed. V. Standen & R.A. Foley, pp. 1–36. Boston: Blackwell Scientific Publications.
Crook, J.H., Ellis, J.E. & Goss-Custard, J.D. (1976). Mammalian social systems: structure and function. *Anim. Behav.* **24**, 261–74.
Crook, J.H. & Gartlan, J.S. (1966). Evolution of primate societies. *Nature* **210**, 1200–3.
Dawkins, M.S. (1989). The future of ethology: how many legs are we standing on? In *Perspectives in Ethology*, vol. 8, *Whither Ethology*? ed. P.P.G. Bateson & P.H. Klopfer, pp. 47–54. New York: Plenum Press.
Dillon, J.E., Raleigh, M.J., McGuire, M.T., Bergin-Pollack, D. & Yuwiler, A. (1992). Plasma catecholamines and social behavior in male vervet monkeys (*Cercopithecus aethiops sabaeus*). *Physiol. Behav.* **51**, 973–7.
Dixson, A.F. (1983). Observations on the evolution and behavioral significance of "sexual skin" in female primates. In *Advances In The Study of Behavior*, vol. 13, ed. J.S. Rosenblatt, R.A. Hinde, C. Beer & M.-C. Busnel, pp. 63–106. New York: Academic Press.
Dixson, A.F., Bossi, T. & Wickings, E.J. (1993). Male dominance and genetically determined reproductive success in the mandrill (*Mandrillus sphinx*). *Primates* **34**, 525–32.
Dunbar, R.I.M. (1984). *Reproductive Decisions. An Economic Analysis Of Gelada Baboon Social Strategies*. Princeton: Princeton University Press.
Dunbar, R.I.M. 1989. Ecological modelling in an evolutionary context. *Folia primatol.* **53**, 235– 46.
Dunbar, R.I.M. (1995). The price of being at the top. *Nature* **373**, 22–3.
Eaton, C.G. & Resko, J. A. (1974). Plasma testosterone and male dominance in a Japanese macaque (*Macaca fuscata*) troop compared with repeated measures of testosterone in laboratory males. *Horm. Behav.* **5**, 251–9.
Eberhardt, J.A., Keverne, E.B. & Meller, R.E. (1980). Social influences on plasma testosterone levels in male talapoin monkeys. *Horm. Behav.* **14**, 247–66.
Eberhardt, J.A., Keverne, E.B. & Meller, R.E. (1983). Social influences on circulating levels of cortisol and prolactin in male talapoin monkeys. *Physiol. Behav.* **30**, 361–9.
Ehardt, C.L. (1988). Absence of strongly kin-preferential behavior by adult

female sooty mangabeys (*Cercocebus atys*). *Am. J. phys. Anthropol.* **76**, 233–43.

Eisenberg, J.F., Muckenhirn, N.A. & Rudran, R. (1972). The relation between ecology and social structure in primates. *Science* **176**, 863–74.

Fabre-Nys, C., Martin, G.B. & Venier, G. (1993). Analysis of the hormonal control of female sexual behavior and the preovulatory LH surge in the ewe: roles of quantity of estrogen and duration of its presence. *Horm. Behav.* **27**, 108–21.

Fawcett, D.W., Neavres, W.B. & Flores, M.N. (1973). Comparative observations on intertubular lymphatics and the organization of interstitial tissue of the mammalian testis. *Biology Reprod.* **9**, 500–12.

Fentress, J.C. (1991). Analytical ethology and synthetic neuroscience. In *The Development and Integration of Behavior. Essays in Honour of Robert Hinde*, ed. P. Bateson, pp. 77–120. New York: Cambridge University Press.

Garcia-Segura, L.M., Chowen, J.A., Parducz, A. & Naftolin, F. (1994). Gonadal hormones as promotors of synaptic plasticity: Cellular mechanisms. *Prog. Neurobiol.* **44**, 279–307.

Gordon, T.P., Bernstein, I.S. & Rose, R.M. (1978). Social and seasonal influences on testosterone secretion in the male rhesus monkey. *Physiol. Behav.* **21**, 623–7.

Gordon, T.P., Gust, D. A., Busse, C.D. & Wilson, M.E. (1991). Hormones and sexual behavior associated with postconception perineal swelling in the sooty mangabey (*Cercocebus torquatus atys*). *Am. J. Primatol.* **12**, 585–97.

Gordon, T.P., Rose, T.M. & Bernstein, I.S. (1976). Seasonal rhythm in plasma testosterone levels in the rhesus monkey (*Macaca mulatta*): a three year study. *Horm. Behav.* **7**, 229–43.

Gordon, T.P., Rose, T.M., Grady, C.L. & Bernstein, I.S. (1979). Effects of increasing testosterone secretion on the behavior of adult male rhesus living in a social group. *Folia primatol.* **32**, 149–60.

Goss-Custard, J.D., Dunbar, R.I.M. & Aldrich-Blake, F.P.G. (1972). Survival, mating, and rearing strategies in the evolution of primate social structure. *Folia primatol.* **17**, 1–19.

Gouzoules, H. & Goy, R.W. (1983). Physiological and social influences on mounting behavior of troop-living female monkeys. (*Macaca fuscata*). *Am. J. Primatol.* **5**, 39–49.

Green, S. (1981). Sex differences and age gradations in vocalizations of Japanese and liontailed monkeys. *Am. Zool.* **21**, 165–83.

Gunnar, M.R. & Mangelsdorf, S. (1989). The dynamics of temperament-physiology relations: a comment on biological processes in temperament. In *Temperament in Childhood*, ed. G. A. Kohnstamm, J. E. Bates & M. K. Rothbart, pp. 145–52. New York: John Wiley and Sons.

Gust, D.A. & Gordon, T.P. (1994). The absence of a matrilineally based dominance system in sooty mangabeys, *Cercocebus torquatus atys*. *Anim. Behav.* **47**, 589–94.

Gust, D.A., Gordon, T.P., Hambright, M.K. & Wilson, M.E. (1993). Relationship between social factors and pituitary-adrenocortical activity in female rhesus monkeys (*Macaca mulatta*). *Horm. Behav.* **27**, 318–31.

Hamilton, W.J. (1984). Significance of paternal investment by primates to the evolution of male-female associations. In *Primate Paternalism*. ed. D.M. Taub. pp. 309–35. New York: Van Nostrand Reinhold.

Harcourt, A.H. (1987). Dominance and fertility among female primates. *J. Zool.* **213**, 471–87.

Harcourt, A.H. Harvey, P.H., Larson, S.G. & Short, R.V. (1981). Testis weight, body weight and breeding system in primates. *Nature* **293**, 55–7.
Harvey, P.A. & Harcourt, A.H. (1984). Sperm competition, testes size, and breeding systems in primates. In *Sperm Competition and the Evolution of Animal Mating Systems*, ed. R.L. Smith, pp. 589–600. New York: Academic Press.
Higley, J.D., Mehlman, P.T., Taub, D.M., Higley, S.B., Suomi, S.J., Vickers, J.H. & Linnoila, M. (1992). Cerebrospinal fluid monoamine and adrenal correlates of aggression in free-ranging rhesus monkeys. *Arch. Gen. Psychiatry* **49**, 436–41.
Higley, J.D. & Suomi, S.J. (1989). Temperamental reactivity in non-human primates. In *Temperament in Childhood*, ed. G.A.Kohnstamm, J.E. Bates & M.K. Rothbart, pp. 153–68. New York: John Wiley and Sons.
Higley, J.D., Suomi, S.J. & Linnoila, M. (1991). CSF monoamine metabolite concentrations vary according to age, rearing, and sex, and are influenced by the stressor of social separation in rhesus monkeys. *Psychopharmacol.* **103**, 551–6.
Higley, J.D., Thompson, W.W., Champoux, M., Goldman, D., Hasert, M.F., Kraemer, G.W., Scanlan, J.M., Suomi, S.J. & Linnoila, M. (1993). Paternal and maternal genetic and environmental contributions to cerebrospinal fluid monoamine metabolites in rhesus monkeys (*Macaca mulatta*). *Arch. gen. Psychiatry.* **50**, 615–23.
Hinde, R.A. (1979). *Towards Understanding Relationships*. London: Academic Press.
Hinde, R.A. (1982). *Ethology: Its Nature and Relations with other Sciences*. New York: Oxford University Press.
Hooks, B.L. & Green, P.A. (1993). Cultivating male allies: a focus on primate females, including *Homo sapiens. Hum. Nature* **4**, 81–107.
Hrdy, S.B. (1977). *The Langurs Of Abu.* Cambridge, MA: Harvard University Press
Hrdy, S.B. (1981). *The Woman That Never Evolved.* Cambridge, MA: Harvard University Press.
Hrdy, S.B. (1988). The primate origins of human sexuality. In *The Evolution of Sex*, ed. R. Bellig & G. Stevens, pp. 101–36. San Francisco: Harper and Row.
Hrdy, S.B. & Whitten, P.L. (1987). Patterning of sexual activity. In *Primate Societies*, ed. B.B. Smuts, D.L. Cheney, R.M. Seyfarth, R.W. Wrangham & T.T. Struhsaker, pp. 370–84. Chicago: University of Chicago Press.
Insel, T.R. & Carter, C.S. (1995). The monogamous brain. *Nat. Hist.* **104,** 12–14.
Insel, T.R., Winslow, J.T., Williams, J.R., Hastings, N., Shapiro, L.E. & Carter, C.S. (1993). The role of neurohypophyseal peptides in the central mediation of complex social processes – evidence from comparative studies. *Reg. Peptides* **45,** 127–31.
Jégou, B. & Sharpe, R.M. (1993). Paracrine mechanisms in testicular control. In *Molecular Biology of the Male Reproductive System*, ed. D. de Kretser, pp. 271–310. New York: Academic Press.
Johnson, T.J.M. (1985). Lion-tailed macaque behavior in the wild. In *The Lion-tailed Macaque: Status and Conversation*, ed. P.G. Heltne, pp. 41–63. New York: Alan R. Liss.
Kaplan, J.R. (1986). Psychological stress and behavior in nonhuman primates. In *Comparative Primate Biology,* Volume 2A*: Behavior, Conservation, and Ecology*, ed. G. Mitchell & J. Erwin, pp. 455–92. New York: Alan R. Liss.
Kaplan, J.R., Manuck, S.B.& Gatsonis, C. (1990). Heart rate and social status

among male cynomologous monkeys (*Macaca fascicularis*) housed in disrupted social groupings. *Am. J. Primatol.* **21**, 175–87.
Kaplan, J.R., Fontenot, M.B., Berard, J., Manuck, S. B. & Mann, J.J. (1995). Delayed dispersal and elevated monoaminergic activity in free-ranging rhesus monkeys. *Am. J. Primatol.* **35**, 229–34.
Kaplan, J.R., Heise, E.R., Manuck, S.B., Shively, C.A., Cohen, S., Rabin, B.S. & Kasprowicz, A.L. (1991). The relationship of agonistic and affiliative behavior patterns to cellular immune function among cynomologous monkeys (*Macaca fasciciularis*) living in unstable social groups. *Am. J. Primatol.* **25**, 157–73.
Kenagy, G.L. & Tombulak, S.C. (1986). Size and function of mammalian testes in relation to body size. *J. Mammal.* **67**, 1–22.
Ketterson, E.D. & Nolan, V., Jr. (1992). Hormones and life histories: an integrative approach. *Am. Nat.* **140**, S33–S62.
Keverne, E.B., Eberhart, J.A. & Meller, R.E. (1982). Social influences on behaviour and neuroendocrine responsiveness of talapoin monkeys. *Scand. J. Psychol.* (*Suppl.*) **1**, 37–47.
Krebs, J.R. & Davies, N.B. (1991). Preface. In *Behavioural Ecology. An Evolutionary Approach*, 3rd edn., ed. J.R. Krebs & N.B. Davies, pp. ix–x. Boston: Blackwell Scientific Publications.
Kling, A. & Dunne, K. (1976). Social environmental factors affecting behavior and plasma testosterone in normal and amygdala-lesioned *Macaca speciosa*. *Primates* **17**, 23–42.
Lancaster, J.B. & Lee, R.B. (1965). The annual reproductive cycle in monkeys and apes. In *Primate Behavior: Field Studies of Monkeys and Apes*, ed. I. DeVore, pp. 486–513. New York; Holt.
Lindburg, D. (1987). Seasonality of reproduction in primates. In *Comparative Primate Biology*, vol. 2B, ed. G. Mitchell & J. Erwin. pp. 167–218. New York: Alan R. Liss.
Longcope, C., Billiar, R.B., Takaoka, Y., Reddy, P.S., Richman, D. & Little, B. (1983). Tissue sites of aromatization in the female rhesus monkey. *Endocrinol.* **113**, 1679–82.
Lott, D.F. (1984). Intraspecific variation in the social systems of wild vertebrates. *Behaviour* **88**, 266–325.
Lott, D.F. (1991). *Intraspecific Variation in the Social Systems of Wild Vertebrates*. New York: Cambridge University Press.
McCamant, S.K., Klosterman, L.L.. Goldman, E.S., Murai, J.T. & Siiteri, P.K. (1987). Conversion of androgens to estrogens in the male squirrel monkey (*Saimiri sciureus*). *Steroids* **50**, 549–57.
McEwen, B.S., Davis, P.G., Parsons, B. & Pfaff, D.W. (1979). The brain as a target for steroid hormone action. *Ann. Rev. Neuroscience* **2**, 65–112.
McEwen, B.S. (1991). Non-genomic and genomic effects of steroids on neural activity. *Tr. pharmacol. Sci.* **12**, 141–7.
McGuire, M.T., Raleigh, M.J. & Pollack, D.B. (1994). Personality features in vervet monkeys: the effects of sex, age, social status, and group composition. *Am. J. Primatol.* **33**, 1–13.
McGuire, M.T., Brammer, G.L. & Raleigh, M.J. (1986). Resting cortisol levels and the emergence of dominant male status in vervet monkeys. *Horm. Behav.* **20**, 106–17.
McGuire, M.T. & Raleigh, M.J. (1985). Serotonin-behavior interactions in vervet monkeys. *Psychopharmacol. Bull.* **21**, 458–63.
Marshall, G.R., Jockenhövel, F., Lüdecke, D. & Nieschlag, E. (1986).

Maintenance of complete but quantitatively reduced spermatogenesis in hypophysectomized monkeys by testosterone alone. *Acta Endocrinol.* **113**, 424–31.

Martensz, N.D., Vellucci, S.V., Fuller, L.M., Everitt, B.J., Keverne, E.B. & Herbert, J. (1987). Relationship between aggressive behaviour and circadian rhythms in cortisol and testosterone in social groups of talapoin monkeys. *J. Endocrinol.* **115**, 107–20.

Mason, W.A. (1990). Premises, promises, and problems of primatology. *Am. J. Primatol.* **22**, 123–38.

Mendoza, S.P., Lowe, E.L., Davidson, J.M. & Levine, S. (1978). Annual cyclicity in the squirrel monkey (*Saimiri sciureus*): the relationship between testosterone, fatting, and sexual behavior. *Horm. Behav.* **11**, 295–303.

Michael, R.P. & Zumpe, D. (1993). A review of hormonal factors influencing the sexual and aggressive behavior of macaques. *Am. J. Primatol.* **30**, 213–41.

Møller, A.P. (1988). Ejaculate quality, testes size and sperm competition in primates. *J. Hum. Evol.* **17**, 479–88.

Moore, J. (1984). Female transfer in primates. *Int. J. Primatol.* **5**, 537–89.

Moore, M. C. (1991). Application of organization-activation theory to alternative male reproductive strategies: a review. *Horm. Behav.* **25**, 154–79.

Naftolin, F. (1994). Brain aromatization of estrogens. *J. reprod. Med.* **39**, 257–61.

Oates, J.F. (1994). The natural history of African colobines. In *Colobine Monkeys: Their Ecology, Behaviour and Evolution*, ed. A.G. Davies & J.F. Oates, pp.75–128. Cambridge: Cambridge University Press.

Packer, C., Collins, D.A., Sindlmwo, A. & Goodall, J. (1995). Reproductive constraints on aggressive competition in female baboons. *Nature* **373**, 60–3.

Packer, C. & Pusey, A. (1987). Dispersal and philopatry. In *Primate Societies*, ed. B.B. Smuts, D.L. Cheney, R.M. Seyfarth, R.W. Wrangham & T.T. Struhsaker, pp. 250–66. Chicago: University of Chicago Press.

Pagel, M. (1994). The evolution of conspicuous oestrous adverstisement in Old World monkeys. *Anim. Behav.* **47**, 1333–41.

Phillips-Conroy, J.E., Jolly, C.J., Nystrom, P.I. & Hemmalin, H.A. (1992). Migration of male hamadryas baboons into anubis groups in the Awash National Park, Ethiopia. *Int. J. Primatol.*, **13**, 455–76.

Popp J.L. & DeVore, I. (1979). Aggressive competition and social dominance theory: synopsis. In *The Great Apes*, ed. D Hamburg & E. R. McCown, pp. 317–38. Menlo Park, CA: Benjamin/Cummings.

Raleigh, M.J., Yuwiler, A., Brammer, G.L., McGuire, M.T., Geller, E. & Flannery, J.W. (1981). Peripheral correlates of serotonergically-influenced behaviors in vervet monkeys (*Cercopithecus aethiops sabaeus*). *Psychopharmacol.* **72**, 241–6.

Raleigh, M.J., Brammer, G.L., Ritvo, E.R., Yuwiler, A., McGuire, M.T. & Geller, E. (1986). Effects of chronic fenfluramine on blood serotonin, cerebrospinal fluid metabolites, and behavior in monkeys. *Psychopharmacol.* **90**, 503–8.

Raleigh, M.J. & McGuire, M.T. (1989). Female influences on male dominance acquisition in captive vervet monkeys (*Cercopithecus aethiops sabaeus*). *Anim. Behav.* **37**, 59–67.

Raleigh, M.J., McGuire, M.T., Brammer, G.L., Pollack, D.B. & Yuwiler, A. (1991). Serotonergic mechanisms promote dominance acquisition in adult male vervet monkeys. *Brain Res.* **559**, 181–90.

Ray, J.C. & Sapolsky, R.M. (1992). Styles of male social behavior and their endocrine correlates among high-ranking wild baboons. *Am. J. Primatol.* **28**, 231–50.

Rejeski, W.J., Brubaker, P.H., Herb, R.A., Kaplan, J.R. & Kaltnik, D. (1988). Anabolic steroids and aggressive behavior in cynomologous monkeys. *J. behav. Med.* **11**, 95–105.
Richard, A. (1981). Changing assumptions in primate ecology. *Am. Anthropol.* **83**, 517–33.
Richard, A. (1985). *Primates in Nature*. New York: W.H. Freeman & Co.
Rodman, P.S. (1988). Resources and group sizes of primates. In *The Ecology of Social Behavior*, ed. C. N. Slobodchikoff, pp. 83–108. New York: Academic Press.
Rose, R.M., Holady, J.W. & Bernstein, I.S. (1971). Plasma testosterone, dominance rank and aggressive behavior in male rhesus monkeys. *Nature* **231**, 366–8.
Rosenblum, L.A. & Kaufmann, I.C. (1967). Laboratory observations of early mother–infant relations in pigtail and bonnet macaques. In *Social Communication Among Primates*, ed. S.A. Altmann, pp. 33–41. Chicago: University of Chicago.
Rowell, T.E. (1970). Reproductive cycles of two *Cercopithecus* monkeys. *J. Reprod. Fertil.* **22**, 321–38.
Rowell, T.E. (1988). The social systems of guenons, compared with baboons, macaques and mangabeys. In *A Primate Radiation: Evolutionary Biology of the African Guenons*, ed. A. Gautier-Hion, F. Bourliere, J.-P. Gautier & J. Kingdon, pp. 439–51. Cambridge: Cambridge University.
Rowell, T.E. (1994). Reification of social systems. *Evol. Anthropol.* **2**, 135
Sapolsky, R.M. (1983). Endocrine aspects of social instability in the olive baboon. *Am. J. Primatol.* **5**, 365–72.
Sapolsky, R.M. (1986). Endocrine and behavioral correlates of drought in the wild baboon. *Am. J. Primatol.* **11**, 217–24.
Sapolsky, R.M. (1991). Testicular function, social rank and personality among wild baboons. *Psychoneuroendocrinol.* **16**, 281–93.
Sapolsky, R.M. (1993). Endocrinolgy alfresco: psychendocrine studies of wild baboons. *Rec. Prog. Hormone Res.* **48**, 437–68.
Sapolsky, R.M. & Ray, J.C. (1989). Styles of dominance and their endocrine correlates among wild olive baboons (*Papio anubis*). *Am. J. Primatol.* **18**, 1–13.
Sassenrath, E.N. (1970). Increased adrenal responsiveness related to social stress in rhesus monkeys. *Horm. Behav.* **1**, 283–94.
Scallet, A.C., Sumoi, S.J. & Bowman, R.E. (1981). Sex differences in adrenocortical response to controlled agonistic encounters in rhesus monkeys. *Physiol. Behav.* **26**, 385–90.
Shideler, S.E. & Lasley, B.L. (1982). A comparison of primate ovarian cycles. *Am. J. Primatol. (Suppl.)* **1**, 1171–80.
Shideler, S.E., Ortuno, A.M., Moran, F.M., Moorman, E.A. & Lasley, B.L. (1993). Simple extraction and enzyme immunoassays for estrogen and progesterone metabolites in the feces of *Macaca fascicularis* during non-conceptive and conceptive ovarian cycles. *Biol. Reproduct.* **48**, 1290–8.
Shiveley, C., Clarke, S., King, N., Schapiro, S. & Mitchell, G. (1982). Patterns of sexual behavior in male macaques. *Am. J. Primatol.* **2**, 373–84.
Short, R.V. (1981). Sexual selection and great apes. In *Reproductive Biology of the Great Apes*, ed. G.E. Graham, pp. 319–41. New York: Academic Press.
Sillén-Tullberg, B. & Møller, A.P. (1993). The relationship between concealed ovulation and mating systems in anthropoid primates: a phylogenetic analysis. *Am. Nat.* **141**, 1–25.

Small, M.F. (1993). *Female Choices: Sexual Behavior of Female Primates*. Ithaca: Cornell University Press.
Smuts, B.B. (1985). *Sex and Friendship in Baboons*. Hawthorne: Aldine.
Smuts, B.B. (1987). Sexual competition and mate choice. In *Primate Societies*, ed. B.B. Smuts, D.L. Cheney, R.M. Seyfarth, R.W. Wrangham & T.T. Struhsaker, pp. 385–99. Chicago: University of Chicago.
Smuts. B.B. & Smuts, R.W. (1993). Male aggression and sexual coercion of females in nonhuman primates and other mammals: evidence and theoretical implications. In *Advances in the Study of Behavior*, vol. 22, ed. P. J.B. Slater, J.S. Rosenblatt, M. Milinski & C.T. Snowdon, pp. 1–63. New York: Academic Press.
Stavisky, R.C., Russell, E., Stallings, J., Smith, E.O., Worthman, C. & Whitten, P.L. (1995). Fecal steroid analysis of ovarian cycles in free-ranging baboons. *Am. J. Primatol.* **36**, 285–97.
Steklis, H.D., Brammer, G.L., Raleigh, M.J. & McGuire, M.T. (1985). Serum testosterone, male dominance and aggression in captive groups of male vervet monkeys (*Cercopithecus aethiops sabaeus*). *Horm. Behav.* **19**, 154–63.
Strassman, B.I. (1981). Sexual selection, paternal care, and concealed ovulation in humans. *Ethol. Sociobiol.* **2**, 31–40.
Strier, K.B. & Ziegler, T.E. (1994). Insights into ovarian function in wild Muriqui monkeys (*Brachyteles arachnoides*). *Am. J. Primatol.* **32**, 31–40.
Thierry, B. (1990). Feedback loop between dominance and kinship: the macaque model. *J. theoret. Biol.* **45**, 511–21.
van Schaik, C.P. (1989). The ecology of social relationships amongst female primates. In *Comparative Socioecology: The Behavioral Ecology of Humans and Other Animals*, ed. V. Standon & R.A. Foley, pp. 195–218. Oxford. Blackwell Scientific.
de Waal, F.B.M & Luttrell, L.M. (1987). Toward a comparative socioecology of the genus *Macaca*: Different dominance styles in rhesus and stumptail monkeys. *Am. J. Primatol.* **19**, 83–109.
Wallen, K. (1982). Influence of female hormonal state on rhesus sexual behavior varies with space for social interaction. *Science* **217**, 375–6.
Wallen, K. (1990). Desire and ability: hormones and the regulation of female sexual behavior. *Neurosci. Biobehav. Revs.* **14**, 233–41.
Wallen, K. & Goy, R.W. (1977). Effects of estradiol benzoate, estrone, and propionates of testosterone or dihydrotestosterone on sexual and related behaviors of ovariectomized rhesus monkeys. *Horm. Behav.* **9**, 228–48.
Wallen, K., Winston, L.A., Gaventa, S., Davis-DaSilva, M. & Collins, D.C. (1984). Periovulatory changes in female sexual behavior and patterns of ovarian steroid secretion in group-living female rhesus monkeys. *Physiol. Behav.* **36**, 369–75.
Wasser, S.K. (1983). Reproductive competition and cooperation among female yellow baboons. In *Social Behavior of Female Vertebrates*, ed. S.K. Wasser, pp. 349–90. New York: Academic Press.
Wasser, S.K., Monfort, S.L. & Wildt, D.E. (1991). Rapid extraction of faecal steroids for measuring reproductive cyclicity and early pregnancy in free-ranging yellow baboons (*Papio cynocephalus cynocepalus*). *J. Reprod. Fertil.* **73**, 133–8.
Wehling, M. (1994). Nongenomic actions of steroid hormones. *Tr. Endocrinol. Metabol.* **5**, 347– 53.
Weinbauer, G.F., Göckeler, E. & Nieschlag, E. (1988). Testosterone prevents complete suppression of spermatogenesis in the gonadotropin-releasing

hormone antagonist-treated nonhuman primate (*Macaca fascicularis*). *J. Clin. Endocrinol. Metabol.* **67**, 284–90.
Weinbauer, G.F. & Nieschlag, E. (1993). Hormonal control of spermatogenesis. In *Molecular Biology of the Male Reproductive System*, ed. D. de Krester, pp. 99–142. New York: Academic Press.
Whitten, P.L. (1987). Males and infants. In *Primate Societies*, ed. B.B. Smuts, D.L. Cheney, R.M. Seyfarth, R.W. Wrangham & T.T. Struhsaker, pp. 343–57. Chicago: University of Chicago Press.
Whitten, P.L. & Russell, E. (1996). Information content of sexual swellings and fecal steroids in sooty mangabeys (*Cercocebus torquatus atys*). *Am. J. Primatol.* **40**, 67–82.
Wickings, E.J., Bossi, T. & Dixson, A.F. (1993). Reproductive success in the mandrill, *Mandrillus sphinx*: correlations of male dominance and mating success with paternity, as determined by DNA fingerprinting. *J. Zool.* **231**, 563–74.
Wickings, E.J. & Dixson, A.F. (1992). Testicular function, secondary sexual development, and social status in male mandrills (*Mandrillus sphinx*). *Physiol. Behav.* **52**, 909–16.
Wilson, E.O. (1975). *Sociobiology*. Cambridge, MA: Harvard University Press.
Wilson, M.E., Gordon, T.P. & Collins, D.C. (1982). Variation in ovarian steroids associated with the annual mating period in female rhesus (*Macaca mulatta*). *Biol. Reprod.* **27**, 530–9.
Wingfield, J.C. (1994). Hormone-behavior interactions and mating systems in male and female birds. *The Differences Between the Sexes*, ed. R.V. Short & E. Balaban, pp. 303–33. New York: Cambridge.
Wingfield, J.C., Hegner, R.E., Dufty, A.M., Jr. & Ball, G.F. (1990). The "challenge hypothesis": theoretical implications for patterns of testosterone secretion, mating systems, and breeding strategies. *Am. Nat.* **136**, 829–46.
Worthman, C. (1990). Socioendocrinology: key to a fundamental synergy. In *Socioendocrinology of Primate Reproduction*, ed. F. Bercovitch & T. Ziegler, pp. 187–212. New York: Wiley-Liss.
Wrangham, R.W. (1980). An ecological model of female-bonded groups. *Behaviour* **75**, 262– 300.
Wrangham, R.W. (1993). The evolution of sexuality in chimpanzees and bonobos. *Hum. Nature* **4**, 47–79.
Zuckerman, S. (1932). *The Social Life of Monkeys and Apes*. New York: Harcourt, Brace, and Co.
Zumpe, D. & Michael, R.P. (1996). Social factors modulate the effects of hormones on sexual and aggressive behavior of macaques. *Am. J. Primatol.* **38**, 233–61.

11

Behavioral ecology and socioendocrinology of reproductive maturation in cercopithecine monkeys

FRED B. BERCOVITCH

Introduction

This contribution explores reproductive maturation from both evolutionary and physiological perspectives, in order to describe and evaluate the forces that affect the timing of puberty and first birth in cercopithecine monkeys. Evidence from both field and laboratory studies is used. Although approximately 40 cercopithecine species are recognized (see Smuts *et al.*, 1987, appendix 1), most of these data are derived from fewer than a dozen species. They are applied to three related themes: the effect of weight or fatness on reproductive maturation; the influence of dominance status on reproductive development; and the role of sex differences in reproductive strategies on the pace of reproductive maturation.

Perspectives on reproductive maturation

Behavioral ecology and reproductive maturation

While nutritional status and reproductive success are undoubtedly closely linked, the relationship between them is not simple. Malnutrition suppresses reproduction, but the compensatory effects of supplementary feeding have been exaggerated (Loy, 1988). As Darwin (1859: 33) astutely observed, "on the one hand, we see domesticated animals and plants, though often weak and sickly, breeding freely under confinement; and . . . on the other hand, we see individuals, though taken young from a state of nature perfectly tamed, long-lived and healthy . . . having their reproductive system . . . fail to act." Ecological, nutritional, social, and demographic factors mold the timing of reproductive development and onset of reproduction.

A key tenet of behavioral ecology (Krebs and Davies, 1987) is that foraging strategies link nutrition and fertility by translating food acquisition and processing into reproduction (Gaulin and Konner, 1977). Foraging strategies are molded by predator pressure, food distribution, abundance, and renewal rate, and by constraints imposed by population structure, such as group size and levels of intra- and inter-group competition (Richard, 1985; Garber, 1987; Dunbar, 1988). The fundamental problem to be solved by primate foraging strategies is to avoid predation while traveling with a troop, or a subgroup, whose members differ in metabolic demands and activity levels (Altmann, 1974; Lindburg, 1977). Analyses of the behavioral ecology of primates have tended to emphasize interspecific differences in troop size and structure, rather than intraspecific and inter-individual differences in foraging strategies that could influence the tempo of reproductive maturation.

In herbivores, phytochemistry influences foraging strategies and has the potential to directly or indirectly affect reproductive endocrinology. Phytoestrogens are associated with ovarian cycle abnormalities, acting either as antagonists or as functional mimics of estrogen (Labov, 1977). Phytoestrogens also adversely influence spermatogenic function in laboratory rats (Awoniyi *et al.*, 1996), but similar effects on primates have yet to be systematically evaluated. Food chemistry may affect reproduction indirectly, through its influence on health. Chimpanzees (*Pan troglodytes*), for instance, might partially regulate fertility by ingesting medicinal plants (Huffman and Wrangham, 1994) and leaves and fruits of the *Balanites* tree eaten by hamadryas baboons (*Papio hamadryas hamadryas*) might lower the incidence of schistosome parasites (Phillips-Conroy, 1986). However, nutrients probably have a greater bearing on reproductive maturation in cercopithecines than either secondary compounds or phytoestrogens.

Feeding strategies are closely intertwined with reproductive strategies, and will be affected by group living. No cercopithecine monkey is solitary in habits, and about 35% of species live in large multimale/multifemale groups (Smuts *et al.*, 1987, appendix 1). In the case of social primates, optimal foraging models are of limited use in explaining how food is converted into offspring, partly because of the extent of social constraints and partly as a consequence of the large dietary diversity characteristic of most primate species (Post, 1984; Richard, 1985; Garber, 1987). The modal food of most cercopithecine species is fruit, making up 40–80% of the diet in all except in some baboon (*Papio* and *Theropithecus*) populations (Richard, 1985). Dietary composition and social system interact in constructing life history strategies, which allocate resources among growth, maintenance, and reproduction (Gadgil and Bossert, 1970; Wilson, 1975; Stearns, 1992). Although malnutrition hampers maturation, the extent to which differential access to food within a population affects differences in rates of reproductive maturation is little known. Variation in resource availability among three troops of vervet monkeys (*Cercopithecus aethiops*) living in Amboseli National Park, Kenya influenced age at first birth (Cheney *et al.*, 1988) and differences in resource availability across time modified the age at onset of puberty among members of a single olive baboon (*P. cynocephalus anubis*) troop at Gilgil, Kenya (Bercovitch and Strum, 1993).

Metabolic rate and reproductive maturation

Foraging strategies and metabolic rate form a bidirectional pathway. Body size, fat level, growth rate, and activity patterns influence metabolism, which, in turn, has repercussions for reproductive endocrinology (Steiner, 1987; Bronson, 1989; I'Anson *et al.*, 1991). Metabolic cues late in juvenile life play a key role in triggering rhythmic GnRH release and the activation of puberty (Cameron, 1989, 1991; Plant, 1994). This physiological transformation begins with nocturnal increases in hypothalamic pulsatile discharges of luteinizing hormone-releasing hormone (LHRH). These trigger LH bursts which, in turn, stimulate the gonads to produce sex steroids (Bercovitch and Goy, 1990; Hotchkiss and Knobil, 1994; Plant, 1994; Terasawa, 1995). The hypothalamic–pituitary–gonadal axis is fine-tuned during adolescent development, which is longer in cercopithecines than in most mammals of similar size (Periera and Altmann, 1985; Caine, 1986; Bercovitch and Goy, 1990; see under section ' Sexual Bimaturism'). The elevated caloric demand of the prepubertal brain possibly limits LHRH output, hampering the onset of reproductive maturation (Winterer *et al.*, 1984; Steiner, 1987). Both cerebral and basal metabolic rates decline at puberty (Winterer *et al.*, 1984). According to this hypothesis, the diversion of resources from somatic to neural development essentially starves reproductive tissues of the stimulation necessary for their maturation.

Nutritional intake and metabolic level affect reproductive maturation in at least three essential ways: first, energy restriction dampens LH levels; second, glucose, amino acids, and insulin can augment LH secretion; and, third, dietary lipids have a direct connection to levels of various neurotransmitters (Steiner, 1987; I'Anson *et al.*, 1991). The effect of diet on reproductive maturation depends upon reproductive stage (Fig. 11.1). Despite the link between metabolism and maturation, detailed field studies

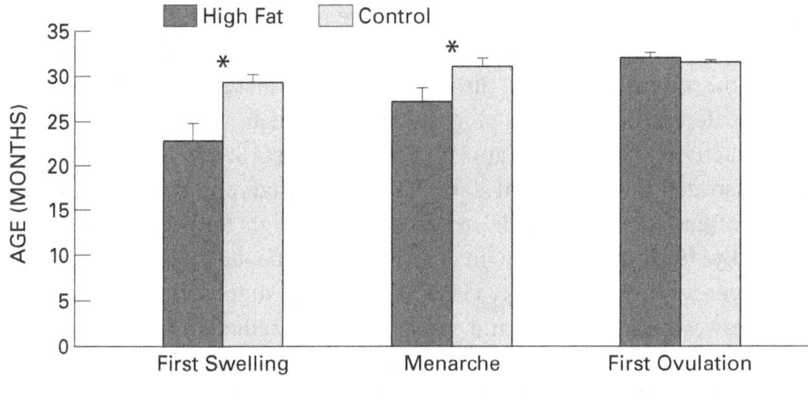

Fig. 11.1. The influence of diet on reproductive maturation in female rhesus monkeys. In the high fat diet, 31% of calories were derived from fat, whereas the control diet consisted of 12% of calories derived from fat. Ovulation was determined by comparison of progesterone levels across sequential samples. Five females received the high fat diet, whereas ten females received the control diet, but sample sizes decreased to four individuals in each group by the age of first ovulation. $*p<0.05$. Data from Schwartz *et al.*, 1988.

concentrating on associations between activity levels, dietary intake, and rates of reproductive development are lacking.

Retarded growth early in life can be offset if subsequent growth is accelerated by increased feeding activity (Bronson, 1989), so that different juvenile growth trajectories might result in the same age at onset of puberty. Males are less affected by undernutrition than females because reduced food intake has a more pronounced impact on steroidogenesis than on spermatogenesis (Bronson, 1989), and the two processes are independent (Bercovitch and Goy, 1990; Bercovitch and Nürnberg, 1996). Due to sex differences in reproductive roles, the nutritional modulation of reproductive endocrinology might exhibit greater differences between males and females of the same species than between same-sex samples of different species (I'Anson *et al.*, 1991).

Socioendocrinology and reproductive maturation

Social factors also impinge upon the two-way relationship between hormones and behavior (Ziegler and Bercovitch, 1990). Social subordination can affect the timing of reproductive maturation in primates, but the endocrine mechanisms are unclear and the data are inconsistent (see under section 'Dominance rank and reproductive maturation'). A logical candidate for an inhibitor of reproductive development is elevated cortisol levels, given the connections that link acute stress, increased glucocorticoid output, depressed secretion of gonadotropins such as LH, and impaired reproductive functioning (Sapolsky, 1994). Several studies, however, have demonstrated that low social status and suppressed reproductive hormone levels often do not coincide in primates (Raleigh and McGuire, 1990; Sapolsky, 1992, 1993; Saltzman *et al.*, 1994; Bercovitch and Clarke, 1995; Ziegler *et al.*, 1995). Sapolsky (1992, 1993, 1994) suggests that the connection between cortisol level and social status is mediated by at least three parameters: individual personality; stability of the hierarchy; and whether the subject is falling or rising in rank. Stress, social subordination, and high cortisol concentrations do not necessarily form a triangle that constrains primate reproductive maturation.

Elevated cortisol is only part of the endocrine stress response. Endogenous opioids released in response to stress inhibit GnRH release, but interfering with opioid action in prepubescent primates does not delay the onset of puberty (Plant, 1994). The capacity to cope with stress could result from pre- and neo-natal stress levels that modify the development of the neuroendocrine feedback system (Clarke *et al.*, 1994). It may be the inability to cope with stressful predicaments, rather than stress itself, that dampens reproductive function.

Social behavior triggers neuroendocrine responses that can have significant effects on male reproduction. For example, male-biased dispersal is a key event associated with reproductive maturation in almost all cercopithecine species (Pusey and Packer, 1987), and is probably associated with heightened stress and reduced immune function. Raleigh and McGuire, (1990) have even speculated that increased susceptibility to disease during dispersal, brought about by altered stress-related immune function, could equal predation in its effect upon survivorship.

In an experimentally simulated dispersal test, adolescent male vervet monkeys that had the greatest difficulty integrating into a group experienced increases in natural killer cell activity, whereas males that eased into a new social group showed the opposite response (Raleigh and McGuire, 1990). Natural killer cell activity is probably detrimental to immune function. A young adult male immigrant yellow baboon (*P. c. cynocephalus*) that experienced social difficulties and was very aggressive had remarkably high cortisol and testosterone concentrations, coupled with low lymphocyte counts (Alberts *et al.*, 1992). Male rhesus monkeys (*Macaca mulatta*) reintroduced to a social group after a year's absence showed the same pattern, with increased levels of both testosterone and cortisol, as well as significant decreases in T lymphocyte counts (Gust *et al.*, 1993). When male long-tailed macaques (*M. fascicularis*) are exposed to females, cortisol concentrations increase, with a more pronounced elevation accompanying exposure to strange females (Glick, 1984). In summary, reproductive hormones such as testosterone can increase concomitantly with stress hormones such as cortisol when males immigrate into new troops.

The pace of reproductive maturation

Interspecific differences in the timing of reproductive development will be channeled by the species' heritage, but intra-populational variation can arise from differences in foraging strategies, metabolic profiles, and social circumstances. How these variables interact to mold the pace of reproductive maturation is the subject of the remainder of this chapter.

Fatness, weight, and reproductive maturation

The "critical weight/fat threshold model" posits that crossing a minimum weight or fatness threshold is a necessary, but not sufficient, condition to trigger and sustain ovarian function (Frisch and McArthur, 1974; Frisch, 1984). Criticism of the model (Bronson and Manning, 1991) has been directed at the presumed directional effect, rather than the correlation itself. Threshold fat levels and the onset of ovulation might both be associated with metabolic mechanisms. Although both food restriction and intense activity suppress pulsatile GnRH release, no known physiological mechanism links fat levels with GnRH pulse pattern (Bronson and Manning, 1991). Testing the critical weight/fat model in the field presents a logistical nightmare. It requires that individuals be accurately weighed on the first day of evidence of puberty, and that controls be established for possible seasonal effects. Data from semi-free ranging and captive cercopithecines suggest that although a precise correlation between onsct of puberty and body weight often cannot be established, achieving a minimum body weight seems necessary for the onset of reproductive maturation in both males and females.

In females, the critical fat threshold was reasoned to be an adaptive mechanism to ensure a source of energy for maintaining pregnancy and lactation despite the vagaries of the food supply (Frisch, 1984). Frisch (1984) also suggested that female weight was important to reproductive success by virtue of its correlation with infant birth weight, which, in turn, is associated with infant survivorship. This model does not consider weight or fat levels in males, which might also contribute importantly to their reproductive success (Bercovitch, 1992; Andersson, 1994; Bercovitch and Nürnberg, 1996). Testis function in male rhesus macaques is dependent upon testis size, which itself is correlated with body weight (Bercovitch, 1993; Bercovitch and Nürnberg, 1996). The "fatted male" phenomenon enables males in seasonally breeding species to channel their effort into reproductive rather than feeding behavior (Bercovitch, 1992), but whether fat levels affect the timing of reproductive maturation in males is unknown.

Some early studies indicated that a lower age at menarche is correlated with greater body weight in rhesus macaques (van Wagenen and Catchpole, 1956), but more recent studies have failed to confirm a relation between weight or fat and onset of puberty (Schwartz *et al.*, 1988). In male rhesus macaques, spermatogenesis was initiated at 3.5 years of age, but only in individuals exceeding 5 kg in weight (Nadler *et al.*, 1993). In long-tailed macaques, however, body weight was poorly correlated with onset of spermatogenesis (Dang and Meusy-Desolle, 1984). In stump-tailed macaques (*M. arctoides*) age at testicular descent was not correlated with body weight, but descent did not begin until a male exceeded 5 kg in body weight (Nieuwenhuijsen *et al.*, 1987). Heavier female stump-tailed macaques ovulated at younger ages than lighter females (Nieuwenhuijsen *et al.*, 1987). Body weight corresponded with occurrence of first sexual swelling in female gray-cheeked mangabeys, *Lophocebus albigena*, while in males testicular descent appeared to coincide with peak weight increase (Deputte, 1992). Within an age cohort of female pigtail macaques (*M. nemestrina*), first sexual swelling tended to occur earlier in heavier individuals (Erwin and Erwin, 1976). Artificial feeding of Japanese macaques (*M. fuscata*) appears to increase body weight and accelerate age at first birth, but effects on onset of puberty are not reported (Sugiyama and Ohsawa, 1982).

Thus, metabolic signals initiate activity on the hypothalamic–pituitary–gonadal axis, triggering the onset of reproduction. These signals depend upon body composition, activity levels, and energy balance, with a minimum level of weight or fat probably a necessary, but not sufficient, condition. Crossing a weight/fat threshold does not ignite puberty in either sex, but in both sexes it tends to be associated with the timing of reproductive maturation.

Dominance rank and reproductive maturation

High social status provides a potential avenue for accelerating reproductive maturation, but few studies of cercopithecine monkeys have actually found such an effect. High rank is associated with a younger age at first birth among Barbary macaques (*M. sylvanus*; Paul and Kuester, 1988), but no such connection was found in bonnet macaques (*M. radiata*; Silk *et al.*, 1981), stump-tailed macaques (Nieuwenhuijsen *et al.*, 1985), Japanese macaques (Gouzoules *et al.*, 1982; Fedigan *et al.*, 1986), hamadryas baboons (Chalyan *et al.*, 1991), or vervet monkeys (Cheney *et al.*, 1988). Among provisioned Japanese macaques, high ranking females eat more of the artificial food than low ranking females do (Soumah and Yokota, 1991), and provisioning accelerates age at first birth among high ranking but not low ranking females (Sugiyama and Ohsawa, 1982). In provisioned populations of rhesus macaques, high population density (or lowered per capita access to food) reduced the probability of rapid reproductive maturation compared to low density conditions (Fig.11.2a). Furthermore, under high density conditions, high ranking females were significantly more likely to experience accelerated reproductive maturation (Fig.11.2b). Drickamer (1974) reported that low ranking female rhesus macaques produced their first offspring later than middle and high ranking females. Bercovitch and Berard (1993) found that over a 30-year period high ranking females on Cayo Santiago, off the coast of Puerto Rico, generally bore first offspring at the same age as middle and low ranking females, but were disproportionately represented among the 6.6% of females that experienced first parturition at three rather than four years of age. Dominance rank accounted for only 5% of the variance in age at first birth among female olive baboons at Gombe National Park, Tanzania (Packer *et al.*, 1995), it was not associated with age at first birth among olive baboons at Gilgil, Kenya (Bercovitch and Strum, 1993), but seemed to accelerate age at first conception by about six months among yellow baboons at Amboseli National Park, Kenya (Altmann *et al.*, 1988).

A younger age at first birth could extend the breeding lifespan and thus influence lifetime reproductive success. However, longitudinal data have shown age at first birth to be uncorrelated with lifetime reproductive success in Japanese macaques (Fedigan *et al.*, 1986), rhesus macaques (Bercovitch and Berard, 1993), olive baboons (Packer *et al.*, 1995; but see Altmann, 1991), or, probably, in Barbary macaques (Paul and Kuester, 1996). In captive baboons, age at first birth has a very high heritability (Williams-Blangero and Blangero, 1995), a charactersitic of traits with little effect on fitness (Falconer, 1989). Offspring survival has a greater influence upon the lifetime reproductive success of female primates than does rate of offspring production (Harcourt, 1987; Cheney *et al.*, 1988; Bercovitch and Goy, 1990) or age at first birth, and dietary effects on the chances of conception are weaker than dietary effects on maternal ability to

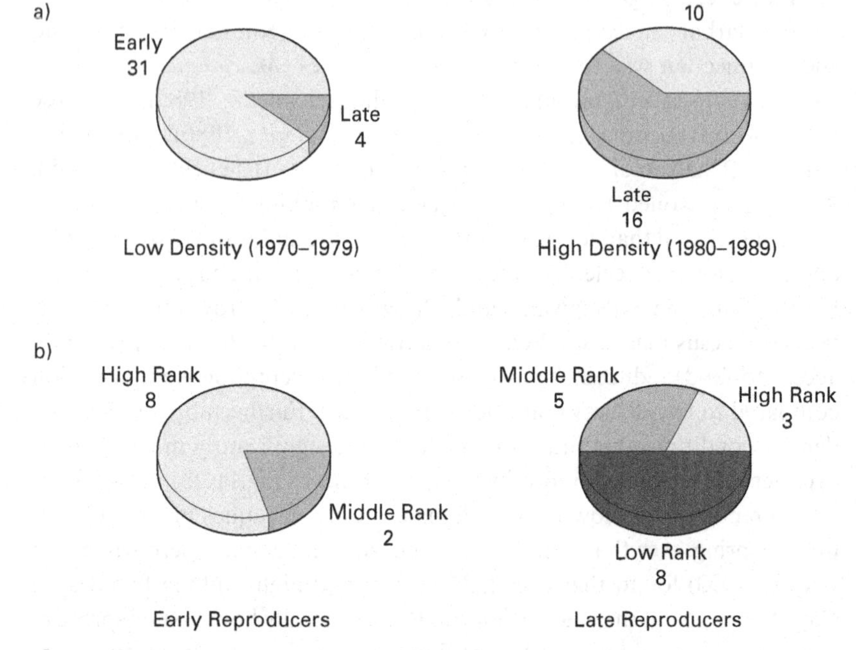

Fig. 11.2. The effect of population density and dominance rank on age at first parturition among the rhesus monkeys of Cayo Santiago, Puerto Rico. (a). The effect of population density on the timing of first parturition ($\chi^2 = 17.00$, df = 1, $p < 0.001$). Cayo Santiago had a population of monkeys that doubled in number between the low density decade and the high density decade. Average age at first parturition in the population is four years, with those bearing first offspring at three years of age referred to as "early", and those producing first offspring at five years of age or later referred to as "late". Sample sizes are indicated in the appropriate places. (b). The effect of dominance rank on age at first parturition under high density conditions (G = 13.380, df = 2, $p < 0.01$). Data from Bercovitch and Berard (1993).

nourish and rear offspring successfully (Lee, 1987). In summary, female dominance rank has little impact on the timing of reproductive maturation. Feeding strategies are more likely to influence rearing than production of offspring, and longitudinal data decouple age at first birth from lifetime reproductive success.

Variance in age at first sirehood among multimale cercopithecines remains to be established, but among captive rhesus macaques high ranking males are more likely than are their lower ranking age mates to sire offspring at younger ages (Bercovitch and Goy, 1990; Bercovitch and Nürnberg, 1996). Accelerated testicular development occurs among high ranking male rhesus macaques (Bercovitch, 1993), mandrills (*Mandrillus sphinx*; Wickings and Dixson, 1992), and yellow baboons (Alberts and Altmann, 1995). Such acceleration need not correspond to earlier first reproduction (Alberts and Altmann, 1995), but such a link has been established for rhesus macaques (Bercovitch and Nürnberg, 1996). Young sons of high ranking female Barbary macaques have a higher mating success than those of low ranking females (Paul and Kuester, 1990), and mating success reliably mirrors relative reproductive success in this population (Paul *et al.*, 1993). These findings suggest that in cercopithecine monkeys dominance rank influences reproductive maturation of males more than that of females.

Subordinate social status might sometimes negatively affect reproductive development and success in female cercopithecines, but is unlikely to inhibit reproduction completely. Hausfater (1975) reported that a middle ranking female yellow baboon continuously cycled without conceiving, but evidence of barrenness in post-pubescent females is rare. Over a six-year period of observation of rhesus macaques at the Sabana Seca Field Station, Puerto Rico, only one middle ranking female had not produced offspring at eight years of age. In the same troop, some low ranking males fail to sire offspring by seven years of age, while some of their high ranking age-mates have sired over a dozen infants between four and seven years of age (Bercovitch and Nürnberg, unpubl. data).

The endocrine stress response, feeding competition, and metabolic efficiency are three mechanisms that could be influenced by social status and exert an impact on the timing of reproductive maturation. Acute stress can disrupt reproductive functioning, but, as described above, the facile explanation that social stress retards reproductive maturation has been extensively challenged by empirical studies. Subordinate monkeys might be subject to more stress than dominant, but the intensity of stress is not great enough to generate an endocrine response sufficient to disrupt the timing of reproductive maturation.

Privileged access to specific food items is implicated as a major factor enabling high ranking females to accelerate reproductive rate (Fedigan, 1983; Whitten, 1983; Harcourt, 1987; Bercovitch, 1991). Feeding competition among female olive baboons intensifies in the wild during the dry season (Barton and Whiten, 1993), but, among many cercopithecines, low ranking females mitigate the effects of feeding competition by adopting strategies that minimize the chances of interference (Iwamato, 1987; van Noordwikj and van Schaik, 1987; Gilleau and Pallaud, 1988). Young adult female rhesus macaques (Johnson *et al.*, 1991) and yellow baboons

(Muruthi *et al.*, 1991) spend more time feeding than older adult females, but whether differences among younger females in feeding time or nutrient intake directly corresponds to variation in the timing of reproductive maturation is unestablished. Given the diverse diet of cercopithecines, one could reason that alternative, but equally productive, feeding strategies could be used by females occupying different rank positions in a troop, resulting in comparable nutrition and hence rates of reproductive maturation. Rank effects on reproductive maturation are more likely in a seasonally breeding species, where dominant females can control access to key food items (Bercovitch and Goy, 1990).

Metabolic efficiency buffers individuals against external challenges to their rhythmic endocrine profiles, and enables them to convert food more rapidly into growth or reproductive effort. Schwartz *et al.* (1985) found that high ranking female rhesus macaques were more likely to ovulate at a younger age than lower ranking conspecifics, despite the lack of rank-related differences in time spent feeding, body weight, fat levels, or growth rates. It was suggested that metabolic efficiency could account for the differences. In captive female vervets, high ranking individuals, though they are no heavier than low ranking females, produce more offspring (Fairbanks and McGuire, 1984). Similarly, body mass among immature female yellow baboons in Amboseli National Park is unrelated to dominance rank (Altmann and Alberts, 1987), although high ranking females have accelerated reproductive maturation (Altmann *et al.*, 1988). In all three studies, there were no rank-related differences in body weight, but rank-related differences in reproduction occurred. This suggests that, for a given body weight, the endocrine system of high ranking females is more efficient.

In summary, both social stress and feeding competition appear to have less influence on the tempo of reproductive maturation than metabolic efficiency. Whether metabolic efficiency does differ by dominance rank, and the causes of any such differences, remain to be investigated. Thus, dominance rank can regulate the timing of reproductive maturation, but such an effect is a conditional probability, not a predictable correlation (Bercovitch, 1991). Few studies of cercopithecines have shown suppressed development in subordinate individuals. The pace of female reproductive maturation seems to depend less upon stress than metabolic signals, and age at first parturition is not a valid indicator of lifetime reproductive success. Among males, dominance rank guides the trajectory of reproductive maturation more powerfully, perhaps due to sexual bimaturism.

Sexual bimaturism in cercopithecine monkeys

Sexual bimaturism – the existence of divergent, sex-specific rates and trajectories of growth and development within a species – evolves when age-dependent expenditures on reproductive effort differ by sex (Wiley, 1974). In females, energetic constraints suppress growth in favor of allocating resources to reproduction (Demment, 1983; Strum, 1991; Martin *et al.*, 1994) because female reproductive effort focuses on nurturing young. Male reproductive effort focuses on mating with multiple females. Postponement of his reproductive maturation enables a male to gain body weight that is advantageous in competition with others. A larger body size in males could be advantageous during dispersal, by reducing susceptibility to predation and buffering against starvation in unfamiliar terrain. Large body size may also enhance endurance rivalry, the ability to control access to females despite reductions in feeding (Andersson, 1994; Bercovitch, 1997). Food availability directly effects the timing of reproductive maturation among females within a population (Bercovitch and Strum, 1993), but food abundance has less effect on body weight in females than in males (Dunbar, 1990; Strum, 1991; but see Altmann *et al.*, 1993). Female reproductive energetics are more narrowly channeled.

Although most explanations of delayed maturation in males concentrate on the advantages of larger body size in male–male competition, a male's access to receptive females is mediated by complex social strategies and alternative reproductive tactics (Dunbar, 1982, 1983; Strum, 1982, 1994; Bercovitch, 1988, 1991; Berard *et al.*, 1994). For example, body weight influences the consort success of 9- to 10-year-old male olive baboons, but lighter males with strong affiliative bonds with females have consort success comparable to their heavier peers (Strum, 1994). In both gelada baboons (Dunbar, 1984) and olive baboons (Bercovitch, 1989) the outcome of aggressive competition is rarely predictable on the basis of male body weight.

Retarded reproductive maturation also extends the time available for honing social skills (Bronson, 1989). Sexual and social maturity are not synchronous in primates (Gautier-Hion and Gautier, 1976). Delayed maturation can be advantageous by allowing time to learn complex skills (Periera and Fairbanks, 1993), and extended growth in males prolongs this immature stage, enabling them to learn the intricacies of social manipulation. In many cercopithecine species, male social strategies rely heavily upon manipulations and alliances with non-kin, whereas female social strategies concentrate on obtaining assistance from kin (Western and Strum, 1983).

Discussions of sexual bimaturism tend to focus on delayed maturation in males and accelerated maturation in females, but postponing reproduction can be advantageous to females if it increases the chances of offspring survival. Primiparous mothers often have lower infant survival rates than multiparae (Bercovitch and Goy, 1990), but it is unknown whether the offspring of primiparae of different ages differ in survival prospects if one controls for dominance rank and sex of progeny. Among low ranking primiparous Barbary macaques, those who produced first offspring at a young age lost more of these infants to mortality than those who produced first offspring at older ages (Paul and Kuester, 1996).

Sexual bimaturism characterizes cercopithecines, but a systematic assessment of rates of reproductive maturation is difficult because ambiguities and discrepancies exist in the criteria for adolescence and adulthood (see Bercovitch and Goy, 1990; Strum, 1991). A proper analysis of sexual bimaturism should incorporate comparable benchmarks in the course of reproductive maturation in both sexes.

The data presented in Table 11.1 were analyzed to investigate the degree to which sexual bimaturism accompanies development in polygynous cercopithecines. Each species was considered an independent data point, since all belong to the same subfamily, obviating major problems arising from phylogenetic bias (Harvey *et al.*, 1987). Average values for each species were derived by combining field and laboratory work, emphasizing the former. Site locality was not considered. Male life history patterns, and paternal uncertainty, preclude comparing variation in actual age at first birth between males and females, but variance in achieving full body size could be documented.

Reproductive maturation commences significantly earlier in females than males (paired $t = -4.261$, $df = 10$, $p = 0.002$), and adult body size is also achieved at significantly younger ages (paired $t = -7.834$, $df = 11$, $p < 0.001$). Furthermore, the interval between these two landmarks is significantly longer in males than in females (paired $t = -2.939$, $df = 8$, $p = 0.019$). The longer growth phase of males presumably renders them more susceptible to perturbations and influences affecting reproductive maturation than females.

Male age at the onset of reproductive maturation is not significantly different from female age at first parturition (paired $t = 1.838$, $df = 10$, $p = 0.096$). In two studies, the youngest males to sire offspring were the same age as the average female giving birth for the first time (Bercovitch and

Nürnberg, 1996; Paul and Kuester, 1996). Interestingly, the weight of male rhesus macaques at the time of testicular descent – 5 kg – (see Nadler *et al.*, 1993) is close to the weight of females of this species when they conceive for the first time (Bercovitch *et al.*, 1998).

Although the chances of conception depend upon body weight, the sex of first offspring produced is independent of maternal weight (Bercovitch *et al.*, 1998) or body condition (Berman, 1988). In general, female cercopithecine monkeys reach full body size about one year after their first birth (paired $t = -3.251$, $df = 10$, $p = 0.009$), while males are unlikely to impregnate females until achieving full size. Female rhesus macaques attain a plateau in body weight at about six years of age (see Table 11.1), with weight at three years highly predictive of weight at six years (Fig. 11.3).

Thus, reproductive maturation in males not only begins later, but takes longer and is more variable in duration, than the comparable period among females. This pattern of sexual bimaturism allows males time to increase their body size and enables them to improve their social skills. However, in some circumstances, delayed reproductive maturation could enhance female reproductive success, while accelerated reproductive maturation could elevate male reproductive success.

Conclusions

Feeding strategies are closely intertwined with reproductive strategies, but the connection between nutrition and fertility among wild cercopithecines is unclear. Competition for food, metabolic efficiency, and the endocrine stress response are three mechanisms that can mediate differences in the pace of reproductive maturation. High stress levels do not seem to thwart reproductive development in low ranking animals, but elevated stress and suppressed immune function are associated with the dispersal of males from their natal group. Alternative feeding strategies, coupled with a diverse diet, probably reduce rank-related variation in rates of reproductive maturation. Most studies of cercopithecine monkeys have failed to find an effect of rank on patterns of reproductive maturation in females, but high ranking males are often characterized by accelerated reproductive maturation. Better nutrition fosters rapid growth, but females seem to have a less labile growth trajectory than males.

Crossing a minimum weight/fat threshold does not itself trigger reproductive maturation, but is correlated with initial reproductive potential. Age at first reproduction in females is not a valid index for calculating lifetime reproductive success. The period of reproductive maturation occupies

Table 11.1. *Sexual bimaturism in cercopithecines*

	Females			Males	
Species	Puberty*	First birth	Full size	Puberty	Full size
Cercocebus albigena	3.6	–	5.0	4.5	7.0
C. atys	3.0	4.5	5.0	5.0	6.0
Cercopithecus aethiops	2.8	5.0	–	5.0	–
Erythrocebus patas	–	3.0	3.0	4.0	5.0
Macaca arctoides	3.7	4.9	–	3.3	–
M. fascicularis	2.5	4.0	6.5	4.0	8.0
M. fuscata	3.5	5.5	7.0	4.5	10.0
M. mulatta	2.5	4.0	6.0	3.5	8.0
M. nemestrina	3.0	3.9	6.0	–	8.5
M. radiata	3.5	4.1	5.5	3.5	9.0
M. sylvanus	3.5	5.0	5.5	4.0	7.5
Miopithecus talapoin	4.0	4.5	5.0	–	6.0
Papio cynocephalus	4.8	6.9	7.0	5.7	9.0
P. hamadryas	4.3	6.1	5.6	5.8	10.0

Notes:
* Puberty in females is recorded as age at first sexual swelling, first menses, or first ovulation, with investigators adopting different criteria across species. Puberty in males is recorded as age at testes descent, production of fertile sperm, eruption of canines, or performing complete ejaculatory activity. In general, testes descent, sperm production, and canine emergence coincide in age (Nigi *et al.*, 1989; Bercovitch and Goy, 1990), so the use of different criteria probably influences cell values for males less than for females.
Source: Data are given in years and present an approximate mean value derived from the following references: Nadler and Rosenblum (1971); Rowell (1977); Mori (1979); Packer (1979); Takahata (1980); Altmann *et al.*, (1981); Matsubayashi and Mochizuki (1982); Sigg *et al.* (1982); Sugiyama and Ohsawa (1982); Bercovitch and Strum (1993); Mann *et al.* (1983); Sly *et al.* (1983); Chism *et al.* (1984); Horrocks (1986); Nieuwenjuisen *et al.* (1987); Paul and Kuester (1988, 1996); Cheney *et al.* (1988); Nigi *et al.* (1989); Turnquist and Kessler (1989); Turnquist (pers. comm.); Bercovitch and Goy (1990); Kahumbu and Eley (1991); Strum (1991); Deputte (1992); Itiogawa *et al.* (1992); Ménard and Vallet (1993); Alberts and Altmann (1995); Packer *et al.* (1995)

a greater proportion of a male's life history timetable than a female's. Average age at first birth among female cercopithecine monkeys is approximately the same as the youngest age at first sirehood. Sexual bimaturism can be attributed not only to the potential advantage of large size in males, but also to constraints imposed on female reproductive energetics and to increasing the time for males to improve social skills necessary for reproductive success.

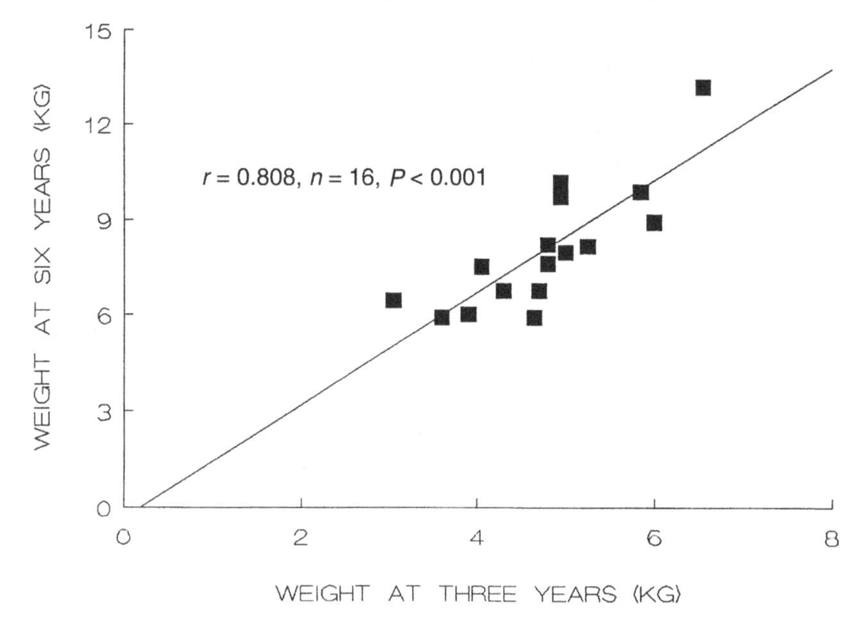

Fig. 11.3. The relationship between weight at three years of age and weight at six years of age among female rhesus macaques residing in a troop of about 150 monkeys that lives in a one-acre outdoor enclosure at the Sabana Seca Field Station, Puerto Rico. Weights were obtained from a battery operated scale about one month prior to onset of the mating season. Each box represents one female.

Acknowledgements

Suggestions for improvement in the text were offered by Cliff Jolly, Paul Whitehead, and Toni Ziegler. The CPRC is supported by NIH Grant RR03640.

References

Alberts, S.C. & Altmann, J. (1995). Preparation and activation: determinants of age at reproductive maturity in male baboons. *Behav. Ecol. Sociobiol.* **36**, 397–406.

Alberts, S.C., Sapolsky, R.M. & Altmann, J. (1992). Behavioral, endocrine, and immunological correlates of immigration by an aggressive male into a natural primate group. *Horm. Behav.* **26**, 167–78.

Altmann, J. & Alberts, S.C. (1987). Body mass and growth rates in a wild primate population. *Oecologia* **72**, 15–20.

Altmann, J., Altmann, S.A. & Hausfater, G. (1981). Physical maturation and age estimates of yellow baboons, *Papio cynocephalus*, in Amboseli National Park, Kenya. *Amer. J. Primatol.* **1**, 389–99.

Altmann, J., Hausfater, G. & Altmann, S.A. (1988). Determinants of reproductive

success in savannah baboons, *Papio cynocephalus*. In *Reproductive Success*, ed. T. H. Clutton-Brock, pp. 403–18. Chicago: University of Chicago Press.
Altmann, J., Schoeller, D., Altmann, S.A., Muruthi, P. & Sapolsky, R.M. (1993). Body size and fatness of free-living baboons reflect food availability and activity levels. *Am. J. Primatol.* **30**, 149–61.
Altmann, S.A. (1974). Baboons, space, time, and energy. *Am. Zool.* **14**, 221–48.
Altmann, S.A. (1991). Diets of yearling female primates (*Papio cynocephalus*) predict lifetime fitness. *Proc. Natl. Acad. Sci.* **88**, 420–3.
Andersson, M. (1994). *Sexual Selection*. Princeton: Princeton University Press.
Awoniyi, C.A., Roberts, D., Chandrashekar, V., Tucker, K.E., Hurst, B.S. & Schlaff, W.D. (1996). Effects of a phytoestrogen, coumestrol, on gonadotropin subunits gene expression and spermatogenesis. *Biol. Reprod.*, **54** (Suppl. 1), 136.
Barton, R.A. & Whiten, A. (1993). Feeding competition among female olive baboons, *Papio anubis. Anim. Behav.* **46**, 777–89.
Berard, J.D., Nürnberg, P., Epplen, J.T. & Schmidtke, J. (1994). Alternative reproductive tactics and reproductive success in male rhesus macaques. *Behaviour* **129**, 177–201.
Bercovitch, F.B. (1988). Coalitions, cooperation, and reproductive tactics among adult male savanna baboons. *Anim. Behav.* **34**, 1198–209.
Bercovitch, F.B. (1989). Body size, sperm competition, and determinants of reproductive success in male savanna baboons. *Evolution* **43**, 1507–21.
Bercovitch, F.B. (1991). Social stratification, social strategies, and reproductive success in primates. *Ethol. Sociobiol.* **12**, 315–33.
Bercovitch, F.B. (1992). Estradiol concentrations, fat deposits, and reproductive strategies in male rhesus macaques. *Horm. Behav.* **26**, 272–82.
Bercovitch, F.B. (1993). Dominance rank and reproductive maturation in male rhesus macaques (*Macaca mulatta*). *J. Reprod. Fert.* **99**, 113–20.
Bercovitch, F.B. (1997). Reproductive strategies in rhesus macaques. *Primates* **38**, 247–63.
Bercovitch, F.B. & Berard, J.D. (1993). Life history costs and consequences of rapid reproductive maturation in female rhesus macaques. *Behav. Ecol. Sociobiol.* **32**, 103–9.
Bercovitch, F.B. & Clarke, A.S. (1995). Dominance rank, cortisol concentrations, and reproductive maturation in male rhesus macaques. *Physiol. Behav.* **58**, 215–21.
Bercovitch, F.B. & Goy, R.W. (1990). The socioendocrinology of reproductive development and reproductive success in macaques. In *Socioendocrinology of Primate Reproduction*, ed. T.E. Ziegler & F.B. Bercovitch, pp. 59–93. New York: Wiley-Liss.
Bercovitch, F.B., Lebron, M.R., Martinez, H.S. & Kessler, M.J. (1998). Primigravidity, body weight and costs of rearing first offspring in rhesus macaques. *Amer. J. Primatol.* **46**, 135–44.
Bercovitch, F.B. & Nürnberg, P. (1996). Socioendocrine and morphological correlates of paternity in rhesus macaques. *J. Reprod. Fert.* **107**, 59–68.
Bercovitch, F.B. & Strum, S.C. (1993). Dominance rank, resource availability, and reproductive maturation in female savanna baboons. *Behav. Ecol. Sociobiol.* **33**, 313–18.
Berman, C.M. (1988). Maternal condition and offspring sex ratio in a group of free-ranging rhesus monkeys: an eleven year study. *Am. Nat.* **131**, 307–28.
Bronson, F.H. (1989). *Mammalian Reproductive Biology*. Chicago: University of Chicago Press.

Bronson, F.H. & Manning, J.M. (1991). The energetic regulation of ovulation: a realistic role for body fat. *Biol. Reprod.* **44**, 945–50.

Caine, N.G. (1986). Behavior during puberty and adolescence. In *Comparative Primate Biology*, vol. 2A, ed. G. Mitchell & J. Erwin, pp. 327–61. New York: Alan R. Liss.

Cameron, J.L. (1989). Nutritional and metabolic determinants of GnRH secretion in primate species. In *Control of the Onset of Puberty*, vol. 3, ed. H. A. Delemarre van de Waal, T. M. Plant, G. P. van Reer & J. Schoemaker, pp. 275–84. Amsterdam: Elsevier.

Cameron, J.L. (1991). Metabolic cues for the onset of puberty. *Horm. Res.* **36**, 97–103.

Chalyan, V.G., Meishvili, N. V. & Dathe, R. (1991). Dominance rank and reproduction in female hamadryas baboons. *Primate Report* **29**, 35–40.

Cheney, D.L., Seyfarth, R.M., Andelman, S.J. & Lee, P.C. (1988). Reproductive success in vervet monkeys. In *Reproductive Success*, ed. T.H. Clutton-Brock, pp. 384–402. Chicago: University of Chicago Press.

Chism, J., Rowell, T.E. & Olson, D. (1984). Life history patterns of female patas monkeys. In *Female Primates: Studies by Women Primatologists*, ed. M.F. Small, pp. 175–90. New York: Alan R. Liss.

Clarke, A.S., Wittwer, D.J., Abbott, D.H. & Schneider, M.L. (1994). Long-term effects of prenatal stress on HPA axis activity in juvenile rhesus monkeys. *Develop. Psychobiol.* **27**, 257–69.

Dang, D.C. & Meusy-Desolle, N. (1984). Quantitative study of testis histology and plasma androgens at onset of spermatogenesis in the prepuberal laboratory-born macaques (*Macaca fascicularis*). *Arch. Androl.* **12**, (*Suppl.* 1), 43–51.

Darwin, C.R. (1859). *On the Origin of Species*. London: Murray.

Demment, M.W. (1983). Feeding ecology and the evolution of body size in baboons. *Afr. J. Ecol.* **21**, 219–33.

Deputte, B.L. (1992). Life history of captive gray-cheeked mangabeys: physical and sexual development. *Int. J. Primatol.* **13**, 509–31.

Drickamer, L.C. (1974). A ten-year summary of reproductive data for free-ranging *Macaca mulatta. Folia primatol.* **21**, 61–80.

Dunbar, R.I.M. (1982). Intraspecific variations in mating strategy. In *Perspectives in Ethology*, vol. 5, ed. P. Klopfer & P. Bateson, pp. 385–431. New York: Plenum.

Dunbar, R.I.M. (1983). Life history tactics and alternative strategies of reproduction. In *Mate Choice*, ed. P.P.G. Bateson, pp. 423–33. Cambridge: Cambridge University Press.

Dunbar, R.I.M. (1984). *Reproductive Decisions*. Princeton: Princeton University Press.

Dunbar, R.I.M. (1988). *Primate Social Systems*. Ithaca: Comstock Publishing.

Dunbar, R.I.M. (1990). Environmental determinants of intraspecific variation in body weight in baboons (*Papio* spp.). *J. Zool.* **220**, 157–69.

Erwin, N. & Erwin, J. (1976). Age of menarche in pigtail monkeys (*Macaca nemestrina*): a cross-sectional survey of perineal tumescence. *Theriogenol.* **5**, 261–6.

Fairbanks, L.A. & McGuire, M.T. (1984). Determinants of fecundity and reproductive success in captive vervet monkeys. *Am. J. Primatol.* **7**, 27–38.

Falconer, D.S. (1989). *Introduction to Quantitative Genetics*, 3rd edn. London: Longman.

Fedigan, L.M. (1983). Dominance and reproductive success in primates. *Yrbk. phys. Anthrop.* **26**, 91–129.
Fedigan, L.M., Fedigan, L., Gouzoules, S., Gouzoules, H. & Koyama, N. (1986). Lifetime reproductive success in female Japanese macaques. *Folia primatol.* **47**, 143–57.
Frisch, R.E. (1984). Body fat, puberty, and fertility. *Biol. Rev.* **59**, 161–88.
Frisch, R.E. & McArthur, J. (1974). Menstrual cycles: fatness as a determinant of minimum weight for height necessary for their maintenance or onset. *Science* **185**, 949–51.
Gadgil, M. & Bossert, W.H. (1970). Life historical consequences of natural selection. *Am. Nat.* **104**, 1–24.
Garber, P.A. (1987). Foraging strategies among living primates. *Ann. Rev. Anthropol.* **16**, 339–64.
Gaulin, S.J.C. & Konner, M. (1977). On the natural diet of primates, including humans. In *Nutrition and the Brain*, vol. 1, ed. R.J. Wurtman & J.J. Wurtman, pp. 1–86. New York: Raven Press.
Gautier-Hion, A. & Gautier, J.P. (1976). Croissance, maturite sexuelle et sociale, reproduction chez les cercopithecines forestiers africains. *Folia primatol.* **26**, 165–84.
Gilleau, F. & Pallaud, B. (1988). Activite alimentarie d'un groupe de babouins (*Papio papio*) vivant en enclos. *Cahiers d'Ethol. Appl.* **8**, 1–62.
Glick, B.B. (1984). Male endocrine responses to females: effects of social cues in cynomologus macaques. *Am. J. Primatol.* **6**, 229–39.
Gouzoules, H., Gouzoules, S. & Fedigan, L. (1982). Behavioural dominance and reproductive success in female Japanese monkeys (*Macaca fuscata*). *Anim. Behav.* **30**, 1138–50.
Gust, D.A., Gordon, T.P. & Hambright, M.K. (1993). Response to removal from and return to a social group in adult male rhesus monkeys. *Physiol. Behav.* **53**, 599–602.
Harcourt, A.H. (1987). Dominance and fertility among female primates. *J. Zool.* **213**, 471–87.
Harvey, P.H., Martin, R.D. & Clutton-Brock, T.H. (1987). Life histories in a comparative perspective. In *Primate Societies*, ed. B.B. Smuts, D.L. Cheney, R.M. Seyfarth, R.W. Wrangham & T.T. Struhsaker, pp. 181–96. Chicago: University of Chicago Press.
Hausfater, G. (1975). *Dominance and Reproduction in Baboons (Papio cynocephalus)*. Basel: Karger.
Horrocks, J.A. (1986). Life-history characteristics of a wild population of vervets (*Cercopithecus aethips sabaeus*) in Barbados, West Indies. *Int. J. Primatol.* **7**, 31–47.
Hotchkiss, J. & Knobil, E. (1994). The menstrual cycle and its neuroendocrine control. In *The Physiology of Reproduction*, vol. 2, 2nd edn., ed. E. Knobil & J.D. Neill, pp. 711–49. New York: Raven Press.
Huffman, M.A. & Wrangham, R.W. (1994). Diversity of medicinal plant use by chimpanzees in the wild. In *Chimpanzee Cultures*, ed. R.W. Wrangham, W.C. McGrew, F.B.M. de Waal & P.G. Heltne, pp. 129–48. Cambridge: Harvard University Press.
I'Anson, H., Foster, D.L., Foxcraft, G.R. & Booth, P.J. (1991). Nutrition and reproduction. *Oxford Rev. Reprod. Biol.* **13**, 239–311.
Itiogawa, N., Tanaka, T., Ukai, N., Fujii, H., Kurokawa, T., Koyama, T., Ando, A., Watanabe, Y. & Imakawa, S. (1992). Demography and reproductive parameters of a free-ranging group of Japanese macaques (*Macaca fuscata*) at Katsuyama. *Primates* **33**, 49–68.

Iwamoto, T. (1987). Feeding strategies of primates in relation to social status. In *Animal Societies: Theories and Facts*, ed. Y. Ito, J.L. Brown & J. Kikkawa, pp. 243–52. Tokyo: Japan Science Society Press.
Johnson, R.L., Malik, I. & Berman, C.M. (1991). Age- and dominance-related variation in feeding time among free-ranging female rhesus monkeys. *Int. J. Primatol.* **12**, 337–56.
Kahumbu, P. & Eley, R.M. (1991). Teeth emergence in wild baboons in Kenya and formulation of a dental schedule for aging wild baboon populations. *Amer. J. Primatol.* **23**, 1–9.
Krebs, J.R. & Davies, N.B. (1987). *An Introduction to Behavioral Ecology*, 2nd edn. Oxford: Blackwell Scientific.
Labov, J.B. (1977). Phytoestrogens and mammalian reproduction. *Comp. Biochem. Physiol.* **57A**, 3–9.
Lee, P.C. (1987). Nutrition, fertility, and maternal investment in primates. *J. Zool.* **213**, 409–22.
Lindburg, D.G. (1977). Feeding behavior and diet of rhesus monkeys (*Macaca mulatta*) in a Siwalik Forest in North India. In *Primate Ecology*, ed. T.H. Clutton-Brock, pp. 223–49. New York: Academic Press.
Loy, J. (1988). Effects of supplementary feeding on maturation and fertility in primate groups. In *Ecology and Behavior of Food-Enhanced Primate Groups*, ed. J. Fa & C.H. Southwick, pp. 153–66. New York: Alan R. Liss.
Mann, D.R., Castracane, V.D., McLaughlin, F., Gould, K.G. & Collins, D.C. (1983). Developmental patterns of serum luteinizing hormone, gonadal and adrenal steroids in the sooty mangabey (*Cercocebus atys*). *Biol. Reprod.* **28**, 279–84.
Martin, R.D., Willner, L.A. & Dettling, A. (1994). The evolution of sexual size dimorphism in primates. In *The Differences Between the Sexes*, ed. R.V. Short & E. Balaban, pp. 159–200. Cambridge: Cambridge University Press.
Matsubayashi, K. & Mochizuki, K. (1982). Growth of male reproductive organs with observation of their seasonal morphologic changes in the Japanese monkey (*Macaca fuscata*). *Jpn. J. Vet. Sci.* **44**, 891–902.
Ménard, N. & Vallet, D. (1993). Population dynamics of *Macaca sylvanus* in Algeria: an 8-year study. *Amer. J. Primatol.* **30**, 101–18.
Mori, A. (1979). Analysis of population changes by measurements of body weight in the Koshima Troop of Japanese monkeys. *Primates* **20**, 371–98.
Muruthi, P., Altmann, J. & Altmann, S.A. (1991). Resource base, parity, and reproductive condition affect females' feeding time and nutrient intake within and between groups of a baboon population. *Oecologia* **87**, 467–72.
Nadler, R.D., Manocha, A.D. & McClure, H.M. (1993). Spermatogenesis and hormone levels in rhesus macaques innoculated with simian immunodeficiency virus. *J. med. Primatol.* **22**, 325–9.
Nadler, R.D. & Rosenblum, L.A. (1971). Factors influencing sexual behavior of male bonnet macaques (*Macaca radiata*). *Proc. III Intl. Congr. Primatol.* **3**, 100–7.
Nieuwenhuijsen, K., Lammers, A.J.J., de Neef, K.J.& Slob, A.K. (1985). Reproduction and social rank in female stumptail macaques (*Macaca arctoides*). *Int. J. Primatol.* **6**, 77–99.
Nieuwenhuijsen, K., Borke-Jansen, M., de Neef, K.J., van der Werf ten Bosch, J.J. & Slob, A.K. (1987). Physiological aspects of puberty in group-living stumptail monkeys (*Macaca arctoides*). *Physiol Behav.* **41**, 37–45.
Nigi, H., Hayama, S. & Torii, R. (1989). Rise in age of sexual maturation in male Japanese monkeys at Takasakiyama in relation to nutritional conditions. *Primates* **30**, 571–5.

van Noordwijk, M.A. & van Schaik, C.P. (1987). Male migration and rank acquisition in wild long-tailed macaques (*Macaca fascicularis*). *Anim. Behav.* **33**, 849–61.

Packer, C. (1979). Inter-troop transfer and inbreeding avoidance in *Papio anubis*. *Anim. Behav.* **27**, 1–36.

Packer, C., Collins, D.A., Sindimwo, A. & Goodall, J. (1995). Reproductive constraints on aggressive competition in female baboons. *Nature* **373**, 60–3.

Paul, A. & Kuester, J. (1988). Life-history patterns of Barbary macaques (*Macaca sylvanus*) at Affenberg Salem. In *Ecology and Behavior of Food-Enhanced Primate Groups*, ed. J. Fa & C.H. Southwick, pp. 199–228. New York: Alan R. Liss.

Paul, A. & Kuester, J. (1990). Adaptive significance of sex ratio adjustment in semifree-ranging Barbary macaques (*Macaca sylvanus*) at Salem. *Behav. Ecol. Sociobiol.* **27**, 287–93.

Paul, A. & Kuester, J. (1996). Differential reproduction in male and female Barbary macaques. In *Evolution and Ecology of Macaque Societies*, ed. J.E. Fa & D.G. Lindburg, pp. 293–317. Cambridge: Cambridge University Press.

Paul, A., Kuester, J., Timme, A. & Arnemann, J. (1993). The association between rank, mating effort, and reproductive success in male Barbary macaques (*Macaca sylvanus*). *Primates* **34**, 491–502.

Periera, M. & Altmann, J. (1985). Development of social behavior in free-living nonhuman primates. In *Nonhuman Primate Models for Human Growth and Development*, ed. E.S. Watts, pp. 217–309. New York: Alan R. Liss.

Periera, M. & Fairbanks, L.A. (1993). *Juvenile Primates*. Oxford: Oxford University Press.

Phillips-Conroy, J. (1986). Baboons, diet, and disease: food plant selection and schistosomiasis. In *Current Perspectives in Primate Social Dynamics*, ed. D.M. Taub & F.A. King, pp. 287–304. New York: Van Nostrand Reinhold.

Plant, T. (1994). Puberty in primates. In *The Physiology of Reproduction*, vol. 2, 2nd edn., ed. E. Knobil & J. D. Neill, pp. 453–85. New York: Raven Press.

Post, D.G. (1984). Is optimization the optimal approach to primate foraging? In *Adaptations for Foraging in Nonhuman Primates*, ed. P.S. Rodman & J.G.H. Cant, pp. 280–303. New York: Columbia University Press.

Pusey, A.E. & Packer, C. (1987). Dispersal and philopatry. In *Primate Societies*, ed. B.B. Smuts, D.L. Cheney, R.M. Seyfarth, R.W. Wrangham & T.T. Struhsaker, pp. 250–66. Chicago: University of Chicago Press.

Raleigh, M.J. & McGuire, M.T. (1990). Social influences on endocrine function in male vervet monkeys. In *Socioendocrinology of Primate Reproduction*, ed. T.E. Ziegler & F.B. Bercovitch, pp. 95–111. New York: Wiley-Liss.

Richard, A.F. (1985). *Primates in Nature*. New York: W.H. Freeman.

Saltzman, W., Schultz-Darken, N.J., Scheffler, G., Wegner, F.H. & Abbott, D.H. (1994). Social and reproductive influences on plasma cortisol in female marmoset monkeys. *Physiol. Behav.* **56**, 801–910.

Sapolsky, R.M. (1992). Cortisol concentrations and the social significance of rank instability among wild baboons. *Psychoneuroendocrinol.* **17**, 701–9.

Sapolsky, R.M. (1993). The physiology of dominance in stable versus unstable hierarchies. In *Primate Social Conflict*, ed. W.A. Mason & S.P. Mendoza, pp. 171–204. Albany: SUNY Press.

Sapolsky, R.M. (1994). *Why Zebras Don't Get Ulcers*. New York: W.H. Freeman.

Schwartz, S.M., Wilson, M.E., Walker, M.L. & Collins, D.C. (1985). Social and growth correlates of puberty onset in female rhesus monkeys. *Nutr. Behav.* **2**, 225–32.

Schwartz, S.M., Wilson, M.E., Walker, M.L. & Collins, D.C. (1988). Dietary influences on growth and sexual maturation in premenarchial rhesus monkeys. *Horm. Behav.* **22**, 231–51.
Sigg, H., Stolba, A., Abegglen, J.-J. & Dasser, V. (1982). Life history of hamadryas baboons: physical development, infant mortality, reproductive parameters and family relationships. *Primates* **23**, 473–87.
Silk, J.B., Clark-Wheatley, C.B., Rodman, P.S. & Samuels, A. (1981). Differential reproductive success and facultative adjustment of sex ratios among captive female bonnet macaques (*Macaca radiata*). *Anim. Behav.* **29**, 1106–20.
Sly, D.L., Harbaugh, S.W., London, W.T. & Rice, J.M. (1993). Reproductive performance of a laboratory breeding colony of patas monkeys (*Erythrocebus patas*). *Amer. J. Primatol.* **4**, 23–32.
Smuts, B.B., Cheney, D.L., Seyfarth, R.M., Wrangham, R.W. & Struhsaker, T.T. (ed.) 1987. *Primate Societies*. Chicago: University of Chicago Press.
Soumah, A.G.& Yokota, N. (1991). Female rank and feeding strategies in a free-ranging provisioned troop of Japanese macaques. *Folia primatol.* **57**, 191–200.
Stearns, S.C. (1992). *The Evolution of Life Histories*. Oxford: Oxford University Press.
Steiner, R.A. (1987). Nutritional and metabolic factors in the regulation of reproductive hormone secretion in the primate. *Proc. Nutr. Soc.* **46**, 159–75.
Strum, S.C. (1982). Agonistic dominance in male baboons: an alternative view. *Int. J. Primatol.* **3**, 175–202.
Strum, S.C. (1991). Weight and age in wild olive baboons. *Am. J. Primatol.* **25**, 219–37.
Strum, S.C. (1994). Reconciling aggression and social manipulation as a means of competition. I. Life-history perspective. *Int. J. Primatol.* **15**, 739–65.
Sugiyama, Y. & Ohsawa, H. (1982). Population dynamics of Japanese monkeys with special reference to the effect of artificial feeding. *Folia primatol.* **39**, 238–63.
Takahata, Y. (1980). The reproductive history of a free-ranging troop of Japanese monkeys. *Primates* **21**, 303–29.
Terasawa, E. (1995). Control of luteinizing hormone-releasing hormone pulse generation in nonhuman primates. *Cell. Mol. Neurobiol.* **15**, 141–64.
Turnquist, J.E. & Kessler, M.J. (1989). Free-ranging Cayo Santiago rhesus monkeys (*Macaca mulatta*). I. Body size, proportion, and allometry. *Am. J. Primatol.* **19**, 1–13.
van Wagenen, G. & Catchpole, H.R. (1956). Physical growth of the rhesus monkey (*Macaca mulatta*). *Am. J. phys. Anthrop.* **14**, 245–73.
Western, J.D. & Strum, S.C. (1983). Sex, kinship, and the evolution of social manipulation. *Ethol. Sociobiol.* **4**, 19–28.
Whitten, P.L. (1983). Diet and dominance among female vervet monkeys (*Cercopithecus aethiops*). *Am. J. Primatol.* **5**, 139–59.
Wickings, E.J. & Dixson, A.F. (1992). Testicular function, secondary sexual development, and social status in male mandrills (*Mandrillus sphinx*). *Physiol. Behav.* **52**, 909–16.
Wiley, R.H. (1974). Evolution of social organization and life history patterns among grouse. *Q. Rev. Biol.* **49**, 201–27.
Williams-Blangero, S. & Blangero, J. (1995). Heritability of age at first birth in captive olive baboons. *Am. J. Primatol.* **37**, 233–9.
Wilson, E.O. (1975). *Sociobiology.* Cambridge, MA: Belknap Press of Harvard University.

Winterer, J., Cutler, G.B. & Loriaux, D.L. (1984). Caloric balance, brain to body ratio, and the timing of menarche. *Med. Hypoth.* **15**, 87–91.
Ziegler, T.E. & Bercovitch, F.B. ed. (1990). *Socioendocrinology of Primate Reproduction*. New York: Wiley-Liss.
Ziegler, T.E., Scheffler, G. & Snowdon, C.T. (1995). The relationship of cortisol levels to social environment and reproductive functioning in female cotton-top tamarins, *Saguinus oedipus. Horm. Behav.* **29**, 407–24.

12

Quantitative assessment of occlusal wear and age estimation in Ethiopian and Tanzanian baboons

JANE E. PHILLIPS-CONROY, THORE BERGMAN AND CLIFFORD J. JOLLY

Introduction

Individual age is an important parameter in studies of primate sociobiology, ecology, and population genetics. As well as being a major determinant of behavior (Dunbar, 1988), it is critical for the construction of life tables, as a scale against which to plot measures of growth and maturation, and (if translated into date of birth) to provide a timescale for microevolution. Yet only a few exceptionally long and continuous primate field studies document age directly, from individual birth-to-death life histories of substantial numbers of animals. In other cases, individual age must be estimated from unreliable indicators such as external appearance or deportment. However, dental eruption, and subsequent wear of the occlusal surface, displays progressive change extending over most of an animal's lifetime. Where animals can be caught or restrained, dental evidence provides a basis for estimating the age of those whose life history is otherwise undocumented, as long as the timetable of eruption and wear can be reliably calibrated by reference to animals of known birth date. Elsewhere (Phillips-Conroy and Jolly, 1988), we have documented the sequence and timing of dental eruption in wild yellow baboons (*Papio hamadryas cynocephalus*) and hamadryas baboons (*P. h. hamadryas*), using animals of known natal age. Here, we develop a method of using the exposure of dentine on the occlusal surface of the molar teeth to estimate age, and apply it to adult individuals. By incorporating data from two populations, we are also able to explore interpopulation differences in dental wear rate.

In order to understand the effect of age on structures such as teeth, true longitudinal data, based upon repeated examination of individually identified animals would be ideal. Few studies of wild cercopithecid populations, however, meet this criterion. The next best strategy – cross-sectional sampling of animals whose birth date is known – is now possible for some populations of macaques (*Macaca*; Dittus, 1980; Fedigan and Asquith, 1991) and baboons (Altmann and Altmann, 1970; Phillips-Conroy and Jolly, 1986, 1988; Rhine *et al.*, 1988; Hamilton and Bulger, 1990; Phillips-Conroy *et al.*, 1991, 1992). A few of these populations have been captured and their teeth examined or cast, but only one – the baboons of the Awash hybrid zone (Phillips-Conroy *et al.*, 1991, 1992) – has been repeatedly sampled. The Awash National Park Baboon Research Project was designed initially to investigate the genetic structure of the hybrid zone between anubis (*P. h. anubis*) and hamadryas baboons (Jolly and Brett, 1973). From the beginning, however, other data were also collected, primarily to estimate developmental stability, health, and longevity in the two parental populations and their hybrids. During 1973, a field team led by F.L. Brett captured 531 baboons representing 11 different social groups. Dental eruption status was recorded, and a dental cast of the left maxillary dentition made (Phillips-Conroy, 1978). The same populations were captured and examined in 11 field seasons between 1982 and 1995. Metal or plastic eartags, and palmar dermatoglyphics recorded photographically or by a graphite-transfer technique (Phillips-Conroy *et al.*, 1986), allowed the identities of retrapped individuals to be established. In a separate study (Phillips-Conroy and Jolly, 1988), yellow baboons (*P. h. cynocephalus*) were captured at Mikumi National Park, Tanzania. They included a number of individuals of known birth date.

Tooth wear is detectable as microstructural alteration of the tooth surface ("microwear"), seen as grooves or pits, and as gross surface change ("macrowear"). While microwear records the recent history of the tooth surface, and is continually erased as the tooth is worn (Teaford and Tylenda, 1991), macrowear is cumulative and is suitable for aging the tooth. A number of studies have used mixed-age samples to examine progressive abrasion of the enamel of the occlusal surface (Bramblett, 1969; Welsch, 1967; Gantt, 1979; Nass, 1981; Dean *et al.*, 1992). However, with the exception of some work on howler monkeys (*Alouatta*; Froehlich *et al.*, 1981; Glander *et al,*. 1996; Noble *et al.*, 1996), we are unaware of any study that has examined dental macrowear in wild, non-human primates sampled longitudinally.

The occlusal surface of the cercopithecoid bilophodont molar wears in a predictable sequence. On an upper molar, wear appears earliest on the protocone, then on the hypocone (Fig. 12.1). Dentine is exposed later on the buccal cusps, with the paracone showing wear before the metacone. Cusp blunting progresses to pinpoint dentine exposure, then to triangular or cuneiform dentine facets that meet across the loph or contact their distal partner. Eventually, a single confluent dentine "lake" is formed, waisted at the central basin. In late stages of wear, the occlusal surface appears as a flat dentine plane. From this point, the tooth margins break, the tooth surface is reduced, and finally the tooth is worn to the gum or avulsed.

Bramblett (1969) used this wear sequence to assign ages to yellow baboons captured at Darajani, Kenya, employing Miles' (1963) method of seriating specimens by occlusal wear. We (Phillips-Conroy and Jolly, 1986) applied a similar method to the 1973 sample of Awash baboons, but, in the absence of longitudinal data, were unable to calibrate the resulting stages against the animal's age. This chapter reports occlusal attrition in Awash baboons retrapped during successive field seasons, and in a sample of Mikumi baboons of known age. We apply these data to the following questions:

(1) Can dental wear be used to age individuals, and if so, across what range of ages?
(2) Can dental wear be used to age individuals across populations?
(3) Do males and females wear their teeth at comparable rates?
(4) Do all three molars wear at similar or different rates?
(5) Does wear occur at a constant rate over the life of a tooth?

Materials and methods

The sample consisted of casts of the left maxillary dentition from 126 males and 69 females (Table 12.1). Twelve males and 15 females from Mikumi had known dates of birth (Rhine *et al.*, 1988). The remaining animals (4 Tanzanian females and 106 Ethiopian males and females) had one or more teeth erupting, with part of the crown visible above the gum, when first captured. Their birth date could be estimated from eruption schedules established on known-age Tanzanian yellow baboons (Phillips-Conroy and Jolly, 1988). We assume that the timing of dental eruption is similar in Awash and Mikumi baboon populations, in spite of ecological and taxonomic differences between them, since our earlier study failed to find differences between the Mikumi yellow baboons living in woodland savannah and the hamadryas baboons of Erer-Gota, a semi-desert habitat even more arid than Awash (Sigg *et al.*, 1982; Phillips-Conroy and Jolly, 1988).

Because some of the Awash animals were examined more than once, the number of observations exceeds the number of individuals. Seventy-seven male Awash baboons contribute 114 observations, of which 89 were of 67

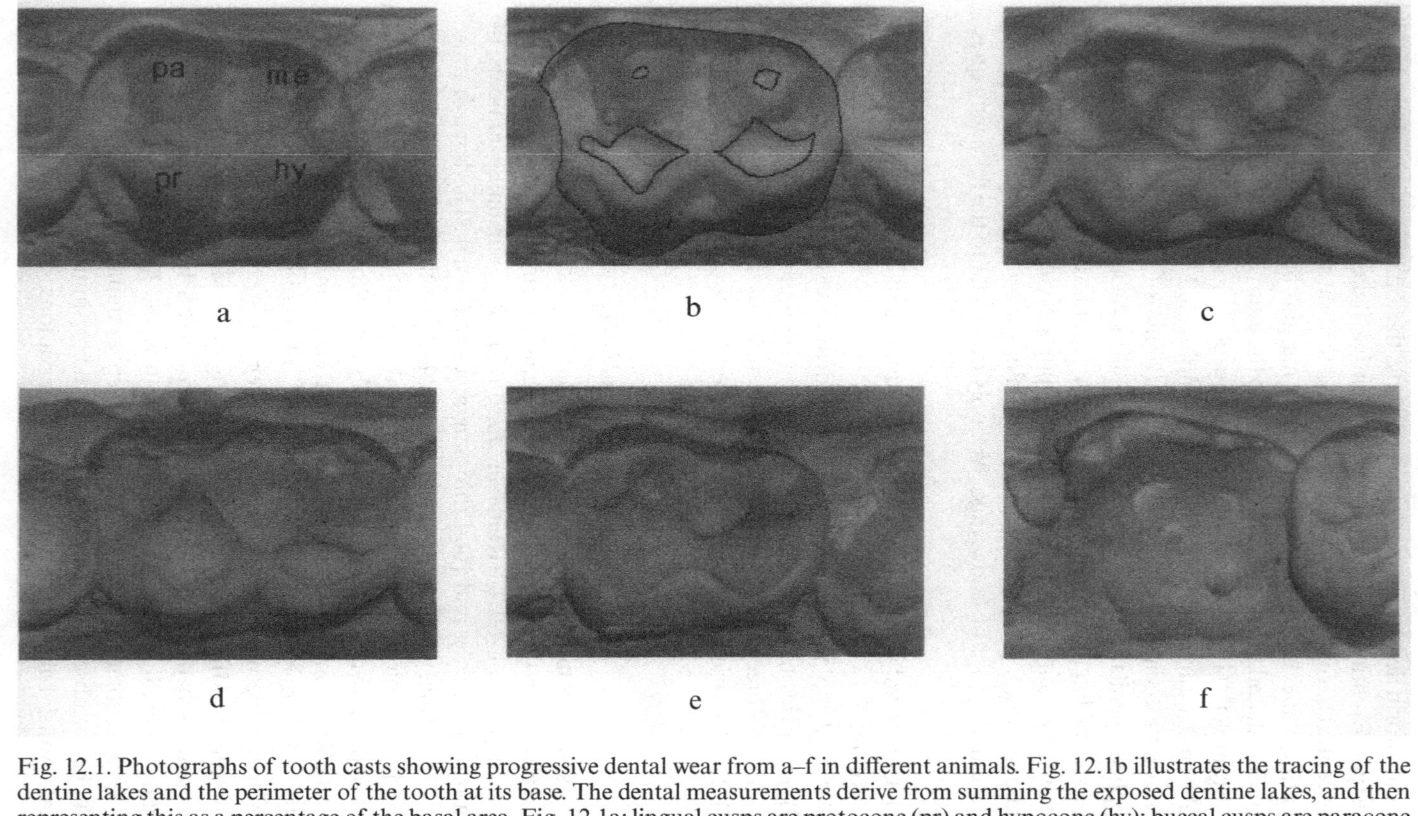

Fig. 12.1. Photographs of tooth casts showing progressive dental wear from a–f in different animals. Fig. 12.1b illustrates the tracing of the dentine lakes and the perimeter of the tooth at its base. The dental measurements derive from summing the exposed dentine lakes, and then representing this as a percentage of the basal area. Fig. 12.1a: lingual cusps are protocone (pr) and hypocone (hy); buccal cusps are paracone (pa) and metacone (me).

Table 12.1. *The sample used in this study*

	Males		Females	
	observations	individuals	observations	individuals
Tanzania	12	12	19	19
Ethiopia				
Known age				
captured once	51	51	31	31
captured >once	38	16	17	8
Total known age	89	67	48	39
Unknown age				
all captured >once	25	10	12	5
Ethiopia Total	114	77	60	44

individuals of known age. Fifty-one of the latter were captured once, and 16 were captured multiple times. Of the 48 observations of known age females, 31 were of individuals captured once, and 17 were of 8 animals captured more than once (Table 12.1). In addition to the animals of known age that were examined more than once, the "recapture" data set also includes 25 observations of 10 males and 12 observations of 5 females that were dentally mature when first caught.

The nature of the data collected – dental casts taken from the living animal – precluded methods of dental wear evaluation dependent upon locating the cemento-enamel junction (Spinage, 1973; Morris, 1978; Hillson, 1986; Hartman, 1995), and the large number of specimens ruled out the accurate but complex, three-dimensional methods that have been applied to the study of tooth wear (Teaford, 1983; van der Bijl *et al.*, 1989). We therefore developed a relatively simple method of measuring wear, by quantifying the exposure of dentine on the occlusal surface.

Each dental cast was mounted in putty, with the occlusal surface of the molar to be photographed perpendicular to the axis of the lens. A scale, situated at the same orientation and distance from the camera, was included in each photograph. Up to three molars could be scored on each animal. When more than one tooth was photographed, the cast was reoriented between exposures. Digital images of the occlusal aspect of the left maxillary molar teeth were recorded and stored using a vertically mounted, digital camera and the NuVista Capture+ program (Truevision, Indianapolis, Indiana, USA) on a Macintosh IIfx computer. For measurement, images were imported into the NIH public domain application,

IMAGE v.1.57. Regions of exposed dentine were outlined, and the area within the outline calculated automatically by IMAGE. Up to four separate dentine exposures (one per cusp) were measured on each molar, and their areas summed. The perimeter of each tooth in occlusal aspect was traced, and the area within it (the basal area) measured in a similar manner. The total area of dentine exposure was divided by the basal area to yield the proportion of tooth surface that consisted of exposed dentine (see Fig. 12.1b). The use of this proportional measure of dentine exposure not only controlled for variation in molar size, it also corrected for the fact that in severely worn teeth the absolute area of exposed dentine is liable to decrease as parts of the crown are lost.

To test whether small, random variations in the orientation of the tooth cast had a significant effect, we re-photographed a number of teeth from various casts in random order. To minimize the potential error in tracing of the dentine lakes, only high quality casts with clearly visible enamel-dentine boundaries were used, and the tooth cast was kept at hand to ensure faithful tracing. Areal measures were highly consistent. Dimensions obtained by two observers tracing the same image repeatedly never varied by more than 5%, and were almost always within 1% of each other.

The functional age of each tooth – its time in occlusion – was estimated by subtracting its eruption age from the animal's age. Based on previous findings (Phillips-Conroy and Jolly, 1988) we used the following eruption ages: for males, 25, 57, and 98 months, and for females, 25, 50, and 92 months for M1, M2, and M3, respectively. In all regressions, tooth wear was set as the independent variable. Data were most numerous for the first molars, fewer for second molars, and for third molars there were too few to generate regression equations, except for the Ethiopian male baboons. We made three sets of comparisons: between populations; between sexes within populations; and among the three molars within each population.

One assumption of this method is that the rate of wear on an occluding molar is unrelated to the animal's age or sex. However, evidence of differential access to food resources by age, sex and rank (Post *et al.*, 1980; Kilgore, 1989) suggests that this assumption may not necessarily be correct. A second assumption is that a molar's degree of wear is directly proportional to its functional age. However, cases where the second molar is more worn than the first, one of which occurred in our sample, suggest that this assumption needs to be evaluated. Finally, the method assumes a constant rate of wear throughout a molar's functional life. However, as later teeth erupt, worn molars may fall below the occlusal plane, reducing their wear rate (Jolly, 1972), and wear on posterior molars may increase as the focus of mastication shifts distally.

Preliminary inspection of the regression plots indicated a nonlinear relationship between age and wear, and a skewed distribution of wear against age. Both of these violations of parametric statistical assumptions were improved by transforming the data to natural logarithms. We then inspected and compared slopes and intercepts for tooth wear against functional age between populations and between sexes. We examined the relationship between the residuals of these regressions and the dependent variables to see whether we might need to correct for bias (Condon *et al.*, 1986).

To evaluate the validity of the regression equations we used a jackknife technique (Sokal and Rohlf, 1995), where each animal is sequentially withdrawn from the sample, the regression recalculated, and the age of the dropped specimen then estimated from the new regression equation. The standard error of the estimate or the standard deviation of the residuals (i.e. the deviations of the estimated age via the jackknife procedure from the known age) was then compared with the known age. The jackknife technique permits error values to be estimated for the parameters of a regression equation when its underlying distributional assumptions are not met by the current data set (Sokal and Rohlf, 1995).

The data from animals of known age were used to derive separate regression equations for male and female, Ethiopian and Tanzanian, baboons, respectively. These equations were used to assign ages to male and female Ethiopian baboons of unknown initial age that were captured more than once, in order to examine differences in prediction when using formulae for the different molars, and to examine the deviation between known and estimated intervals between captures. Then we used the recaptured animals in order to explore the issue of constancy of wear rates within a tooth over time.

In detransforming our data from the logarithmic to the arithmetic scale for interpretation we adopted Smith's (1993) correction for bias by multiplying predicted values by the mean of the detransformed values of the residuals (the "smearing estimate"). The estimated age of the animal, derived from a given molar, was the functional age of that tooth plus its eruption age. A maximum of three estimated ages (one from each molar) could be derived for each animal. The age of the animal was estimated as the mean of these estimates.

Results

The difference between the standard error of the regression (the residual mean square value) and the jackknife standard error was small, except for second molars in Tanzanian males and third molars in Ethiopian males, suggesting that these are poor predictors of age. The plot of residuals against dependent variables for Ethiopian male and female first and second molars indicated no need for further correction.

Table 12.2 presents the regression equations for each of the three molar teeth for males and females in each population, and the relationships are then graphed in Figs. 12.2 and 12.3.

In both males and females, the regression coefficients for M1 wear on age are higher, but the intercept is lower, in the Tanzanian population than the Ethiopian. Analysis of variance reveals a significant locality by wear interaction (*F* ratio 4.02; $p = 0.048$ in males, and *F* ratio of 5.717 and $p = 0.021$ for females). The single Tanzanian male outlier in Fig. 12.2 has a strong influence on the regression; removing this point yields a recalculated slope of 0.75 and an intercept of 2.614, sharpening the differences between the populations. The influence of the three Ethiopian male outliers has a less significant effect on the regression, changing the equation to 3.578 + 0.358*ln(x) of M1 wear.

We further explored this interpopulation difference in a subsample of animals whose lnm1age was <3.5, equivalent to the age at which the second molars erupt. For Ethiopian males, the mean lnm1wear is −0.156 log units versus the Tanzanian male mean of 0.805 ln units. An unpaired t-test shows this difference to be significant ($p = 0.0018$; $t = -3.518$, $df = 23$). Among females, this comparison shows the Ethiopian lnm1wear mean to be −0.724 versus the Tanzanian mean of 0.343 ($p = 0.025$; $t = -2.732$; $df = 8$).

Thus, at a given age, Tanzanian baboons have more worn molars than do Ethiopian baboons. The regression's intercept, however, is significantly lower in Tanzanian baboons of both sexes, with non-overlapping 95% confidence intervals in both the male and female comparisons. This indicates that though the rate of wear is slower, the onset of wear occurs at a younger age in the Tanzanian population.

Comparisons involving the distal molars were limited by small sample sizes. However, a significant relationship between wear and age was demonstrated in the M2 of Ethiopian males and females and Tanzanian females (Table 12.2). There were no significant differences between the populations, in either male or female comparisons, for either M2 or M3.

Inspection of the plots for M1 and M2 in both the Ethiopian and the

Table 12.2. *Regression equations for predicting age from occlusal wear from known age Ethiopian and Tanzanian baboons (95% confidence intervals are given in parentheses)*

	Ethiopian males	Ethiopian females	Tanzanian males*	Tanzanian females
Molar[1]				
Slope	0.346 (0.296–0.397)	0.32 (0.243–0.396)	0.528 (0.261–0.795)	0.459 (0.373–0.547)
Intercept	3.59 (3.513–3.661)	3.77 (3.654–3.902)	2.93 (2.426–3.428)	3.129 (2.927–3.332)
p	<0.0001	<0.0001	<0.0001	<0.0001
n	84	35	12	17
r^2	0.724	0.687	0.66	0.89
Molar[2]				
Slope	0.34 (0.237–0.443)	0.294 (0.184–0.4)	0.257 (0.278–0.798)	0.36 (0.247–0.474)
Intercept	3.439 (3.29–3.586)	3.97 (3.345–4.078)	3.553 (2.774–4.332)	3.621 (3.345–3.898)
p	<0.0001	<0.0001	0.176	<0.0001
n	41	15	4	7
r^2	0.508	0.723	0.68	0.931
Molar[3]				
Slope	0.338			
Intercept	3.261			
p	0.067			
n	7			
r^2	0.52			

Notes:
*Without outlier for Molar 1, Slope = 0.75 (0.363–1.136), Int = 2.614 (1.987–3.241), $r^2 = 0.681$ and $p = 0.0017$

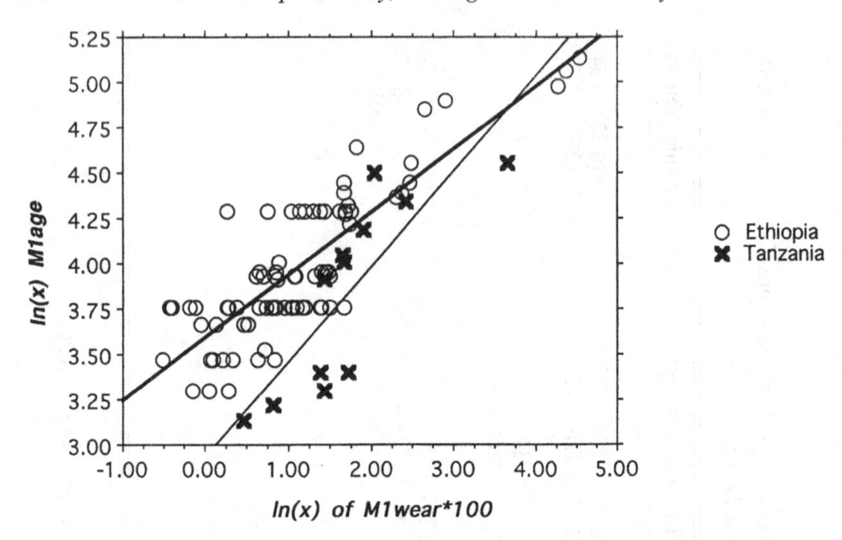

Fig. 12.2. Scatterplot and regression line for regression of lnM1wear on lnM1age in Ethiopian and Tanzanian male baboons. The dark line represents the regression line drawn for the Ethiopian baboons; the light line represents the regression line drawn for the Tanzanian baboons.

Equation for Ethiopian male baboons:

$$\ln(x) \text{ of M1age} = 3.588 + 0.345*\ln(x) \text{ of M1 wear}*100;\ r2 = 0.703$$

Equation for Tanzanian male baboons:

$$\ln(x) \text{ of M1age} = 2.927 + 0.528*\ln(x) \text{ of M1 wear}*100;\ r2 = 0.66$$

Tanzanian populations suggests differences between the sexes that lie in the onset rather than the rate of wear. In each case, the location of the Y intercept indicates that males begin to show molar wear before females. Analysis of variance confirmed this. At neither locality did the slopes differ between the sexes, but for both M1 and M2, Ethiopian male intercepts were significantly lower than those of Ethiopian females (F ratio 8.094 and $p = 0.005$ for M1; F ratio $= 12.39$ and $p = 0.001$ for M2).

The equations presented in Table 12.2 show that within each of the four groups, the slopes of the lines relating wear to age decrease from the more anterior to posterior molars, suggesting that there might be differences in the pattern of wear of the molars. Pending testing of this hypothesis, we suggest that analyses of dental wear should treat the molars separately unless their rates of wear are demonstrably uniform.

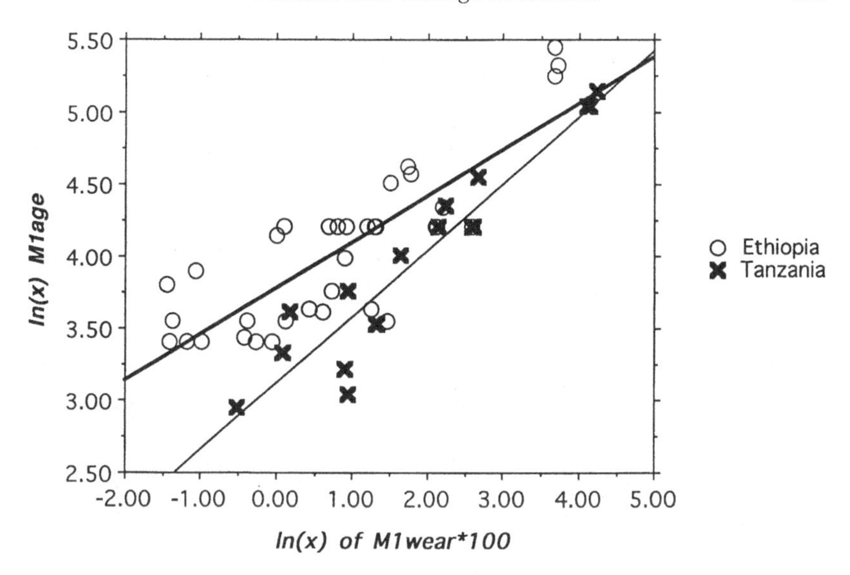

Fig. 12.3. Scatterplot and regression line for regression of lnM1wear and lnM1age in Ethiopian and Tanzanian female baboons. The dark line represents the regression line drawn for the Ethiopian baboons; the light line represents the regression line drawn for the Tanzanian baboons.

Equation for Ethiopian female baboons:

$$\ln(x) \text{ of M1age} = 3.766 + 0.313 * \ln(x) \text{ of M1 wear} * 100; \; r2 = 0.681$$

Equation for Tanzanian female baboons:

$$\ln(x) \text{ of M1age} = 3.129 + 0.46 * \ln(x) \text{ of M1 wear} * 100; \; r2 = 0.894$$

Age estimation in animals of unknown age

Only first and second molars could be used to predict age in females, although all three molars could be used in males. One of the five females was captured three times, the rest twice (11 observations). Of the 10 males, seven were captured twice, one was captured three times and two were captured four times (25 observations). Tables 12.3 and 12.4 show the estimated ages for recaptured animals, the known interval between captures, and the interval estimated from the regression of age on wear.

All the recaptured females were fully adult when originally caught, and their estimated ages ranged from 110 to 190 months (Table 12.4). The average absolute difference between estimates derived from first and second molars was 5.5 months, with a minimum difference of 0.12 months for the youngest animal and a maximum difference of 9.02 months for the oldest animal. Age tends to be overpredicted at the low end of the tooth wear scale

Table 12.3. *Age prediction using equations from Table 12.2 for Ethiopian baboon males*

ID	DS	M1	M2	M3	M1est	M2est	M3est	Xest	Int.a.	Int.b.
83006	10594		6.8	2	–	123.3	137.5	130.4	53	49
88008	10594	13.8	10.8	5.3	119.5	134.7	152.7	135.6	–	–
84021	10599	20.4	5.5	–	133.1	118.8	–	126.0	12	10
83012	10599	13.9	4.2	–	119.7	113.2	–	116.5	–	–
83015	10602	3	0.7	–	80.7	87.4	–	84.1	12	10
84032	10602	5	1.5	–	91.6	96.7	–	94.1	–	–
84026	10604	–	19.2	–	–	151.6	–	151.6	12	26
83017	10604	7.7	17.7	–	102.3	149	–	125.6	–	–
83018	10605	67.8	13.8	–	188.8	141.4	–	165.1	29	17
86028	10605	66	–	16.2	187.2	–	177.9	182.5	–	–
83019	10606	36.3	25.6	1.2	157	161.4	131.1	149.8	35	46
86080	10606	58.5	61.3	42.7	180.7	197.7	208.7	195.7	18	0
88001	10606	71.2	65.6	–	191.6	201	–	196.3	34	0
91051	10606	77.1	64.3	–	196.1	200	–	198.1	–	–
88026	10607	53	19.8	7.6	175.4	152.5	159.8	162.6	41	58
84004	10607	9.6	2.1	–	108.4	101.2	–	104.8	–	–
83021	10608	39.2	–	4.1	160.5	–	148.3	154.4	12	11
84031	10608	56.9	27.7	–	179.1	164.3	–	171.7	18	21
86031	10608	73.4	–	–	193.3	–	–	193.3	36	31
89013	10608	90.5	65.4	–	205.9	200.8	–	203.3	–	–
83029	10616	–	4.2	–	–	113.2	–	113.2	29	28
86034	10616	26.9	10.2	3.3	144	133.1	144.4	140.5	53	36
90039	10616	65.3	–	11	186.6	–	167.9	177.3	–	–
84030	10620	2.1	–	–	74.4	–	–	74.4	70	52
90027	10620	12.6	5.8	2.8	116.7	119.7	141.8	126.1	–	–

Notes:
ID: ID number – first two digits represent year of capture; next three digits represent order of capture; DS: data set number – this number remains the same across years, whereas ID number changes; M1: M1 wear; M2: M2 wear; M1est: estimated age from regression equation for M1 (see Table 12.2); M2 est: estimated age from regression equation for M2 (see Table 12.2); M3 est: estimated age from regression equation for M3 (see Table 12.2); xest: average of M1est and M2 est; Int.a.: known interval between recaptures in months; Int.b.: estimated interval between recaptures in months from tooth wear and age prediction.

and underpredicted at the high end. For example, the known elapsed time between 86003 and her recapture as 88030 is 24 months. The time interval predicted by comparing the estimated ages is 43 months. When this animal was captured over five years later as 93020, she was estimated to be only 31 months older. Similarly, when 90007, also an old female, was captured five years later as 95014, sixteen months, rather than the known 60 months, was estimated to have elapsed.

Table 12.4. *Age prediction using equations from Table 12.2 for Ethiopian baboon females*

ID	DS	M1	M2	M1est	M2est	X est	Int.a.	Int.b.
86003	10659	7.52	1.62	109.9	110	109.9	24	43
88030	10659	30.63	9.21	156.8	149.3	153.1	67	31
93020	10659	60.56	23.39	188.1	180.1	184.1	–	–
86020	10674	10.8	1.91	120.1	112.9	116.5	6	2
86079	10674	9.62	2.85	116.7	120.7	118.7	–	–
86059	10701	11.38	3.37	121.7	124.2	122.9	18	65
88046	10701	56.28	31.87	184.4	192.4	188.4	–	–
90007	10860	46.19	18.45	174.8	171.5	173.2	60	16
95014	10860	67.47	26.33	193.7	184.7	189.2	–	–
88036	10991	15.78	–	132.1	–	–	29	19
90017	10991	28.35	8.81	153.6	148.1	150.8	–	–

Notes:
ID: ID number – first two digits represent year of capture; next three digits represent order of capture; DS: data set number – this number remains the same across years, whereas ID number changes; M1: M1 wear; M2: M2 wear; M1est: estimated age from regression equation for M1 (see Table 12.2); M2 est: estimated age from regression equation for M2 (see Table 12.2); M3 est: estimated age from regression equation for M3 (see Table 12.2); Xest: average of M1est and M2 est; Int.a.: known interval between recaptures in months; Int.b.: estimated interval between recaptures in months from tooth wear and age prediction.

In males, the average difference in age estimates from M1 and M2 was 14.14 months (range 3.0–47.4 months) (Table 12. 4). Close inspection of the individual recaptures reveals some idiosyncratic patterns that explain the high range of estimates for first molars: in 83017 the second molar was considerably more worn than the first; and 83018 had a remarkably steep gradient between first and second molars. When these two animals with highly divergent estimates (47 months) were excluded, the mean difference was 9.78, closer to the female mean. The average difference between estimates based upon first and third molars was 18.7 (range 0.5–33.0) and 16.3 for second and third molars (range 7.3–30.2). Thus, there is less intertooth agreement in age estimates for males than for females. On average, the match between known interval and estimated interval differs by 10.6 months, with a range of 1 to 34 months – the latter representing the old male who showed little progressive wear on his highly worn molars.

In Figs. 12.4 and 12.5, tooth wear is plotted against the months elapsed between captures for all females and for a selected number of males. The initial capture episode is represented on the graph as interval = 0; the

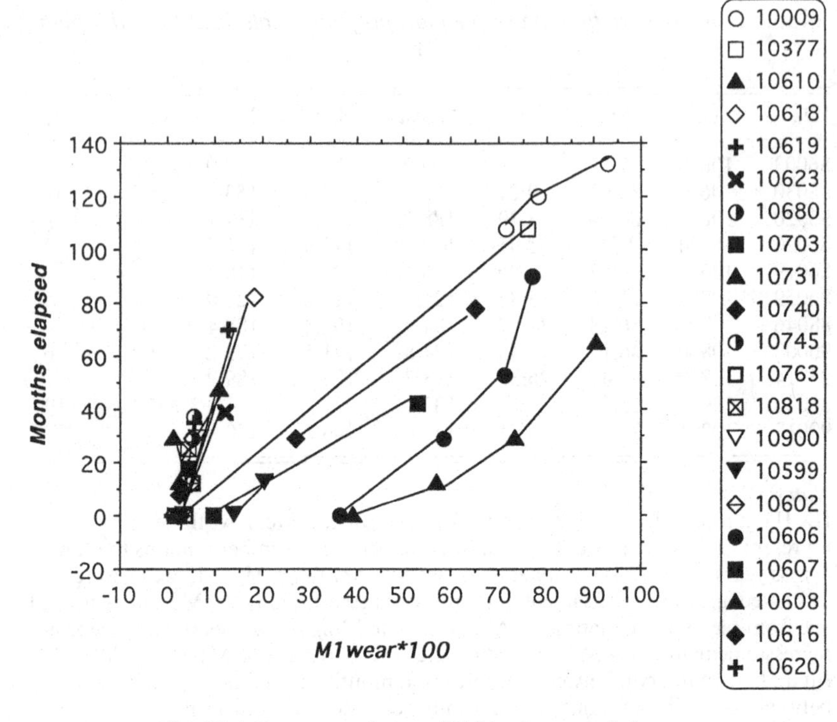

Fig. 12.4. Recapture slopes of Ethiopian male baboons.

recapture intervals then range up to 135 months for males and to 95 months for females.

The plots confirm that the rate at which a tooth wears is not constant over time. Wear, as measured by dentine exposure, is slowest immediately after eruption (that part of the plot where less than 10% of the surface has exposed dentine) and accelerates as dentine is exposed. Once dentine exposure is marked, the rate of wear decelerates. This is seen in both males and females. For example, the rate of wear in baboon 10860 – a female with 45% of the occlusal surface exposed as dentine when initially captured – is far slower than in dentally younger females such as 10659 between first and second capture or even in 10659 between second and third capture (Fig. 12.5). Wear profiles of two males captured four times each, further illustrate this point. While estimated to be of equivalent age in 1983, their wear trajectories diverged over the next five to six years (Fig. 12.4). At his first capture in 1983, 10606 weighed 20 kg, but had already lost four teeth, and was estimated to be 150 months old. In 1991, 10606 had lost 17 teeth and his weight had dropped to 16 kg. Presuming that age estimates at the earlier

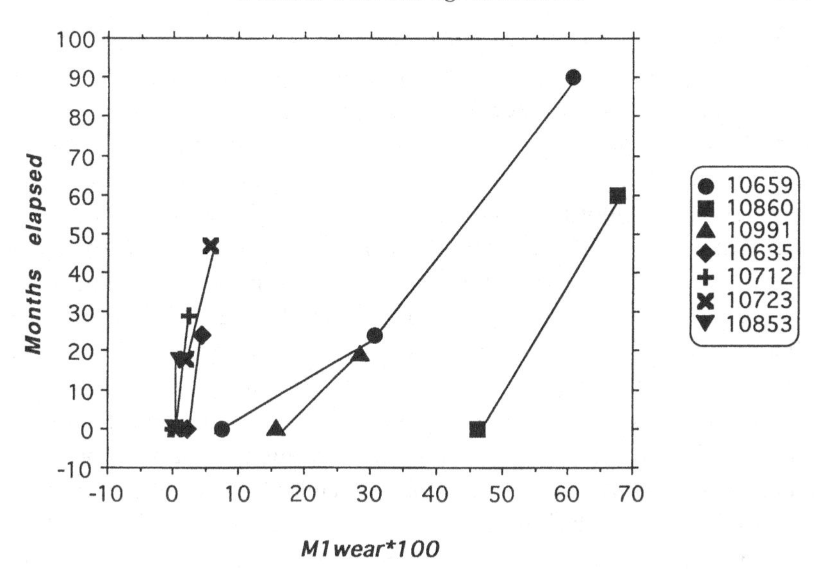

Fig. 12.5. Recapture slopes of Ethiopian female baboons.

end of the range were more accurate, in 1991 he was at least 20 years. Age estimates derived for 10606 on the basis of wear were 150 months in 1983, 195.7 months in 1986, 196.3 months in 1988 and 198.1 months in 1991. The estimated age in 1986 is within a few months of his 1983 age plus the time elapsed. Between 1986 and 1991, little further wear could be measured by our method, and so the subsequent age estimates are clearly poor. On the other hand, 10608 aged at 154.4 months in 1983 was estimated at 171.7 months in 1984, 193.3 months in 1986 and 203.3 months in 1989. The fit between his presumed age (original age plus months elapsed) is also better at the lower end of the range; his estimated age of 203 months in 1989 contrasts with his presumed age of 232 months. Similarly, his weight dropped from 22 to 20 kg in the six years. Despite many highly worn teeth, he lost only three.

Additionally, an animal's pattern of differential wear may change from year to year. In animal 10606, the second molar wore at a faster rate than M1 between 1983 and 1986 (not shown in Fig. 12.4). Between 1986 and 1991, the second molar rate slowed while the first molar rate picked up again.

Discussion

The differences in the onset and the rate of wear between Ethiopian and Tanzanian baboons are consistent across sexes. While the rate of wear is higher in Ethiopian baboons, dentine exposure appears earlier in the Tanzanian population. Although these observations may seem contradictory, they make sense in the ecological context of the two populations.

At Awash, feeding is concentrated on riverine resources and in the thornbush (Nagel, 1973). In the rainy seasons, newly-opened flowers, flush and grass leaves are important components in the diet. Seeds, fruits and shoots of riparian trees and shrubs (*Acacia tortilis, Celtis* sp., *Zizyphus* sp., *Ficus* sp., *Cordia* sp., *Salvadora persica*, etc.) and thornbush shrubs and trees (*Acacia nubica, Balanites aegyptiaca, Grewia* sp., *Dobera glabra*, etc.) form a major component of the diet through the long dry seasons. While much of this dry-season food is gathered after falling onto dusty ground with little herbaceous cover, and therefore inevitably involves the ingestion of abrasive materials, the fact that most foraging occurs within the riparian woodland means that younger (and hence lighter) juveniles have the opportunity to climb and forage for the leaves, fruits, and buds that remain in the treetops.

At Mikumi, overall rainfall is higher, and ground cover is much more continuous in favored feeding areas, presumably making for a diet that is less dusty and abrasive, and hence an overall lower rate of occlusal wear, than at Awash. On the other hand, intense, seasonal exploitation of a particular subterranean resource may explain the earlier onset of occlusal wear at Mikumi. During the dry season, Mikumi baboons feed heavily upon the corms of sedges (*Cyperus*; Rhine and Westlund, 1978). Although highly nutritious, corms require husking and manual rubbing to eliminate attached grit particles. Sedges form extensive beds, so that they are accessible to all animals. Indeed, while a group is occupied at a sedge-patch, little other food is available to its members, and corms form a central element of the diet even of young juveniles. Perhaps the early onset of occlusal wear at Mikumi reflects exploitation of this highly abrasive food item even by very young animals.

Whatever their explanation, the observed, inter-populational differences confirm the need for a population-specific schedule if occlusal wear is to be used to estimate age in nonhuman primates as it is in humans (Miles, 1963; Molnar, 1971; Scott, 1979). There is also evidence suggesting that there may be an influence of niche differentiation between the sexes. It may be possible to locate markers of life history events in the onset and rates of dental wear. Dental wear is a mirror of both elapsed time and of differential access to food resources. The relative lack of fit of individuals to a regression line may well relate to rank, social status, and consequent access to high or low quality resources with differing amount of gritty contaminants. In both anubis and yellow baboons, young males of a group are natal, while the adult males are immigrants, usually from several different natal groups. Heterogeneity among a group's adult males may then reflect both the conditions in the group where they originated, and conditions and their relative status in their new groups. Similarly, variation among females may reflect the influence of individual and matrilineal rank, and their effects on access to resources.

We were unable to show differences in wear rate between molars, but have shown differences in wear rate over the same period of time between individuals with similar initial wear states. To some extent, these may reflect differences in rank. Kilgore (1989) noted that a 15-year-old subordinate female chimpanzee had dental wear at death typical of a far older animal.

These observations indicate the limitations as well as the utility of this measure of dental wear as an age indicator. As shown in Figs. 12.2–12.5, once the dentine is fully exposed, further progressive wear cannot be quantitatively assessed by our technique. Subsequent changes may be of critical functional importance, with reduction in crown height, breakdown of the occlusal surface, and ultimately loss of the tooth itself.

Our data from recaptured animals show that tooth wear is not linearly related to time. Froehlich *et al.* (1981) used recapture intervals to calibrate tooth wear in their study of free-ranging *Alouatta* of Panama, estimating incremental wear from nine animals recaptured once. They recognized differential wear in animals of differing ages and attempted to calibrate their scale to accommodate this. Our recapture slopes suggest that more accurate predictions would result from equations based upon teeth grouped by degree of wear, but we had too few old animals to do so in this study. Among male Awash baboons, the wear rate averages 0.23 units/month when first molars have less than 20% of their surface worn; 0.95 units/month when the surface is worn between 21–40%, and 0.74 units/month when more than 40% of the surface is worn.

In Awash baboons under 16 years, age can be predicted to the nearest year from occlusal wear. Beyond this age, the method is no longer reliable. Even so, it effectively doubles the range across which ages can be assigned, compared to the use of eruption alone. These methods open the prospect of exploring growth trajectories and maturation schedules in the Awash hybrid zone, where cross-sectional data predominate.

Acknowledgements

In Ethiopia, data were collected in collaboration between JP-C and CJJ. We thank Ato Leykun Abunie, Dr. Beyene Petros, Dr. Yalemtsehaye

Makonnen, Dr Seyoum Mengistou, and the Wardens and staff of the Awash National Park for facilitating our work. In Tanzania, data were collected in a study directed by JP-C and Dr. Jeffrey Rogers. We are grateful to the Tanzanian National Scientific Research Organization and Serengeti Research Institute for research permission; and to our collaborators R. Rhine, G. Norton and S. Wasser for generously making available demographic records of the Animal Behavior Research Unit of Mikumi National Park. We acknowledge the contribution of numerous field assistants and game scouts at both research sites, and thank Jim Cheverud and Richard Smith for their generosity in time and statistical advice.

Research in Tanzania was supported by NSF BNS83–03506 (Phillips-Conroy and Rogers) and in Ethiopia by Earthwatch/The Center for Field Research, the Harry Frank Guggenheim Foundation, and New York University.

References

Altmann, S.A. & Altmann, J. (1970). *Baboon Ecology*. Chicago: University of Chicago Press.

Bramblett, C.A. (1969). Non-metric skeletal age changes in the Darajani baboon. *Am. J. phys. Anthrop.* **30**, 161–72.

Condon, K., Charles, D.K., Cheverud, J.M. & Buikstra, J.E. (1986). Cementum annulation and age determination in *Homo sapiens*. II. Estimates and accuracy. *Am. J. phys. Anthrop.* **71**, 321–30.

Dean, M.C., Jones, M.E. & Pilley, J.R. (1992). The natural history of tooth wear, continuous eruption and periodontal disease in wild shot great apes. *J. hum. Evol.* **22**, 23–39.

Dittus, W. (1980). The social regulation of primate populations: a synthesis. In *The Macaques: Studies in Ecology, Behavior and Evolution*. ed. D. G. Lindberg, pp. 263–86. New York:Van Nostrand Reinhold.

Dunbar, R.I.M. (1988). *Primate Social Systems*. Ithaca: Cornell University Press.

Fedigan, L.M. & Asquith, P.J. (1991). *The Monkeys of Arashiyama*. New York: State University of New York Press.

Froehlich, J.W., Thorington, R.W. & Otis, J.S. (1981). The demography of howler monkeys (*Alouatta palliata*) on Barro Colorado Island, Panama. *Int. J. Primatol.* **2**, 207–36.

Gantt, D.G. (1979). Patterns of dental wear and the role of the canine in Cercopithecinae. *Am. J. phys. Anthrop.* **51**, 353–60.

Glander, K.E., Teaford, M.F. & Noble, V.E. (1996). Group differences in *Alouatta palliata* feeding time. *Am. J. phys. Anthrop.* (*Suppl.*) **22**, 113.

Hamilton, W.J. III & Bulger, J.B. (1990). Natal male baboon rank rises and successful challenges to resident alpha males. *Behav. Ecol. Sociobiol.* **26**, 257–62.

Hartman, G.D. (1995). Age determination, age structure and longevity in the mole, *Scalopus aquaticus* (Mammalia, Insectivora). *J. Zool.* **237**, 107–22.

Hillson, S. (1986). *Teeth*. New York: Cambridge University Press.

Jolly, C.J. (1972). The classification and natural history of *Theropithecus*: baboons

of the African Plio-Pleistocene. *Bull. Brit. Mus. Nat. Hist. (Geol.)* **22**, 3–123.
Jolly, C.J. & Brett, F.L. (1973). Genetic markers and baboon biology. *J. med. Primatol.* **2**, 85–99.
Kilgore, L. (1989). Dental pathologies in ten free-ranging chimpanzees from Gombe National Park, Tanzania. *Am. J. phys. Anthrop.* **80**, 219–28.
Miles, A.E.W. (1963). The dentition in the assessment of individual age in skeletal material. In *Dental Anthropology*, ed. D. R. Brothwell, pp. 191–209. Oxford: Pergamon Press.
Molnar, S. (1971). Human tooth wear, tooth function and cultural variability. *Am. J. phys. Anthrop.* **34**, 175–90.
Morris, P. (1978). The use of teeth for estimating the age of wild mammals. In *Development, Function and Evolution of Teeth.* ed. P. M. Butler & K. A. Joysey, pp. 483–94. New York: Academic Press.
Nagel, U. (1973). A comparison of anubis baboons, hamadryas baboons and their hybrids at a species border in Ethiopia. *Folia primatol.* **19**, 104–65.
Nass, G.G. (1981). Sex differences in tooth wear of *Macaca fuscata*, the Arashiyama-A Troop in Texas. *Primates* **22**, 266–76.
Noble, V.E., Teaford, M.F. & Glander, K.E. (1996).Group differences in incisor macrowear of *Alouatta palliata. Am. J. phys. Anthrop.* (*Suppl.*) **22**, 178.
Phillips-Conroy, J.E. (1978). *Dental Variability in Ethiopian Baboons: An Examination of the Anubis-Hamadryas Hybrid Zone in the Awash National Park, Ethiopia.* PhD Dissertation, New York University.
Phillips-Conroy, J.E. & Jolly, C.J. (1986). Changes in the structure of the baboon hybrid zone in the Awash National Park, Ethiopia. *Am. J. phys. Anthrop.* **71**, 337–49.
Phillips-Conroy, J.E. & Jolly, C.J. (1988). Dental eruption schedules of wild and captive baboons. *Am. J. Primatol.* **15**, 17–29.
Phillips-Conroy, J. E., Jolly, C. J. & Brett, F.L. (1991). Characteristics of hamadryas-like male baboons living in anubis baboon troops in the Awash hybrid zone, Ethiopia. *Am. J. phys. Anthrop.* **86**, 353–68.
Phillips-Conroy, J.E., Jolly, C.J., Nystrom, P. & Hemmalin, H. (1992). Migration in male hamadryas baboons of the Awash National Park, Ethiopia. *Int. J. Primatol.* **13**, 455–76.
Phillips-Conroy, J.E., Nystrom, P. & Jolly, C.J. (1986). Palmar dermatoglyphics as a means of identifying individuals in a baboon population. *Int. J. Primatol.* **7**, 435–47.
Post, D., Hausfater, G. & McCuskey, S. (1980). Feeding behavior of yellow baboons (*Papio cynocephalus*): relationship to age, gender, and dominance rank. *Folia primatol.* **34**, 170–95.
Rhine, R.J. & Westlund, B.J. (1978). The nature of a primary feeding habit in different age-sex classes of yellow baboons (*Papio cynocephalus*). *Folia primatol.* **30**, 64–79.
Rhine, R.J., Wasser, S.K. & Norton, G.W. (1988). Eight-year study of social and ecological correlates of mortality among immature baboons of Mikumi National Park. *Am. J. Primatol.* **16**, 199–212.
Scott, E.C. (1979). Dental wear scoring technique *Am. J. phys. Anthrop.* **51**, 203–12.
Sigg, H., Stolba, A., Abegglen, J-J. & Dasser, V. (1982). Life history of hamadryas baboons: physical development, infant mortality, reproductive parameters and family relationships. *Primates* **23**, 473–87.
Smith, R.J. (1993). Logarithmic transformation bias in allometry. *Am. J. phys. Anthrop.* **90**, 215–28.

Sokal, R.R. & Rohlf, F. J. (1995). *Biometry*. New York: W.H. Freeman & Co.
Spinage, C.A. (1973). A review of the age determination of mammals by means of teeth, with especial reference to Africa. *E. Afr. Wildlife J.* **11**, 165–87.
Teaford, M.F. (1983). Differences in molar wear gradient between adult macaques and langurs. *Int. J. Primatol.* **4**, 427–44.
Teaford, M.F. & Tylenda, C.A. (1991). A new approach to the study of tooth wear. *J. Dent. Res.* **70**, 204–7.
Van der Bijl, P., DeWaal, J., Botha, I.A. & Dreyer, W.P. (1989). Assessment of occlusal tooth wear in vervet monkeys by reflex microscopy. *Archiv. Oral Biol.* **34**, 723–9.
Welsch, U. (1967). Tooth wear in living pongids. *J. Dent. Res.* **46**, 989–92.

13

Maternal investment throughout the life span in Old World monkeys

LYNN A. FAIRBANKS

Introduction

Cercopithecid monkey life histories include a period between weaning and sexual maturation when juveniles of both sexes continue to reside with their mothers. In many species, especially among the Cercopithecinae, mature daughters live in their natal group throughout their life span. This social system allows maternal influence on the behavior, development, and reproduction of offspring to extend far beyond early infancy. In such societies, mothers have the opportunity to influence their lifetime reproductive success, not only through production and care of infants, but also through maternal care that promotes survival of juvenile offspring and the reproduction of adult daughters.

Field studies of the 1960s produced detailed descriptions of mother–infant behavior (DeVore, 1963; Jay, 1963; Struhsaker, 1971; Ransom and Rowell, 1972), but the more subtle relationships between mothers and their older offspring were more difficult to detect, and kinship among older animals was rarely known with certainty. Since that time, longitudinal studies of individually recognized animals in captive and free-ranging populations have documented the ongoing nature of the mother–offspring and, particularly, the mother–daughter relationship. Longitudinal data on known individuals are now available for numerous free-ranging populations of macaques (*Macaca*), baboons (*Papio*), vervet monkeys (*Chlorocebus aethiops*, s.l.), and Hanuman langurs (*Semnopithecus entellus*) (e.g. Koyama, 1970; Kurland, 1977; Hasegawa and Hiraiwa, 1980; Horrocks and Hunte, 1983; Altmann *et al.*, 1988; Cheney *et al.*, 1988; Dittus, 1988; Rhine *et al.*, 1988; Mori *et al.*, 1989; Nakamichi, 1989; Smuts and Nicolson, 1989; Borries *et al.*, 1994). Some provisioned captive and semi-free-ranging populations have been managed so as to maintain a natural cross-generational group structure with known kinship (e.g. Sade, 1972; Silk *et al.*,1981; Small and Smith, 1981; Bernstein and Ehardt, 1985; Eaton *et al.*, 1986; Fedigan *et al.*, 1986; Paul and Kuester, 1987; Fairbanks, 1993a; deWaal, 1993).

In spite of the growing body of information on the relationships of mothers with their juvenile and adult offspring in these species, discussions of maternal behavior still focus almost exclusively on infancy (Nicolson, 1987; Pryce *et al.*, 1995). Insights from evolutionary biology about the influence of life history trade-offs and parent–offspring conflict have been applied primarily to maternal behavior toward infants. Kin selection theory has been profitably used to explain altruistic relationships among older kin (Gouzoules and Gouzoules, 1987; Silk, 1987), but these discussions have generally focused on the degree of relatedness between individuals, ignoring differences based on the specific relationship involved, or on the direction of behaviors between generations.

The object of this chapter is to investigate the relationships between cercopithecid monkey mothers and their juvenile and adult offspring, using constructs from parental investment theory. Do mothers interact with their juvenile and adult offspring in ways that increase offspring fitness? Is there evidence that mothers reduce juvenile mortality and increase the chance that their offspring will survive to reproductive age? Do mothers influence the reproductive success of their adult offspring? What are the consequences of being orphaned as a juvenile or young adult? Is there any evidence that grandmothers promote the survival and socialization of their grand-offspring? What are the costs and trade-offs involved in post-weaning maternal investment for the mother?

Data from longitudinal studies of free-ranging, provisioned and captive populations will be used to address these questions and to evaluate the hypothesis that monkey mothers continue to invest reproductive effort in their juvenile and adult offspring long after weaning. Available data do not represent a true cross-section of the Cercopithecidae. Colobines are under-represented, as are forest-living cercopithecines. The extent to which the conclusions drawn here can be generalized to these lesser-known groups will be discussed, but can only be determined when longitudinal information on individuals of known kinship becomes available for more species. (See also Chapter 19.)

Predictions from parental investment theory

Parental investment theory posits that parental care should be allocated among offspring and over time in ways that maximize the parent's expected lifetime reproductive success (Clutton-Brock, 1991). It starts from the assumption that the amount of effort an individual can devote to reproduction is limited and that parental effort that benefits one offspring will reduce the parent's ability to invest in other offspring (Trivers, 1972). Quantitative models then predict how a parent should allocate its effort under different circumstances to maximize its genetic representation in future generations (Stearns, 1993).

Hypothetically, reproductive effort can be directed primarily toward production of offspring, with little or no parental care, or parental effort can be devoted to care that increases the fitness of individual offspring. Effects on offspring fitness can be further differentiated into those that improve the offspring's chance of survival, and those that increase its reproductive success (Blurton-Jones, 1993). The reproductive success of the parent, measured as the number of grandchildren, is the product of the number of offspring produced, their survivorship to reproductive age, and their success at reproduction as adults. The trade-off between offspring number and offspring quality at birth has been extensively modeled and discussed (Stearns, 1993), but less attention has been paid to identifying aspects of parental care that promote survival and offspring reproductive success at later life stages.

Among mammals, internal gestation and lactation dictate that the mother will make a major investment in individual offspring – generally outweighing that of the father – and that maternal care will continue after birth. Determining the optimal allocation of maternal care requires understanding how a particular unit of care benefits the offspring, and how it affects the mother. One might expect that the more energy a mother expends, the greater will be her cost and the greater the benefit to the individual offspring. The relationship among these components is not always so straightforward, however, and would be expected to vary according to the kind of maternal care and the life history stage of both mother and offspring. For example, carrying and nursing a dependent infant exerts a substantial energetic cost on the mother, a measurable delay in future reproduction, and a large benefit to the infant (Altmann, 1980; Altmann and Samuels, 1992). Tolerance of juvenile offspring at a sleeping site might involve no detectable cost to the mother in future reproduction, only a minimal risk to her dependent infant, but a substantial benefit to the juvenile offspring in terms of survival. Calculating exactly how a mother should allocate care among all her offspring involves understanding the complex interactions among the variables that affect parent and offspring fitness.

If maternal care is not shareable, then any care given to one offspring will reduce parental expenditure available for other offspring. Many aspects of maternal care, however, do not seem to conform to this strict definition of investment. When a mother gives an alarm call, the expected cost to her future fitness will be the same, regardless of how many of her offspring benefit. Effort invested in maintaining a dominance position, or in defending a territory, will produce benefits that do not depreciate with the number of offspring who benefit. This distinction has been referred to as shared versus unshared, depreciable versus nondepreciable, or individual versus umbrella care (Lazarus and Inglis, 1986; Clutton-Brock, 1991; Blurton-Jones, 1997). Offspring compete with one another for individual care from the mother, but the benefits received from nondepreciable umbrella care, such as alarm calls, are not discounted by the presence of other beneficiaries. Based on these general principles, and our knowledge of cercopithecid life histories, what would monkey mothers be expected to do for their juvenile and adult offspring? After weaning, juveniles face the challenges of feeding themselves, avoiding illness and predation, integrating into the social system, and growing into competent adults (Fairbanks and Pereira, 1993). The juvenile period, between weaning and sexual maturity, is a time of relatively high mortality in most undisturbed primate populations, and young juveniles are particularly vulnerable to predation (Winkler *et al.*, 1984; Horrocks, 1986; Cheney and Wrangham, 1987; Janson and van Schaik, 1993). Survival to breeding age is a major component of lifetime reproductive success for wild primates and often explains more variance than fertility or adult longevity (Clutton-Brock, 1988). Thus, maternal behaviors that reduce juvenile mortality could potentially have a major effect on the mother's lifetime reproductive success. Theoretically, a mother would benefit more by saving the life of a juvenile son or daughter than by producing a new infant, because the juvenile has a higher reproductive value. If the cost to the mother of saving a juvenile is less than or equal to the cost of producing and raising another infant, then we would expect her to allocate effort to reducing juvenile mortality.

Many Old World monkey societies, particularly those of more terrestrial cercopithecines, are characterized by a relatively high degree of within-group competition (van Schaik, 1989). Individuals compete for dominance rank, and rank influences resource access and reproduction (Harcourt, 1987). Mothers could promote offspring reproductive success in several ways. They could help their young compete for high quality food, allowing them to grow faster, become stronger, and mature earlier. They could assist individual offspring in contests with other group members, and work to maintain intra-group social relationships that are advantageous to all of their juvenile and adult offspring. They could also provide direct support for the reproductive efforts of their offspring, and care and protection for their grand-offspring. In most of the cercopithecid species for which relevant data are available, females are highly philopatric, while males emigrate before breeding. In such societies, we would expect mothers to have more opportunities to influence their daughters' than their sons' reproductive success. When mothers invest effort in individual depreciable care to juvenile and adult offspring, they should assist the offspring who can benefit the most, relative to the cost. We would also expect mothers with more than one offspring to selectively invest in the forms of umbrella care that can simultaneously benefit all family members. The total fitness benefits of nondepreciable maternal investment in activities such as defending a home range, or acquiring and maintaining high rank would increase with the number of offspring available.

Relationships of Old World monkey mothers with juvenile offspring

It is widely accepted that mothers maintain affiliative and supportive relationships with their juvenile offspring after weaning (Pereira and Altmann, 1985), and it is often assumed that these interactions function to promote juvenile survival. Juveniles are typically found close to their mothers much more often than would occur by chance, and young juveniles usually spend more time near their mother than near any other adult female (Kurland, 1977; Hayaki, 1983; Pereira, 1988; Nakamichi, 1989; Janson and van Schaik, 1993; van Noordwijk *et al.*, 1993). Most data on sleeping subgroups rely on inferred relationships, because observation conditions do not permit individual recognition, but the groupings that have been observed suggest that juveniles, and particularly yearlings, huddle with their mothers at night (Anderson and McGrew, 1984; Ansorge *et al.*, 1992; Hammerschmidt *et al.*, 1994). Juveniles also receive considerably more grooming from their mothers than from other group members (Sade, 1972; Missakian, 1974; Kurland, 1977; Cheney, 1978; Walters, 1980, 1981; Pereira and Altmann, 1985; Nakamichi, 1989; Koyama, 1991).

The preferential spatial and grooming relationships of juveniles with their mothers in West African vervet monkeys (*C. ae. sabaeus*), is illustrated in Figure 13.1. Fifty-six, 2-year-old monkeys were observed between 1984 and 1988, in four captive social groups at the UCLA-Sepulveda colony. (For information about the history of the colony and observation methods, see Fairbanks and McGuire, 1985, 1995.) Males remain in their natal group until they are four to six years of age, and females stay in their natal group

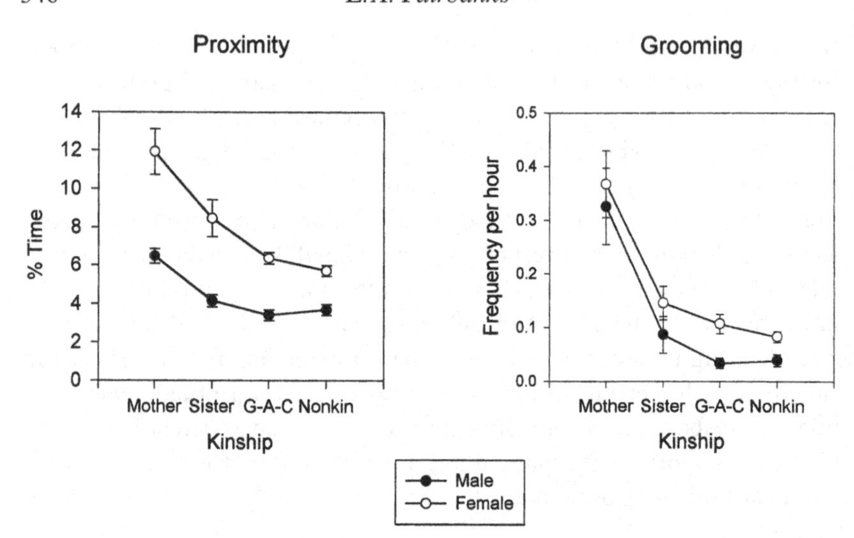

Fig. 13.1. Percentage time in close proximity (within 1 m) to individual adult female group members, and frequency per hour of grooming received from adult females, by matrilineal kinship for two-year-old male and female vervet monkeys. (G–A–C refers to maternal grandmothers, aunts and cousins.) Analysis of variance indicated significant effects of kinship on both proximity ($F_{3,374}=42.40$, $p<0.001$) and grooming received ($F_{3,374}=43.66$, $p<0.001$), with mothers differing significantly from all other classes of adult females in *post-hoc t*-tests at $p<0.05$.

for life. As a result, each juvenile has multiple adult female matrilineal relatives. Figure 13.1 demonstrates that daughters spend more time near their mothers than sons do, but both sons and daughters spend significantly more time near their mother than near any other class of adult females. In addition, both juvenile males and juvenile females receive more than twice as much grooming from their mothers as from their older sisters, while grooming received from adult sisters and more distant adult female kin does not differ significantly from grooming by unrelated adult females. It should be noted that the relationship between degree of relatedness and these affiliative behaviors is not linear (Kurland, 1977; Altmann, 1979). A larger than expected proportion of the effect of kinship on both proximity and grooming received is attributable to the special relationship of juveniles with their mothers.

Mothers in the species surveyed here not only provide proportionally more affiliative support, they are also the principal defenders of juveniles in fights with other group members. In his study of fight interference among rhesus macaques (*M. mulatta*) on Cayo Santiago, Puerto Rico, Kaplan (1978) found that a major portion of the effect of kinship on agonistic aid was due to the high rate of aid given by mothers to their offspring. Mothers aided their offspring at a rate 12 times higher than the rate of aiding between siblings. In a more recent study of agonistic intervention in the same population, Janus (1992) found that of 339 interventions by an adult in conflicts between juveniles, 80% involved the mother of at least one of the participants. Mothers were particularly likely to intervene when their juvenile offspring was the victim of aggression. Rhesus mothers were also more likely to respond to the noisy screams of their immature offspring, which signal greater risk of injury, than from other immature relatives (Gouzoules *et al.*, 1986). Similar high rates of defense and agonistic aiding of offspring by their mothers have been reported for other species and settings (pig-tailed macaques (*M. nemestrina*), Massey, 1977; rhesus macaques, Bernstein and Ehardt, 1985; vervet monkeys, Fairbanks and McGuire, 1985; Hunte and Horrocks, 1987; yellow baboons (*Papio hamadryas cynocephalus*), Pereira, 1989).

When they are injured, juveniles go to their mothers for assistance. Dittus and Ratnayeke (1989) compared social interactions among toque macaques (*M. sinica*), observing the same individuals when they were injured and when they were not. Animals responded to injury by increasing the rate at which they initiated contact and solicited grooming from other group members. Injured females directed 46% of their contact initiations to their mothers, and 44% of the grooming they received was from their mothers. When a four-year-old male Japanese macaque (*M. fuscata*) lost his foot, he began to maintain contact with his mother and was groomed by her every day (Nakamichi, 1989). This pattern contrasted sharply with that of other, uninjured males of the same age who were rarely seen near their mothers.

While such evidence clearly documents close affiliative and supportive relationships between mothers and their juvenile offspring, most studies have demonstrated that it is the juveniles rather than the mothers who are primarily responsible for maintaining the relationship. When juveniles and their mothers are separated, it is usually the juvenile who reestablishes proximity, and the amount of time that mothers and their juvenile offspring spend together is predicted more accurately by the approaches of the juvenile than by the behavior of the mother (Fairbanks and McGuire, 1985; Ehardt, 1987; Rowell and Chism, 1986; Pereira, 1988). A large proportion of the grooming received from mothers is solicited by the juveniles (Cheney, 1978; Walters, 1981; Pereira and Altmann, 1985), and in the examples of special aid described above, the juveniles typically had to solicit the assistance from their mothers.

The active role of yearlings in soliciting maternal care is also evident in studies of sleeping subgroups. Sleeping subgroups including a mother, an infant, and a yearling were common among Barbary macaques (*M. sylvanus*) but mothers did not immediately accept their yearling offspring into their sleeping groups (Hammerschmidt *et al.*, 1994). When yearlings attempted to make contact with their mothers at sleeping sites, the mothers frequently rejected them by turning away, moving, threatening, and sometimes pushing them. A marked increase in calling was recorded at dusk as the yearlings begged to be admitted into a sleeping cluster. Hammerschmidt interpreted the dusk calling as evidence of parent–offspring conflict over maternal care.

The relationship between mothers and juvenile offspring includes a relatively high level of conflict. Bernstein and Ehardt (1986) reported that juvenile rhesus macaques received more aggression from their mothers than from other group members, and the mother's aggression included hitting, pushing, and biting as well as non-contact threats. They concluded that mother–offspring aggression serves a socializing function as mothers try to limit intrusive and unacceptable behavior. The birth of a younger sibling is often a time of accelerated conflict between mothers and juvenile offspring as mothers turn their attention toward their new infants (Lee, 1983a; Chism, 1986, 1991; Ehardt, 1987; Holman and Goy, 1988). In conflicts between siblings, mothers typically side with younger offspring against older siblings (Horrocks and Hunte, 1983; Bernstein and Ehardt, 1986).

Adult females who are pregnant or caring for young infants are under considerable pressure to meet their own needs and those of their infants (Altmann, 1980; Dunbar and Dunbar, 1988). It is likely that a mother's attention and attentiveness have limitations, and that a high level of vigilance directed to the safety of a juvenile offspring would reduce the mother's ability to keep track of her more vulnerable infant (Maestripieri, 1993; Cords, 1995). Predation takes a heavy toll of immature animals in many cercopithecid populations (Cheney and Wrangham, 1987). When predators are sighted, most mothers instantly grab their dependent infants (Hauser, 1988), and may even risk their own lives trying to defend them. Cheney and Wrangham, (1987) recount several instances of red colobus (*Colobus badius*) mothers being killed while attempting to save their offspring from chimpanzees. In all these cases, the offspring were young infants (C. Stanford, pers. comm.). Anecdotes describing mothers actively helping a juvenile offspring to escape from danger are remarkably hard to find. In a study comparing spatial relationships when predators were or were not present, Stanford (1995) reported that juvenile red colobus showed no tendency to be closer to adult females when in the presence of chimpanzee predators, but they were more likely to be near adult males. Other studies have also noted that juveniles seek the protection of adult males, or join larger and more cohesive subgroups, in response to predators (Rhine, 1975; Janson and van Schaik, 1993). Vervet monkey mothers let their juvenile offspring approach and explore potentially dangerous novel spaces and objects, while they hold back themselves and protect their younger infants (Fairbanks and McGuire, 1988, 1993; Fairbanks, 1993b).

The behavioral evidence suggests that mothers do not attend closely to the needs of their offspring after the first year of life, and that juveniles have to actively solicit assistance from their mothers. Mothers generally respond by tolerating juvenile offspring, and by providing active care when it is requested. Juveniles seek the protection of other group members, but not necessarily their mothers, in times of danger.

Survival of juvenile orphans

If mothers are investing parental effort to ensure the safety and survival of their juvenile offspring, then we would expect that juveniles who lose their mothers would suffer higher rates of mortality than juveniles living with their mothers. Reports of Old World monkeys that have been orphaned in the field do not provide much support for this hypothesis, however, and suggest instead that the presence of the mother is not essential for juvenile survival. The most extensive data on survival rates of immature orphans come from a study of free-ranging Japanese macaques at Takagoyama, Japan (Hasegawa and Hiraiwa, 1980). A high mortality rate associated with cessation of provisioning left a large number of immature animals without mothers. While most of the infants less than one year of age died, the majority of the yearlings (5/8) and almost all of the older juveniles (15/17) survived the loss of their mothers. Other reports of young, motherless animals in free-ranging populations also suggest that motherless juveniles are able to survive. Chism *et al.* (1984) mentioned two patas (*Erythrocebus patas*), orphaned at seven and nine months of age, that lived through a subsequent harsh dry season. Infant baboons orphaned or separated from their mothers before 10 months of age usually died within a few days, but five that were orphaned between 10 and 16 months survived (Rhine *et al.*, 1980; Hamilton *et al.*, 1982). Similar results were reported by Lee (1983b) for vervet monkeys (*C. a. pygerthrythrus*) at Amboseli, Kenya. Four orphaned infants less than one-year-old died shortly after their mothers. A 13-month-old orphan was killed by baboons two weeks after his mother's death, but all 12 other orphans between two and five years of age survived.

One reason for the relatively low mortality of orphaned juveniles may be their ability to establish bonds with substitute care-givers. Each of the immature Japanese macaque orphans studied by Hasegawa and Hiraiwa (1980) formed a close relationship with another group member. The new proximity partners varied, including adult males, immature siblings, close relatives, other juveniles and even unrelated adult females, but these new relationships seemed to substitute effectively for the lost mother. The orphaned juveniles did not differ in the amount of grooming they received, and in no case was an orphan observed isolated from the troop. Similar results have also been reported for free-ranging yellow baboons (Walters, 1981) and vervet monkeys (Lee, 1983b) at Amboseli. Two juvenile female baboons who had lost their mothers compensated by forming strong relationships with other group members, one with an older sister and the other with an unrelated adult female. These substitute relationships were so similar to the typical mother–daughter relationship in grooming rates and response to grooming solicitations that they were classified with mother–daughter dyads in a discriminant function analysis of juvenile–adult female interactions. In a comparison of 10 orphaned vervet monkeys varying in age from one to five years, Lee (1983b) found no consistent changes before and after the loss of the mother in the amount of aggression received, time spent grooming, or time in proximity with other group members.

Experimental removal of mothers under more controlled conditions at the UCLA-Sepulveda vervet monkey colony confirmed the finding that orphaned juveniles form substitute relationships that successfully replace the affiliative interactions they had received from their mothers. Three to five adult females were removed from each of four social groups at the colony in 1990, leaving behind 13 two-year-old and 12 three-year-old orphans. In the year that followed, all these orphans were able to form relationships with substitute caretakers, including adult sisters, juvenile peers, and in one case, an unrelated adult female. Figure 13.2 shows rates of social interactions of the two-year-old orphans compared with age- and sex-matched controls. The orphans did not differ from controls in the amount of aggression they received, or in the amount of time they spent alone. There was a slight decrease in the total amount of grooming orphans received from other group members, but the differences were not statistically significant.

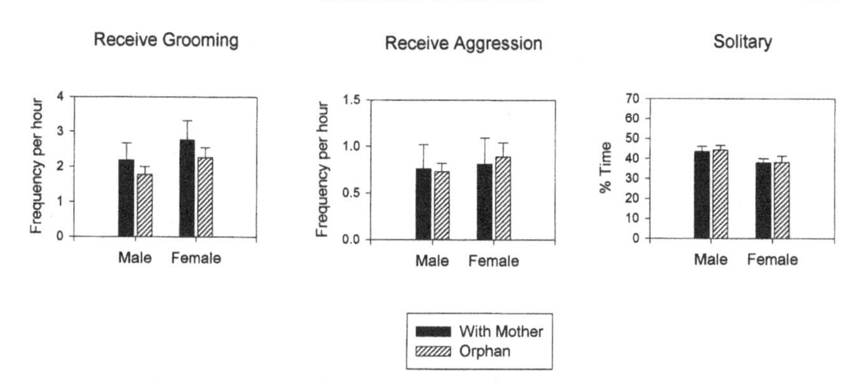

Fig. 13.2. Total frequency per hour of grooming and aggression received from all group members, and percentage time spent solitary, for two-year-old vervet monkey orphans in the year following removal of their mothers, compared with 2-year-olds with mothers in the group. There are no significant differences between orphans and controls in grooming received ($F_{1,15}=1.67$, $p=0.22$), aggression received ($F_{1,15}=0.02$, $p=0.88$), or time spent solitary ($F_{1,15}=0.03$, $p=0.87$).

Mothers' contribution to group membership

Does the fact that juvenile orphans adjust so readily to the loss of their mothers, and seem to survive at the same rate as other juveniles, mean that mothers have little or no influence on juvenile safety and survival? Juveniles of almost all Old World monkey species remain in their mother's troop, even if their mother dies. They are thus able to grow and develop in the context of a familiar home range, with known predation risks, food sources, and social companions, many of whom will be kin. A juvenile that loses its mother still benefits from the time and effort she invested in maintaining the group's home range, and in developing stable and predictable relationships with other group members. Immature animals living in large groups with better territories have been shown to have higher survival rates in free-ranging long-tailed macaques (*M. fascicularis*) and vervet monkeys (van Schaik, 1983; Cheney and Seyfarth, 1987). The importance for juvenile survival of continued residence in the natal troop is often ignored because it is a relatively universal feature of primate social organization, but the few cases where young juveniles are forced to emigrate from the relative safety of their mother's troop demonstrate that the effects of leaving can be disastrous. Following adult-male take-overs in Hanuman langur troops, many juvenile males are ejected from the natal troop. Even though these young males usually emigrate with male peers and often live in groups with their fathers, half of them die in the first year after emigration

(Rajpurohit and Sommer, 1993). Mortality attributable to predation is also high for vervet monkeys that abruptly move into new areas as a result of group fusion (Isbell *et al.*, 1993).

Juveniles orphaned at a young age continue to reside in their natal troop, but the timing of emigration for adolescent males is influenced by the mother's presence. Orphaned male baboons and vervet monkeys leave their natal troop at an earlier age than males with mothers still living (Cheney, 1983; Alberts and Altmann, 1995a).

Offspring reproductive success and the inheritance of dominance rank

In species with clear-cut dominance hierarchies, one of the principal ways that mothers can influence the reproductive success of their offspring is through matrilineal rank inheritance. Females from high-ranking matrilines have been found to mature earlier, have shorter inter-birth intervals, or have lower infant mortality than females from lower ranking matrilines (e.g. Fairbanks and McGuire, 1984; Harcourt, 1987; Altmann *et al.*, 1988; Smuts and Nicolson, 1989).

Matrilineal inheritance of female dominance rank has been extensively documented in macaques, baboons, and vervet monkeys, where virtually all daughters assume the rank position of their mothers relative to females from other matrilines (see Chapais, 1992, and Pereira, 1992, for recent reviews). Early discussions of this phenomenon questioned the importance of active support from the mother in rank acquisition because it is possible for female orphans to assume their birth rank when their mothers are absent (Sade, 1967; Walters, 1980). Matrilineal rank inheritance is aided by other members of a female's kin group and by stabilizing support from other matrilines, but the mother also plays an important role.

Spontaneously occurring rank changes and experimental modifications of group structure have both verified the importance of agonistic support from the mother in dominance acquisition and maintenance. When mothers fall in rank relative to members of other matrilines, their immature daughters almost invariably fall with them (Bernstein, 1969; Koyama, 1970; Gouzoules, 1980; Johnson, 1987; Samuels *et al.*, 1987). By experimentally combining subgroups of juvenile females with or without their mothers, Chapais demonstrated that support from the mother enabled lower-ranking immature animals to rise in rank above higher-ranking age mates (Chapais, 1988a). They were able to maintain their new rank in the absence of their mothers, until the balance of power was shifted by adding the mother or an older sister from the other matriline (Chapais, 1988b).

Studies of orphaned individuals from a variety of species and settings also provide evidence that mothers actively participate in their daughters' rank acquisition and that the mother's support is particularly important for attainment of high rank. In the study of orphaned juvenile Japanese macaques described above, Hasegawa and Hiraiwa (1980) found that the primary disadvantage of being orphaned was the difficulty in maintaining family rank. Some orphans dropped in rank after they lost their mothers, and none of the orphaned daughters of dominant mothers was able to achieve her birth rank. Of 21 two-year-old Japanese monkeys studied by Koyama (1970) at Arashiyama, Japan, 19 shared their mother's rank among peers. The two who did not were orphans. Similar findings have been reported for baboons and vervet monkeys. At Eburru Cliffs, orphaned olive baboons (*P. h. anubis*) more than three years of age were able to keep their matrilineal rank, but all females orphaned before the age of three failed to achieve their birth rank (Johnson, 1987). At Amboseli National Park in Kenya, vervet monkey populations underwent a 10-year period of high mortality and declining group size that left several individuals without close female relatives (Cheney and Seyfarth, 1987). Only 11 rank reversals among unrelated adult females were observed during this period, but eight of them involved individuals with close female kin, including mothers, offspring, and sisters, rising in rank above those who had lost their last close female relative.

Experimental removal of mothers from four social groups at the UCLA-Sepulveda vervet monkey colony, described above, left behind 10 daughters of the highest-ranking females. These daughters ranged in age from 1.5 to 5.5 years at the time their mothers were removed. The female dominance hierarchies underwent major changes with new alpha females emerging in the first few months after the removals in each of the four groups. None of the 10 daughters of the old alpha females was able to assume the new alpha position in her group, and none rose to the alpha position in the four subsequent years.

These data indicate that the active presence of the mother is influential, if not essential, for acquisition of matrilineal rank in cercopithecid monkeys. When a mother is available, her daughters have a greater chance of assuming her position in the hierarchy. The presence of the mother is particularly important if the advantages of high rank are to pass from mothers to daughters. It is very difficult for juvenile daughters of high-ranking families to inherit their mother's rank in her absence.

Mother's rank has also been shown to influence the development of sons in ways that are likely to have an impact on their reproductive success. In rhesus and Japanese macaques, sons as well as daughters tend to inherit their mother's rank (Pereira, 1992). The determinants of male reproductive success, assessed genetically, vary among populations, but recent evidence suggests that high-ranking macaque males father more offspring than lower-ranking males (Bauers and Hearn, 1995; deRuiter *et al.*, 1995; Smith, 1995). In baboons, male rank is usually acquired independent of maternal rank (Johnson, 1987), but high ranking mothers are able to confer other benefits to their sons. Alberts and Altmann (1995b) found that the timing of sexual maturation was influenced by maternal rank. Sons of high-ranking mothers matured physically at an earlier age, and entered the adult dominance hierarchy sooner, than sons of lower ranking mothers.

Mothers and adult offspring

Affiliative social relationships with adult daughters

When individuals have been followed long enough for adult genealogies to be known, it is clear that the strong bond between a mother and her immature offspring continues into adulthood for daughters, but not for sons. For example, in his study of kin relationships among Japanese macaques, Kurland (1977) found that four-year-old daughters, just reaching reproductive maturity, spent 50% of the time near their mothers. Mother–daughter proximity declined with age, but fully mature daughters over six years of age still spent more time near their mothers than more distant relatives. In contrast, proximity between mothers and adult sons dropped precipitously from age three on, and five-year-old sons were not distinguishable from non-kin in the amount of time they spent near their mothers.

Mothers not only spend time near their adult daughters, they also invest effort in them and take risks to promote their welfare. Longitudinal studies of Japanese macaques, vervet monkeys and Hanuman langurs all show the same general pattern (Kurland, 1977; Fairbanks and McGuire, 1986; Koyama, 1991; Nakamichi, 1991; Borries *et al.*, 1994). Mothers groom their adult daughters at higher rates than they groom other adult females, and at higher rates than they groom their adult sons. Mothers also continue to support their adult daughters in agonistic conflicts with other group members.

There is some evidence that maternal investment in adult daughters is strategically placed where it is likely to have the greatest benefit. Young adult females are more likely to need assistance than more mature adults. They are still actively involved in establishing their adult rank relationships and they also have higher rates of pregnancy failure and infant mortality than prime-aged females. When older mothers have more than one adult offspring, they focus their attention on the younger of their adult daughters. Old adult females in the Arashiyama East group spent more time near their youngest adult daughter than near their older daughters, and tended to groom younger daughters more than older ones (Nakamichi, 1991).

When groups of Old World monkeys undergo fission, daughters with living mothers will almost always stay in the same group with their mothers. Dittus (1988) described four fissions in an unprovisioned, free-ranging population of toque macaques. In all four cases, groups split along matrilineal lines. In every case of known relatedness, and all but one case of assumed relatedness, daughters remained with their mothers. Adult daughters also remain with their mothers during group fissions among rhesus monkeys on Cayo Santiago, and the loss of a matriarch increases the probability that a matrilineage will divide (Chepko-Sade and Sade, 1979). Menard and Vallet (1993) described the fission of a wild group of Barbary macaques into three subgroups. Early in the process, 10 individuals from seven different lineages were observed away from their mothers, but at the end only one individual, a seven-year-old male, was not included in the same nucleus as his mother. This tendency for juveniles and adult females to stay with their mothers during group divisions implies that the continuing mother–offspring relationship has value for the offspring.

Reproductive success of adult daughters

A few studies provide direct evidence that surviving mothers can positively influence their daughters' reproductive success. Ten years after the founding of the UCLA-Sepulveda vervet colony, the matrilines were in their third or fourth generation and several of the original matriarchs had died. At that time, young adult females with mothers still living in the group were more likely to reproduce successfully than comparable females who had lost their mothers (Fairbanks and McGuire, 1986). When the grandmother was present, the young adult daughters had slightly higher fertility rates and much lower levels of neonatal mortality, resulting in significantly more surviving offspring. A similar finding has also been reported by Hasegawa and Hiraiwa (1980) for Japanese macaque orphans at Takagoyama. Seven female orphans reached adulthood during the study and produced liveborn infants, but only two of the seven infants survived the first six months of life. In contrast, during the same time period, the five primiparous mothers who had not been orphaned all succeeded in keeping their infants

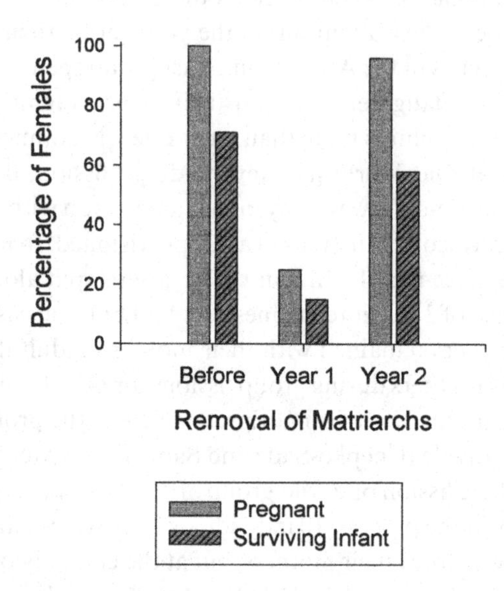

Fig. 13.3. Fertility of young adult female orphans (age 4–6 years) in the two years following removal of their mothers compared with baseline fertility levels for young adult females in the year before removal. In the year following matriarch removal, the rates of both pregnancy and infant survival were significantly lower than baseline (Pregnancy: $X^2=21.4$, $df=1$, $p<0.001$; Surviving infants: $X^2=13.23$, $df=1$, $p<0.001$). In the second year, fertility and infant survival returned to normal levels.

alive to one year. The presence of the young mothers' own mothers increased the probability that they would successfully raise their first infants.

The experimental removal of mothers at the UCLA-Sepulveda colony confirmed the influence of matriarchs on their daughters' reproductive success under controlled circumstances (Fairbanks 1993c). The young adult females in the group, who generally have lower fertility and higher rates of reproductive failure than prime-aged females, were highly sensitive to the stress of social instability caused by the removal of their mothers. In the first year after the removal, there was a dramatic reduction in fertility among the orphaned young adult daughters (Fig. 13.3). As the groups recovered from the disruption, and new relationships have stabilized, fertility and mortality rates returned to pre-existing levels. Matriarchs appear to play an important role in maintaining a safe and predictable social environment for their daughters' early attempts at reproduction. Young adult daughters can cope with the loss of their mother, but at the cost of temporarily reduced reproductive success.

Grandmothers and the next generation

A female monkey can become a grandmother when she is still a relatively young adult. Most grandmothers continue to reproduce, and must balance their own metabolic needs with the costs and benefits of investing in new infants, existing juvenile offspring, adult daughters, and grand-offspring. Grandmothers are more closely related to their offspring than to their grand-offspring by a factor of two, so we would not expect them to provide primary care for a grand-offspring in place of an offspring. The actual costs to the grandmother and the benefits to the grand-offspring should depend on many factors, such as the number and sex of offspring and grand-offspring, and the relative power of the grandmother to influence offspring reproductive success. Because grandmothers would have lower reproductive value and a shorter expected future lifespan than the mother, the costs of high-risk investments to promote grand-offspring survival would be lower for grandmothers than for mothers. The empirical data indicate that there is variability in the level of care that grandmothers give to grand-offspring, with some evidence that grandmothers do contribute to the fitness of their grand-offspring in predictable ways.

Grandmothers promote the survival and socialization of their grand-offspring

Field studies have provided anecdotal evidence that grandmothers will take substantial risks to defend their grand-offspring. Collins *et al.* (1984) observed a baboon grandmother harass an adult male who had attacked her seven month-old granddaughter. Hrdy (1977) recounts several episodes of old female langurs vigorously defending infants from infanticidal males. Exact kin relationships were not known but it is likely that the older females were defending their grand-offspring. Old langur females were also active in defending the group against dogs and humans that threaten the safety of the group. In a study of free-ranging rhesus monkeys on La Parguera, Puerto Rico, Partch (described in Hrdy, 1981) recorded 406 instances of post-reproductive females protecting or defending infants belonging to the old females' matrilines. Post-reproductive females were also more likely than breeding females to attack or chase intruders during intergroup conflicts.

The grandmother can serve as a buffer against the loss of the mother. Chepko-Sade and Sade (1979) described the adoption of an infant rhesus monkey whose mother died within a few days of the birth. The grandmother took the place of the mother in grooming relationships, and the orphaned female assumed the rank position of her grandmother in relationships with other group members.

The contribution of grandmothers to the safety and socialization of their grand-offspring has been discussed by Borries (1988) who performed an in-depth analysis of the relationships of two older females with their grand-offspring in free-ranging Hanuman langurs. Borries found that one of the langur grandmothers had a distinctive relationship with both her infant grandson and her juvenile granddaughter. This grandmother spent more time near her grandson than near other infants and was more likely to respond to his vocalizations and intervene on his behalf. She also maintained higher levels of contact with her juvenile granddaughter than with other juveniles, responded more often to her vocalizations and provided support in interactions with adult females. In contrast, the second langur grandmother did not show any preference for her infant grandson.

Results from the UCLA-Sepulveda vervet monkey colony have demonstrated that, when grandmothers are available, they direct their attention to their grand-infants in ways that reflect the grandmother's ability to influence offspring fitness. In a study of 30 grandmother/grand-infant dyads at the colony, infants spent significantly more time near their grandmothers than near other adult females, kin or non-kin (Fairbanks, 1988). Infants were also groomed by their grandmothers more often than by any adult females other than their mothers. The intensity of the grandmother/grand-infant relationship was influenced by factors associated with the relative benefits that infants would expect to receive, such as the infant's vulnerability to mortality or the grandmother's ability to provide effective social support. High-ranking grandmothers spent more time near their grand-infants and initiated more grooming and caretaking than lower-ranking grandmothers. Grandmothers spent more time near their daughter's first surviving infant than near later-born infants, and when grandmothers had more than one adult daughter, they spent more time near the infant of the younger daughter. These results suggest that grandmothers are actively contributing to the reproductive success of their adult daughters and to the survival of their infant grand-offspring.

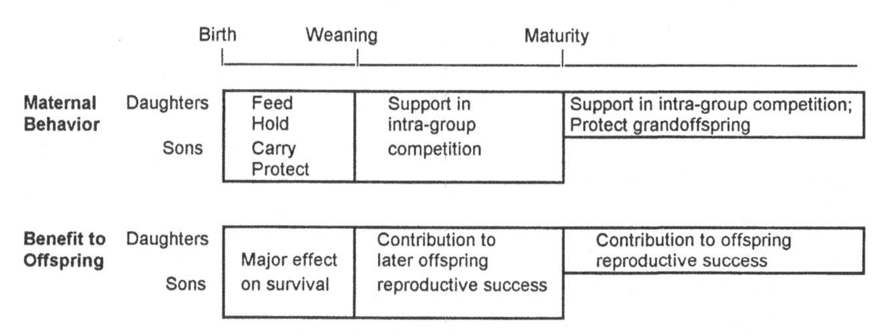

Fig. 13.4. Summary of maternal effects on offspring fitness throughout the life span.

Maternal investment throughout the lifespan

The matrilineal social organization of most of the Old World monkeys creates a context where it is possible for a mother to continue to benefit her offspring long after weaning. The data reviewed here support some of the hypotheses outlined earlier but not others, and suggest that post-weaning maternal investment is limited and selective. Figure 13.4 presents a schematic model of maternal behavior and proposed effects on offspring fitness at different life stages, derived from the above review. First, there is no doubt that mothers invest a large portion of their reproductive effort in care of infants between birth and weaning (Altmann, 1980; Nicolson, 1987; Dunbar and Dunbar, 1988). Maternal care of infants has measurable costs in terms of the delay of future reproduction, and the quality of care has demonstrable effects on infant survival (Fairbanks, 1993d; Fairbanks and McGuire, 1995). Infants that lose their mothers before weaning have almost no chance of surviving.

Examination of maternal behavior with juvenile offspring indicates that mothers do provide affiliative attention and support, but there was not much direct evidence that mothers have a major impact on juvenile survival. The data from natural populations do not indicate that orphaned juveniles have a markedly higher mortality rate than comparable juveniles with mothers. This result is rather surprising considering the high cost of producing a replacement offspring. The failure to find the expected effects of mothers on offspring survival could be explained in several ways. First, the expectation that mothers will attend to and invest in their offspring after weaning may be wrong, and it may be more difficult for mothers to reduce juvenile mortality than we have assumed. By two years of age, most Old World monkey juveniles are capable of looking out for their own safety, and mothers may not have the means of reducing mortality further without risking their own death from predation or disease. Or conversely, mothers may be investing in juvenile survival, but good estimates of mortality rates take much longer and are harder to get than good estimates of behavior. Studies of the social relationships of orphans would naturally be biased toward populations with high rates of orphan survival. Estimates of survival to maturity are needed for juveniles with and without mothers in a number of natural populations. Given the importance of juvenile mortality in Old World monkey lifetime reproductive success, this is certainly an area that deserves to be the focus of attention in future research.

In contrast to the lack of evidence that mothers play a major role in promoting offspring survival after weaning, in species with clear-cut dominance hierarchies, there is convincing support for the hypothesis that mothers actively promote offspring reproductive success by supporting their offspring in intra-group competition and by influencing rank attainment of their juvenile and adult daughters. Juveniles without mothers seem to be able to replace the relatively inexpensive benefits of affiliative associations and grooming time, but orphaned females are less likely to find an effective ally for the more costly interventions that promote rank acquisition. Active support by the mother helps to assure the transmission of family rank and the inheritance of advantage from one generation of females to the next. In species without strict hierarchies, mothers may also provide support that influences the competitive ability and later reproductive success of their sons and daughters. In situations that involve within- or between-group competition over food or over safe positions, advantages that are gained by the mother can be conferred to her offspring. The relative benefit of maternal support in intra-group competition is likely to be greater for juvenile daughters than for sons, and greater for species where resource access and reproductive success are influenced by dominance rank.

There is a growing body of evidence from matrilocal societies that mothers positively influence the reproductive success of their adult daughters. Mothers continue to associate with and support their daughters in intra-group conflicts. The presence of the mother contributes to group stability and the maintenance of a predictable environment for young adult daughters beginning their reproductive careers. There is also preliminary evidence from a few sites that grandmothers play an active part in promoting the reproductive success of their adult daughters, and the survival and socialization of their grand-offspring. The degree of influence that maternal investment of this type would have on offspring reproductive success should depend on the social system, with greater benefits being associated with higher levels of within-group competition.

More data from a greater variety of species are needed to confirm and refine these conclusions. Information is sorely lacking on post-weaning relationships between mothers and offspring for most of the arboreal Cercopithecinae and Colobinae (Cords, 1987; Newton and Dunbar, 1994). Thus, features of social organization that would be expected to have an impact on parental investment decisions, such as female transfer or the higher paternity certainty in one-male groups, have not been examined. Long-term studies of a platyrhine, the red howler monkey (*Alouatta seniculus*), have demonstrated that the presence of the mother is an important factor in determining whether a female will be able to stay and breed in the natal troop or be forced to bear the higher costs of emigration (Crockett and Pope, 1993). A similar effect might be expected among Old World monkeys with facultative female transfer (Moore, 1984). When paternity certainty is high, as in mountain gorilla groups, the effect of the loss of the mother for juveniles is buffered by the presence of their fathers (Watts and Pusey, 1993). A greater role of male parental investment may also be characteristic of species with one-male social units, such as hamadryas baboons and many colobines and forest cercopithecines (Cords, 1987; Stammbach, 1987; Struhsaker and Leland, 1987). More information from species with different social systems will help us identify the factors that promote the extension of maternal care beyond infancy, and will contribute to our understanding of the evolution of extended maternal care in the primates.

References

Alberts, S.C. & Altmann, J. (1995a).Balancing costs and opportunities: dispersal in male baboons. *Am. Nat.* **145**, 279–306.

Alberts, S.C. & Altmann, J. (1995b). Preparation and activation: determinants of age at reproductive maturity in male baboons. *Behav. Ecol. Sociobiol.* **36**, 397–406.

Altmann, J. (1980). *Baboon Mothers and Infants*. Cambridge, MA: Harvard University Press.

Altmann, J., Hausfater, G. & Altmann, S.A. (1988). Determinants of reproductive success in savannah baboons, *Papio cynocephalus.* In *Reproductive Success*, ed. T.H. Clutton-Brock, pp. 403–18. Chicago: University of Chicago Press.

Altmann, J. & Samuels, A. (1992). Costs of maternal care: infant-carrying in baboons. *Behav. Ecol. Sociobiol.* **29**, 391–8.

Altmann, S. (1979). Altruistic behavior: the fallacy of kin deployment. *Anim. Behav*. **7**, 958–9.

Anderson, J. R. & McGrew, W.C. (1984). Guinea baboons (*Papio papio*) at a sleeping site. *Am. J. Primatol.* **6**, 1–14.

Ansorge, V., Hammerschmidt, K. & Todt, D. (1992). Communal roosting and formation of sleeping clusters in Barbary macaques (*Macaca sylvanus*). *Am. J. Primatol.* **28**, 271–80.

Bauers, K.A. & Hearn, J.P. (1995). Patterns of paternity in relation to male social rank in the stumptailed macaque, *Macaca arctoides*. *Behaviour* **129**, 149–76.

Bernstein, I.S. (1969). Stability of the status hierarchy in a pigtail monkey troop (*Macaca nemestrina*). *Anim. Behav.* **17**, 452–8.

Bernstein, I.S. & Ehardt, C.L. (1985). Agonistic aiding: kinship, rank, age, and sex influences. *Am. J. Primatol.* **8**, 37–52

Bernstein, I.S. & Ehardt, C. (1986). The influence of kinship and socialization on aggressive behaviour in rhesus monkeys (*Macaca mulatta*). *Anim. Behav.* **34**, 739–47.

Blurton-Jones, N. (1993). The lives of hunter–gatherer children: effects of parental behavior and parental reproductive strategy. In *Juvenile Primates*, ed. M.E. Pereira & L.A. Fairbanks, pp. 309–26. New York: Oxford University Press.

Blurton-Jones, N. (1997). Too good to be true? Is there really a trade-off between number and care of offspring in human reproduction? In *Human Nature: A Critical Reader*, ed. L. Betzig, pp. 83–6. New York: Oxford University Press.

Borries, C. (1988). Patterns of grandmaternal behaviour in free-ranging Hanuman langurs (*Presbytis entellus*). *Human Evol.* **3**, 239–60.

Borries, C., Sommer, V. & Srivastava A. (1994). Weaving a tight social net: allogrooming in free-ranging female langurs (*Presbytis entellus*). *Int. J. Primatol.* **15**, 421–43.

Chapais, B.(1988a). Experimental matrilineal inheritance of rank in female Japanese macaques. *Anim. Behav.* **36**, 1025–37.

Chapais, B. (1988b). Rank maintenance in female Japanese macaques: experimental evidence for social dependency. *Behaviour* **104**, 41–59.

Chapais, B. (1992). The role of alliances in social inheritance of rank among female primates. In *Coalitions and Alliances in Humans and other Animals*, ed. A. Harcourt & F. deWaal, pp. 29–59. New York: Oxford University Press.

Cheney, D.L. (1978). Interactions of immature male and female baboons with adult females. *Anim. Behav.* **26**, 389–408.

Cheney, D.L. (1983). Proximate and ultimate factors related to the distribution of male migration. In *Primate Social Relationships*, ed. R.A. Hinde, pp. 241–9 Sunderland, MA: Sinauer.

Cheney, D.L. & Seyfarth, R. M. (1987). The influence of intergroup competition on the survival and reproduction of female vervet monkeys. *Behav. Ecol. Sociobiol.* **21**, 375–86.

Cheney, D.L., Seyfarth, R.L., Andelman, S.J. & Lee, P.C. (1988). Reproductive success in vervet monkeys. In *Reproductive Success*, ed. T.H. Clutton-Brock, pp. 227–39. Chicago: University of Chicago Press.

Cheney, D.L. & Wrangham, R.W. (1987). Predation. In *Primate Societies*, ed. B.B. Smuts, D.L. Cheney, R.M. Seyfarth, R.W. Wrangham & T.T. Struhsaker, pp. 227–39. Chicago: University of Chicago Press.

Chepko-Sade, D.B. & Sade, D.S. (1979). Patterns of group splitting within matrilineal kinship groups. *Behav. Ecol. Sociobiol.* **5**, 67–86.

Chism, J. (1986). Development and mother-infant relations among captive patas monkeys. *Int. J. Primatol.* **7**, 49–82.

Chism, J. (1991). Ontogeny of behavior in humans and nonhuman primates: the search for common ground. In *Understanding Behavior: What Primate Studies Tell Us About Human Behavior*, ed. J. D. Loy & C. B. Peters, pp. 90–120. New York: Oxford University Press.

Chism, J., Rowell, T. & Olson, D. (1984). Life history patterns of female patas monkeys. In *Female Primates: Studies by Women Primatologists*, ed. M.F. Small, pp. 175–90. New York: Alan R. Liss.

Clutton-Brock, T.H. (Ed.) (1988). *Reproductive Success*. Chicago: University of Chicago Press.

Clutton-Brock, T.H. (1991). *The Evolution of Parental Care*. Princeton, NJ: Princeton University Press.

Collins, D.A. Busse, C.D. & Goodall, J. (1984). Infanticide in two populations of savannah baboons. In *Infanticide: Comparative and Evolutionary Perspectives*, ed. G. Hausfater and S.B. Hrdy, pp. 173–91. Hawthorne, NY: Aldine.

Cords, M. (1987). Forest guenons and patas monkeys. In *Primate Societies*, ed. B.B. Smuts, D.L. Cheney, R.M. Seyfarth, R.W. Wrangham & T.T. Struhsaker, pp. 98–111. Chicago:University of Chicago Press.

Cords, M. (1995). Predator vigilance costs of allogrooming in wild blue monkeys. *Behaviour*, **132**, 559–69.

Crockett, C.M. & Pope, T.R. (1993). Consequences of sex differences in dispersal for juvenile red howler monkeys. In *Juvenile Primates*, ed. M.E. Pereira & L.A. Fairbanks, pp. 104–18. New York: Oxford University Press.

deRuiter, J.R. van Hooff, J.A.R. A.M. & Scheffrahn, W. (1995). Social and genetic aspects of paternity in wild long-tailed macaques (*Macaca fascicularis*). *Behaviour* **129**, 203–24.

DeVore, I. (1963). Mother–infant relations in free-ranging baboons. In *Maternal Behavior in Mammals*, ed. H. L. Reingold, pp. 305–35. New York: John Wiley.

deWaal, F.B.M. (1993). Codevelopment of dominance relations and affiliative bonds in rhesus monkeys. In *Juvenile Primates*, ed. M. E. Pereira & L. A. Fairbanks, pp. 259–70. New York: Oxford University Press.

Dittus, W.P.J. (1988). Group fission among toque macaques as a consequence of female resource competition and environmental stress. *Anim. Behav*. **36**, 1626–45.

Dittus, W.P.J. & Ratnayeke, S.M. (1989). Individual and social behavioral responses to injury in wild toque macaques (*Macaca sinica*). *Int. J. Primatol.* **10**, 215–34.

Dunbar, R.I.M. & Dunbar, P. (1988). Maternal time budgets of gelada baboons. *Anim. Behav*. **36**, 970–80.

Eaton, G.G., Johnson, D.F., Glick, B.B. & Worlein, J.M. (1986). Japanese macaque (*Macaca fuscata*) social development: gender differences in juvenile behavior. *Primates* **27**, 141–50.

Ehardt, C. (1987). Birth-season interactions of adult female Japanese macaques (*Macaca fuscata*) without newborn infants. *Int. J. Primatol.* **8**, 245–59.

Fairbanks, L.A. (1988). Vervet monkey grandmothers: interactions with infant grandoffspring. *Int. J. Primatol.* **9**, 425–41.

Fairbanks, L.A. (1993a). Juvenile vervet monkeys: establishing relationships and practicing skills for the future. In *Juvenile Primates*, ed. M.E. Pereira & L.A. Fairbanks, pp. 211–27. New York: Oxford UniversityPress.

Fairbanks, L.A. (1993b). Risk-taking by juvenile vervet monkeys. *Behaviour* **124**, 57–72.

Fairbanks, L.A. (1993c). Loss of matriarchs inhibits reproduction of young adult daughters in vervet monkeys (*Cercopithecus aethiops sabaeus*): an experimental verification. *Am. J. Primatol.* **30**, 309.

Fairbanks, L.A. (1993d). What is a good mother? Adaptive variation in maternal behavior of primates. *Curr. Dir. Psychol. Sci.* **2**, 179–83.

Fairbanks, L.A. & McGuire, M.T. (1984). Determinants of fecundity and reproductive success in captive vervet monkeys. *Am. J. Primatol.* **7**, 27–38.
Fairbanks, L.A. & McGuire, M.T. (1985). Relationships of vervet mothers with sons and daughters from one through three years of age. *Anim. Behav.* **33**, 40–50.
Fairbanks, L.A. & McGuire, M.T. (1986). Age, reproductive value, and dominance-related behaviour in vervet monkey females: cross-generational influences on social relationships and reproduction. *Anim. Behav.* **34**, 1710–21.
Fairbanks, L.A. & McGuire, M.T. (1988). Long-term effects of early mothering behavior on responsiveness to the environment in vervet monkeys. *Develop. Psychobiol.* **21**, 711–24.
Fairbanks, L.A. & McGuire, M.T. (1993). Maternal protectiveness and response to the unfamiliar in vervet monkeys. *Am. J. Primatol.* **30**, 119–29.
Fairbanks, L.A. & McGuire, M.T. (1995). Maternal condition and the quality of maternal care in vervet monkeys. *Behaviour* **132**, 733–54.
Fairbanks, L.A. & Pereira, M.E. (1993). Juvenile primates: dimensions for future research. In *Juvenile Primates*, ed. M.E. Pereira & L. A. Fairbanks, pp. 359–66. New York: Oxford University Press.
Fedigan, L.M., Fedigan, L. Gouzoules, S., Gouzoules, H. & Koyman, N. (1986). Lifetime reproductive success in female Japanese macaques. *Folia Primatol.* **47**, 143–57.
Gouzoules, H. (1980). A description of genealogical rank changes in a troop of Japanese monkeys (*Macaca fuscata*). *Primates* **21**, 262–7.
Gouzoules, S. & Gouzoules, H. (1987). Kinship. In *Primate Societies*, ed. B. B. Smuts, D.L. Cheney, R.M. Seyfarth, R.W. Wrangham & T.T. Struhsaker, pp. 299–305. Chicago: University of Chicago Press.
Gouzoules, H., Gouzoules, S. & Marler, P. (1986). Vocal communication: a vehicle for the study of social relationships. In *The Cayo Santiago Macaques*, ed. R.G. Rawlins & M.J. Kessler, pp. 111–29. Albany, NY: SUNY Press.
Hamilton, W.J., Busse, C. & Smith, K.S. (1982). Adoption of infant orphan chacma baboons. *Anim. Behav.* **30**, 29–34.
Hammerschmidt, K., Ansorge, V., Fischer, J. & Todt, D. (1994). Dusk calling in Barbary macaques (*Macaca sylvanus*): demand for social shelter. *Am. J. Primatol.* **32**, 277–89.
Harcourt, A.H. (1987). Dominance and fertility among female primates. *J. Zool.* **213**, 471–87.
Hasegawa, T. & Hiraiwa, M. (1980). Social interactions of orphans observed in a free-ranging troop of Japanese monkeys. *Folia primatol.* **33**, 129–58.
Hauser, M.D. (1988). Variation in maternal responsiveness in free-ranging vervet monkeys: a response to infant mortality risk. *Am. Nat.* **131**, 573–87.
Hayaki, H. (1983). The social interactions of juvenile Japanese monkeys on Koshima Islet. *Primates* **24**, 139–53.
Holman, S.D. and Goy, R.W. (1988). Sexually dimorphic transitions revealed in the relationships of yearling rhesus monkeys following the birth of siblings. *Int. J. Primatol.* **9**, 113–33.
Horrocks, J.A. (1986). Life-history characteristics of a wild population of vervets (*Cercopithecus aethiops sabaeus*) in Barbados, West Indies. *Int. J. Primatol.* **7**, 31–47.
Horrocks, J.A. & Hunte, W. (1983). Maternal rank and offspring rank in vervet monkeys: an appraisal of the mechanisms of rank acquisition. *Anim. Behav.* **31**, 772–82.

Hrdy, S.B. (1977). *The Langurs of Abu*. Cambridge, MA: Cambridge University Press.

Hrdy, S.B. (1981). "Nepotists" and "Altruists": The behavior of old females among macaque and langur monkeys. In *Other Ways of Growing Old*, ed. P. T. Amoss & S. Harell, pp. 59–76, Stanford: Stanford University Press.

Hunte, W. & Horrocks, J.A. (1987). Kin and non-kin interventions in the aggressive disputes of vervet monkeys. *Behav. Ecol. Sociobiol.* **20**, 257–63.

Isbell, L.A., Cheney, D.L. & Seyfarth, R.M. (1993). Are immigrant vervet monkeys, *Cercopithecus aethiops*, at greater risk of mortality than residents? *Anim. Behav*. **45**, 729–34.

Janson, C.H. & van Schaik, C.P. (1993). Ecological risk aversion in juvenile primates: slow and steady wins the race. In *Juvenile Primates*, ed. M.E. Pereira and L. A. Fairbanks, pp. 57–74. New York: Oxford University Press.

Janus, M. (1992). Interplay between various aspects in social relationships of young rhesus monkeys: dominance, agonistic help, and affiliation. *Am. J. Primatol.* **26**, 291–308.

Jay, P. (1963). Mother-infant relations in langurs. In *Maternal Behavior in Mammals*, ed. H.L. Reingold, pp. 282–304. New York: John Wiley and Sons.

Johnson, J.A. (1987). Dominance rank in juvenile olive baboons, *Papio anubis*: the influence of gender, size, maternal rank and orphaning. *Anim. Behav*. **35**, 1694–708.

Kaplan, J.R. (1978). Fight interference and altruism in rhesus monkeys. *Am. J. phys. Anthropol.* **49**, 241–50.

Koyama, N. (1970). Changes in dominance rank and division of a wild Japanese monkey troop in Arashiyama. *Primates* **11**, 335–90.

Koyama, N. (1991). Grooming relationships in the Arashiyama group of Japanese monkeys. In *The Monkeys of Arashiyama*, ed. L.M. Fedigan & P.J. Asquith, pp. 276–85. Albany, NY: SUNY Press.

Kurland, J.A. (1977). *Kin Selection in the Japanese Monkey*, Contrib. Primatol, vol 12. Basel: S.Karger.

Lazarus, J. & Inglis, I.R. (1986). Shared and unshared parental investment, parent–offspring conflict and brood size. *Anim. Behav*. **34**, 1791–804

Lee, P.C. (1983a). Effects of parturition on the mother's relationship with older offspring. In *Primate Social Relationships*, ed. R.A. Hinde, pp. 134–9. Sunderland, MA.: Sinauer

Lee, P.C. (1983b). Effects of the loss of the mother on social development. In *Primate Social Relationships*, ed. R.A. Hinde, pp. 73–9. Sunderland, MA: Sinauer

Maestripieri, D. (1993). Vigilance costs of allogrooming in macaque mothers. *Am. Nat*. **141**, 744–53.

Massey, A. (1977). Agonistic aids and kinship in a group of pigtail macaques. *Behav. Ecol. Sociobiol.* **2**, 31–40.

Menard, N. & Vallet, D. (1993). Dynamics of fission in a wild Barbary macaque group (*Macaca sylvanus*). *Int. J. Primatol.* **14**, 479–500.

Missakian, E.A. (1974). Mother–offspring grooming relations in rhesus monkeys. *Arch. Sexual Behav*. **3**, 135–41.

Moore, J. (1984). Female transfer in primates. *Int. J. Primatol.* **5**, 537–89.

Mori, A., Watanabe, K. & Yamaguchi, N. (1989). Longitudinal changes of dominance rank among the females of the Koshima group of Japanese monkeys. *Primates* **30**, 147–73.

Nakamichi, M. (1989). Sex differences in social development during the first 4 years in a free-ranging group of Japanese monkeys, *Macaca fuscata*. *Anim. Behav*. **38**, 737–48.

Nakamichi, M. (1991). Behavior of old females: comparisons of Japanese monkeys in the Arashiyama East and West groups. In *The Monkeys of Arashiyama*, ed. L. M. Fedigan & P. J. Asquith, pp. 175–93. Albany NY, SUNY Press.

Newton, P.N. and Dunbar, R.I.M. (1994). Colobine monkey society. In *Colobine Monkeys*, ed. A. G. Davies & J. F. Oates, pp 311–46, Cambridge: Cambridge University Press.

Nicolson, N.A. (1987). Infants, mothers, and other females. In *Primate Societies*, ed. B.B. Smuts, D.L. Cheney, R.M. Seyfarth, R.W. Wrangham & T.T. Struhsaker, pp. 330–42. Chicago: University of Chicago Press.

Paul, A. & Kuester, J. (1987). Dominance, kinship and reproductive value in female Barbary macaques (*Macaca sylvanus*) at Affenberg Salem. *Behav. Ecol. Sociobiol*. **21**, 323–31.

Pereira, M.E. (1988). Effects of age and sex on intra-group spacing behaviour in juvenile savannah baboons, *Papio cynocephalus cynocephalus*. *Anim. Behav*. **36**, 184–204.

Pereira, M.E. (1989). Agonistic interactions of juvenile savanna baboons. II. Agonistic support and rank acquisition. *Ethology* **80**, 152–71.

Pereira, M.E. (1992). The development of dominance relations before puberty in cercopithecine societies. In *Aggression and Peacefulness in Humans and Other Primates*, ed. J. Silverberg & P. Gray, pp. 117–49. New York: Oxford University Press.

Pereira, M.E. & Altmann, J. (1985). Development of social behavior in free-living nonhuman primates. In *Nonhuman Primate Models for Human Growth and Development*, ed. E.S. Watts, pp. 217–309. New York: Liss.

Pryce, C., Martin, R. & Skuse, D. (1995). *Motherhood in Human and Nonhuman Primates: Biological and Social Determinants*. Basel: Karger

Rajpurohit, L.S. & Sommer, V. (1993). Juvenile male emigration rom natal one-male troops in Hanuman langurs. In *Juvenile Primates*, ed. M.E. Pereira & L.A. Fairbanks, pp. 86–103. New York: Oxford University Press.

Ransom, T.W. & Rowell, T.E. (1972). Early social development of feral baboons. In *Primate Socialization*, ed. F. E. Poirier, pp. 105–44. New York: Random House.

Rhine, R.J. (1975). The order of movement of yellow baboons (*Papio cynocephalus*) *Folia primatol*. **23**, 73–104.

Rhine, R.J., Norton, G.W., Roertgen, W.J. & Klein, H.D. (1980). The brief survival of free-ranging baboon infants (*Papio cynocephalus*) after separation from their mothers. *Int. J. Primatol*. **4**, 401–9.

Rhine, R.J., Wasser, S.K. & Norton, G.W. (1988). Eight-year study of social and ecological correlates of mortality among immature baboons of Mikumi National Park, Tanzania. *Am. J. Primatol*. **16**, 199–212.

Rowell, T.E. & Chism, J. (1986). The ontogeny of sex differences in the behavior of patas monkeys. *Int. J. Primatol*. **7**, 83–107.

Sade, D.S. (1967). Determinants of dominance in a group of free-ranging rhesus monkeys. In *Social Communication among Primates*, ed. S.A. Altmann, pp. 99–114. Chicago: University of Chicago Press.

Sade, D.S. (1972). A longitudinal study of social behavior of rhesus monkeys. In *The Functional and Evolutionary Biology of Primates*, ed. R. Tuttle, pp. 378–98. Chicago: Aldine-Atherton.

Samuels, A. Silk, J.B. & Altmann, J. (1987). Continuity and change in dominance relations among female baboons. *Anim. Behav.* **35**, 785–93.
Silk, J.B. (1987). Social behavior in evolutionary perspective. In *Primate Societies*, ed. B.B. Smuts, D.L. Cheney, R.M. Seyfarth, R.W. Wrangham & T.T. Struhsaker, pp. 330–42. Chicago: University of Chicago Press.
Silk, J.B., Samuels, A. & Rodman, P. (1981). The influence of kinship, rank, and sex on affiliation and aggression between adult female and immature bonnet macaques (*Macaca radiata*). *Behaviour* **78**, 111–37.
Small, M.F. & Smith, D.G. (1981). Interactions with infants by full siblings, paternal half-siblings, and nonrelatives in a captive group of rhesus macaques (*Macaca mulatta*). *Am. J. Primatol.* **1**, 91–4.
Smith, D.G. (1995). Male dominance and reproductive success in a captive group of rhesus macaques (*Macaca mulatta*). *Behaviour* **129**, 225–42.
Smuts, B.B. & Nicolson, N. (1989). Reproduction in wild female olive baboons. *Am. J. Primatol.* **19**, 229–46.
Stammbach, E. (1987). Desert, forest and montane baboons: multilevel societies. In *Primate Societies*, ed. B.B. Smuts, D.L. Cheney, R.M. Seyfarth, R.W. Wrangham and T.T. Struhsaker, pp. 112–20, Chicago: University of Chicago Press.
Stanford, C.B. (1995). The influence of chimpanzee predation on group size and anti-predator behaviour in red colobus monkeys. *Anim. Behav.* **49**, 577–87.
Stearns, S.C. (1993). *The Evolution of Life Histories*. Oxford: Oxford University Press.
Struhsaker, T.T. (1971). Social behavior of mother and infant vervet monkeys (*Cercopithecus aethiops*). *Anim. Behav.* **19**, 233–50.
Struhsaker, T.T. & Leland, L. (1987). Colobine infanticide. In *Primate Societies*, ed. B.B. Smuts, D.L. Cheney, R.M. Seyfarth, R.W. Wrangham & T.T. Struhsaker, pp. 83–97. Chicago: University of Chicago Press.
Trivers, R.L. (1972). Parental investment and sexual selection. In *Sexual Selection and the Descent of Man 1871–1971*, ed. B. G. Campbell, pp. 136–79. Chicago: Aldine.
van Noordwijk, M.A., Hemelrijk, C.K., Herremans, L.A.M. & Streck, E.H.M. (1993). Spatial position and behavioral sex differences in juvenile long-tailed macaques. In *Juvenile Primates*, ed. M.E. Pereira & L.A. Fairbanks, pp. 77–85. New York: Oxford University Press.
van Schaik, C.P. (1983). Why are diurnal primates living in groups? *Behav.* **87**, 120–44.
van Schaik, C. P. (1989). The ecology of social relationships amongst female primates. In *Comparative Socioecology*, ed. V. Standon & R. A. Foley, pp. 195–218. Oxford: Blackwell.
Walters, J. (1980). Interventions and the development of dominance relationships in female baboons. *Folia primatol.* **34**, 61–89.
Walters, J. (1981). Inferring kinship from behaviour: maternity determinations in yellow baboons. *Anim. Behav.* **29**, 126–36.
Watts, D.P. & Pusey, A.E. (1993). Behavior of juvenile and adolescent great apes. In *Juvenile Primates*, ed. M.E. Pereira & L.A. Fairbanks, pp. 148–67. New York: Oxford University Press.
Winkler, P., Loch, H., Vogel, C. (1984). Life history of Hanuman langurs (*Presbytis entellus*). *Folia primatol.* **43**, 1–23.

14

Cognitive capacities of Old World monkeys based on studies of social behavior

IRWIN S. BERNSTEIN

Introduction

Psychology is defined as the study of the mind (Psyche), and comparative psychology as the study of animal minds. Comparative psychology began by testing animals individually on tasks presumed to reflect general cognitive abilities, or a general intelligence factor, and taxa were compared by their achievements on these tasks. Ethologists were less interested in capacities and more interested in species-typical behavior.

Having been schooled in both traditions, I was quick to accept Mason's suggestions in 1960 that the study of primates living in groups would yield more insight into their cognitive capacities than would individual testing. Whereas Chance and Mead (1952) and Jolly (1966), theorized that primate intelligence evolved in part because of the selective pressures of group living, it was a quarter of a century later that Cheney and Seyfarth (1985a,b) suggested that primate intelligence is extraordinary only in the social domain. Today, investigators regularly examine social interactions and their functions in a search for evidence of extraordinary cognitive capacities in nonhuman primates.

As early as 1880, Lauder hypothesized the existence of infanticide, reciprocity, and teaching in animals. He argued that animals understand death, express anger, jealousy, rage and revenge, commit suicide, suffer from moral insanity, and recognize and respond to criminal behavior. We may describe such writings as quaint, but similar terms, arguments, and supporting evidence are used in recent accounts of the abilities of nonhuman primates. These are often presented against a background of sophisticated evolutionary theory, but the evidence is often distressingly sparse.

Attempts to identify the biological basis for human behavior often link nonhuman primate behavior with complex human activities, but the claim of genetic contributions precludes the assumption of analogy. Ellis (1986) looks for criminal behavior in animals and indicates that rape, child abuse, spouse abuse, infanticide, and even murder take place. He does not argue solely on functional outcomes, but claims that monkeys have murderous intentions. Van Lawick-Goodall (1973) writes that chimpanzees (*Pan*) hunt cooperatively. Although several may pursue the same prey animal, and some meat sharing among hunters and non-participants takes place, there is little evidence of a division of labor or organization in these "cooperative" hunts. The use of terms like "warfare" to describe intertroop fighting must be taken strictly as an analogy.

This chapter will review some of the current evidence concerning the cognitive abilities of nonhuman primates.

Sociobiological theory

Reproductive strategies

Sociobiological theory has focused attention on reproductive strategies. Male primates are said to use a variety of means to assess the probability that a female's offspring is theirs, and females are said to deceive a male into believing that he is the father of her offspring, in order to get him to invest in it (Lancaster, 1984a,b). Males are said to estimate the probability of their being the father of an infant and regulate their behavior accordingly (van Noordwijk and van Schaik, 1988). Hector *et al.* (1989) show that male vervet monkeys (*Cercopithecus aethiops*) affiliate more with infants that they could have fathered, and females are said to select fathers for their offspring.

Is this evidence that nonhuman primates understand that infants have fathers? Hauser (1986) suggests that males calculate the percentage of copulations that a female had with them to estimate the probability of paternity. Even if only metaphor, this implies a prodigious memory, an enormous ability to track a female's behavior and some computational skill. Jolly (1985) states that "males that consort with particular females may have a high probability of knowing their own young" (p. 327), and "pregnant langurs (*Presbytis entellus*) may be particularly prone to mate with new harem lords, which may fool the male into accepting the subsequent offspring as his own." (p. 264). Anderson (1992) indicates that a male baboon (*Papio*), witnessing a female copulate with another male, gives the subsequent infant less care.

All of this implies long-term memory and distal association. Moreover, it implies that a male knows that he is a father and that sexual activity relates to paternity. Of course, one may argue that this is only metaphor and that the behavior was selected to serve the described functions. Even so, the proximate mechanisms would require considerable cognitive ability. In order to favor offspring that he might have sired, a male would need to be able to recognize a female he had copulated with one gestation period ago plus her infant's age, and distinguish her from females that he had copulated with either too recently, or too long previously, to have fathered their current offspring.

The use of certain patterns of courtship behavior by males, and the proposed arrangements for exchange of services, is also often couched in terms suggesting conscious awareness. Carrying of infants by male tamarins (*Saguinus*) is interpreted to be "aimed at demonstrating their competence in infant care to the female to encourage her to accept them as mates" (Price, 1990). A similar theory is presented by Smuts (1987), who writes that female baboons mate with males "in exchange for protection for themselves and their offspring against other males." (p. 395) and that a male may "signal the female that he will take good care of her infant if she mates with him" (p. 396). These authors may be describing functional outcomes of behavior, but their language connotes proximal cause and cognitive intentionality.

Infanticide

Male protection seems to be important in species in which male reproductive strategies include infanticide, and infanticide has been described to serve a number of social as well as biological functions. For some authors, infanticide signifies intentional behavior. For example, Sommer (1987:189) mentions "infanticidal intentions of males", and these are recognized by the females, in whom "expectation of infanticide" is said to induce abortion. According to Itani (1982), infanticide is used by chimpanzees to eliminate infants produced incestuously, and as an intentional and biologically sophisticated strategy by male macaques (*Macaca*) to return a female to receptivity – "males may know that the loss of the infant is followed by physiological changes in the female who had been nursing her infant, causing her return to sexual arousal."(p. 362). Paul and Kuester (1988:224) agree that incest avoidance may be the proximal cause of behavior. Nishida and Kawanaka (1985:274), however, suggest that infanticide by chimpanzees may be used to punish or correct a female's reproductive behavior – "the infanticidal male might have reason to suspect the paternity of her infant." Takahata (1985:168) elaborates on this, "may have awakened the

M group males' suspicion towards the infant's paternity, then the males may have fostered the erroneous idea that the infant had been sired by the males of other groups, or may have wanted to remove the slightest possibility that infants sired by the males of other groups would remain in their own unit group." Hamai *et al.* (1992) see infanticide as a form of male–male competition, reporting that adult male chimpanzees killed a female's infant because she had mated primarily with subadult males. This infanticide was unusual in that the males carried the infant ventrally, competed for the infant, and the infant later died as they were eating it. Newton (1987) suggests that infanticide may serve to weaken the position of a resident male, thereby making deposing him easier and Rijksen (1981) writes that infanticide may serve to improve a male's dominance.

These authors may be stressing functional outcomes without any attempt to describe proximal mechanism, but when Pirta and Singh (1981) say that a male monkey may kill an infant in order to get a female to follow him, they explain that his increased reproductive success is his reinforcement for infanticide. "Reinforcement" must be being used in a different sense than in discussions of learning, where reinforcement is said to increase the probability of the response in an individual in the future. It is not unusual for authors to equate function with evolutionary cause, but in writing about infanticide many authors (e.g. Caljan *et al.*, 1987) totally confound function and proximal cause. Even the existence of infanticide is sometimes assumed from the advantages that might in theory accrue to animals showing the behavior.

Cognitive aggression?

Aggression, lethal and non-lethal, is reported to serve a variety of biological and social functions when directed towards females. Wrangham (1979) reports that male chimpanzees attack females to show their superior defensive ability against other males. One may, therefore, ask if a female then mates with the male who injures her most severely? Fossey (1981) suggests that a young gorilla (*Gorilla*) male killed an old female to improve his status and genetic fitness, but exactly how is unclear. Itani (1982) states that males may use aggression against a female to intensify her sexual arousal, but again the mechanism is unclear.

Wasser and Starling (1986:352) suggest true Machiavellian intelligence in female monkeys of high ranking families, who incite aggression among the members of lower ranking families by provoking intrafamilial redirection – "females appeared to instigate target females of the next lowest matriline by deliberately initiating such attacks on targets when the target female was in close proximity with kin." Harcourt (1992) believes that only primates have the special intellectual abilities required to plan and execute patterns that actively block the formation of alliances among subordinates. This special skill of apes and monkeys extends to the selection of partners – "primates deliberately try to establish supportive relationships with, for example, powerful potential allies" (Harcourt, 1989:223). Harcourt (1989:224) draws parallels to human practices – "giving and exchanging of services in a manner very reminiscent of exchange of goods or services in human societies." The cognitive abilities implied are explicit because "primates are capable of such second-order social calculations", in that "the aim of such indirect aggression is presumably to inhibit future aggression" and, with reference to supporting the juvenile offspring of high ranking females, the "aim of indirect support is to promote future support" (Harcourt, 1989:233).

Reciprocity

There are widespread reports of reciprocity and the exchange of services and favors among nonhuman primates. Wrangham (1993) writes that bonobos (*Pan paniscus*) exchange sexual behavior for favors in other currencies. His explanation suggests that the animals are acting in a goal-directed manner in order to achieve the biological benefits, rather than due to any immediate hedonic feedback.

Seyfarth (1977) presents the most widely examined hypothesis of the exchange of services. He theorizes that nonhuman primates compete to groom the highest ranking females in order to obtain later agonistic support from the most powerful allies. This is further elaborated in Cheney *et al.* (1986), where reciprocation of aiding for grooming is used as one example of intelligence in the social domain. A number of anecdotes suggesting deception are also presented, but the authors call for rigorous field tests of these hypotheses.

Cheney *et al.* (1987) state that monkeys trade grooming for aid, and Moore (1984) claims that monkeys consciously use grooming to form useful alliances. Van Hooff and van Schaik (1992) do not speak of conscious intent, but report the regular exchange of services such as grooming. Vogel (1985) interjects an element of learning in this exchange, stating that monkeys ally with adjacently ranked partners because they learn that higher ranking partners do not reciprocate as well as lower ranking partners. A slight variation on the exchange theme is reported by Ransom and

Ransom (1971), who write that baboons will steal an infant in order to force the mother to groom them.

Grooming, which Tinklepaugh (1931) called "fur picking as an act of adornment," has been analyzed in the search for evidence of exchanges of services and favors in something other than a straight reciprocation or mutualistic relationship. Although this is hotly contested, and although grooming is often confounded with variables such as long-term association patterns (kinship) and other generally positively affiliated associations, some authors believe that grooming patterns support hypotheses of trading of services or reciprocal altruism. Noe (1990) concludes that none of the proposers had data to support their theories, and uses veto game theory to show that such complex reciprocity is unlikely to result from natural selection.

Complex exchanges and reciprocation connote some degree of sophisticated memory and cognitive skills, including a system of accounting. Cheney and Seyfarth (1985a:188) state "monkeys are readily able to deduce a dominance hierarchy among conspecifics from their observation of dyadic interactions." Further, "within the social group, the behavior of monkeys suggests an understanding of causality, transitive inference and the notion of reciprocity" (Cheney and Seyfarth, 1985a:197); and "primates seem to both remember past interactions and to adjust their cooperative acts depending on who has previously behaved affinitively towards them" (p. 195). Similarly, they "appear to be able to remember who has behaved altruistically towards them in the past." (Cheney and Seyfarth, 1985b:152). To understand "altruism", the monkeys would have to appreciate not only the benefits that they received, but also the costs experienced by the donor. In later writings, Cheney and Seyfarth (1990a) adopt a more conservative stance, denying any ability on the part of a monkey to appreciate how another monkey experiences the world. They do, however, reaffirm their belief that monkeys keep track of reciprocity.

De Waal (1992) writes that chimpanzees keep mental records of exchanges in maintaining reciprocity. Other authors extend this bookkeeping ability to Old World monkeys. Johnson (1989) states that baboons know relative rank differences, and Cheney and Seyfarth (1982:520) conclude that a monkey "recognizes the precise rank which each other individual holds" as well as "exact ranks and degrees of relatedness." A "precise rank" signifies an ability to count, and "degrees of relatedness" suggests at least ordinal skill. Another type of numerical ability is implied by Silk *et al.* (1981), when they write that females produce more female offspring in a year following a year when a disproportionately high number of female infants survive. They report the differences in their data to be nonsignificant, but nonetheless believe that the phenomenon exists.

Deception

Consideration of deception has provoked some of the most far-reaching discussions of cognitive abilities. Menzel (1973) implies that chimpanzees can "lie", and de Waal (1985) writes that a chimpanzee may suppress excitement, hide its own behavior, "tell on" another, feign interest, hide its own face to mask involuntary facial expressions, and deceive by ignoring. Cheney and Wrangham (1987) report that monkeys practice deception by silence, while Byrne and Whiten (1985) state that monkeys use false alarms and other deceptive tactics.

The evidence for lying and deception in these cases is variable. Menzel (1973) presents systematic studies of false signaling according to context, and de Waal (1985) reports a series of observations that are less compelling. The evidence in the case of monkeys is either strictly anecdotal (Byrne and Whiten, 1987) or a functional description of the consequence of behavior or the absence of a signal.

The argument is not whether monkeys are sometimes misled by others, but whether the monkeys are actively deceiving others. Byrne and Whiten (1987) suggest, from the results of their questionnaire survey, that deliberate deception is a widespread social tactic in monkeys and apes. They conclude that conscious intention, rather than learned association of outcome with action, is the more likely explanation. While interpreting the anecdotes in this framework, Byrne and Whiten acknowledge that what they call "kill-joy" explanations are also possible. Hauser (1992) reports that monkeys who deceive others, by not giving food calls on finding food, are punished for this deception by other group members. Punishing a deceptive animal upon detecting the deception suggests that an element of conscious awareness is involved. Presumably, the animal that attempted the deception is expected to learn not to attempt to do so again.

The accumulation of anecdotes suggesting active conscious deception on the part of Old World monkeys and apes generated a ready enthusiasm to hypothesize the existence of the underlying cognitive processes and awareness. The plural of anecdote, however, is not data. Experimental verification of awareness and intentional deception on the part of monkeys was not forthcoming and Whiten (1993) later decided that the monkey data were inconclusive.

Deception, in an active sense, implies that a monkey knows what another monkey knows, and how it knows it. The fact that a monkey is deceived by the action of another is not proof of intentional deception, any more than camouflage coloration or stealthy hunting patterns reflect an intention to hide. Ristau (1986) tries to equate intention and thinking with selective response and preception, but this confuses the issue. Producing one of several possible responses in a situation indicates a process of choice, but not necessarily one involving knowledge of how a response produces that consequence. Perception involves processing and interpretation of sensory data, but not necessarily thinking, conscious awareness, or goal-seeking.

Perceptual processes may involve identification of an object, but it is often difficult to know how the object is being identified and what class concepts exist for the animal doing the identification. Perceptual processing into cognitive classes implies the existence of the concepts, and demonstration of the existence of concepts in an animal requires carefully designed studies. The level of cognitive classification is often unclear.

Learning, cognition and science

The existence of a large cognitive component in primate behavior is undeniable. The variability of behavior, within and between taxa, and the flexibility of individual behavior preclude any assumption of rigidly preprogrammed behavior. Learning and problem-solving abilities, however, are sometimes described in terms loaded with excess meaning. Terms such as "strategies" and "tactics" suggest that nonhuman primates are anticipating and seeking functional consequences with adaptive value. Some insist that such usage is only metaphorical, while for others it correctly implies that a degree of cognitive awareness and intentionality is involved in learning to perform responses selectively as a function of differential consequences. This mixture of evolutionary selective pressures operating to select cognitive processing abilities, based on the specific consequences produced, and learning to respond to a situation with responses that have been reinforced in the past, runs the risk of seeing a consequence in the future as the proximate cause of behavior leading to that future outcome. The confusion of function and proximal cause is teleological.

Learning need not be teleological if one identifies the stimuli to which an animal responds, the responses elicited, and the resultant consequences and their associations. Rescorla (1987) argues that animals learn to associate a stimulus with a response and a response with its usual consequence. Stimuli are associated with consequences, in that the presence of the stimulus indicates the potential of a particular response to produce the consequence with which it is associated – a form of classical conditioning. It is important to note that the association is with the usual consequence that occurred in the past. Observing the consequence of a particular response does not mean that the animal was seeking that consequence. I may have suffered a flat tire driving to my office this morning, and the flat tire was a consequence of my driving, but I had no intention nor any expectation of having to change a tire when I set out to work this morning.

Assessing animal intentions and expectations (associations) remains a difficult task, that is not eliminated by observing the functional consequences of behavior. Moreover, an animal is likely to associate a response with an immediate consequence, such as a pleasurable or noxious sensations, rather than with distant and more biologically meaningful events such as pregnancies, nutritional states, or infant survival. Natural selection acts by way of these consequences, but individuals engage in behavior based on immediate stimuli and the usual hedonic consequences of their responses. For example, natural selection favors individuals that breed in the right season, regardless of the cue to which they respond. One must avoid explaining seasonal breeding by the successful production of offspring in the future. It is especially misleading to imply proximate causality in terms such as "in order to", when describing functional outcomes.

Imitation

Primates are noted for their ability to make new associations and modify their behavior based on past experiences. Mechanisms range from simple individual trial-and-error learning and conditioning, through several complex layers including various forms of social learning. Watson (1908) reported imitation in monkeys, and this was considered well-established by Warden and Jackson (1935). Cambefort (1981) presented evidence that monkeys learn about bitter tastes by watching others. Colvin (1985) believed that monkeys learn about social relationships through observational learning. Chamove (1980) wrote that monkeys learned aggression by imitation. Berman (1990) indicated that the rate at which female monkeys reject their infants was a consequence of imitation of their own mothers' rejection rate of themselves when they were infants. Imitation was said to be important in rank acquisition where an infant imitating its mother's behavior acquires its mother's rank (Johnson, 1985). Walters (1980) suggested that infants and group members remember an infant's mother's rank at its birth, and aid the infant in its attempts to acquire that rank in adolescence, even if the mother had subsequently lost rank or died.

The range of abilities described as imitation extends from behavior such as avoiding an object or location when a conspecific indicates distress at that location or on contact with that object, to processes requiring memory of events occurring as long as a lifetime ago. It also includes extrapolation of relationships between others to oneself, which implies using another as a representation of self, and as a model for one's own behavior.

The transmission of information between monkeys by imitation is assumed in the folk saying "Monkey see, monkey do". Imitation, however, must be differentiated from social facilitation, seen in almost every social animal, local site enhancement (directing attention at the location of the activities of another), mimicking, and learning to perform a specific motor act by modeling one's own actions on another's action. Imitation can be viewed as the modeling of one's own motor action on another individual's motor action, to achieve a goal. No evidence for imitation, in this last sense, is found in New or Old World monkeys (Visalberghi and Fragaszy, 1990; Cheney and Seyfarth, 1990b). A clear interest in the activities of others is apparent, and attention is often directed to the same object or to the site of anothers' activity, but motor patterns are not reproduced. Mason and Mendoza (1993) indicate that the existence of observational learning is not demonstrated. Beck (1978) concludes that it is rare, and Strayer (1976) believes that dominance may preclude observational learning even where it is possible. Whiten (1989) clearly distinguishes problem-solving, following attention directed towards a stimulus, from imitation and finds little evidence for the latter. Jouventin *et al.* (1977), present clear evidence of avoidance of an object or location by monkeys after witnessing another monkey's negative response to it. This, however, is not imitation. The widespread notion that monkeys must learn much of their behavior by modeling after other monkeys seems less likely following Milton's (1993) description of typical social behavior, social organization and complex social interactions in a spider monkey troop derived from a collection of juveniles that matured in the total absence of adults and without any possibility of modeling.

Teaching

Whereas discussions of social learning focus on the acquisition of information by a subject, discussions of the model from which information is obtained often involve the concept of "teaching". Teaching, as an active process, denotes that the teacher is acting in a manner contingent on whether the pupil had learned. A mother's activities may provide an infant with an opportunity to learn, but without feedback between mother's and infant's behavior, "teaching" cannot be inferred. Nicolson (1991) reviews the evidence for active teaching among monkeys and finds little to suggest that it occurs.

Caro and Hauser (1992) define teaching functionally, denying any need for intentionality or knowledge of what another knows, but they nevertheless require the teacher to distinguish naive from sophisticated pupils. Without feedback and assessment, any individual that by its behavior affords another the opportunity to learn can be considered a teacher. In this sense, a predator teaches prey and prey teach predators whenever one shows a relatively long-term change in behavior as a consequence of a particular interaction.

There are, nonetheless, reports of "teaching" by nonhuman primates. Nishida (1983) believes that chimpanzee allomothers teach immature animals what is edible. Burton (1972) posits that Barbary macaque (*M. sylvanus*) males teach infants most of the essential social skills and communication forms. Operational definitions of teaching, as with imitation, are seldom provided and empirical tests are rare.

Assumptions of cognitive abilities

Seyfarth (1987) suggests that monkeys deceive, and that kinship and rank exist in an animal's mind, but finding a means to test these hypotheses has been extraordinarily difficult. Reynolds (1986) suggests intentionality in chimpanzees and deduces that they think, worry, are stressed, and act "in order to." There have been few attempts to test such abilities, even in chimpanzees. It will be even more difficult to demonstrate "ostracism" – a "socially induced exclusion from vital resources necessary for life and reproduction" (Lancaster 1986:68); or the "process of individual 'decisions' and 'strategies' to optimize personal reproductive success." (p. 67) in nonhuman primates.

Impressive social mechanisms are suggested by Weisbard and Goy (1976), who claim that a female monkey receives reinforcement from the group for providing good care for her infant. A more elaborate group potential is suggested by Itani's (1982:368) discussion of chimpanzee cannibalism, "Or did it signify the acceptance of the fact of killing? Or was it a communal feast to eliminate the fact of the act from group memory?" O'Hara (1985) indicates that a mystical mechanism is described by Watson

(1979), who is quoted as writing "Let us say, for argument's sake that the number (of potato washers) was 99 and that . . . one further convert was added . . . (this) carried the number across some sort of threshold . . . because by that evening . . . the habit seems to have jumped natural barriers and to have appeared spontaneously . . . in colonies on other islands and on the mainland".

Scientific logic

Compared to the above, inferring grief from the carrying of a dead infant (Nicolson, 1991) seems conservative. The over-interpretation of anecdotes, however, has been widespread. Whereas all good science begins with anecdotes, the process does not end there. Anecdotes are used to generate hypotheses that then must be tested against new data. Just because a hypothesis or theory could explain an observation does not mean it is true. I can explain how a boy hits an oncoming ball with a bat in terms of mechanics, trigonometry, and algebra involving estimating the velocity of the ball, deceleration due to air friction, gravitational attraction, distance, the acceleration force on the bat, and the distance and time to the point of impact. Small boys, however, bat balls without any knowledge of physics and mathematics. Accepting an explanation because it explains an event is called the "Error of Affirming the Consequent".

Science requires hypotheses to be falsifiable, and identification of the null may be a problem. De Waal (1991) defends attribution of human thought processes to chimpanzees, arguing that, since chimpanzees are so much like humans, there is no reason to discard the null hypothesis of no mental differences. He believes that chimpanzees keep mental records but indicates that anecdotal evidence must be restated as testable hypotheses. It is not obvious that the null hypothesis of "no difference from humans" is more appropriate than the null hypothesis of "no ability in the chimpanzee can be said to exist until you demonstrate that a chimpanzee's behavior can be accounted for only by the existence of such an ability". Moreover, failing to reject a null that chimpanzees and humans have identical mental abilities does not prove the hypothesis that they do.

Mirrors, awareness and intentions

Although a difference between apes and monkeys is generally acknowledged, there is no general agreement about the nature of the difference. Gallup's (1970) work on mirror recognition provides some evidence of self-referencing in chimpanzees and other great apes, but such tests generally fail to find such evidence in Old World monkeys. This ability is not universal in chimpanzees (Povinelli *et al.*, 1993), but this ability would make it possible for a chimpanzee to differentiate itself from others. This differentiation would permit the development of a "theory of mind", i.e. the idea that what one knows, another may not, and that this difference may be due to how information is obtained (Premack, 1983).

Although there is still some debate about the significance of the mirror test, and exactly what monkeys can and cannot do, Povinelli *et al.* (1991) present data that indicate that monkeys do not differentiate between informants who could and could not possibly know the location of food. Menzel (1991) indicates that monkeys, like young human infants, are more likely to search for food where they last found it, than where they last saw it hidden. Cheney and Seyfarth (1990a,b) have adopted a much more cautious stance than previously, and state that monkeys do not know what they know, nor what others know and therefore cannot tell the difference. Infants attend to mother's orientation and attention, but do not know what mother knows and do not imitate, nor do mothers teach.

Although Cheney and Seyfarth take the position that monkeys cannot tell what another monkey knows and, presumably, do not know how another monkey knows things, they believe that alarm calls (presumed to be elicited as a response to a specific stimulus or stimulus class) are deliberately withheld as an act of deception (Cheney and Seyfarth, 1990a). Alarm calls responding to a specific stimulus may be emitted, or not, according to the social context. It is more parsimonious, however, to interpret the failure to signal as an example of "audience effect" (in which the presence of a specific class of other is required before a stimulus elicits a response in the subject), rather than as intentionally deceptive suppression of a response. Pigtailed macaques (*M. nemestrina*) respond to the same stimuli differently according to the social context (Bernstein, 1966). A "control animal" either approached a threatening human intruder or fled, depending upon whether the social group was present or absent. Group members fled the human intruder or ran towards the control animal, depending on whether the control animal was free or confined to a small cage in a fixed location. A difference in response to stimuli as a function of social context is to be expected, and implies no intention to deceive.

Proximal cause and function

Much of the argument about Old World monkey cognitive capacities may stem from differences in terminology and the use of metaphor. Writers risk confusing proximal cause with function when they state that a monkey performs an act *because of* its adaptive function, which implies at least that a knowledge of the function explains the animal's motivation. This problem led to a plea by Sugiyama (1987) for investigators to study proximal causes instead of theorizing about function. If they did so, authors would not state that, as a male monkey ages, he seeks out older female partners and does not compete for female kin *in order to* avoid incest (Simpson and Barton, 1992). Kummer *et al.* (1990) clearly point to the difficulties that arise when behavior is named by its function, especially the implication that behavior is consciously directed toward the achievement of such functional consequences. Eddy *et al.* (1993) say that this focus on intent reveals a basic anthropomorphism; when behavior seems structurally similar in a monkey or ape and a human subject, authors assume that the nonhuman primate has the same intentions as the human performer. Mason (1986) declares that the problem is deeper, in that descriptions of behavior are incorrigibly mentalistic and treat the actor as agent with motives, goals, perceptions, and memories.

The combination of mentalistic terminology and a functional focus can lead to implications about cognitive abilities that would be startling, even in chimpanzees (Hasegawa, 1989). Descriptions often confuse wanting and knowing with working towards a goal and learning (Mason, 1982). A particular behavior pattern may improve an individual's Darwinian fitness, and thus be favored by natural selection. But there is no reason to deduce, from the fact that individuals perform the advantageous behavior pattern, that they know about natural selection and the consequences for their fitness. Although cercopithecoid monkeys learn from the consequences of their behavior, they cannot learn from its effect upon their lifetime reproductive success, nor is there any reason to believe they could assess such an effect.

Limited knowledge

Welker and Schaeffer-Witt (1992) agree that primates are smart and that intelligence may be heritable, but state that primate intelligence is expressed in learning when to attack, when to avoid, and when to submit. Although humans may be claimed to be especially intelligent, other primates show levels of sociality and intelligence well within the range of other mammals. Eisenberg (1973) exemplifies scientists who dispute any primate uniqueness in social complexity, in displays, or in the means used to transmit information.

Cheney and Seyfarth (1990b) and Dasser (1988) clearly demonstrate that group members associate an infant with its mother, but this association does not imply that monkeys understand the complex concept of kinship (Bernstein, 1991), either in the sense of genes in common or descent from a common ancestor. Wu *et al.* (1980) suggest some type of kinship recognition system in monkeys independent of experience, but efforts at replication indicate that monkeys respond preferentially to familiar associates regardless of their degree of relatedness or recency of an ancestor in common (Fredrickson and Sackett, 1984; Sackett and Fredrickson, 1987).

Theories concerning paternity have also made unwarranted assumptions about a knowledge of kinship and parentage. Hamilton (1984) hypothesizes that sexual swellings stimulate male–male competition, thereby increasing a male's confidence in paternity. The term "confidence" is surely not meant cognitively, and the exact mechanism is implied rather than described. Andelman (1987) develops a more elaborate theory involving male paternity confidence, concealed ovulation, infanticidal behavior, and male parental care in vervet monkeys. This complex of variables seems to explain almost any observed outcome, but the data supporting the proposed linkages are flimsy. Rhesus macaque (*M. mulatta*) females rarely show sexual swellings, but I do not agree that ovulation is "concealed" since color and behavioral and chemical changes clearly concentrate male attention at the time of ovulation.

Busse (1985) discusses "confidence in paternity", but within the context of possible functions, and he concentrates on patterns of male and female affiliation as proximate mechanisms. Burt (1992) points out that arguments about the evolution of concealed ovulation may be moot if advertisement rather than its lack is the derived condition.

The social inheritance of dominance rank is usually explained without invoking a heritable component, and the mechanism is usually described as preferential aid to daughters by age. Walters (1980) proposes a more complex system involving observational learning of ranks and long-term memory of mother's status at the birth of the subject. Chapais (1992) and Datta (1986) critique Walters' theory of observational learning and targeting, showing that ranks were attained by active aiding by a mother or older relative. Harcourt (1992), nonetheless, believes that only primates are intelligent enough to choose allies and manipulate alliances in order to achieve status.

Popp and DeVore (1979) indicate that primates fight for status (and therefore have this concept?). De Waal (1987) also believes that primates engage in intentional status striving, and that this causes fights. Mason and Mendoza (1993), however, conclude that a transactional analysis of monkey aggressive encounters fails to provide any evidence for status striving as a proximal cause of aggression.

Data supporting social cognition

The reaction against richly interpreted anecdotes and field observation as evidence of extraordinary cognitive abilities in Old World monkeys does not diminish the significance of cognitive abilities demonstrated by systematic observation and experimental techniques. Kummer and his colleagues (Kummer, 1968; Kummer *et al.*, 1978; Bachmann and Kummer, 1980) reported field experiments and observations indicating that hamadryas baboon males respond to behavioral interactions of familiar males with previously unfamiliar females as a signal not to compete for the female, regardless of dominance relationships among the males. Behavior described as "notification" is used to coordinate male behavior and modify competition among familiar males. These data are in accord with long-acknowledged abilities of Old World monkeys to recognize each individual in their social unit, remember their dominance relationship with each individual, and to recognize the usual associates of each group member. Occasional errors by immature animals and newly-immigrant adult males in recognizing dominance relationships and alliances are to be expected (Bernstein and Ehardt, 1986).

Kummer (1973) also showed that the recognition of a relationship between other baboons is extended to recognition of an association between another individual and items in its immediate vicinity. Competition for incentives is inhibited if a familiar group member is close to, and directly acting upon, an object. Because proximity and active interaction are required for "ownership" to be respected, the concept of ownership was not extrapolated to a notion of "property". Further tests stressed the active participation of both members in social relationships (Kummer, 1978). Kummer *et al.* (1978) report more extensively on this work and their experiments using cans containing food and associations of individuals with feeding sites. Sigg and Falett (1985) also found sex differences in this behavior, and Kummer and Cords (1991) found ownership to be independent of drive levels.

Conclusions

Although data are lacking to support the existence of teaching or imitation (in the strict sense) among Old World monkeys, directed attention and local site enhancement are evidently powerful tools for transmitting information. An infant whose attention is directed to what its mother is manipulating, and to where and to what she is eating, can readily learn what is edible, where to find it, and even how to process it using its own problem-solving techniques. No deliberate teaching or imitation is required. Local site enhancement can also lead to similar discoveries, as in sweet potato washing, placer mining of wheat, etc., in Japanese macaques (*M. fuscata*; Kawai, 1965). This transmission of information between generations is facilitated by the long period of dependency in Old World monkeys and prolonged generational overlap, and may meet the basic requirements for culture (Kawamura, 1959).

Attention to social interactions has included aggression and the aftermath of aggression, and complex cognitive processes are involved in the explanations of reconciliation in Old World monkeys. This pattern in rhesus monkeys was carefully operationalized and documented by De Waal and Yoshihara (1983). Similar patterns are found in patas monkeys (*Erythrocebus patas*; York and Rowell, 1988), sooty mangabeys (*Cercocebus atys*; Gust and Gordon, 1993), pigtailed macaques (*M. nemestrina*; Judge, 1991), Japanese macaques (Aureli *et al.*, 1992), long tailed macaques (*M. fascicularis*; Cords, 1992), tonkean macaques (*M. tonkeana*; Thierry, 1985), and stumptailed macaques (*M. arctoides*; De Waal and Johanowicz, 1993). The last authors demonstrate a learned component in reconciliation patterns by creating a mixed group of rhesus and stumptailed macaques. Rhesus monkeys, which when normally reared show much lower rates of reconciliation than do stumptail macaques, showed significantly higher rates of reconciliation when reared with stumptails.

Cords and Thurnheer (1993) show reconciliation to occur more frequently when partners cooperate to obtain food. Aureli *et al.* (1993), while demonstrating that operational definitions for redirection and reconciliation could be satisfied in Japanese macaques, find no evidence for consolation behavior (an increase in affiliative behavior with the victim of aggression by individuals not involved in the agonistic episode). Such consolation would have suggested empathy, and the ability to understand the difference between your own feelings and those of another. On the other hand, Aureli *et al.* (1992), present evidence that Japanese macaques recognize the usual associates of another by showing that a victim, or its associates, are likely to redirect aggression against the usual associates of an attacker.

In line with the suggestion that primate intelligence is especially well-developed in the social domain, De Waal (1989) proposes that primates are especially skillful at restoring peaceful relationships. Peacemaking may be of special importance because "Nonsocial ecological techniques are poorly developed in primates. Their specialization must be sought in the way they act as groups" (Kummer, 1971:38).

Maintenance of social relationships with a 100 or more other individuals must require enormous cognitive abilities on the part of Old World monkeys. Not only must each individual keep track of social relationships and maintain them, but it must know how to use each potential partner to solve the problems posed by the physical and social ecology within which it exists. The reliability, abilities, and personal preferences of each partner must be assessed and remembered. Moreover, whether friend or foe, it is essential to know the friends and associates of your interaction partners, their likelihood of interference and their abilities. A heavy reliance on social knowledge selects for social skills. As Schultz (1969) has suggested, the primary adaptive attributes of the primates may be that: (1) they are smart (they can modify their behavior as a function of experience); (2) they are social; and (3) they have long periods of biological dependency. From these three attributes all else may follow, including an understanding of our own complex social lives.

References

Andelman, S.J. (1987). Evolution of concealed ovulation in vervet monkeys (*Cercopithecus aethiops*). *Am. Nat.* **129**, 785–99.

Anderson, C.M. (1992). Male investment under changing conditions among chacma baboons at Siukerbosrand. *Am. J. phys. Anthrop.* **87**, 479–96.

Aureli, F., Cozzolino, R., Cordischi, C. & Scucchi, S. (1992). Kin-oriented redirection among Japanese macaques: an expression of a revenge system? *Anim. Behav.* **44**, 283–91.

Aureli, F., Veenama, H.C., van Pathaleon van Eck, C.J. & van Hooff, J.A.R.A.M. (1993). Reconciliation, consolation, and redirection in Japanese macaques (*Macaca fuscata*). *Behaviour* **124**, 1–21.

Bachmann, D. & Kummer, H. (1980). Male assessment of female choice in hamadryas baboons. *Behav. Ecol. Sociobiol.* **6**, 315–21.

Beck, B.B. (1978). Ontogeny of tool use by non-human animals. In *The Development of Behavior: Comparative and Evolutionary Aspects*, ed. G.M. Burghardt & M. Bekoff, pp. 406–19. New York: Garland STPM Press.

Berman, C.M. (1990). Intergenerational transmission of maternal rejection rates among free-ranging rhesus monkeys. *Anim. Behav.* **39**, 329–37.

Bernstein, I.S. (1966). An investigation of the organization of pigtail monkey groups through the use of challenges. *Primates* **7**, 471–80.
Bernstein, I.S. (1991). The correlation between kinship and behaviour in non-human primates. In *Kin Recognition*, ed. P.G. Hepper, pp. 6–29. Cambridge: Cambridge University Press.
Bernstein, I.S. & Ehardt, C.L. (1986). Selective interference in rhesus monkey (*Macaca mulatta*) intragroup agonistic episodes by age-sex class. *J. comp. Psychol.* **100**, 380–4.
Burt, A. (1992). "Concealed ovulation" and sexual signals in primates. *Folia primatol.* **58**, 1–6.
Burton, F.D. (1972). The integration of biology and behavior in the socialization of *Macaca sylvanus* of Gibraltar. In *Primate Socialization*, ed. F.E. Poirier, pp. 29–62. New York: Random House.
Busse, C.D. (1985). Paternity recognition in multi-male primate groups. *Am. Zool.* **25**, 873–81.
Byrne, R.W. & Whiten, A. (1985). Tactical deception of familiar individuals in baboons (*Papio ursinus*). *Anim. Behav.* **33**, 669–72.
Byrne, R. & Whiten, A. (1987). The thinking primate's guide to deception. *New Scientist* **116**, 54–7.
Caljan, V.G., Meisvili, N.V. & Vancatova, M.A. (1987). Infanticide and primate evolution. In *Behaviour As One of the Main Factors of Evolution*, ed. V. Leonovicova & V.J.A. Novak, pp. 321–30. Prague: Czechoslovak Academy of Sciences.
Cambefort, J.P. (1981). A comparative study of culturally transmitted patterns of feeding habits in the chacma baboon *Papio ursinus* and the vervet monkey *Cercopithecus aethiops*. *Folia primatol.* **36**, 243–63.
Caro, T.M. & Hauser, M.D. (1992). Is there teaching in nonhuman animals? *Q. Rev. Biol.* **67**, 151–74.
Chamove, A.S. (1980). Nongenetic induction of acquired levels of aggression. *J. abnorm. Psychol.* **89**, 469–88.
Chance, M.R.A. & Mead, A.P. (1952). Social behavior in primate evolution. In *Evolution: Symp. Evol. Soc. exper. Biol.* **7**, 395–439. Cambridge: Cambridge University Press.
Chapais, B. (1992). The role of alliances in social inheritance of rank among female primates. In *Coalitions and Alliances in Humans and Other Animals*, ed. A.H. Harcourt & F.B.M. deWaal, pp. 29–59. Oxford: Oxford University Press.
Cheney, D.L. & Seyfarth, R.M. (1982). Recognition of individuals within and between groups of free-ranging vervet monkeys. *Am. Zool.* **22**, 519–29.
Cheney, D.L. & Seyfarth, R.M. (1985a). Social and non-social knowledge in vervet monkeys. *Phil. Trans. Roy. Soc.* B **308**, 187–201.
Cheney, D.L. & Seyfarth, R.M. (1985b). Vervet monkey alarm calls: Manipulation through shared information? *Behaviour* **94**, 150–66.
Cheney, D., Seyfarth, R. & Smuts, B. (1986). Social relationships and social cognition in nonhuman primates. *Science* **234**, 1361–6.
Cheney, D.L., Seyfarth, R.M., Smuts, B.B. & Wrangham, R.W. (1987). The study of primate societies. In *Primate Societies*, ed. B.B. Smuts, D.L. Cheney, R.M. Seyfarth, R.W. Wrangham & T.T.Struhsaker, pp. 1–8. Chicago: University of Chicago Press.
Cheney, D.L. & Seyfarth, R.M. (1990a). *How Monkeys See the World. Inside the Mind of Another Species*. Chicago: University of Chicago Press.

Cheney, D.L. & Seyfarth, R.M. (1990b). Attending to behaviour versus attending to knowledge: examining monkey's attribution of mental states. *Anim. Behav.* **40**, 742–53.
Cheney, D.L. & Wrangham, R.W. (1987). Predation. In *Primate Societies*, ed. B.B. Smuts, D.L. Cheney, R.M. Seyfarth, R.W. Wrangham & T.T. Struhsaker, pp. 227–39. Chicago: University of Chicago Press.
Colvin, J.D. (1985). Breeding-season relationships of immature male rhesus monkeys with females: I. Individual differences and constraints on partner choice. *Int. J. Primatol.* **6**, 261–87.
Cords, M. (1992). Post-conflict reunions and reconciliation in long-tailed macaques. *Anim. Behav.* **44**, 57–61.
Cords, M. & Thurnheer, S. (1993). Reconciling with valuable partners by long-tailed macaques. *Ethology* **93**, 315–25.
Dasser, V. (1988). A social concept in Java monkeys. *Anim. Behav.* **36**, 225–30.
Datta, S.B. (1986). The role of alliances in the acquisition of rank. In *Primate Ontogeny, Cognition and Social Behaviour*, ed. J.G. Else & P.C. Lee, pp. 219–25. New York: Cambridge University Press.
de Waal, F.B.M. (1985). Deception in the natural communication of chimpanzees. In *Deception: Perspectives on Human and Nonhuman Deceit*, ed. R.W. Mitchell & N.S. Thompson, pp. 221–44. Albany: SUNY Press.
de Waal, F.B.M. (1987). Dynamics of social relationships. In *Primate Societies*, ed. B.B. Smuts, D.L. Cheney, R.M. Seyfarth, R.W. Wrangham & T.T. Struhsaker, pp. 421–9. Chicago: University of Chicago Press.
de Waal, F.B.M. (1989). *Peacemaking Among Primates*. Cambridge, MA: Harvard University Press.
de Waal, F.B.M. (1991). Complementary methods and convergent evidence in the study of primate social cognition. *Behaviour* **118**, 297–320.
de Waal, F.B.M. (1992). Coalitions as part of reciprocal relations in the Arnhem chimpanzee colony. In *Coalitions and Alliances in Humans and Other Animals*, ed. A.H. Harcourt & F.B.M. de Waal, pp. 233–57. Oxford: Oxford University Press.
de Waal, F.B.M. & Johanowicz, D.L. (1993). Modification of reconciliation behavior through social experience: an experiment with two macaque species. *Child Devel.* **64**, 897–908.
de Waal, F.B.M. & Yoshihara, D. (1983). Reconciliation and redirected affection in rhesus monkeys. *Behaviour* **85**, 224–41.
Eddy, T.J., Gallup, G.G. Jr. & Povinelli, D.J. (1993). Attribution of cognitive states to animals: Anthropomorphism in comparative perspective. *J. Soc. Issues* **49**, 87–101.
Eisenberg, J.F. (1973). Mammalian social systems: Are primate social systems unique? In *Symp. 4th internat. Congr. Primatol.*, vol. 1: *Precultural Primate Behavior*, ed. E.W. Menzel, pp. 232–49. Basel: S. Karger.
Ellis, L. (1986). Evolution and the nonlegal equivalent of aggressive criminal behavior. *Agress. Behav.* **12**, 57–71.
Fredrickson, W.T. & Sackett, G.P. (1984). Kin preferences in primates (*Macaca nemestrina*): Relatedness or familiarity? *J. comp. Psychol.* **98**, 29–34.
Fossey, D. (1981). The imperiled mountain gorilla: a grim struggle for survival. *Nat. Geographic* **159**, 501–23.
Gallup, D., Jr. (1970). Chimpanzees' self-recognition. *Science* **157**, 86–7.
Gust, D.A. & Gordon, T.P. (1993). Conflict resolution in sooty mangabeys. *Anim. Behav.* **46**, 685–94.

Hamai, M., Nishida, T., Takasaki, H. & Turner, L.A. (1992). New records of within-group infanticide and cannibalism in wild chimpanzees. *Primates* **33**, 151–62.

Hamilton, W.J., III (1984). Significance of paternal investment by primates to the evolution of adult male-female associations. In *Primate Paternalism*, ed. D.M. Taub, pp. 309–35. New York: Van Nostrand Reinhold Co.

Harcourt, A.H. (1989). Social Influences on Competitive Ability. Alliances and their consequences. In *Comparative Socioecology. The Behavioural Ecology of Humans and Other Mammals*, ed. V. Standen & R.A. Foley, pp. 223–42. Oxford: Blackwell Scientific Publications.

Harcourt, A.H. (1992). Coalitions and alliances: are primates more complex than non-primates? In *Coalitions and Alliances in Humans and Other Animals*, ed. A.H. Harcourt & F.B.M. deWaal, pp. 445–71. Oxford: Oxford University Press.

Hasegawa, T. (1989). Sexual behavior of immigrant and resident female chimpanzees at Mahale. In *Understanding Chimpanzees*, ed. P.G. Heltne & L.A. Marquardt, pp. 90–103. Cambridge, MA: Harvard University Press.

Hauser, M.D. (1986). Male responsiveness to infant distress calls in free-ranging vervet monkeys. *Behav. and Ecological Sociobiol.* **19**, 65–71.

Hauser, M.D. (1992). Costs of deception: Cheaters are punished in rhesus monkeys (*Macaca mulatta*). *Proc. Natl. Acad. Sci.* **89**, 12137–9.

Hector, A.C.K., Seyfarth, R.M. & Raleigh, M.J. (1989). Male parental care, female choice and the effect of an audience in vervet monkeys. *Anim. Behav.* **38**, 262–71.

Itani, J.(1982). Intraspecific killing among non-human primates. *J. Soc. Biol. Struct.* **5**, 361–8.

Johnson, J.A. (1989). Supplanting by olive baboons: dominance rank difference and resource value. *Behav. Ecol. Sociobiol.* **24**, 277–83.

Johnson, P. (1985). Comparison of maternal-infant and dyad-troop interactions in lion-tailed macaques. In *The Lion-Tailed Macaque: Status and Conservation*, ed. P.G. Heltne, pp. 265–8. New York: Liss.

Jolly, A. (1966). Lemur social behavior and primate intelligence. *Science* **153**, 501–6.

Jolly, A. (1985). *The Evolution of Primate Behavior*, 2nd edn. New York: MacMillan.

Jouventin, P., Pasteur, C. & Cambefort, J.P. (1977). Observational learning of baboons and avoidance of mimics: Exploratory tests. *Evolution* **31**, 214–18.

Judge, P.G. (1991). Dyadic and triadic reconciliation in pigtail macaques (*Macaca nemestrina*). *Am. J. Primatol.* **23**, 225–37.

Kawamura, S. (1959). The process of sub-culture propagation among Japanese macaques. *Primates* **2**, 43–60.

Kawai, M. (1965). Newly-acquired pre-cultural behavior of the natural troop of Japanese monkeys on Koshima Islet. *Primates* **6**, 1–30.

Kummer, H. (1968). *Social Organization of Hamadryas Baboons*. Chicago: University of Chicago Press.

Kummer, H. (1971). *Primate Societies. Group Techniques of Ecological Adaptation*. Chicago: Aldine-Atherton.

Kummer, H. (1973). Dominance versus possession: an experiment on hamadryas baboons. In *Symp. 4th internat. Congr. Primatol.*, vol. 1: *Pre-cultural Behavior*, ed E.W. Menzel, pp. 226–31. Basel: S. Karger.

Kummer, H. (1978). Value of social relationships to nonhuman primates: Heuristic scheme. *Soc. Sci. Informat.* **17**, 687–705.

Kummer, H. Abegglen, J.J., Bachman, C.H., Falett, J. & Sigg, H. (1978). Grooming relationship and object competition among hamadryas baboons. In *Recent Advances in Primatology*, vol 1, *Behaviour*, ed. D.J. Chivers & J. Herbert. pp. 31–8. London: Academic Press.

Kummer, H. & Cords, M. (1991). Cues of ownership in longtailed macaques (*Macaca fascicularis*). *Anim. Behav.* **42**, 529–49.

Kummer, H., Dasser, V. & Hoyningen-Huene, P. (1990). Exploring primate social cognition: some critical remarks. *Behaviour* **112**, 84–98.

Kummer, H., Götz, W. & Angst, W. (1974). Triadic differentiation: an inhibitory process protecting pair bonds in baboons. *Behaviour* **49**, 62–87.

Lancaster, J.B. (1984a). Introduction. In *Female Primates: Studies by Women Primatologists*, ed. M.F. Small, pp. 1–10. New York: Alan R. Liss Inc.

Lancaster, J.B. (1984b). Evolutionary perspectives on sex differences in the higher primates. In *Gender and Life Course*, ed. A.S. Rossi, pp. 3–27. New York: Aldine.

Lancaster, J.B. (1986). Primate social behavior and ostracism. *Ethol. Sociobiol.* **7**, 215–25.

Lauder, W. (1880). *Mind in the Lower Animals in Health and Disease*, vol. II, *Mind in Disease*, pp. 571. New York: D. Appleton and Company.

Mason, W.A. (1982). Primate social intelligence: contributions from the laboratory. In *Animal Mind-Human Mind, Report of the Dahalem Workshop Konferenzen*, ed. D.R. Griffin, pp. 131–43. Berlin: Springer-Verlag.

Mason, W.A. (1986). Behavior implies cognition. In *Integrating Scientific Disciplines: Case Studies from the Life Sciences*, ed. W. Bechtel, pp. 297–302. Dordrecht. The Netherlands: M. Nijhoff.

Mason, W.A. & Mendoza, S.P. (1993). Primate social conflict: an overview of sources, forms and consequences. In *Primate Social Conflict*, ed. W.A. Mason & S.P. Mendoza, pp. 1–11. Albany: State University of New York Press.

Menzel, C.R. (1991). Cognitive aspects of foraging in Japanese monkeys. *Anim. Behav.* **41**, 397–402.

Menzel, E.W., Jr. (1973). Leadership and communication in young chimpanzees. *In Symp. 4th internat. Congr. Primatol.*, vol. 1: *Precultural Primate Behavior*. pp. 192–225. Basel: S. Karger.

Milton, K. (1993). Diet and social organization of a free-ranging spider monkey population: the development of species-typical behavior in the absence of adults. *In Juvenile Primates: Life History, Development, and Behavior.* ed. M.E. Pereira & L.A. Fairbanks, pp. 173–81. New York: Oxford University Press.

Moore, J. (1984). Age and grooming in langur male bands (*Presbytis entellus*). In *Current Primate Researches*, ed. M.L. Roonwal, S.M. Mohnot & N.S. Rathore, pp. 381–7. Jodhpur, India: University of Jodhpur.

Newton, P.N. (1987). The social organization of forest hanuman langurs (*Presbytis entellus*). *Int. J. Primatol.* **8**, 199–232.

Nicolson, N.A.(1991). Maternal behavior in human and nonhuman primates. *In Understanding Behavior. What Primate Studies Tell Us About Human Behavior*, ed. J.D. Loy & C.B. Peters, pp. 17–50. New York: Oxford University Press.

Nishida, T. (1983). Alloparental behavior in wild chimpanzees of the Mahale Mountains, Tanzania. *Folia primatol.* **41**, 1–33.

Nishida, T. & Kawanaka, K. (1985). Within-group cannabalism by adult male chimpanzees. *Primates* **26**, 274–84.

Noe, R. (1990). A Veto game played by baboons: a challenge to the use of the Prisoner's Dilemma as a paradigm for reciprocity and cooperation. *Anim. Behav.* **39**, 78–90.

O'Hara, M. (1985). Of myths and monkeys: A critical look at a theory of critical mass. *J. humanistic Psychol.* **25**, 61–78.

Paul, A. & Kuester, J. (1988). Life-history patterns of Barbary macaques (*Macaca sylvanus*) at Affenberg Salem. In *Ecology and Behavior of Food-Enhanced Primate Groups*. ed. J.E. Fa & C.H. Southwick, pp. 199–228. New York: Alan R. Liss, Inc.

Pirta, R.S. & Singh, M. (1981). Forcible snatching and probable killing of infants by a rhesus (*Macaca mulatta*) alpha male in a wild habitat. *Behaviour Analysis Letters* **1**, 339–44.

Popp, J.L. & DeVore, I. (1979). Aggressive competition and social dominance theory: a synopsis. In *The Great Apes*, ed. D.A. Hamburg & E.R. McCown, Menlo Park, CA: Benjamin/Cummings.

Povinelli, D.J., Parks, D.A. & Novak, M.A. (1991). Do rhesus monkeys (*Macaca mulatta*) attribute knowledge and ignorance to others? *J. comp. Psychol.* **105**, 318–25.

Povinelli, D.J., Rulf, A.B., Landau, K.R. & Bierschwale, D.T. (1993). Self-recognition in chimpanzees (*Pan troglodytes*): distribution, ontogeny, and patterns of emergence. *J. comp. Psychol.* **107**, 347–72.

Premack, D. (1983). Animal cognition. *Ann. Rev. Psychol.* **34**, 351–62.

Price, E.C. (1990). Infant carrying as a courtship strategy of breeding male cotton-top tamarins. *Anim. Behav.* **40**, 784–6.

Ransom, T.W. & Ransom, B.S. (1971). Adult male-infant relations among baboons (*Papio anubis*). *Folia primatol.* **16**, 179–95.

Rescorla, R.A. (1987). A Pavlovian analysis of goal-directed behavior. *Am. Psychologist* **42**, 119–29.

Reynolds, V. (1986). Primate social thinking. In *Primate Ontogeny, Cognition and Social Behavior*, ed. J.G. Else & P.C. Lee, pp. 53–60. New York: Cambridge University Press.

Rijksen, H.D. (1981). Infant killing: A possible consequence of a disputed leader role. *Behaviour* **78**, 138–67.

Ristau, C.A. (1986). Do animals think? In *Animal Intelligence: Insights into the Animal Mind*, ed. R.J. Hoage & L. Goldman, pp. 165–85. Washington, DC: Smithsonian Institution Press.

Sackett, G.P. & Fredrickson, W.T. (1987). Social preferences by pigtailed macaques: familiarity versus degree and type of kinship. *Anim. Behav.* **35**, 603–6.

Schultz, A.H. (1969). *The Life of Primates*. London: Weidenfeld & Nicolson.

Seyfarth, R.M. (1977). A model of social grooming among adult female monkeys. *J. theoret. Biol.* **65**, 671–98.

Seyfarth, R.M. (1987). Vocal communication and its relation to language. In *Primate Societies*. ed. B.B. Smuts, D.L. Cheney, R.M. Seyfarth, R.W. Wrangham & T.T. Struhsaker, pp. 440–51. Chicago: University of Chicago Press.

Sigg, H. & Falett, J. (1985). Experiments on respect of possession and property in hamadryas baboons (*Papio hamadryas*). *Anim. Behav.* **33**, 978–84.

Silk, J.B., Clark-Wheatley, C.B., Rodman, P.S. & Samuels, A. (1981). Differential reproductive success and facultative adjustment of sex ratios among captive female bonnet macaques (*Macaca radiata*). *Anim. Behav.* **29**, 1106–20.

Simpson, A.J. & Barton, R.A. (1992). Dominance and mating success: avoiding incest and artefact. *Anim. Behav.* **44**, 1164–5.

Smuts, B.B. (1987). Sexual competition and mate choice. In *Primate Societies*, ed. B.B. Smuts, D.L., Cheney, R.M. Seyfarth, R.W. Wrangham & T.T. Struhsaker, pp. 385–99. Chicago: University of Chicago Press.

Sommer, V. (1987). Infanticide among free-ranging langurs (*Presbytis entellus*) at Jodhpur (Rajasthan/India): recent observations and a reconsideration of hypotheses. *Primates* **28**, 163–97.

Strayer, F.F. (1976). Learning and imitation as a function of social status in macaque monkeys (*Macaca nemestrina*). *Anim. Behav.* **24**, 835–48.

Sugiyama, Y. (1987). A review of infanticide among Hanuman langurs and other primates. *J. Bombay Nat. Hist. Soc.* **83**, 7–11.

Takahata, Y.(1985). Adult male chimpanzees kill and eat male newborn infant: newly observed intragroup infanticide and cannibalism in Mahale National Park, Tanzania. *Folia primatol.* **44**, 161–70.

Thierry, B.(1985). Patterns of agonistic interactions in three species of macaque (*Macaca mulatta, M. fascicularis, M. tonkeana*). *Agress. Behav.* **11**, 223–33.

Tinklepaugh, O.L. (1931). Fur picking in monkeys as an act of adornment. *J. Mammalogy* **12**, 430–1.

van Hooff, J.A.R.A.M. & van Schaik, C.P. (1992). Cooperation in competition: The ecology of primate bonds. In *Coalitions and Alliances in Human and Other Animals*. ed. A.H. Harcourt & F.B.M. de Waal, pp. 357–89. Oxford: Oxford University Press.

van Lawick-Goodall, J. (1973). Behavior of chimpanzees in their natural habitat. *Am. J. Psychiat.* **130**, 1–12.

van Noordwijk, M.A. & van Schaik, C.P. (1988). Male careers in Sumatran long-tailed macaques (Macaca fascicularis). *Behav.* **107**, 24–43.

Visalberghi, E. & Fragaszy, D.M. (1990). Do monkeys ape? In *Language and Intelligence in Monkeys and Apes. Comparative Developmental Perspectives*, ed. S.F. Parker & K.R. Gibson, pp. 247–73. New York: Cambridge University Press.

Vogel, C. (1985). Helping, cooperation, and altruism in primate societies. *Fortschr. Zool.* **31**, 375–89.

Walters, J. (1980). Interventions and the development of dominance relationships in female baboons. *Folia primatol.* **34**, 61–89.

Wasser, S.K. & Starling, A.K. (1986). Reproductive competition among female yellow baboons. In *Primate Ontogeny, Cognition and Social Behaviour*, ed. J.G. Else & P.C. Lee. pp. 343–54. New York: Cambridge University Press.

Warden, C.J. & Jackson, T.A. (1935). Imitative behavior in the rhesus monkey. *J. Genetic Psychol.* **46**, 103–25.

Watson, J.B. (1908). Imitation in monkeys. *Psychol. Bull.* **5**, 169–78.

Watson, L. (1979). *Lifetide: The Biology of the Unconscious*. New York: Simon & Schuster.

Weisbard, C. & Goy, R.W. (1976). Effect of parturition and group composition on competitive drinking order in stumptail macaques (*Macaca arctoides*). *Folia primatol.* **25**, 95–121.

Welker, C. & Schaefer-Witt, C. (1992). The need of long-term studies to interpret actual behaviour patterns observable in social groups, the crab-eating monkey *Macaca fascicularis* as an example. *Primate Report* **32**, 31–47.

Whiten, A. (1989). Transmission mechanisms in primate cultural evolution. *Tr. Ecol. Evol.* **4**, 61–2.

Whiten, A. (1993). Evolving a theory of mind: the nature of non-verbal mentalism in other primates. In *Understanding Other Minds: Perspectives From Autism*, ed. S. Baron-Cohen, H. Tager-Flushery and D.S. Cohen, pp. 367–96. Oxford: Oxford University Press.

Wrangham, R.W. (1979). On the evolution of ape social systems. *Soc. Sci. Informat.* **18**, 335–68.
Wrangham, R.W. (1993). The evolution of sexuality in chimpanzees and baboons. *Human Nature* **4**, 47–93.
Wu, H.M.H., Holmes, W.G., Medina, S.R. & Sackett, G.P. (1980). Kin preference in infant *Macaca nemestrina. Nature* **285**, 225–7.
York, A.D. & Rowell, T.E. (1988). Reconciliation following aggression in patas monkeys, *Erythrocebus patas. Anim. Behav.* **36**, 502–9.

15

The effects of predation and habitat quality on the socioecology of African monkeys: lessons from the islands of Bioko and Zanzibar

THOMAS T. STRUHSAKER

Introduction

Explanations for the evolution of primate social groups and interspecific associations have invoked phylogenetic history and selective pressures exerted by predation and habitat upon individual fitness (e.g. Crook and Gartlan, 1966; Struhsaker, 1969, 1981; Wrangham, 1987; Gautier-Hion, 1988). The problem has been debated as to whether predation or foraging advantages exclusively shaped primate social systems (Wrangham, 1980; Van Schaik, 1983). Such a dichotomy is too simple, however, as shown by the great variation in primate behavioral ecology described over the past 30 years (e.g. Hall, 1965; Struhsaker, 1969; Van Schaik and Van Hooff, 1983; Wrangham, 1987). It is now generally accepted that primate social associations are formed and maintained in response to numerous interacting variables, most of which are not mutually exclusive in their effect. The challenge is to understand more fully how each factor contributes to the variation observed in nature.

This chapter considers two related issues, that of intraspecific and interspecific associations in African monkeys. Monkeys apparently associate for various reasons, including reproduction, foraging advantages, resource defense, predator avoidance, and hygiene (grooming). Any of these potential advantages may be realized in both intra- and interspecific associations (the latter involving occasional hybridization; Struhsaker *et al*., 1988). Here, I examine how the abundance of a major predator, the crowned hawk-eagle (*Stephanoaetus coronatus*), and gross habitat quality correlate with the frequency of polyspecific associations, the abundance of solitary monkeys, social group size, adult sex ratios within social groups, and terrestriality of African forest monkeys. The islands of Bioko and Zanzibar, where the crowned hawk-eagle has never been recorded, provide important test cases for understanding the relative importance of predation in shaping primate associations.

Study sites and methods

Bioko, politically part of Equatorial Guinea, is a volcanic island approximately 2020 km^2 in area and 2886 m in maximum elevation. It is separated from the coast of Cameroon by a 50–60 m deep channel (Butynski and Koster, 1989; Castelo, 1994) established by rising ocean levels about 10000–15000 years ago (Hamilton, 1982). Of the 10 native primate species, five are considered to be subspecifically distinct from the mainland forms (Butynski and Koster, 1994). There are no records of crowned hawk-eagle or leopard from Bioko.

In March 1992, I spent 16 days conducting systematic censuses and general surveys for primates near the mouth of the Rio Epola, on the southwest coast of Bioko (3° 18′N), a few kilometers northwest of Punta Oscura. The cover of lowland, tropical rainforest is strongly influenced by the severe storms and heavy rainfall (approximately 3500 mm per year; Fa, 1989, cited in Butynski and Koster, 1989). In overall structure, particularly its relatively low height and vine-dominated under-story, it recalls Caribbean hurricane forests. Tree-falls were common, and the ground was dominated by basalt rocks and boulders. The entire southern half of Bioko has a very sparse human population. No one lived within 15 km of the study site, and hunting was still at very low levels. Most of the primates at Rio Epola seemed to be naïve of humans and did not flee from us.

The two census routes, 3173 m and 2213 m in length, ran inland from the ocean, approximately 350 to 400 m apart, along escarpments flanking the Rio Epola on either side. The longer route was sampled on six days and the shorter on four days. Data on polyspecific associations were collected only during the systematic censuses, whereas information on group size and composition was gathered *ad libitum*.

Zanzibar (or Unguja) lies off the east coast of Tanzania. It is 1657 km^2 in areas, with a maximum elevation of about 100 m. The channel between Zanzibar and the mainland is 35 m deep and also formed 10000–15000 years ago (Hamilton, 1982). Zanzibar's substrate is largely an old coral reef and sandbar. Five primates occur on the island, of which one, the Zanzibar red colobus (*Procolobus* or *Colobus kirkii*), is endemic. There are no records of the crowned hawk-eagle, and leopards have been virtually exterminated.

Data from Zanzibar were collected during five trips of three to four weeks each during 1992–1995. Information on polyspecific associations was largely collected *ad libitum*, while making group counts or conducting surveys. Most data were collected in and around the Jozani Forest (6° 16′S). This is a ground-water forest dominated by relatively few species of trees and vines (*Calophyllum, Vitex, Syzgium, Elaeis, Phoenix, Pandanus*). Its history is poorly known, but virtually all has been greatly influenced by human activities, such as pitsawing. Adjacent to the forest are three other habitats: coral rag scrub (dry substrate, supporting low-stature, sclerophytic, evergreen trees and shrubs); shambas (perennial gardens dominated by exotic trees, e.g. *Mangifera indica, Terminalia catappa, Cocos nucifera*); and mangrove swamp. The red colobus are not hunted in this area, but the Sykes monkeys (*Cercopithecus mitis albogularis*) are occasionally shot, trapped, or harassed as agricultural pests. Human population density is relatively high.

The other study sites are described in the references cited within this chapter. Data on polyspecific associations were again collected during systematic censuses and surveys, while information on group size and composition was gathered *ad libitum*.

The relative abundance of crowned hawk-eagles was classified into three categories: (1) *common* (seen at least once per week, often daily); (2) *rare* (seen approximately once every two to three years or no records, but within the species' broad range; and (3) *absent* (no sightings and outside known range).

Habitat was classified into two broad categories: (1) large blocks (>15 km^2) of relatively aseasonal, evergreen forest; and (2) small, fragmented evergreen blocks and highly seasonal, semi-deciduous forest, often lower plant diversity. Category 2 forests were often heavily disturbed by human activities. Although many other habitat variables might be considered (e.g. Struhsaker, 1981; Cords, 1987; Chapman and Chapman, 1996), these were selected because they are some of the most obvious, and because information on them is available from all sites compared.

Tests for significant associations between species relied on chi-squared tests and Fager's (1957) index of affinity. The models of chance developed by Waser (1982, 1987) and Whitesides (1989) were not used, because the information they require was unavailable for most sites or species, and because many, if not most, of their underlying assumptions are not valid.

Polyspecific associations

Island versus mainland sites

Polyspecific associations among monkey species are common and widespread throughout the rainforests of Africa. These associations are operationally defined as the spatial intermingling of individuals of two or more species, so that members of different species are usually within 20 m of one another (Gartlan and Struhsaker, 1972). In general, over 60% of all social groups encountered are with other monkey species (Table 15.1). Notable exceptions are the islands of Bioko and Zanzibar, the heavily logged areas of Kibale in Uganda, and Tiwai Island, Sierra Leone (Table 15.1). Crowned hawk-eagles are absent from Bioko and Zanzibar and extremely rare at Tiwai. Only two probable or possible sightings were made of these eagles at Tiwai during more than five person-years of field work there (J.F. Oates and G.H. Whitesides, pers. comm.). The absence or rarity of crowned hawk-eagles cannot, however, explain the relatively low occurrence of polyspecific associations in the heavily logged areas of Kibale, where eagles are common (Skorupa, 1989; Struhsaker, unpub. data).

African monkey species differ strikingly in their tendency to form polyspecific associations (Struhsaker, 1975, 1981; Gautier-Hion, 1988), and therefore, a meaningful comparison between island and mainland communities requires that they have similar species compositions. For Bioko, the best comparison is with Cameroon, where all monkey species found on Bioko are at one or more of the study sites. The mainland site most comparable and closest to Jozani, Zanzibar is Magombera, Tanzania. Both of these sites have a species of red colobus and subspecies of Sykes monkey, but Magombera also has *Colobus angolensis.* Comparison of these two pairs of sites shows the same trend as does the pan-African comparison (Table 15.2). The island sites, lacking crowned hawk-eagles, have significantly fewer polyspecific associations than do the mainland sites.

The analysis can be further refined by comparing populations of the same species or closely related species. This was possible for five species or superspecies (Table 15.3). Red colobus (*C. badius*, s.l.) social groups on islands were in polyspecific associations less (Zanzibar) or no more (Bioko) often than expected by chance. In contrast, those at mainland sites associated with other species more than expected. The one exception was the heavily logged part of Kibale where red colobus social groups occurred as often alone as with other species. The other species showed similar patterns (Table 15.3). Island populations either occurred significantly more often in

Table 15.1. *Polyspecific associations: percentage of encounters with social groups*

Site	*n*	Number of anthropoid species	Percentage polyspecific	Percentage Monospecific
*** Zanzibar Island[1]**	**360**	**2**	**30.6**	**69.4**
Magombera, Tanzania[2]	28	3	78.6	21.4
Kanyawara, Kibale (unlogged)[2]	498	8	71.7	28.3
*** Kanyawara, Kibale (heavily logged)[2]**	**309**	**8**	**39.5**	**60.5**
Ngogo, Kibale (unlogged)[2]	133	8	65.4	34.6
Afarama, Ituri, Zaire[3]	55	11	70.9	29.1
Lamako, Zaire[4]	389	6	60.4	39.6
Idenau, Cameroon[2,5]	151	8	57.0	43.0
S. Bakundu, Cameroon[2,5]	671	6	79.6	20.4
W. & E. Cameroon (10 other sites)[5]	104	6–8	79.8	20.2
Douala-Edea Reserve, Cameroon[6]	217	8	82.0	18.0
Korup National Park, Cameroon[7]	75	8	68.0	32.0
Gabon (4 sites)[2,8]	251	8	70.9	29.1
*** Bioko Island, Equatorial Guinea[9]**	**69**	**6**	**39.0**	**61.0**
*** Bioko Island, Equatorial Guinea[10]**	**86**	**7**	**22.1**	**77.9**
Tai National Park, Cote d'Ivoire[11]	439	7	80.0	20.0
*** Tiwai Island, Sierra Leone[12]**	**349**	**8**	**42.1**	**57.9**

Notes:
* Exceptions
1: Struhsaker (1992–1995, unpub. data); 2: Struhsaker (1981 and 1973–1976 unpub. data); 3: Struhsaker & L. Leland (1988, unpub. data); 4: McGraw (1994); 5: Gartlan & Struhsaker (1972); 6: Whitesides (1981); 7: Edwards (1992); 8: Gautier & Gautier-Hion (1969); 9: Struhsaker (1992, unpub. data from Rio Epola area); 10: Butynski & Koster (1994); 11. Galat & Galat-Luong (1985); 12: Oates & Whitesides (1990).

Table 15.2. *Polyspecific associations on islands versus nearest mainland sites (All Species Combined)*

Location	Number and Percentage of Social Groups in[1]:		
	Polyspecific associations	Monospecific associations	Totals
West Africa			
Bioko Island[2]	27 (51.4), 39%	42 (17.6), 61%	69
Cameroon[3]	932 (907.6), 76.5%	286 (310.4), 23.5%	1218
Totals	959	328	1287
East Africa			
Zanzibar Island[4]	110 (122.5), 30.6%	250 (237.5), 69.4%	360
Magombera, Tanzania[5]	22 (9.5), 78.6%	6 (18.5), 21.4%	28
Totals	132	256	388

Notes:
West African section: $\chi^2 = 41.52$, $df = 1$, $p < 0.001$
East African section: $\chi^2 = 26.83$, $df = 1$, $p < 0.001$
1: Expected values in parentheses; 2: Sample includes 6 species (Struhsaker, 1992, unpub. data); 3: 13 sites from Gartlan & Struhsaker (1972), plus Douala–Edea (Whitesides, 1981), & Korup (Edwards 1992). Sample includes 6–8 species at each site; 4: Sample includes 2 species (Struhsaker, 1992–95, unpub. data); 5: Sample includes 3 species (Struhsaker, 1981)

monospecific associations, or showed no significant difference from expectation in association tendencies, while most mainland populations associated with other species significantly more than expected by chance. Again, the heavily logged parts of Kibale proved an exception, with *C. mitis* and *C. ascanius* social groups showing no significant difference in their occurrence in mono- or polyspecific associations. The same was true for *C. mitis* at Magombera (Table 15.3).

Finally, I compared indices of affinity (association) between pairs of species (Table 15.4) at island and mainland sites. (The various forms (species and/or subspecies) of red colobus are combined as *C. badius*, s.l.) Species pairs on the islands show no significant affinities, whereas those on the mainland do. The only mainland exceptions were at the heavily logged sites in Kibale.

All four comparisons consistently suggest that the occurrence of the crowned hawk-eagle strongly predicts whether or not social groups of monkeys will form polyspecific associations more often than expected by chance. The most notable exception is the heavily logged part of Kibale.

Table 15.3. *Percentage of encounters with social groups: islands versus mainland*

Species & location	*n*	Polyspecific (%)	Monospecific (%)	χ^{2*}
Colobus badius				
Bioko Island, Equatorial Guinea	13	38.5	61.5	0.69
Bioko Island, E.G. (several sites)***	10	40	60	0.4
Zanzibar Island (all habitats), Tanzania	218	25	75	54.75*
Korup National Park, Cameroon	13	92	8	9.17*
Magombera, Tanzania	10	90	10	6.4*
Kibale (unlogged), Uganda**	228–266	55.2–64.5	35.5–44.7	2.95*–19.10*
Kibale (heavily logged), Uganda**	98–114	48.2–56.1	43.9–51.8	0.14–1.47
Cercopithecus mitis				
Zanzibar Island (all habitats), Tanzania	142	38.7	61.3	7.72*
Magombera, Tanzania	9	55.6	44.4	0.5
Kibale (unlogged), Uganda**	107–112	89.3–93.5	6.5–10.7	69.1*–80.8*
Kabale (heavily logged), Uganda**	32–45	53.3–75	25–46.7	0.2–8.0*
Cercopithecus pogonias				
Bioko Island, Equatorial Guinea	30	30	70	4.8*
Bioko Island, E.G. (several sites)***	15	13.3	86.7	8.07*
Korup, Cameroon	9	88.9	11.1	5.44*
13 sites Cameroon	60	96.7	3.3	52.27*
Douala–Edea, Cameroon	29	96.6	3.4	25.14*
4 sites Gabon	49	77.6	22.4	14.88
Cercopithecus erythrotis/cephus				
Bioko Island, Equatorial Guinea	16	50	50	0
Bioko Island, E.G. (several sites)***	41	14.6	85.4	20.5*
13 sites Cameroon	275	80.7	19.3	103.86*

Table 15.3. (*cont.*)

Species & location	*n*	Polyspecific (%)	Monospecific (5)	χ^{2*}
Douala–Edea, Cameroon	44	93.2	6.8	32.8*
4 sites Gabon	59	88.4	11.6	61.73*
Cercopithecus ascanius				
Kibale (unlogged), Uganda**	182–209	74.2–85.2	14.8–25.8	48.8*–90.02*
Kibale (heavily logged), Uganda**	56–75	46.7–62.5	37.5–53.3	0.33–3.5*

Notes:
* $\chi^2 > 2.71$, $p < 0.10$, $\chi^2 > 3.84$, $p < 0.05$.
** Note: there was a bias in favor of scoring polyspecific associations because they were scored even if their complete species composition was not known, whereas monospecifics were only scored if it was certain no other species was present. For Kibale, two percentages are given. The lowest for monospecific associations is the more conservative and the highest is when all monospecific associations were included even if there was some uncertainty about the presence or absence of other species. Unlogged Kibale includes Kanyawara and Ngogo sites.
*** Butynski & Koster 1994. Other sources given in Table 15.2.

Table 15.4. *Interspecific associations (Fager's Index of Affinity)*[a]

Species pairs	Mainland	Island
Colobus badius–Cercopithecus pogonias	0.73*[1]	0.17[2]
Cercopithecus pogonias–Cercopithecus erythrotis	0.74*[3]	0.29[2]
	0.55*[4]	
Colobus badius–Cercopithecus mitis	0.42*[5]	0.31[7]
	0.52*[6]	
	0.20[8]	

Notes:
[a] 2J/A + B, J = number of joint occurrences, A&B = totals of each species
* Significant associations $p<0.05$ (see Fager, 1957)
[1] Korup National Park, Cameroon (Edwards, 1992)
[2] Bioko Island, Equatorial Guinea (Struhsaker, unpub. data, 1992)
[3] Idenau, Cameroon (Gartlan & Struhsaker, 1972)
[4] Douala–Edea Reserve, Cameroon (Whitesides, 1981)
[5] Magombera, Tanzania (Struhsaker, 1981)
[6] Kanyawara, Kibale (unlogged), Uganda (Struhsaker, 1975 and unpub. data, 1974–76)
[7] Zanzibar Island (all habitats), Tanzania (Struhsaker, unpub. data, 1992–95)
[8] Kanyawara, Kibale (heavily logged), Uganda (Struhsaker, 1975 and unpub. data 1973–76)

Variation within sites

Although polyspecific associations are common in the forests of mainland Africa, there is often considerable variation within populations in the frequency of these associations. Variation occurs between groups of the same species and between subpopulations in the same forest, and temporal and spatial variation occurs within social groups (Struhsaker, 1975, 1981; Cords, 1987; Lwanga, 1987; Gautier-Hion, 1988; Whitesides, 1989; Oates and Whitesides, 1990). In most cases, these differences appear to be related to spatial and/or temporal variation in food (diversity, density, distribution, species composition) and/or species composition of the primate community within the group's home range.

One example is provided by heavily logged and unlogged sites at Kibale, referred to above. These are separated by only 3.5–7 km within the same forest, and yet the differences in interspecific associations are profound (Struhsaker, 1975, 1981). In the heavily logged forest, no species associated with other species more than expected nor were there any significant affinities between any pairs of species. By contrast, in the unlogged part, three of six species for which there were adequate samples, occurred in polyspecific associations more than expected by chance and three pairs of species had significant affinities (see discussion below).

The second example comes from Zanzibar. The four distinct habitats at the Jozani study site (coral rag/dry forest; plantations of *Calophyllum inophyllum*; ground-water forest; shambas) are contiguous and some groups of monkeys move between all four. Habitat type appears to have very little effect on the propensity of red colobus to associate with the Sykes monkeys. In contrast, the Sykes monkeys tend to associate with the red colobus much more than expected when they are in the shambas (Figs. 15.1 and 15.2).

For each habitat type, the frequency of monospecific social groups were compared to the frequency of bispecific groups for both red colobus (Fig.15.1) and Sykes monkeys (Fig.15.2). Two-cell, chi-squared tests were run for each of the two species in each of the four habitats. Red colobus formed monospecific groups significantly more often than they associated with Sykes in all habitats ($\chi^2 = 5.4$ to 54.8, $df = 1$, $p < 0.02$ to 0.001) except the coral rag, where there was no significant difference ($\chi^2 = 2.45$, $df = 1$, $0.2 > p > 0.1$). Sykes groups were alone significantly more often than they associated with red colobus in the ground-water forest and *Calophyllum* plantations ($\chi^2 = 25.1$ and 4.6, $p < 0.001$ and $0.05 > p > 0.02$, respectively). There was no significant difference in association tendencies of Sykes in the coral rag ($\chi^2 = 0.33, p > 0.05$).

In marked contrast to all other results for Zanzibar, when Sykes groups were in shambas, they were with red colobus groups significantly more often than they were alone ($\chi^2 = 10.7$, $df = 1$, $0.01 > p > 0.001$). It is in the shambas that the Sykes are most prone to hunting and harassment by humans, because, unlike the red colobus, they are considered to be pests. The shamba associations of Sykes with red colobus are probably adaptive for the Sykes. Not only are there more individual monkeys to detect humans, but the large and noisy groups of red colobus (rarely hunted or harassed by humans on Zanzibar) may act as a screen for the Sykes, which tend to forage lower and in denser vegetation than the colobus. It may be harder for a person defending crops to detect the Sykes when they are with the red colobus. A similar, but less pronounced, phenomenon may account for associations in the coral rag, where humans hunting for duikers with guns and nets may pose a threat. The association of olive colobus and diana monkeys at Tiwai Island may have an analogous explanation. These are the only two species at Tiwai that associate with one another significantly more than expected. Oates and Whitesides (1990) suggest that as the colobus feeds quietly in the dense understory, where it is comparatively vulnerable

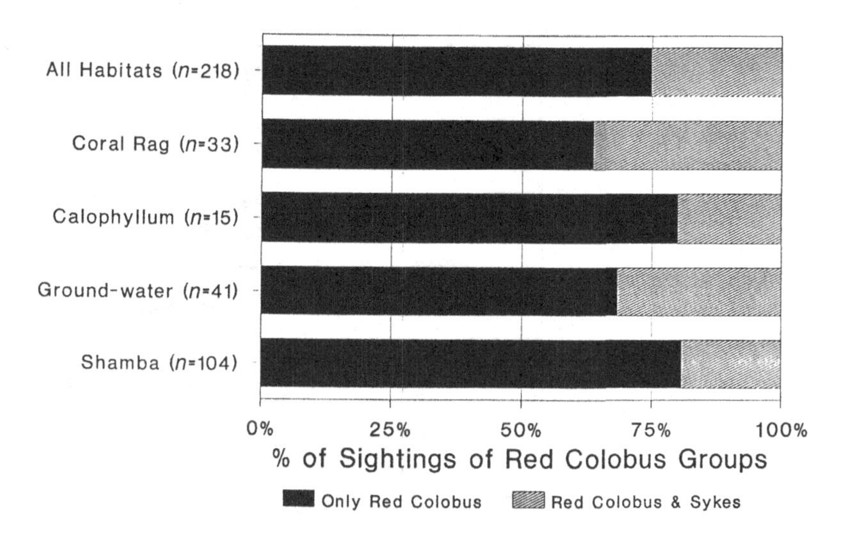

Fig. 15.1. The percentage of sightings of red colobus social groups that were alone or with social groups of Sykes monkeys in four different habitats. All habitats includes these four habitats, as well as others such as swamp and mangrove. Red colobus were significantly more often alone than with Sykes in all habitats except coral rag where there was no significant difference in association patterns. See text for statistics.

to predation by chimpanzees, it is using the conspicuous and noisy diana monkey in the upper canopy as a distracting screen.

Discussion of polyspecific associations

Reviews of polyspecific associations in primates (e.g. Struhsaker, 1981; Cords, 1987; Waser, 1980, 1987; Gautier-Hion, 1988; Whitesides, 1989; Terborgh, 1990) generally argue that they occur by chance, or because they benefit one or more of the species involved, or are simply aggregations at common resources. Chance occurrences depend on population densities, group spread and rates and direction of travel. Benefits include defense against predation and foraging advantages.

The data presented here support the hypothesis that predation is an extremely important force favoring polyspecific associations. In the absence of crowned hawk-eagles, there are significantly fewer associations between

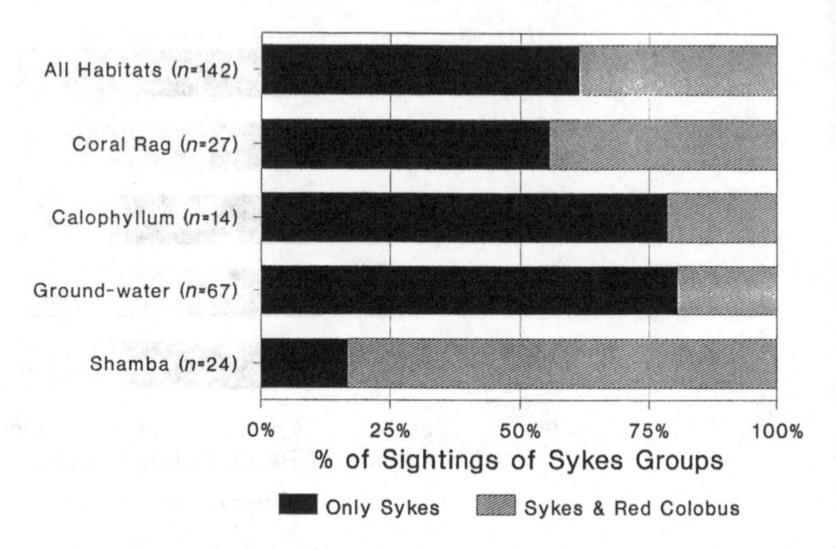

Fig. 15.2. The percentage of sightings of Sykes monkey social groups that were alone or with social groups of red colobus monkeys in four different habitats. Sykes were alone significantly more often in the *Calophyllum* plantations and ground-water forest, but there was no significant difference in association patterns when they were in coral rag. In contrast, they were with red colobus significantly more than expected when in the shambas (perennial gardens). See text for statistics.

species than when these eagles are common. This is consistent with the conclusions of a pan tropical review (Struhsaker, 1981). Polyspecific associations occur more often in African rainforests, where monkey-eating eagles are common, while in South America they are much less common and primarily involve the smaller species. The larger platyrhines (*Alouatta, Ateles, Lagothrix*) rarely form polyspecific associations, and I believe this is because the harpy eagle (*Harpia harpyja*), the only avian predator large enough to kill them, is absent or extremely rare in most areas that have been studied. Smaller platyrhines such as *Cebus, Saimiri,* and *Saguinus* frequently form polyspecific associations, perhaps because they are prone to predation by a wide range of smaller raptors. Small monkeys such as *Callicebus*, that live as mated pairs, do not commonly form polyspecific associations (T.T. Struhsaker, unpub. data), having, apparently, adopted a strategy against avian predators that depends more upon concealment and stealth.

The situation in Southeast Asia is, in many ways, a test case analogous to the African islands. Raptors capable of taking adult monkeys are absent from most of the area, and no such attacks are recorded. As expected, primate polyspecific associations are much less common than in Africa or the neotropics (Struhsaker, 1981).

Data from Kibale suggest that, although predation by eagles is an important selective pressure favoring interspecific associations, foraging costs and benefits may also influence them. In the logged areas, polyspecific associations are much less common than in the nearby unlogged areas, although crowned hawk-eagles are abundant in both (Skorupa, 1989; Struhsaker and Leakey, 1990). The greatest difference between them is in vegetation. At least 50% of the basal area of trees was removed from the logged areas, and their failure to regenerate is reflected in significantly reduced tree density and diversity (Skorupa, 1986, 1988). The striking differences in polyspecific associations at these two sites is probably related to food, with the benefits of polyspecific associations against predation being outweighed by the costs to foraging efficiency incurred by such associations where food diversity and density are low. Lwanga (1987) has presented data suggesting that high levels of dietary overlap (food competition?) between primate species may actually lead to a decrease in associations, in spite of high predation risks.

Could similar habitat variables explain the infrequent occurrence of polyspecific associations on Bioko and Zanzibar? This seems unlikely. The Bioko data come from many different areas and forest types of the island (Butynski and Koster, 1994; Struhsaker, this chapter), and the forests sampled appeared similar in structure to many of the forests surveyed on the mainland. On Zanzibar, the pattern of association was consistent in radically different habitats. The sole exception, the greater than expected occurence of associations in the shambas, is likely to be a response to harassment and predation by humans. Although there were differences in plant species composition, the ground-water forest and shamba habitats had similar densities and diversities of red colobus food species (Siex, 1995). Furthermore, the monodominant *Calophyllum* plantations were very different in plant species composition from the adjacent ground-water forest and yet the two monkey species showed identical patterns of association in these two habitats.

If hunting and harassment by humans explain why Sykes monkeys on Zanzibar associate more with red colobus in shambas than elsewhere, this would seem to contrast with the mainland situation. Waser (1982) proposed that differences in hunting pressure by humans may account for an apparently greater occurrence of polyspecific associations in hunted areas of Cameroon and Gabon than in unhunted Kibale. This perception, however, is an artifact of comparing census and survey data with longitudinal, all-day observations of specific social groups (Struhsaker, 1981). Census and survey data are biased in favor of polyspecific associations. When census data from Kibale are compared with similar data from hunted areas of West Africa, there are no apparent differences in the frequency of polyspecific associations (Struhsaker, 1981).

Attempts to correlate habitat variables with the frequency of polyspecific associations have had mixed results. Social groups of redtail and blue monkeys associate more frequently when uncommon foods that are favored by both species are available (Struhsaker, 1981; Cords, 1987). A group of redtail monkeys in Kibale associated more with other monkey species, especially *C. mitis*, when fruit constituted a high proportion of the diet (Struhsaker, 1981), but this was not the case in the Kakamega Forest (Cords, 1987). Cords (1987) found that the frequency of associations between a specific group of blue monkeys and redtail monkeys was significantly correlated with habitat type, but a study in Kibale (Chapman and Chapman, 1996)) found no correlations between food density or distribution and the frequency of polyspecific associations. The contrasting outcome probably results from the broader and coarser-grained approach of the latter study, which combined the majority of plant foods regardless of abundance or importance in the diet, and measured food available only as present or absent rather than on a scale of relative abundance.

Wasser (1993) attempted to explain differences in association patterns between two species of colobus monkeys at Udzungwa, Tanzania and Kibale, Uganda on the basis of contrasting patterns in dietary overlap and presumed feeding competition. He speculated that these differences in interspecific association tendencies and dietary overlap were ultimately due to differences in plant-species composition of the two forests. In the Udzungwas, the two colobus (*C. angolensis* and *C. badius gordonorum*) occur together about 50–55% of the time (Wasser 1993), whereas in an unlogged part of Kibale, two comparable species (*C. guereza* and *C. badius tephrosceles*) are rarely together (0.7%) (Struhsaker, 1981). Wasser (1993) suggests that the two colobus can associate with one another more frequently at Udzungwa than at Kibale because, although they show great overlap in plant *species* eaten (51%), they have less dietary overlap at the level of plant *parts* (48%) (figures 13.2 and 3.3 in Wasser, 1993). However, overlap between the two colobus at Kibale is identical (47.5%) in plant parts and somewhat less (33%) in plant species (Struhsaker, 1975; 1978; Oates, 1977). Even more important, dietary overlap between the two colobus in species-specific plant parts is greater at Udzungwa (27%, fig. 13.3 in Wasser, 1993) than at Kibale (7.1%; Struhsaker, 1978). Contrary to Wasser's (1993) contention, competition between the two colobus is potentially greater at Udzungwa than Kibale, yet the two colobus associate more often at Udzungwa. It is possible that the relatively high dietary overlap at Udzungwa may not concern foods that are limiting either of the colobus populations. Moreover, crowned hawk-eagles are very abundant at Udzungwa (Wasser, 1993) and the benefits of interspecific associations against this predator may outweigh the costs of any feeding competition that might exist.

To what extent might chance explain the differences between island and mainland populations? The models developed by Waser (1987 review) and Whitesides (1989) incorporate assumptions such as consistent and known configuration and spatial dimensions of group spread, rates of group movement, and random direction of movement. Even if most of the assumptions were valid, the data necessary to test these models were rarely reported in the studies used here.

In any event, chance is an unlikely explanation for the low frequency of polyspecific associations on Bioko and Zanzibar because anthropoid species richness (Table 15.1) and population densities there were at least as high as on the mainland (Butynski and Koster, 1994; Siex, 1995; Struhsaker, unpub. data). Furthermore, at least some of the species on Bioko and some groups of red colobus in the ground-water forest of Zanzibar had social groups that frequently divided into separate foraging parties, which increases the probability of polyspecific associations by chance and yet this did not happen. Finally, it is important to emphasize a point made by Whitesides (1989) that an interspecific association can still be biologically significant even if it occurs by chance.

The new data and those reviewed support the argument that predation is a major determinant of polyspecific associations among African forest monkeys, but that anti-predation benefits may be outweighed by the costs of foraging with other species in areas of low food density and/or diversity. The benefits of interspecific associations to foraging efficiency and predator avoidance are, of course, not mutually exclusive (Struhsaker, 1981; Cords, 1987; Gautier-Hion, 1988).

Table 15.5. *Abundance of solitary monkeys: islands versus mainland. Perentage of sightings that were solitary* (n = *solitaries* + *group sightings*)

Species	Location			
	Bioko (island)[1]	Korup (mainland)[2]	Zanzibar (island)[3]	Kibale (K30) (mainland)[4]
Red colobus[5]	23.5% (17)[A]	0 (36)[B]	5.2% (233)[C]	1.8% (219)[B]
Cercopithecus erythrotis/ascanius[6]	20% (20)[A]	10% (20)[A,B]	NA	5.2% (172)[B]
Cercopithecus pogonias	0 (31)[A]	0 (47)[A]	NA	NA
Cercopithecus mitis	NA	NA	4% (152)[A]	3.8% (106)[A]

Notes:
within a species row, shared letters (A,B,C) indicate no significant difference, $\chi^2 < 2.71$, $p > 0.10$
[1] Struhsaker, unpub. data (1992)
[2] Edwards (1992)
[3] Struhsaker, unpub. data (1992–95)
[4] Struhsaker, unpub. data from censuses 1970–76
[5] Includes *Colobus/Procolobus* forms of *pennanti, preussi, kirkii,* & *tephrosceles*
[6] Very closely related, same superspecies, perhaps conspecific

Group composition and size

Solitary monkeys

Solitary monkeys (i.e. animals separated by at least 50 m from conspecifics, but not necessarily isolated socially) are more common in rainforest than in open savannas. This fact has been attributed to the higher risk of predation in more open habitats, as a result of more predators and fewer refuges in trees or cover (Struhsaker, 1969). These data on solitaries may reflect the extent of ecological forces selecting for spatial cohesion.

In the absence of crowned hawk-eagles and leopards, one would predict more solitary monkeys in the forests of Bioko and Zanzibar than in mainland forests. This seems to be true for red colobus, somewhat the case for *C. erythrotis/ascanius* but not so for *C. pogonias* and *C. mitis* (Table 15. 5). In addition to the sightings made during censuses (Table 15.5), three more solitary female red colobus were seen (two adults – one with large perineal swelling – and a medium juvenile – 35 m from edge of group). While the solitary red colobus on Bioko were all females (adult, subadult, medium juvenile), those on Zanzibar were mostly males (adult, subadult, large juvenile), suggesting that factors in addition to predation may be involved, such as competition for mates.

Table 15.6. *Abundance of solitary monkeys in Kibale: effect of intense logging. Percentage of sightings during systematic censuses that were solitary (*n *=solitaries + group sightings)*

Species	Location			
	K30(UL)	Ngogo(UL)	K28/29 (UL/ML)*	K12,13,17 (HL)
Red colobus[1]	1.8% (219)[A]	1.9% (52)[A]	2.2% (45)[A]	4.2% (119)[A]
Cercopithecus ascanius	5.2% (172)[A]	8.0% (50)[A]	9.1% (33)[A]	7.4% (81)[A]
Cercopithecus mitis	3.8% (106)[A]	0 (10)[A]	26.7% (19)[B]	10% (50)[A]

Notes:
UL: unlogged; UL/ML*: unlogged for most of censuses, then moderately logged; HL: heavily logged.
Data from 1970–78 (Struhsaker unpub. data).
[1] *Procolobus/Colobus badius/rufomitratus tephrosceles.*
Note: within a species row, shared letters (A,B) indicate no significant difference, $\chi^2 < 2.71, p > 0.10$.

This comparison suggests that there are important interspecific differences in the frequency of solitary monkeys, some of which can be related to the degree of variability and/or flexibility in social systems. Red colobus social groups are extremely variable in size and age-sex composition, whereas most *Cercopithecus* species are much less variable or flexible in these respects. For example, most *Cercopithecus* social groups contain only one fully adult male on a full-time basis. Correspondingly, in the absence of predators, the red colobus showed the most dramatic increase in solitary individuals compared to the *Cercopithecus*.

Logging in Kibale appeared to lead to a slight increase in the incidence of solitary monkeys (Table 15.6). However, in only one case was there a statistically significant difference (*C. mitis* in K28/29) and this may not have been due to logging, which began after most of the data for this compartment were collected. Predation pressure by eagles remained high in the logged parts of Kibale and this probably explains why there were only slight differences in the numbers of solitaries. Factors other than predation, such as density, diversity and dispersion of food, and intrasexual competition, may influence the abundance of solitary monkeys. For example, the density of solitary male blue monkeys (*C. mitis*) based on detailed studies of individuals was estimated to be seven times greater at Ngogo than at the Kanyawara study site of Kibale, even though the two sites are only 10 km

Table 15.7. *Group size*

Species & location	x̄ Number in group[A] (range)	*n* (number groups or party counts)
Cercopithecus pogonias		
Bioko Island, Equatorial Guinea[1]	11.4 ($<6\leq15$)	7
Cameroun (13 sites)[2]	14.6 (11–19)	9
Makokou, Gabon[3]	15.5 (13–18)	1
Colobus satanas		
Bioko Island, Equatorial Guinea[1]	2.6* (2–3)	5
Lac Tisongo, Douala–Edea, Cameroun[2]	6.7 (2–10)	5
Douala–Edea, Cameroun[4]	15–16*	1
Lope, Gabon[5]	9–14*	1

Notes:
[A] All entries are reliable estimates unless followed by an asterisk (*), which indicates complete counts.
[1] Struhsaker (unpub. data, 1992). Rio Epola.
[2] Struhsaker (1969).
[3] Gautier-Hion and Gautier (1974).
[4] McKey and Waterman (1982).
[5] M. Harrison in Oates (1994).

apart and both are comprised of unlogged, mature forest (Struhsaker *et al.*, 1988), and experience similar predation pressure from eagles.

Group size

If predation favors aggregation, a decrease in the size of social groups, or at least of foraging parties, might be expected in the absence of predators. For example, group size of long-tailed macaques in Indonesia was significantly smaller on the island of Simeulue (felid predators absent) than on Sumatra (felid predators present) (van Schaik and van Noordwijk, 1985). A comparative study of several primate species found that estimated rates of predation were inversely related to the log of group size (Isbell, 1994).

Reliable group counts or estimates were obtained for three species on Bioko, and for red colobus on Zanzibar (Tables. 15.7 and 15.8). *Cercopithecus pogonias* and *Colobus satanas* occurred in significantly smaller groups or foraging parties on Bioko island than on the mainland (Table 15.7; $U=13, p=0.05$ and $U=4, p=0.015$, 1-tail, respectively). Some of the counts on Bioko may have been repeats of the same party, but on different days. Furthermore, on at least two occasions it was clear that small parties of *Cercopithecus pogonias* moved 150–200 m to join other conspecific parties. No aggression was detected between parties during these unions, and they appeared to belong to one social group that had split into temporary foraging parties. Compared to the mainland populations, *C. pogonias* on Bioko may have a greater propensity to divide into small, temporary foraging parties, rather than to form smaller social groups.

Red colobus group size is extremely variable both within and between populations (Table 15.8), and may change over time. On the Tana River, for example, mean group size declined significantly, from 18 to 12, over an 11 to 15-year period (Table 15.8, $U=38, p<0.05$, 2-tail) (see Decker, 1994a).

Although only two complete group counts (9 and 10 individuals) were made of red colobus on Bioko, reliable estimates indicated distinctly smaller groups there than on the mainland (Table 15.8). In contrast, the red colobus groups on Zanzibar are similar in size to some of the mainland populations where crowned hawk-eagles are common, for example Tai, Magombera, and Mbisi. A comparison of group size between Zanzibar and the nearest mainland site, Magombera, revealed no significant difference (Table 15.8, $U=37, p>0.10$, 2-tail).

It appears that for red colobus, group size is influenced by more variables than predation. Gross habitat quality is likely to be one such variable. When crowned hawk-eagles are common, red colobus groups are bigger in large rainforest blocks than they are in blocks that are small, fragmented and/or seasonal (Figure 15.3a). A similar trend is apparent even when eagles are rare, but note that even in the large blocks of rainforest, group size may be slightly smaller than where eagles are common (Figure 15.3b). In the absence of eagles, there is no clear relationship between group size and gross forest type (Figure 15.3c).

Further evidence that group and/or foraging party size is likely influenced by habitat quality comes from Kibale. In the heavily logged parts, where hawk-eagles are common, red colobus group size is similar to that in nearby unlogged forest (47 vs. 50; Struhsaker, 1975; Skorupa, 1988), however, the main study group in heavily logged forest divided into two or more subgroups or foraging parties on 33% of the sample days ($n=74$ over 15 months; Skorupa, 1988). These subgroups were sometimes separated by several hundred meters for more than one day (Skorupa, person. comm.). This type of fusion–fission behavior has never been seen in the unlogged tracts of Kibale, during thousands of hours of observation. Similar fusion–fission behavior has been reported in *Presbytis melalophos*, in response to heavy logging (Johns, 1983), and in *Macaca fascicularis* as a response to extensive damage after a major forest fire (Berenstain, 1986).

Table 15.8. *Red colobus group size*

Location	$\bar{x}$ Number in group[A] (range)	*n* (number groups or party counts)
West Africa		
Bioko Island, Equatorial Guinea[1]	≤14 (5–≤30)	14–15[B]
Korup, Cameroon[2]	≥47(>24≥80)	7
Tai, Cote d'Ivoire[2]	~40 (>27<55)	10
Tai, Cote d'Ivoire[3]	32*	1
Tiwai, Sierra Leone[4]	35 (NA)	NA
Senegal & Gambia[2]	24.5* (12–34)	6
Fathala, Senegal[5]	24.3* (3–62)	20
East Africa		
Zanzibar Island, Tanzania[6]	31* (11–62)	12
Magombera, Tanzania[7]	26.6 (23–33)	4
Magombera, Tanzania[8]	34.3* (26–50)	8
Tana River, Kenya[9]	18.0* (13–30)	13
Tana River, Kenya[10]	12.0* (4–23)	13
Gombe, Tanzania[11]	55–58.8 (30–82)	5
Mbisi, Tanzania[12]	≥24 (~≤14–≥29)	5
Kibale, Uganda[13]	≥44 (9–80)	20
Central Africa		
Badane, CAR[14]	11.7* (3–18)	3
Salonga, Zaire[15]	>60	1

Notes:

[A] All entries are reliable estimates unless followed by an asterisk (*), which indicates complete counts.

[B] In one case it was uncertain if the count was of 1 or 2 parties, so it was treated as 1.5. Some of the Bioko counts probably involved recounts of the same individuals.

[1] Struhsaker (unpub. data, 1992). Rio Epola area.

[2] Struhsaker (1975).

[3] Galat et Galat-Luong (1985).

[4] Oates *et al.* (1990).

[5] Gatinot (1975).

[6] Siex 1995; Struhsaker (unpub. data, 1992–95).

[7] Struhsaker and Leland (1980).

[8] Decker (1994b).

[9] Marsh (1979).

[10] Decker (1989 and 1994a).

[11] Clutton-Brock (1972).

[12] Rodgers *et al.* (1984).

[13] Struhsaker (1975) plus unpub. data from Ngogo (a mix of complete counts and estimates).

[14] Galat-Luong & Galat (1979). Study of <1 month.

[15] Maisels *et al.* (1994).

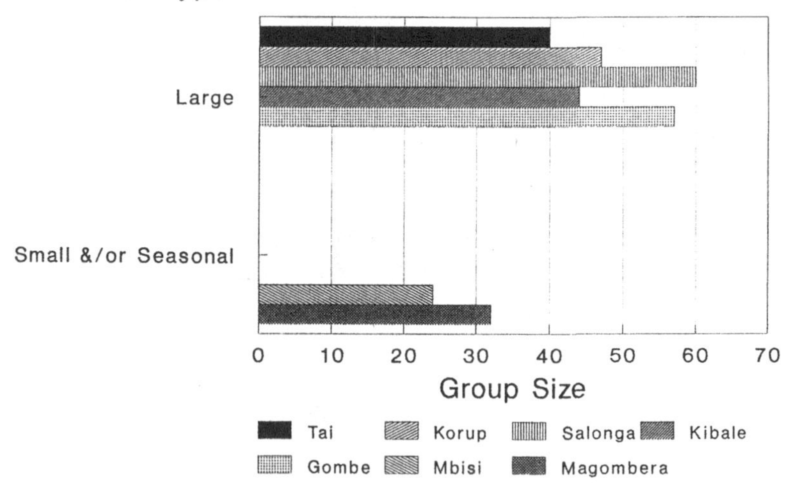

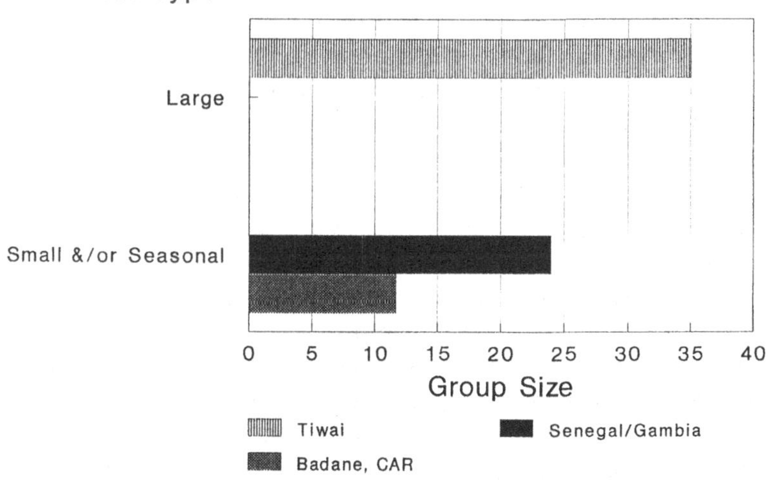

Fig. 15.3. The relation between red colobus social group size, the relative abundance of predaceous eagles, and gross habitat type in red colobus. Both predators and habitat appear to influence group size. Note differences in x axis scale.

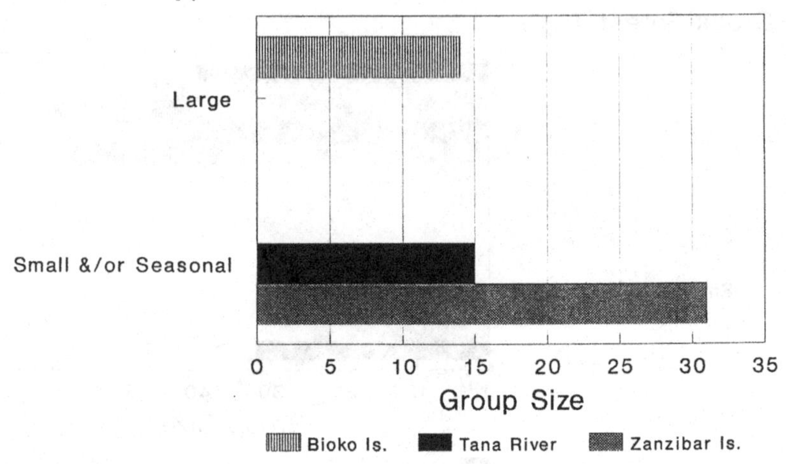

Fig. 15.3. (*cont.*)

Even light logging can influence group size in some species. In Kibale, for example, black and white colobus groups are 25% smaller in lightly logged than unlogged forest, while blue monkey groups in the same logged forest are approximately half the size of those in the adjacent unlogged area (summarized in Struhsaker, 1997). Decrease in the size of foraging parties and social groups presumably reflect declines in density and diversity of food species, as well as a more patchy distribution of food associated with habitat disturbance.

Fusion–fission social groups have also been observed among some of the red colobus living in the ground-water forest at Jozani, Zanzibar (Struhsaker and Siex, unpub. data). Some of these social groups were divided into at least two parties approximately 49% of the time (Siex, 1995). This temporary division of social groups into foraging parties has not been observed in the red colobus groups living in the adjacent shambas. Siex (1995) has shown that food resources are more evenly distributed (less clumped) and that home ranges are smaller among the main shamba study groups than the forest groups. A similar situation may prevail on Bioko. The prevalence of small groups of all species, but particularly the red colobus, may, in fact, reflect small foraging parties of larger, fusion–fission social groups.

It has been suggested that fusion–fission societies, as well as smaller groups, are most likely to occur when predators are absent and food is scarce and/or widely dispersed in a clumped manner (Struhsaker and Leland, 1979). This may well be the case on Bioko and some parts of Zanzibar. In the heavily logged parts of Kibale, where predators are still abundant, the advantages of large social groups to predator avoidance are apparently outweighed by foraging costs, leading to smaller groups or fusion–fission societies.

While the data presented here support the idea that group size is greatly influenced by the complex interplay of predation and food, it would be misleading to suggest that these are the only variables involved. Group size and, especially, group composition are also affected by intrasexual competition. Intolerance amongst adult males, so pronounced in *Cercopithecus*, can result in one-male groups, which, in turn, affects group size. The extent of intrasexual tolerance often closely parallels phylogenetic relationships (Struhsaker and Leland, 1979).

Adult sex ratio in social groups

The presence or absence of predators has no apparent effect on adult sex ratios among the social groups of forest *Cercopithecus* species. In none of the surveys or detailed studies were there deviations from the typical pattern, in which each social group has only one adult male most of the time. It is not known what variables are most important in determining the frequency of the temporary multi-male influxes described for *C. ascanius* and *C. mitis* (Cords, 1988; Struhsaker, 1988).

In contrast, adult sex ratios in red colobus groups are extremely variable between and within study sites (Table 15.9). In those populations with sufficient samples, the coefficient of variation in adult sex ratio of social groups ranged from 23% to 48%. This great variation occurred regardless of the abundance of crowned hawk-eagles; the coefficient of variation was 42% at Magombera (eagles common) and 45–48% on the Tana (eagles absent).

Even within the same social group, the ratio of adult females to adult males varied from 1.5 to 10 over a period of nearly 18 years (Table 15.9, Kibale). Furthermore, these sex ratios can change for entire populations over time. For example, in the Tana River population, the number of adult females per adult male in social groups declined from 7.3 to 5.1 over an 11 to 15 year-period ($U=60$, $p<0.05$, 2-tail, Table 15.10) (Marsh, 1979; Decker, 1994a). The highest ratio of females to males occurred after compression of the population as a result of forest cutting by people. The ratio

Table 15.9. *Adult sex ratio in red colobus*

Species/ subspecies	Location	*n*	$\bar{x}$ no. adult females	$\bar{x}$ no. adult males	$\bar{x}$ ratio: female/male	Crowned hawk-eagle	Source
Colobus badius temminckii	E. Gambia	2	6.5 (3–10)	3 (2–4)	2.2 (1.5–2.5)	Rare	Struhsaker (1975)
C. b. temminckii	Fathala, Senegal	12	13.3 (5–27)	6.7 (3–13)	1.92 (1–2.7)	Rare	Gatinot (1975)
C. b. temminckii	Abuko, Gambia	2	10.5 (8–13)	2.5 (2–3)	4.2	Rare	Starin (1991) (cited in Oates, 1994)
C. b. badius	Tiwai, Sierra Leone	1	13	7	1.8	Rare	A.G. Davies (cited in Oates, 1994)
C. b. badius	Tai, Cote d'Ivoire	1	13	3	4.3	Common	Galat & Galat-Luong (1985)
C. b. tephrosceles	Kibale, Uganda	1	9.3 (2–15)	3.3 (1–6)	3.3 (1.5–10)	Common	Struhsaker (1970–88, unpub. data)*
C. b. tephrosceles	Kibale, Uganda	1	13.7 (8–21)	6.5 (5–9)	2.1 (1.3–2.9)	Common	Struhsaker (1970–88, unpub. data)*
C. b. tephrosceles	Kibale, Uganda	1	17–25	7	2.4–3.6	Common	Struhsaker (1995, unpub. data)**
C. b. tephrosceles	Kibale, Uganda	2	26.5 (21–32)	11.5 (10–13)	2.4 (1.6–3.2)	Common	Clutton-Brock (1972)
C. b. tephrosceles	Gombe, Tanzania	1	28	11	2.5	Common	Clutton-Brock (1972)
C. (b.) gordonorum	Magombera, Tanzania	6	14 (9–26)	1.8 (1–3)	8.5 (3.7–13)	Common	Decker (1994b)
C. b. rufomitratus	Tana River, Kenya	13	9.6 (5–18)	1.5 (1–2)	7.3 (3–11)	Absent	Marsh (1979)

C. b. rufomitratus	Tana River, Kenya	18	5.9 (2–18)	1.1 (0–1)	5.1 (2–11.3)	Absent	Decker (1994a)
C. kirkii	Jozani, Zanzibar, Tanzania	7	8.9 (6–16)	4.9 (3–9)	2.2 (1.1–5.3)	Absent	Silkiluwasha (1981)
C. kirkii	Jozani, Zanzibar, Tanzania	3	17.1 (14–21)	3.0 (2–5)	5.7 (4.7–7.7)	Absent	Mturi (1991)
C. kirkii	Jozani, Zanzibar, Tanzania	4	13.7 (9–18)	3.5 (2–5)	4.1 (3–7.3)	Absent	Struhsaker & Siex (unpub. data)***
C. kirkii	Jozani, Zanzibar, Tanzania	11	8.9 (2–27)	1.5 (1–5)	5.9 (3–12)	Absent	Struhsaker & Siex (unpub. data)****

Notes:

* Struhsaker data for Kibale are for two groups. The first, CW, was studied from 1970–1988 at Kanyawara. The numbers and sex ratios are weighted means computed from counts during 200 months. The second group, RUL, was studied at Ngogo from 1978–83. Numbers and sex ratios are weighted means computed from counts over 64 months.

** A group of 54 was counted once at Kanyawara, Kibale. 17 adult females were definitely identified. In addition, there were three to five subadult/young adult females.

*** Groups in Jozani Forest Reserve.

**** Groups in perennial gardens mixed with secondary scrub adjacent to Jozani Forest.

Table 15.10. *Comparison of red colobus monkey adult sex ratios*[A]. *Results of Mann-Whitney U Test; no = does not support hypothesis that predation pressure leads to lower ratio of females per male; yes = supports hypothesis*

Sites compared, (females/male)	U value
(1) *Eagle rare versus common:* Senegal, Gambia, and Tiwai ($\bar{x}=2.5$, $n=16$) versus Tai, Kibale, and Gombe ($\bar{x}=2.9$, $n=7$)	22** no
(2) *Eagle common versus common:* Magombera ($\bar{x}=8.5$, $n=6$) versus Kibale, and Gombe ($\bar{x}=2.7$, $n=6$)	0*** no
(3) *Eagle common versus rare:* Magombera ($\bar{x}=8.5$, $n=6$) versus Senegal, Gambia, and Tiwai ($\bar{x}=2.5$, $n=16$)	1*** no
(4) *Eagle absent versus absent:* Tana River (Marsh 1979, $\bar{x}=7.3$, $n=13$) versus Tana River (Decker 1994a, $\bar{x}=5.1$, $n=17$)	60** no
(5) *Eagle common versus absent:* Magombera (Decker 1994b, $\bar{x}=8.5$, $n=6$) versus Tana (Decker 1994a, $\bar{x}=5.1$, $n=17$)	22** no
(6) *Eagle common versus absent:* Magombera ($\bar{x}=8.5$, $n=6$) versus Tana (Marsh 1979, $\bar{x}=7.3$, $n=13$)	30 (ns) no
(7) *Eagle absent versus absent:* Zanzibar (Silkiluwasha 1981, $\bar{x}=2.2$, $n=7$) versus Zanzibar (Mturi, 1991; Struhsaker and Siex, unpub. data, $\bar{x}=5.5$, $n=18$)	10*** no
(8) *Eagle absent versus absent:* Jozani Forest, Zanzibar ($\bar{x}=4.1$, $n=4$) versus Jozani Shambas, Zanzibar ($\bar{x}=5.9$, $n=11$)	9_a no
(9) *Eagle absent versus absent:* Tana River ($\bar{x}=6.2$, $n=30$) versus Zanzibar ($\bar{x}=5.5$, $n=18$)	243 (ns) yes
(10) *Eagle common versus absent:* Magombera ($\bar{x}=8.5$, $n=6$) versus Zanzibar ($\bar{x}=5.5$, $n=18$)	27* no
(11) *Eagle common versus absent:* Kibale and Gombe ($\bar{x}=2.7$, $n=6$) versus Zanzibar ($\bar{x}=5.5$, $n=18$)	6*** yes
(12) *Eagle common versus absent:* Kibale and Gombe ($\bar{x}=2.7$, $n=6$) versus Zanzibar (Silkiluwasha 1981, $\bar{x}=2.2$, $n=7$)	11 (ns) no
(13) *Eagle common versus absent:* Kibale and Gombe ($\bar{x}=2.7$, $n=6$) versus Tana River ($\bar{x}=6.2$, $n=30$)	18*** yes
(14) *Eagle rare versus absent:* Senegal, Gambia, Tiwai ($\bar{x}=2.5$, $n=16$) versus Zanzibar ($\bar{x}=5.5$, $n=18$)	6*** yes?
(15) *Eagle rare versus absent:* Senegal, Gambia, Tiwai ($\bar{x}=2.5$, $n=16$) versus Tana River ($\bar{x}=6.2$, $n=30$)	25*** yes?

Notes:
A: details found in Table 15.9
all tests 2-tail: * $p<0.10$, ** $p<0.05$, *** $p<0.002$, ns: $p>0.10$, a: $p<0.10$, 1-tail.

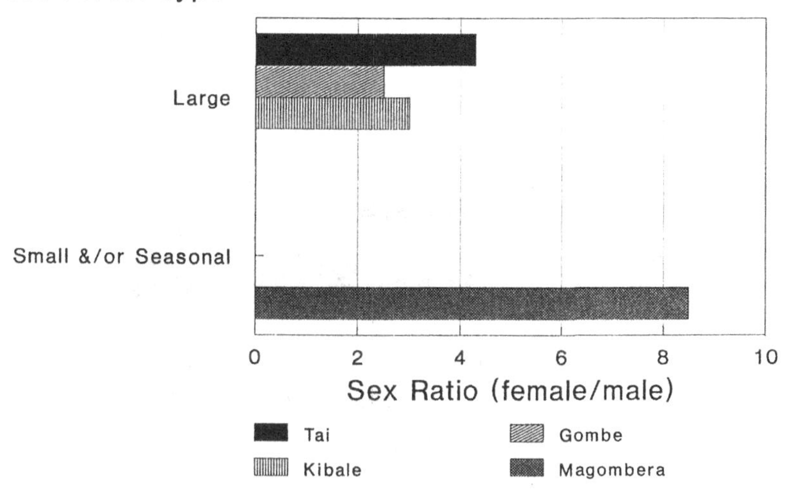

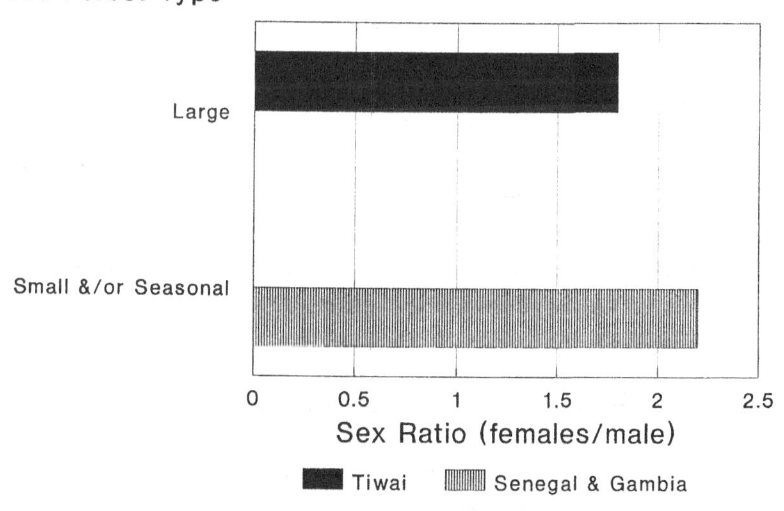

Fig. 15.4. The relation between adult sex ratios in red colobus social groups and the relative abundance of predators and gross habitat quality. Neither predators nor gross habitat quality are strong predictors of sex ratios in groups. Note differences in x axis scale.

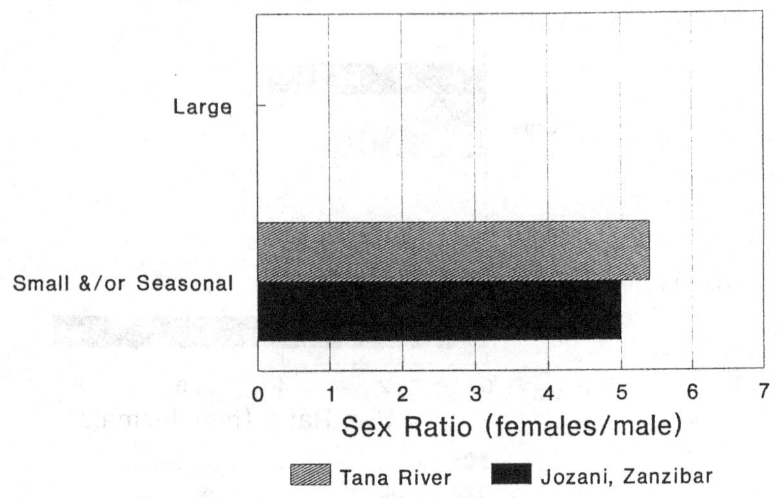

Fig. 15.4. (*cont.*)

declined as the population crashed over the next 11–15 years, apparently because of proportionately greater mortality among females. Indeed, there were significantly fewer adult females in the Tana study groups at the end of the 15-year period than at the beginning (5.6/group vs. 9.6, $U=27$, $p<0.001$, 2-tail).

According to the argument of van Schaik and Horstermann (1994), one would expect to find a greater representation of adult males in red colobus social groups where predation by eagles is present, because males are presumed to defend the females. Contrary to this expectation, however, no consistent correlation is seen between adult sex ratios and the relative abundance of crowned hawk-eagles (Table 15.9 and Figure 15.4).

Unfortunately, no counts of red colobus groups on Bioko island were comprehensive in terms of age–sex composition. Many of the groups on Bioko appeared to have only one adult male, but recall that these groups were unusually small and so the sex ratio may have also been low. At least one group definitely had two adult males and on a few other occasions two sources of male loud calls were heard near to one another, suggesting that there was more than one male in a group. On another occasion, a group of 10–15 red colobus on Bioko had at least three adult males and two adult females (one with a clinging young infant and the other with a sexual swelling). Contrary to the predation hypothesis, the adult sex ratio in this group may have been between 1.5 and 2.5, not unlike mainland populations where eagles are common.

Two populations have sex ratios that could be interpreted as supporting the predation hypothesis. Eagles are absent from the Tana River and Zanzibar, and the adult sex ratios of these populations are generally higher than sites where eagles are present, except for Magombera (Tables 15.9 and 15.10). Unfortunately, the comparison is confounded because these two sites are also characterized by small, fragmented, seasonal, and relatively simple habitats, which, as the Tana River case shows, can influence sex ratios through differential mortality.

In contrast, many examples are inconsistent with the predation hypothesis (Tables 15.9, 15.10, and Figure 15.4). For example, very high ratios occur at Magombera and in the Tai forest, both sites where eagles are common (Table 15.9; T.T. Struhsaker, unpublished data), while the higher ratios expected where eagles are rare are not seen at Tiwai, Senegal, or Gambia. When all sites were statistically compared with one another (Table 15.10), fewer than one-third of the 15 tests support the hypothesis that adult sex ratios should decline with predation pressure from eagles. In fact, the number of adult males in a red colobus group is highly correlated with the number of adult females, regardless of whether crowned hawk-eagles were common, rare, or absent ($r = 0.85$, $p < 0.01$, $df = 14$; Table 15.9, Figure 15.5). Furthermore, the residuals of the regression of male to female numbers show no consistent pattern in relation to eagle abundance (Figure 15.5). The great intrapopulational variance in group sex ratios (see above) also indicates low selective pressure by eagles in this regard. These results are consistent with interspecific comparisons showing that the number of adult males in social groups is strongly corrrelated with the number of adult females (Mitani *et al.*,1996).

Perhaps one of the greatest weaknesses of the predation hypothesis is the assumption that adult male monkeys in the rainforest are effective deterrents against eagle attacks and that more males provide greater protection. Indeed, adult male monkeys will charge and displace eagles *after* they have attacked, but there is little evidence that males prevent eagle attacks. An interspecific comparison of primates found no significant correlation between estimated predation rates and the number of adult males per group, nor were any differences found in predation rates between species with multimale versus one male social groups (Cheney and Wrangham, 1987).

In Kibale, crowned hawk-eagles prey less upon red colobus than

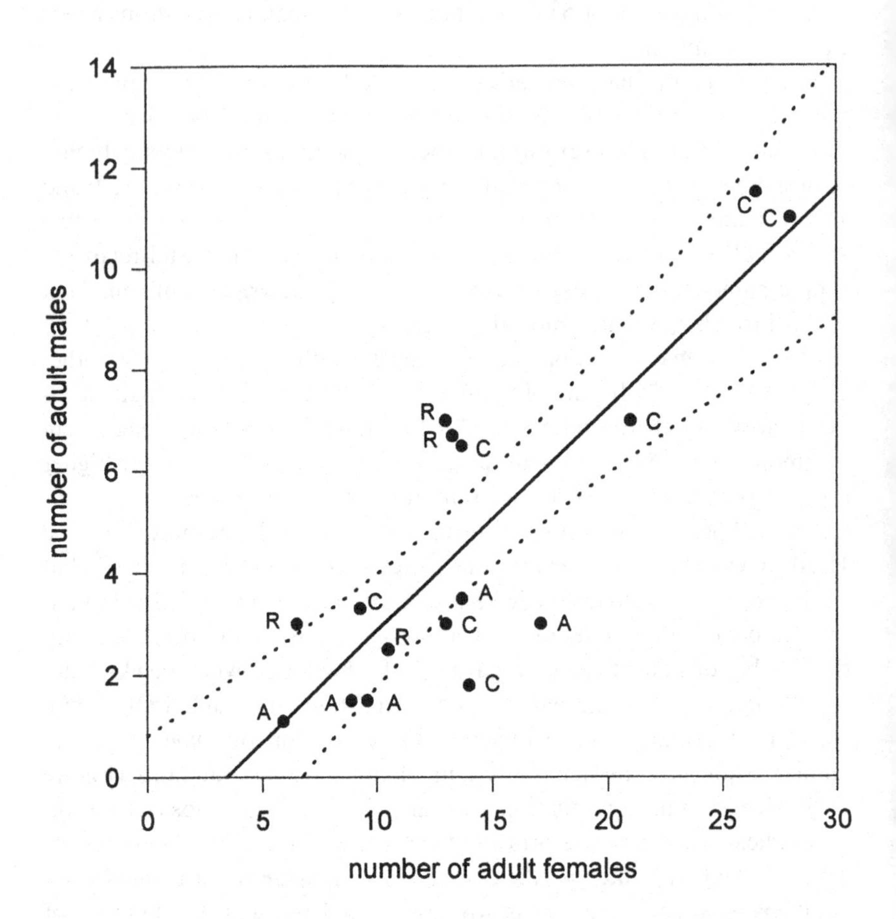

Fig. 15.5. Significant regression of numbers of red colobus adult males to adult females and 95% confidence limits (data from Table 15.9, excluding Silkiluwasha, 1981) regardless of eagle abundance.

expected (Struhsaker and Leakey, 1990), but it is unclear to what extent this is due to the unusually aggressive nature of male red colobus, or their high frequency of association with other primates, particularly redtail monkeys, which are among the first to detect eagles. Certainly, prey selection by eagles is not strictly determined by adult sex ratios of groups, because both

Colobus guereza in their one-male groups and *Cercocebus albigena* in multi-male groups were preyed upon more than expected. Although predation may not select for greater numbers of males in a group, male-biased predation by eagles contributes directly to female-biased sex ratios in these two species (Struhsaker and Leakey, 1990). Even if males are to some extent effective in deterring eagle attacks, it must be determined what constitutes a biologically significant difference in sex ratio. Is, for example, a sex ratio of 1:3 any more effective in avoiding eagle attacks than a ratio of 1:4?

Although predation by eagles may influence adult sex ratios in social groups, other variables are equally if not more important. Habitat quality is one of these, causing differential mortality between the sexes, especially in highly seasonal, fragmented, and small blocks of forest. The extent of this differential mortality also varies. Male mortality was apparently high at Magombera, but low in Senegal and Gambia (Tables 15.9 and 15.10). On the Tana, mortality was higher in males than females, but during the population crash of the late 1970s and early 1980s, female mortality increased and the sex ratio decreased (Tables 15.9 and 15.10). Even within populations there are differences in sex ratio that can be related to habitat quality. On Zanzibar, red colobus adult sex ratios were slightly higher in the shamba habitat than in adjacent ground-water forest (Table 15.10) in spite of similarly low predation pressures.

Perhaps two of the most important variables in determining sex ratios in social groups are the extent of competition and tolerance between conspecific males. These are often closely correlated with phylogeny. Sex ratios in groups will also be affected by the degree to which males can monopolize females and mating opportunities and whether males form coalitions (e.g. Struhsaker and Leland, 1979). These variables, in turn, will be influenced by habitat quality and demography. Finally, sex ratios in social groups are probably influenced by the ratio between reproductive and foraging benefits and costs to the fitness of group members (Struhsaker and Leland, 1979).

In summary, the variation in adult sex ratios among red colobus groups cannot be explained by a single factor, but appears to be dynamically dependent on the complex interplay of several variables, including predation, habitat quality, phylogeny, male–male tolerance, mating opportunities, and demographic trends. The importance of predation in determining adult sex ratios in social groups has been exaggerated in other species besides red colobus. Van Schaik and Horstermann (1994), for example, suggest that predation by harpy eagles leads to the formation of multimale groups in howler monkeys (*Alouatta*), although for at least half the sites at which they imply that harpy eagles are present, there are, in fact, no records of this eagle.

Terrestriality

Monkeys were often seen foraging and walking on the ground on both Bioko and Zanzibar islands, where leopards are absent. During 16 days of field work on Bioko, monkeys were seen on the ground 13 times (6 red colobus; 4 *Cercopithecus erythrotis*; 3 *C. pogonias*). On Zanzibar, red colobus and Sykes were seen on the ground many times every day, especially in the shambas, in spite of the fact that they were occasionally chased by dogs in this habitat. Dogs probably constitute a far less serious threat than do leopards.

On the mainland, where leopards are present or only recently exterminated, the same or closely related monkey species rarely come to the ground. Although red colobus at Kibale did come to the ground, they were never seen to do so in more than 418 encounters during systematic censuses over an 18-year period. Social groups of *C. ascanius* (same superspecies as *C. erythrotis*) were encountered more than 317 times during the same censuses, but never on the ground. *Cercopithecus pogonias* was never seen on the ground during nearly three years of field work in Cameroon.

Most, if not all, species occasionally travel and forage on the ground in mainland forests, but, compared to the island monkeys, they do so very rarely. By contrast, much larger anthropoid primates such as baboons (*Papio*), drills (*Mandrillus leucophaeus*), and chimpanzees (*Pan troglodytes*) spend a great deal of their time on the ground, whether leopards are in the forest or not. Presumably, their larger body size makes them less vulnerable.

As well as the absence of leopards, the distribution of food may also be a factor influencing terrestriality, but this seems unlikely on Bioko for reasons already described. It is clear, however, that in the absence of leopards, the monkeys more freely use foods on the ground.

Summary and conclusions

New data from the islands of Bioko and Zanzibar, combined with published and unpublished data from the African mainland, provide a more realistic understanding of the relative importance of predation and habitat quality to monkey socioecology. Aside from humans, the most important predator of African forest monkeys is the crowned hawk-eagle. The presence and relative abundance of this predator is the single most important

Interspecific Associations Between Social Groups

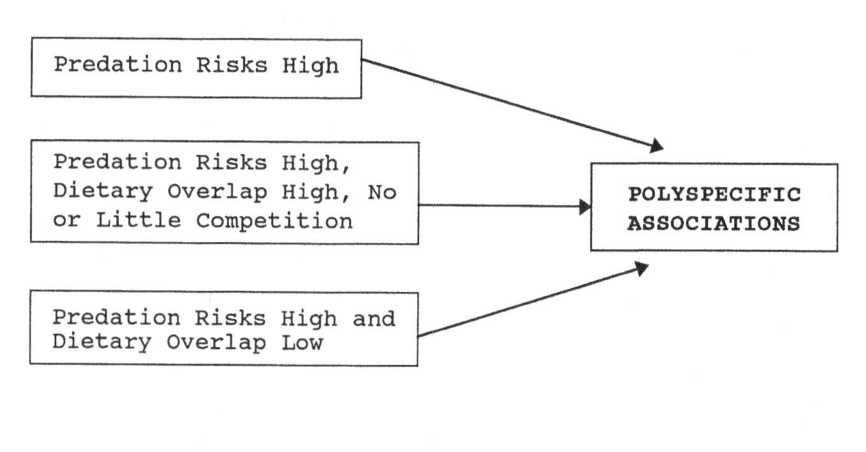

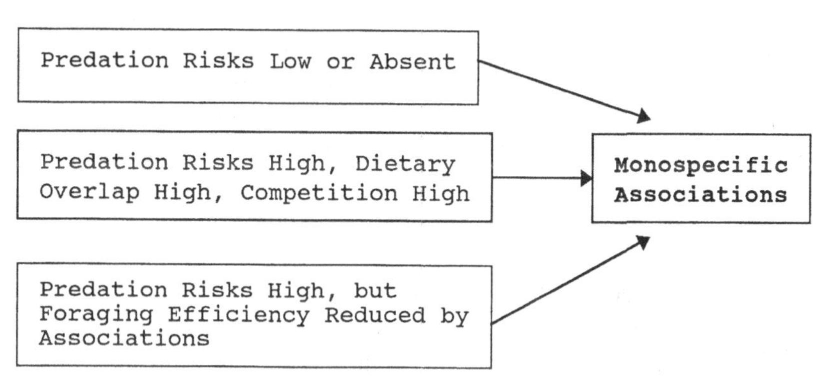

Fig. 15.6. Summary of conditions leading to polyspecific and monospecific associations.

predictor of polyspecific associations, although other variables such as habitat quality, and foraging costs and benefits to a lesser extent also influence the species composition, frequency, and duration of mixed-species associations (Figure 15.6).

The presence of crowned hawk-eagles also appears to be very important in determining the frequency of solitary monkeys. Red colobus populations, especially, include fewer solitary monkeys where predation pressure is higher. Amongst *Cercopithecus*, however, phylogeny apparently dictates that one-male groups prevail whether predators are present or not, and the ratio of solitary monkeys (usually males) to social groups also varies less.

Social group size is also affected both by eagle predation and habitat quality, groups tending to be smaller when eagles are absent, or when the habitat consists of small and fragmented blocks of highly seasonal forest. Small groups and fusion–fission societies sometimes develop in degraded forest even when eagles are abundant. Again, the costs and benefits of foraging with other monkeys are apparently balanced.

Although the adult sex ratio in social groups of monkeys has been considered to be a function of eagle predation, the data from African forest monkeys, and *Alouatta* of Central and South America, do not support this idea. The great variance in sex ratios within populations, within social groups and over time at a given site suggests that predation by eagles is but one variable influencing sex ratios, along with habitat quality, intrasexual tolerance, phylogeny, and demographic trends. Finally, forest monkeys travel and forage on the ground much more often in the absence of leopards.

These findings support the idea that most, if not all, aspects of primate socioecology can best be understood and described as the result of a complex interplay of several variables, an interplay in which the relative contribution of each of the interacting parameters to the observed variance can and does change with time. Attempts at univariate explanations contribute relatively little to a realistic understanding of primate socioecology.

Acknowledgements

Thanks are extended to the following organizations and individuals who made this research possible. Research on Zanzibar was funded by the National Geographic Society and FinnIDA. The Zanzibar Forest Department gave permission and assistance. Valuable assistance on Zanzibar was provided by Mr. Leroy Duvall, Ms. Anna-Lisa Raunio, Ms. Merja Makela, Messrs. Hilal Juma, Timo Vihola, Amour Bakari Omar, and Hamza Rijal. Research on Bioko was funded by the Species Survival Commission of IUCN and Zoo Atlanta. I am particularly grateful to Dr. Dietrich Schaaf for arranging the trip to Bioko and dealing with all the logistic and organization details in the field. The U.S. Ambassador and his staff provided invaluable assistance and hospitality on Bioko. Saturnino Malest Ballovera generously shared his knowledge of the forest and the sea with us. Drs. Carolyn Ehardt, John Mitani, and John Oates and an anonymous reviewer are thanked for valuable comments on the manuscript.

References

Berenstain, L. (1986). Responses of long-tailed macaques to drought and fire in eastern Borneo: a preliminary report. *Biotropica* **18**, 257–62.

Butynski, T.M. & Koster, S.H. (1989). The status and conservation of forests and primates on Bioko Island (Fernando Poo), Equatorial Guinea. (World Wildlife Fund unpub. report.)

Butynski, T.M. & Koster, S.H. (1994). Distribution and conservation status of primates in Bioko island, Equatorial Guinea. *Biodivers. Conservat.* **3**, 893–909.

Castelo, R. (1994). Biogeographical considerations of fish diversity in Bioko. *Biodivers. Conservat.* **3**, 808–27.

Chapman, C.A. & Chapman, L.J. (1996). Mixed-species primate groups in the Kibale Forest: ecological constraints on association. *Int. J. Primatol.* **17**, 31–50.

Cheney, D.L. & Wrangham, R.W. (1987). Predation. In *Primate Societies*, ed. B.B. Smuts, D.L. Cheney, R.M. Seyfarth, R.W. Wrangham & T.T. Struhsaker, pp. 227–39. Chicago: University of Chicago Press.

Clutton-Brock, T.H. (1972). *Feeding and Ranging Behaviour of the Red Colobus Monkey.* PhD thesis, University of Cambridge.

Cords, M. (1987). Mixed-species association of *Cercopithecus* monkeys in the Kakamega forest. *University of California Publications in Zoology* **117**, 1–109.

Cords, M. (1988). Mating systems of forest guenons: a preliminary review. In *A Primate Radiation: Evolutionary Biology of the African Guenons*, ed. A. Gautier-Hion, F. Bourliere, J.-P. Gautier & J. Kingdon, pp. 323–39. Cambridge: Cambridge University Press.

Crook, J.H. & Gartlan, J.S. (1966). Evolution of primate societies. *Nature* **210**, 200–3.

Decker, B.S. (1989). *Effects of Habitat Disturbance on the Behavioural Ecology and Demographics of the Tana River Red Colobus* Colobus badius rufomitratus. PhD thesis, Emory University.

Decker, B.S. (1994a). Effects of habitat disturbance on the behavioral ecology and demographics of the Tana River red colobus (*Colobus badius rufomitratus*). *Int. J. Primatol.* **15**, 703–37.

Decker, B.S. (1994b). Endangered primates in the Selous Game Reserve and an imminent threat to their habitat. *Oryx* **28**, 183–90.

Edwards, A.E. (1992). *The Diurnal Primates of Korup National Park, Cameroon: Abundance, Productivity and Polyspecific Associations.* MSc thesis, University of Florida, Gainesville.

Fager, E.W. (1957). Determination and analysis of recurrent groups. *Ecology* **38**, 586–95.

Galat, G. & Galat-Luong, A. (1985). La communaute de primates diurnes de la foret de Tai, Cote-D'Ivoire. *Rev. Ecol. (Terre Vie)* **40**, 3–32.

Galat-Luong, A. & Galat, G. (1979). Quelques observations sur l'ecologie de *Colobus pennanti oustaleti* en Empire Centrafricain. *Mammalia* **43**, 309–12.

Gartlan, J.S. & Struhsaker, T.T.(1972). Polyspecific associations and niche separation of rain forest anthropoids in Cameroon, West Africa. *J. Zool. Soc. Lond.* **168**, 221–66.

Gatinot, B.L. (1975). *Ecologie d'un Colobe Bai* (Colobus badius temmincki, *Kuhn 1820) dans un Milieu Marginal au Senegal.* PhD thesis, Universite de Paris VI, France.

Gautier, J.-P. & Gautier-Hion, A.(1969). Les associations polyspecifiques chez les Cercopithecidae du Gabon. *La Terre et la Vie.* **23**, 164–201.

Gautier-Hion, A. (1988). Polyspecific associations among forest guenons: ecological, behavioural and evolutionary aspects. In *A primate radiation: evolutionary biology of the African guenons*, ed. A. Gautier-Hion, F.

Bourliere, J.-P. Gautier & J.Kingdon, pp. 452–76. Cambridge: Cambridge University Press.

Gautier-Hion, A. & Gautier, J.-P. (1974). Les associations polyspecifiques des Cercopitheques du plateau de M'passa, Gabon. *Folia primatol.* **22**, 134–77.

Hall, K.R.L. (1965). Social organization of the old world monkeys and apes. *Symp. Zool. Soc. Lond.* **14**, 265–89.

Hamilton, A. C. (1982). *Environmental History of East Africa.* London: Academic Press.

Isbell, L.A. (1994). Predation on primates: ecological patterns and evolutionary consequences. *Evol. Anthropol.* **3**, 61–71.

Johns, A.D. (1983). *Ecological Effects of Selective Logging in a West Malaysian Rain-forest*. PhD thesis, Cambridge University, Cambridge.

Lwanga, J.S. (1987). *Group Fission in Blue Monkeys* (Cercopithecus mitis stuhlmanni): *Effects on the Socioecology in Kibale Forest, Uganda.* MSc thesis, Makerere University, Kampala, Uganda.

Maisels, F., A. Gautier-Hion, & J.-P. Gautier. (1994). Diets of two sympatric colobines in Zaire: more evidence on seed-eating in forests on poor soils. *Int. J. Primatol.* **15**, 681–701.

Marsh, C.W. (1979). Comparative aspects of social organization in the Tana River red colobus, *Colobus badius rufomitratus. Z. Tierpsychol.* **51**, 337–63.

McGraw, S. (1994). Census, habitat preference, and polyspecific associations of six monkeys in the Lomako Forest, Zaire. *Am. J. Primatol.* **34**, 295–307.

McKey, D.B. & Waterman, P.G. (1982). Ranging behavior of a group of black colobus (*Colobus satanas)* in the Douala-Edea Reserve, Cameroun. *Folia primatol.* **39**, 264–304.

Mitani, J., Gros-Louis, J. & Manson, J.H. (1996). Number of males in primate groups: comparative tests of competing hypotheses. *Am. J. Primatol.* **38**, 315–32.

Mturi, F.A. (1991). *The Feeding Ecology and Behaviour of the Red Colobus Monkey* (Colobus badius kirkii). PhD thesis, University of Dar es Salaam, Tanzania.

Oates, J.F. (1977). *The Ecology and Behaviour of the Black and White Colobus Monkey* (Colobus guereza *Ruppell) in East Africa.* PhD thesis, University of London.

Oates, J.F. (1994). The natural history of African colobines. In *Colobine Monkeys: Their Ecology, Behaviour and Evolution*, ed. A.G. Davies & J.F. Oates, pp. 75–128. Cambridge: Cambridge University Press.

Oates, J.F. & Whitesides, G.H. (1990). Association between olive colobus (*Procolobus verus)*, Diana guenons (*Cercopithecus diana)* and other forest monkeys in Sierra Leone. *Am. J. Primatol.* **21**, 129–46.

Oates, J.F., Whitesides, G.H., Davies, A.G., Waterman, P.G., Green, S.M., Dasilva, G.L. & Mole, S. (1990). Determinants of variation in tropical forest primate biomass: new evidence from West Africa. *Ecology* **71**, 328–43.

Rodgers, W.A., Struhsaker, T. T. & West, C. C. (1984). Observations on the red colobus (*Colobus badius tephrosceles)* of Mbisi Forest, south-west Tanzania. *Afr. J. Ecol.* **22**, 187–94.

van Schaik, C.P. (1983). Why are diurnal primates living in groups? *Behav.* **87**, 120–44.

van Schaik, C.P. & van Hooff, J.A.R. A.M. (1983). On the ultimate causes of primate social systems. *Behaviour* **85**, 91–117.

van Schaik, C. P. & Horstermann, M. (1994). Predation risk and the number of adult males in a primate group: a comparative test. *Behav. Ecol. Sociobiol.* **35**, 261–72.

van Schaik, C.P. & van Noordwijk, M.A. (1985). Evolutionary effect of the absence of felids on the social organization of the macaques on the island of Simeulue (*Macaca fascicularis fusca*, Miller 1903). *Folia primatol.* **44**, 138–147.

Siex, K.S. (1995). *The Zanzibar Red Colobus Monkey* (Procolobus kirkii): *Ecology, Demography and Use of Cocos nucifera.* MSc thesis, University of Florida, Gainesville.

Silkiluwasha, F. (1981). The distribution and conservation status of the Zanzibar red colobus. *Afr. J. Ecol.* **19**, 187–94.

Skorupa, J.P. (1986). Responses of rainforest primates to selective logging in Kibale Forest, Uganda: a summary report. In *Primates, The Road to Self-Sustaining Populations*, ed. K. Benirschke, pp 57–70. New York: Springer-Verlag.

Skorupa, J.P. (1988). *The Effects of Selective Timber Harvesting on Rainforest Primates in Kibale Forest, Uganda.* PhD thesis, University of California, Davis.

Skorupa, J.P. (1989). Crowned eagles *Stephanoaetus coronatus* in rainforest: observations on breeding chronology and diet at a nest in Uganda. *Ibis* **131**, 294–8.

Struhsaker, T.T. (1969). Correlates of ecology and social organization among African cercopithecines. *Folia primatol.* **11**, 80–118.

Struhsaker, T.T. (1975). *The Red Colobus Monkey.* Chicago: University of Chicago Press.

Struhsaker, T.T. (1978). Food habits of five monkey species in the Kibale Forest, Uganda. In *Recent Advances in Primatology*, vol. 1, ed. D.J. Chivers & J. Herbert, pp. 225–48. New York: Academic Press.

Struhsaker, T.T. (1981). Polyspecific associations among tropical rainforest primates. *Z. Tierpsychol.* **57**, 268–304.

Struhsaker, T.T. (1988). Male tenure, multi-male influxes, and reproductive success in redtail monkeys (*Cercopithecus ascanius*). In *A Primate Radiation: Evolutionary Biology of the African Guenons*, ed. A. Gautier-Hion, F. Bourliere, J.-P. Gautier & J. Kingdon. pp. 340–63. Cambridge:Cambridge University Press.

Struhsaker, T.T. (1997). *Ecology of an African Rainforest: Logging in Kibale and the Conflict Between Conservation and Exploitation.* Gainesville: University Press of Florida.

Struhsaker, T.T., Butynski, T M. & Lwanga, J.S. (1988). Hybridization between redtail (*Cercopithecus ascanius schmidti*) and blue (*C. mitis stuhlmanni*) monkeys in the Kibale Forest, Uganda. In *A Primate Radiation: Evolutionary Biology of the African Guenons.* ed. A. Gautier-Hion, F. Bourlier, J.-P. Gautier & J. Kingdon. pp. 477–97. Cambridge: Cambridge University Press.

Struhsaker, T.T. & Leakey, M. (1990). Prey selectivity by crowned hawk-eagles on monkeys in the Kibale Forest, Uganda. *Behav. Ecol. Sociobiol.* **26**, 435–43.

Struhsaker, T.T. & Leland, L. (1979). Socioecology of five sympatric monkey species in the Kibale Forest, Uganda. In *Advances in the study of behavior*, vol. 9, ed. J. S. Rosenblatt, R.A. Hinde, C. Beer & M.C. Busnel, pp. 159–228. New York: Academic Press.

Struhsaker, T.T. & Leland, L. (1980). Observations on two rare and endangered populations of red colobus monkeys in East Africa: *Colobus badius gordonorum* and *Colobus badius kirkii. Afr. J. Ecol.* **18**, 191–216.

Terborgh, J. (1990). Mixed flocks and polyspecific associations: costs and benefits of mixed groups to birds and monkeys. *Am. J. Primatol.* **21**, 87–100.

Waser, P.M. (1980). Polyspecific associations of *Cercocebus albigena*: geographic variation and ecological correlates. *Folia primatol.* 33:57–76.
Waser, P.M. (1982). Primate polyspecific associations: do they occur by chance? *Anim. Behav.* **30**, 1–8.
Waser, P.M. (1987). Interactions among primate species. In *Primate Societies*, ed. B. B. Smuts, D.L. Cheney, R.M. Seyfarth, R.W. Wrangham & T T. Struhsaker, pp. 210–26. Chicago: University of Chicago Press.
Wasser, S.K. (1993). The socioecology of interspecific associations among the monkeys of the Mwanihana rain forest, Tanzania: a biogeographic perspective. In *Biogeography and Ecology of the Rain Forests of Eastern Africa*, ed. J.C. Lovett & S.K. Wasser, pp. 267–80. Cambridge: Cambridge University Press.
Whitesides, G.H. (1981). *Community and Population Ecology of Non-Human Primates in the Douala-Edea Forest Reserve.* MSc thesis, The Johns Hopkins University, Baltimore.
Whitesides, G.H. (1989). Interspecific associations of Diana monkeys, *Cercopithecus diana*, in Sierra Leone, West Africa: biological significance or chance? *Anim. Behaviour* **37**, 760–76.
Wrangham, R.W. (1980). An ecological model of female-bonded primate groups. *Behaviour.* **75**, 262–300.
Wrangham, R.W. (1987). Evolution of social structure. In *Primate Societies*, ed. B. B. Smuts, D. L. Cheney, R. M. Seyfarth, R. W. Wrangham & T.T. Struhsaker, pp. 282–96. Chicago: University of Chicago Press.

16

The loud calls of black-and-white colobus monkeys: their adaptive and taxonomic significance in light of new data

JOHN F. OATES, CAROLYN M. BOCIAN AND CARL J. TERRANOVA

Introduction

Writing in the original *Old World Monkeys* volume, Struhsaker (1970) demonstrated for the first time how the loud calls of primates, analyzed with sound spectrograms, can be used in phylogenetic analysis, particularly to understand better the relationships within a genus or species-group. Struhsaker's study built on the work of Marler (1957) and others who had used vocalizations in the study of non-primate taxa. Struhsaker's analysis has been followed by many other primate studies, including those of Wilson and Wilson (1975) on leaf-monkeys (*Presbytis*), Marshall and Marshall (1976) and Mitani (1987) on gibbons (*Hylobates*), Oates and Trocco (1983) on black-and-white colobus, Snowdon *et al.* (1986) on lion tamarins (*Leontopithecus*), Gautier (1988) on guenons (*Cercopithecus*), Zimmermann (1990) on galagos (Galagonidae), and Whitehead (1995) on howler monkeys (*Alouatta*).

These spectrographic studies have found that primate vocalizations, particularly male loud calls, are both relatively stable characters within species, and also measurably different between species. However, most of these previous studies have used limited samples (a few individuals from a few sites) and rarely have they been repeated, or their conclusions tested by examining samples from additional animals or new sites.

The work we present here is a reappraisal of the analysis and conclusions of the study of black-and-white colobus male loud calls by Oates and Trocco (1983). It involves both a new analysis of the original sample of recordings using different methods, and the addition of new recordings from several sites not included in the earlier analysis. It also incorporates new information on black-and-white colobus social behavior and ecology, particularly from a study by one of us (C.B.) on sympatric *Colobus angolensis* and *C. guereza* in the Ituri Forest, Democratic Republic of Congo (former Zaire). The Ituri is the only known area where two black-and-white colobus species still co-occur.

Oates and Trocco (1983) found that black-and-white colobus grouped into three clusters based on the temporal patterning ("tempo"), frequency ("pitch") and modulation of male loud calls, or "roars". These clusters were:

- *C. guereza* and *C. vellerosus*, whose roars had a slow tempo and a low pitch.
- *C. angolensis* and *C. polykomos*, whose roars had an intermediate tempo and a high or intermediate pitch.
- *C. satanas*, whose roars had a fast tempo and a high pitch, as well as often exhibiting modulation.

Combining this information with data on cranial morphology (e.g. Hull, 1979) and pelage (e.g. Rahm, 1970), Oates and Trocco, (1983) concluded that the black-and-white colobus monkeys are best regarded as belonging to five species: *Colobus polykomos* (Zimmerman, 1780); *C. vellerosus* (Geoffroy, 1831); *C. guereza* (Rüppell, 1835); *C. satanas* (Waterhouse, 1838); and *C. angolensis* (Sclater, 1860). Oates and Trocco (1983) also concluded that *C. satanas* retains more primitive features than other members of the group, and that *C. guereza* and *C. vellerosus*, in their low-frequency loud call and other features, are the most derived species.

Although some other, non-vocal, evidence (e.g. Groves *et al.*, 1993) has tended to support the taxonomic conclusions drawn by Oates and Trocco, the sample sizes used in that earlier analysis were small, thus weakening the force of the conclusions reached. *Colobus guereza* was represented by bouts of roaring from five different animals at four different sites, and *C. angolensis* by roaring from three animals at two sites; but *C. polykomos*, *C. satanas* and *C. vellerosus* were each represented by only a single animal at a single site.

We therefore wondered whether consideration of a larger sample would produce any different conclusions, particularly if more data were included from species poorly-represented in the original analysis. We were also intrigued by the fact that on an initial visit to the Ituri Forest, where *C. angolensis* and *C. guereza* live sympatrically without interbreeding, one of us (C.B.) found it difficult to discriminate immediately between the roaring calls of males of the two species. This raised the possibility that the conclusions from the original analysis were biased by a small sample size; perhaps loud calls were not as reliable a tool in taxonomic and phylogenetic analysis as we had supposed.

Here we present analysis of a larger sample of roars, including material from five new sites. We have also used new methods to analyze both the new data, and the recordings used by Oates and Trocco (1983). Our statistical analysis focuses on two quantitative and readily measured features of the male roar that were found by Oates and Trocco to clearly discriminate between three clusters of species: pulse rate, and the median of the fundamental (or "dominant") frequency; these terms are defined below. We consider whether this new analysis points to the same conclusions reached by Oates and Trocco, and we also use the new information on the behavior of black-and-white colobus populations in the Ituri Forest to speculate on the evolution of patterns of variation in roaring across species.

Materials and methods

Terminology

Hill and Booth (1957) expressed the sound of an adult male black-and-white colobus loud call as "rurr rurr rurr rurr." Using the terminology of Marler (1972) this can be thought of as a *sequence* of roaring, made up of individual *phrases* (a single "rurr" is therefore one phrase). Roar phrases are usually delivered in a series, or sequence, and sequences are often organized into *bouts*. Oates and Trocco (1983) defined a bout as a series of roaring sequences produced by a single animal, separated from other bouts by a non-roaring period of at least one minute. A spectrographic examination of roar phrases reveals that each phrase is made up of a number of closely-spaced repeated units, or *pulses* (see Fig. 16.1); Oates and Trocco found a range of 3–31 pulses per phrase, with a mean of 11.5.

Oates and Trocco (1983) used the term "fundamental frequency" to refer to the lowest horizontal dark tracing on sound spectrograms, and called the bands above it "harmonics." This was common terminology for primate calls at that time. J.M. Whitehead (1995, and pers. comm.) has pointed out that "fundamental frequency" in the description of human speech refers to vibration that is a direct function of the action of the glottis, and that it is not equivalent to the lowest frequency on most spectrograms of nonhuman primate calls. Whitehead therefore uses the term "emphasized frequency" to describe the zone on a spectrogram where most energy is concentrated. We accept Whitehead's point about the use of "fundamental frequency" but, because the spectrograms of black-and-white colobus roaring show emphasis on several energy bands, we prefer to use the term "dominant frequency" (Masters, 1991) to describe the lowest emphasized frequency.

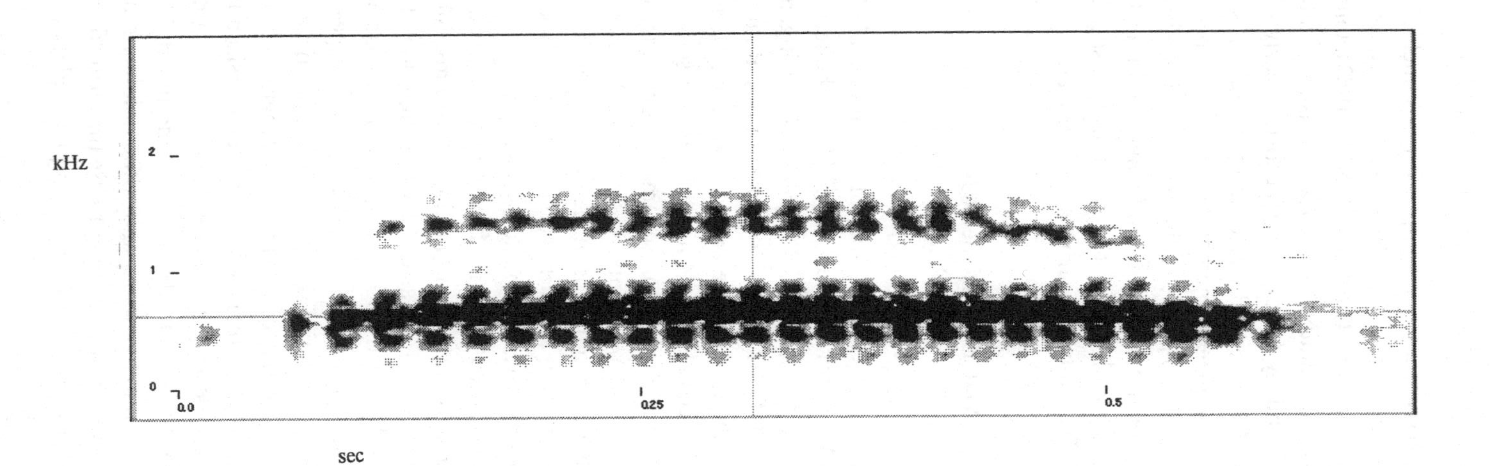

Fig. 16.1. Sound spectrogram of a single *C. angolensis* roar generated by the *Sonogram* program. Horizontal line (at left) is the position of median dominant frequency as judged by visual inspection (see text). Comparative Kay spectrograms from different species are shown by Oates and Trocco (1983).

The vocalization sample set

Twenty-one tape-recorded bouts of roaring were used in this new analysis, collected between 1964 and 1995 from 15 different populations of black-and-white colobus monkeys (Table 16.1). Each species of black-and-white colobus is represented by at least two samples. We used 10 of the 11 recordings used by Oates and Trocco (1983), details of which are given in that paper. Samples 1 and 2 in the earlier analysis consisted of two individuals roaring at the same time at Sango Bay in Uganda; in this new analysis we used the call of just one of these males. New recordings added to the analysis were of *C. angolensis* and *C. guereza* in the Ituri Forest, D.R. Congo, by Bocian; of *C. guereza* in Nigeria and Tanzania, *C. vellerosus* in Ghana, and *C. polykomos* in Sierra Leone by Oates; and of *C. satanas* on Bioko, Equatorial Guinea, by D. Schaaf and T. Struhsaker. The new recordings by Bocian and Oates used a Sony TCD-5000EV, Sony TCM-564V or Marantz PMD-221 cassette tape recorder, and either a Sennheiser ME-80 microphone head with a KD-3U powering module, or an ME-66 head with a K-6 module; these microphones are highly directional, "shotgun" types. More details of the recordings are given in Table 16.1

Analytical and statistical methods applied to vocalizations

The analysis of Oates and Trocco (1983) used measurements taken from sound spectrograms of roaring produced by a Kay Sona-Graph spectrum analyzer (model 7029A), an analog system. The new analysis reported here used a digital system. The 21 recordings were converted from analog to digital format with an audio digitizer (Metaresearch Digital Ears) at a sampling rate of 44.1 kHz, and processed with a digital signal processing program (Metaresearch SoundWorks, release 2.0) stored on a NeXT computer.

In an initial inspection of the processed calls, indistinct roar phrases (such as those made by several animals roaring together) were rejected. From the remaining phrases in each bout, 10 were then selected randomly for spectrographic analysis. Sound spectrograms were produced and analyzed with Sonogram (version 0.9), an acoustic signal analyzer/editor developed by H. Momose at the University of California, Davis. We used a Fourier transform with a frequency resolution of 43.1 Hz and a time resolution of 1.1 ms.

From the spectrograms we measured: (1) number of pulses in the phrase; (2) phrase duration, measured from the beginning of the first pulse to the end of the last pulse; and (3) median dominant frequency. To measure

Table 16.1. *Origins of colobus (*Colobus *spp.) roaring bouts analyzed in this chapter*

Sample no.	Place recorded (coordinates)	Date	Recorded by*	Population classification
1	Sango Bay, Uganda (0°55′S, 31°40′E)	12 January 1972	JO	*C. angolensis ruwenzorii*
3	Nyungwe Forest, Rwanda (2°20′S, 29°20′E)	19 September 1980	TS	*C. angolensis ruwenzorii*
12	Ituri Forest, D.R. Congo (former Zaire) (1°32′N, 28°32′E)	13 March 1993	CB	*C. angolensis cottoni*
13	Ituri Forest, D.R. Congo (former Zaire) (1°32′N, 28°32′E)	21 February 1994	CB	*C. angolensis cottoni*
4	Momela, Mt. Meru, Tanzania (3°13′S, 36°54′E)	15 August 1964	TS	*C. guereza caudatus*
22	Ngare Sero, Mt. Meru, Tanzania (3°20′S, 36°49′E)	19 October 1995	JO	*C. guereza caudatus*
5	Ishasha River, Uganda (0°37′S, 29°40′E)	4 April 1965	PM	*C. guereza occidentalis*
6	Budongo Forest, Uganda (1°45′N, 31°29′E)	16 October 1964	PM	*C. guereza occidentalis*

7	Budongo Forest, Uganda (1°45′N, 31°29′E)	24 April 1965	PM	*C. guereza occidentalis*
8	Kibale Forest, Uganda (0°34′N, 30°21′E)	13 June 1971	JO	*C. guereza occidentalis*
14	Ituri Forest, D.R. Congo (former Zaire) (1°32′N, 28°32′E)	5 April 1993	CB	*C. guereza occidentalis*
15	Ituri Forest D.R. Congo (former Zaire) (1°32′N, 28°32′E)	14 February 1994	CB	*C. guereza occidentalis*
16	Gashaka-Gumpti N. Park, Nigeria (7°20′N, 11°35′E)	8 January 1995	JO	*C. guereza occidentalis*
9	Tai National Park, Côte d'Ivoire (5°50′N, 7°20′W)	31 August 1979	JO	*C. polykomos*
17	Tiwai Island, Sierra Leone (7°33′N, 11°21′W)	30 January 1983	JO	*C. polykomos*
18	Tiwai Island, Sierra Leone (7°33′N, 11°21′W)	10 July 1983	JO	*C. polykomos*
10	Kade, Ghana (6°05′N, 0°52′W)	27 May 1968	TS	*C. vellerosus*

Table 16.1. (*cont.*)

Sample no.	Place recorded (Coordinates)	Date	Recorded by*	Population classification
19	Boabeng-Fiema Sanctuary, Ghana (7°41′N, 1°42′W)	27 August 1993	JO	*C. vellerosus*
11	Douala-Edéa Forest, Cameroon (3°34′N, 9°53′E)	13 February 1974	TS	*C. satanas anthracinus*
20	Gran Caldera de Luba, Bioko (3°20′N, 8°30′E)	18 March 1990	DS	*C. satanas satanas*
21	Gran Caldera de Luba, Bioko (3°20′N, 8°30′E)	20 March 1992	TS	*C. satanas satanas*

Notes:
Samples 1–11 are identical to those analyzed by Oates and Trocco (1983).
* CB: C. Bocian; JO: J. Oates; PM: Peter Marler; DS: D. Schaaf; TS: T. Struhsaker.

median dominant frequency (in Hz), we used the analytical program to place a horizontal line through the middle (determined by visual inspection) of the lowest band of pulses on each spectrogram (and see Fig. 16.1). To obtain pulse rate we divided the number of pulses in a phrase by the phrase duration in milliseconds.

We used these measurements of call structure in three separate analyses. First, we compared measurements on 10 samples (numbers 1 and 3–11) with the data obtained by Oates and Trocco (1983) from the same set of recordings. This was done to check whether differences in data collection and analytical procedures (especially machinery) would result in different interpretations. Second, we asked whether the new, larger data set still allowed us to distinguish the same species clusters recognized by Oates and Trocco. And third, we examined patterns of intraspecific variation.

Patterns of interspecific and intraspecific variation in these samples were examined with a variety of parametric and nonparametric tests. Parametric and nonparametric methods produced similar results, and therefore we present only the results of our nonparametric tests. Comparisons of mean values were made with the Mann-Whitney U test, and analyses of variance with a Kruskal-Wallis test (Sokal and Rohlf, 1981). We used the Systat program (Wilkinson, 1990) for analyses and to produce graphics.

New Observations of Colobus Behavior

C.B. studied niche separation in *C. guereza* and *C. angolensis* in the Central Ituri Forest, D.R. Congo. The study site, Basakwe, was a 900 ha mosaic of mature rainforest and swamp forest within the Okapi Faunal Reserve. At this site the density of *C. angolensis* was estimated at 12.5–16.7 individuals/km^2 and of *C. guereza* at 2.4–9.6 individuals/km^2 (Bocian, unpublished observations; ranges are estimates from different census methods). Systematic observations on one habituated social group of each species were made during monthly 5-day dawn to dusk follows from March 1993 through March 1994, and all occurrences of *Colobus* roaring were noted during 118 days of follows. Additional groups of each species were followed periodically to obtain a clearer picture of each population's social organization; at least eight *C. angolensis* and five *C. guereza* groups used the Basakwe site.

Results

Table 16.2 summarizes the results of our new measurements of dominant frequency and pulse rate for the entire sample of roars.

Table 16.2. *Summary of measurements of dominant frequency and pulse rate of black-and-white colobus (*Colobus *spp.) roars. Sample numbers correspond to Table 1, which gives locality details. In each sample, 10 phrases were analyzed*

		Median dominant frequency (Hz)				Pulse rate (Units/sec)			
Sample no.	Species	Range	Mean	S.D.	C.V.	Range	Mean	S.D.	C.V.
1	*angolensis*	640–700	669	19.51	0.03	34.8–37.5	36.0	1.03	0.03
3	*angolensis*	581–640	614	17.51	0.03	35.2–46.1	42.9	3.44	0.08
12	*angolensis*	510–700	619	64.17	0.10	29.2–36.2	33.8	1.93	0.06
13	*angolensis*	629–723	671	26.91	0.04	33.0–57.4	40.6	6.53	0.16
4	*guereza*	593–652	621	20.31	0.03	13.8–24.6	21.5	3.41	0.16
22	*guereza*	652–700	669	20.29	0.03	20.0–24.5	22.5	1.42	0.06
5	*guereza*	486–593	556	30.83	0.06	19.4–24.1	21.3	1.62	0.08
6	*guereza*	522–605	562	23.83	0.04	16.7–23.1	18.7	1.78	0.10
7	*guereza*	484–553	523	17.72	0.03	15.6–25.1	19.9	3.42	0.17
8	*guereza*	451–569	523	40.87	0.08	20.2–24.4	22.9	1.29	0.06
14	*guereza*	439–498	481	17.86	0.04	13.3–19.4	17.1	2.38	0.14
15	*guereza*	486–592	562	32.13	0.06	20.2–22.6	21.5	0.84	0.04
16	*guereza*	510–546	524	12.25	0.02	17.6–23.5	22.0	1.65	0.08
9	*polykomos*	640–700	672	17.72	0.03	30.5–39.8	34.5	2.94	0.09
17	*polykomos*	629–747	710	36.83	0.05	27.0–38.5	33.5	3.11	0.09
18	*polykomos*	581–771	676	70.72	0.11	24.8–31.6	27.6	1.94	0.07
10	*vellerosus*	534–581	554	17.70	0.03	18.0–22.7	20.6	1.64	0.08
19	*vellerosus*	454–593	525	44.84	0.09	16.3–26.0	22.7	2.82	0.12
11	*satanas*	617–759	690	42.86	0.06	59.6–81.5	69.3	7.55	0.11
20	*satanas*	688–747	716	19.53	0.03	46.0–54.6	49.2	2.37	0.05
21	*satanas*	747–818	780	22.22	0.03	50.9–64.0	55.0	4.23	0.08

Comparison of new and old measurements

The comparison of old (Kay analysis) with new (NeXT analysis) data is presented in Figure 16.2; there is very little difference in roar pulse rates measured by the two machines, but dominant frequencies measured from the Kay spectrograms were consistently slightly higher than frequencies measured on the NeXT (mean difference 33 Hz; the differences are statistically significant at the level of $p < 0.005$ in samples 1 and 8, and at $p < 0.05$ in 6, 7, 10, and 11). It is unclear whether these differences result from the machines themselves or from differences in the techniques of the researchers making the measurements. However, the differences are small and do not affect the overall pattern of the data.

Species differences

Figure 16.3 displays our measurements of roar frequency and pulse rate according to the five species recognized by Oates and Trocco (1983). The same clusters are evident as in the earlier analysis, but the clusters (using a larger sample) show more overlap. Calls in one cluster, from *C. guereza* and *C. vellerosus*, have a slow pulse rate (<26 units/sec) and generally a low dominant frequency (439–700 Hz). Calls in a second cluster, from *C. angolensis* and *C. polykomos*, have an intermediate pulse rate (25–57 units/sec) and an intermediate frequency (510–771 Hz). Calls in the third cluster, from *C. satanas* only, have a fast pulse rate (>46 units/sec) and a high dominant frequency (617–818 Hz).

Kruskal-Wallis tests performed to examine differences in the distribution of variance in pulse rate and dominant frequency among these three clusters found that each differs significantly ($p < 0.001$) from the other two in both parameters, with the exception that clusters 1 (*C. guereza* and *vellerosus*) and 2 (*C. angolensis* and *C. polykomos*) are not significantly different in frequency ($p = 0.06$). Within cluster 1, *C. guereza* and *C. vellerosus* do not differ significantly in pulse rate or frequency ($p < 0.001$) and nor do *C. angolensis* and *C. polykomos* in cluster 2 ($p < 0.001$).

Intraspecific variation

The scatter of points in Figure 16.3 demonstrates the considerable variation in measured roar features not only between but also within species. However, such variation appears to be more pronounced in some species than in others. For instance, there is a greater range in roar pulse rate in one

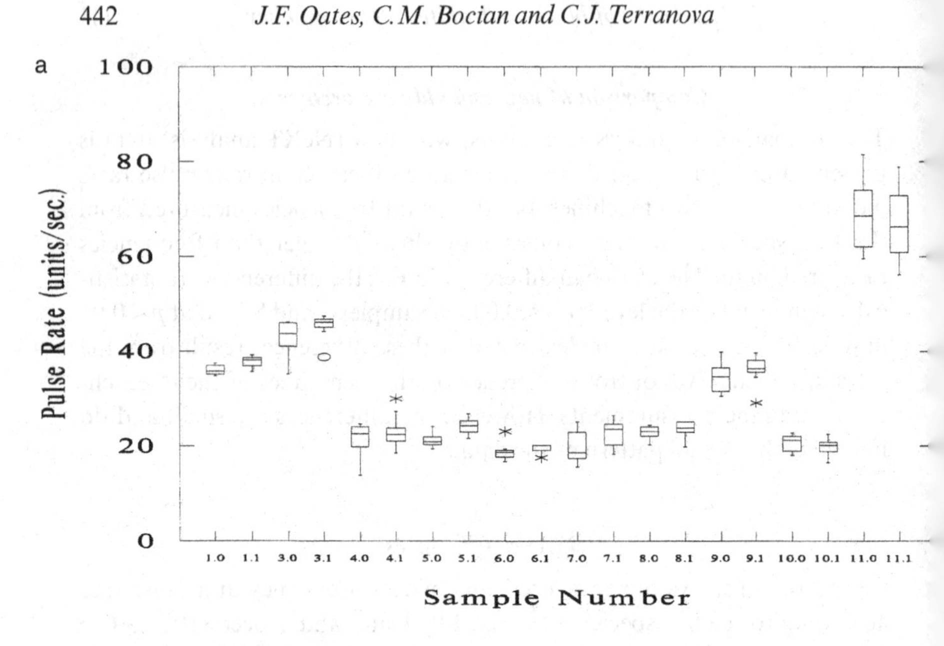

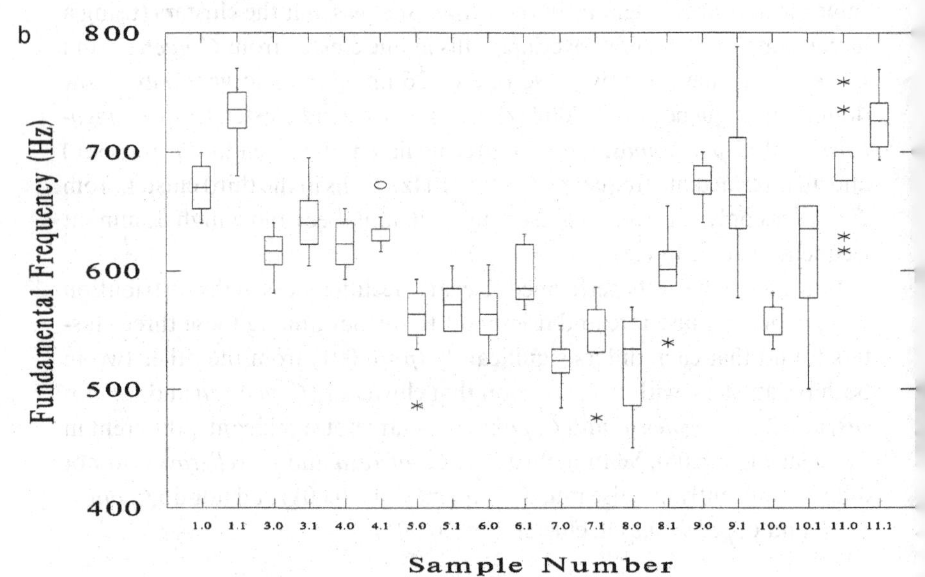

Fig. 16.2. Box plots of pulse rate (a) and dominant frequency (b) analyzed by two different systems. The central horizontal line in each box denotes the median value, hinges split the remaining values in half again. Outliers are represented by asterisks and open circles. The notation "1.0" refers to sample 1 (Table 16.1) analyzed on a Kay Sona-Graph, "1.1" refers to sample 1 analyzed by SoundWorks on a NeXT computer.

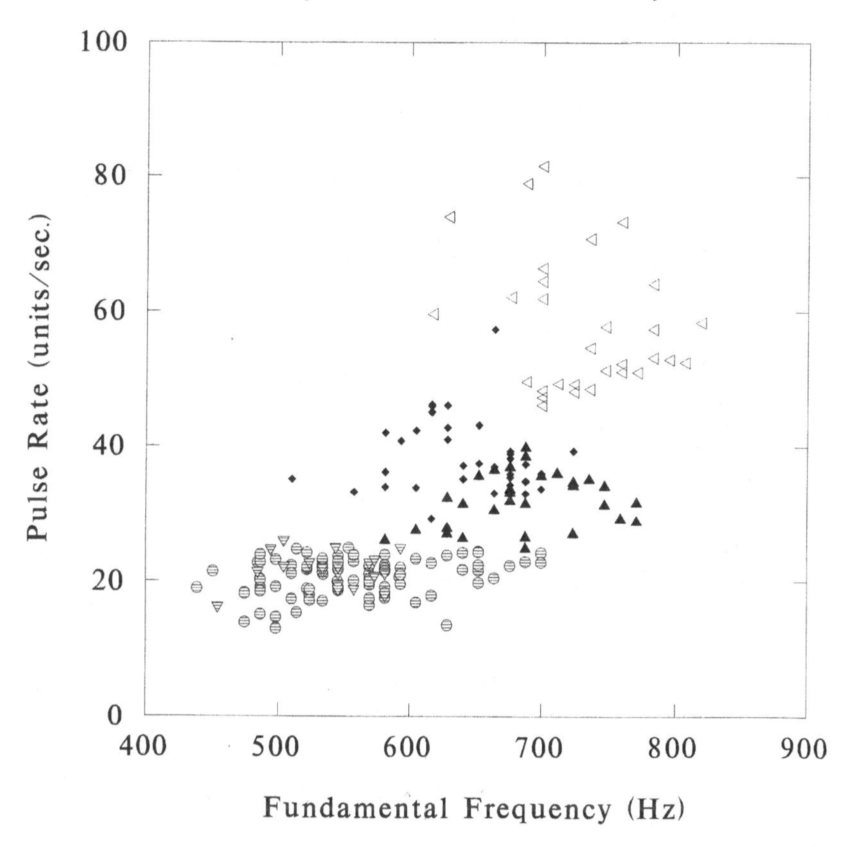

Fig. 16.3. Scatterplot of roar pulse rate on dominant frequency. Symbols: *Colobus guereza*, hatched circles; *C. vellerosus*, hatched triangles; *C. angolensis*, solid diamonds; *C. polykomos*, solid triangles; *C. satanas*, open triangles.

sample of *C. satanas* roaring (no. 11) than across the entire set of nine *C. guereza* samples (see Table 16.2 and Fig. 16.4). Indeed, *C. guereza* shows very little variation in pulse rate across the entire range of the species.

In general, however, we found a great deal of overlap in samples from the same species recorded in different habitats. For instance, the range of variation in pulse rate and dominant frequency found in the recordings of *C. angolensis* from Nyungwe and Sango Bay is embedded within the range found in recordings of the same species from Ituri (Table 16.2). Similarly, the range in pulse rate and frequency in the recording of *C. vellerosus* from Kade is embedded within the range of the same species recorded at Boabeng-Fiema.

Some striking exceptions to this general pattern of overlap are found in

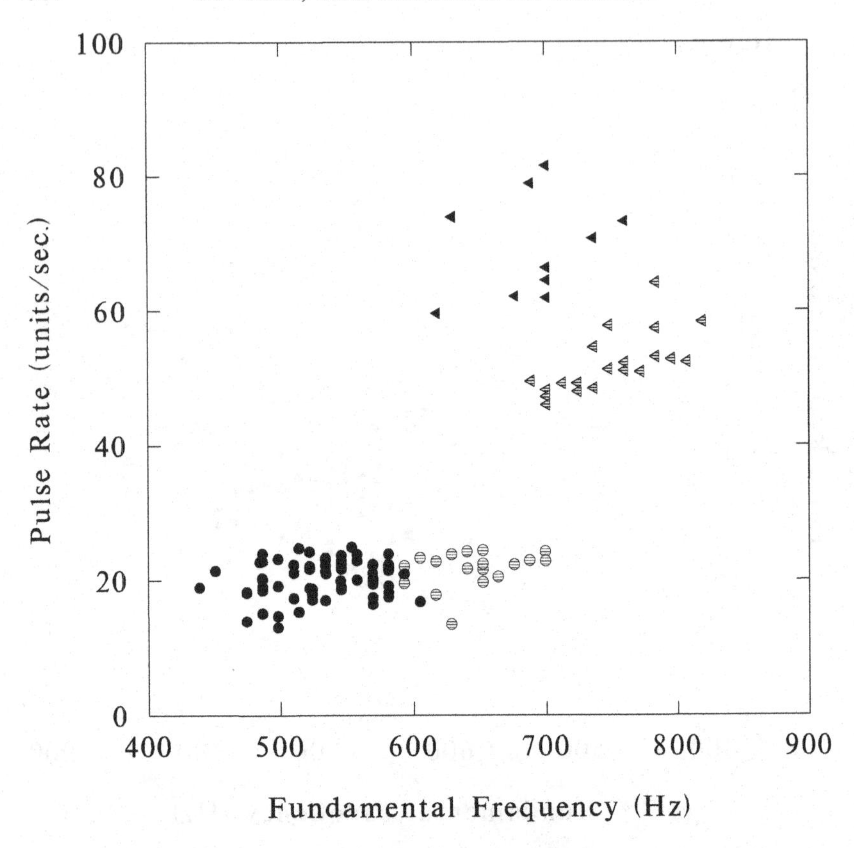

Fig. 16.4. Scatterplot of pulse rate on dominant frequency, indicating differences between *Colobus guereza* and *C. satanas* subspecies. Symbols: *Colobus g. occidentalis*, solid circles; *C. g. caudatus* (from Meru), hatched circles; *C. satanas anthracinus*, solid triangles; *C. satanas satanas* (from Bioko), hatched triangles.

the recordings of *C. guereza* from Meru, Tanzania, compared with other locations, and of *C. satanas* from Bioko Island compared to Cameroon (Fig. 16.4). The dominant frequencies of roars of *C. guereza* from Mt. Meru are significantly different ($p < 0.001$) from those in other populations (yet there is considerable overlap in this measure between Meru animals and West African *C. polykomos*). The roar pulse rate in Bioko samples is also significantly different ($p < 0.001$) from that in mainland *C. satanas*. The Mt. Meru guerezas, together with those on nearby Kilimanjaro, are usually classified as a distinct subspecies, *C. g. caudatus*, isolated by at least 200 km of arid country from the nearest population of *C. g. occidentalis*, the subspecies from which all our other recordings derive. The Bioko black colobus are separated by a 90 km sea channel from the Cameroon population, and are often regarded as belonging to a different subspecies (*C. s. satanas* on Bioko, *C. s. anthracinus* on the mainland).

Social organization in Ituri Forest black-and-white colobus

Colobus angolensis and *C. guereza* groups in the Ituri Forest differed in their composition. *C. angolensis* groups were larger (size range 6–20, mean = 14, $n = 8$) and typically contained several adult males, while *C. guereza* groups (size range 5–11, mean = 8, $n = 5$) contained only one or two adult males. In March 1994, the *C. angolensis* study group consisted of 20 animals, including five adult males, eight adult females, one subadult male, five juveniles, and one infant; the *C. guereza* study group consisted of 10 animals, including two adult males, three adult females, one subadult female, two juveniles, and two infants.

The social behavior of adult females and of young animals was similar in both species. Interspecific differences were apparent, however, in the nature of interactions between adult males within the same social group. *Colobus guereza* adult males were generally intolerant of other adult males; within the study group, interactions between the two males were infrequent, and were usually agonistic in nature. Furthermore, the dominant male had exclusive mating access to females and was the only individual who roared. The subordinate male became increasingly peripheral to the group during the course of the study. Relationships were less clearly defined among the five adult males in the *C. angolensis* study group. While two of these males appeared to share a dominant position, they did not monopolize access to all eight of the adult females, and at least four adult males participated in jumping/roaring displays. Aggression between males was infrequent, while affiliative interactions, such as grooming and resting together were commonly observed.

Patterns of range use were strikingly different between the two study groups. The *C. guereza* group intensively used 14 ha of contiguous forest within a home range of approximately 100 ha. Other *C. guereza* groups were neither observed nor heard within this core area during study group follows. During 60 days of group follows over a period of 12 months, only two *C. guereza* intergroup encounters were observed. In contrast to the *C. guereza* pattern, the *C. angolensis* study group did not intensively use one particular area, and shared their home range of over 370 ha with several other *C. angolensis* groups. *Colobus angolensis* intergroup encounters were common, occurring on at least 25 of 62 days of group follows. In terms of duration, these encounters were of two types: brief meetings in the forest of several minutes to an hour; or longer associations lasting for several hours to more than one day. Associations were generally characterized by long periods of feeding, resting, and travelling in proximity (i.e. members of different groups were within 50 m of each other, and were sometimes mixed together in the same trees); peaceful associations were sometimes interrupted by aggressive interactions between adult males from the two groups.

Roaring behavior in Ituri Forest black-and-white colobus

We noted earlier that Bocian initially had difficulty distinguishing distant roars of *C. angolensis* from those of *C. guereza*. However, experience very soon overcame this problem. By the time systematic data collection began after five months in the field, Bocian felt confident in her ability to identify the roaring species.

During the period of data collection (118 days spread over one year), a total of 303 roars were heard; of these, 231 (76.2%) were identified as coming from *C. guereza*, 67 (22.1%) as coming from *C. angolensis*, and for 5 (1.7%) the species was undetermined. Ninety-two of these roars occurred before dawn; of these, 87 (94.5%) were *C. guereza*; 2 (2.2%) were *C. angolensis*; and 3 (3.3%) were undetermined. During 60 days of *C. guereza* follows, 60 roaring bouts were noted from the study group, all from the dominant adult male. Of these 60 roaring bouts, 17 were predawn and 43 were after dawn. During 58 days of *C. angolensis* follows, nine roaring bouts were noted from the group's males; eight of these roars were after dawn, while only one was a predawn roar.

Most predawn roaring by *C. guereza* involved choruses of individual adult males calling from several different locations. Roars during daylight hours seemed to be associated with a heightened state of arousal, and perhaps alarm, as they were often heard after tree-falls, duiker alarm calls, or the vocalizations of other primate species. Most roaring by *C. angolensis* males (which very rarely called before dawn) occurred only in contexts of arousal or alarm, and abbreviated roars were heard during low-intensity agonistic encounters between adult males within the same social group, or during aggressive interactions between group males and extra-group males travelling with the study group.

Discussion

This reanalysis of black-and-white colobus monkey vocalizations, using additional samples and different analytical equipment, leads to the same conclusion as the earlier analysis by Oates and Trocco (1983): simple acoustic properties of male loud calls allow three species-clusters to be distinguished, one containing *C. guereza* and *C. vellerosus*, one containing *C. angolensis* and *C. polykomos*, and the third containing only *C. satanas*. Although larger sample sizes reveal more variation within species than was apparent in the first analysis, overlap between the clusters remains relatively small and single samples of roaring can be reliably assigned to particular clusters. *Colobus polykomos* remains distinct from *C. vellerosus*, supporting the views of Oates and Trocco (1983) and Groves *et al.* (1993) that these are different species.

Not only does this reanalysis confirm the existence of distinct black-and-white colobus species clusters (within which species may be discriminated by cranial morphology and pelage patterns), it also reveals some significant differences between some subspecies. Therefore, the male loud calls of such primates can indeed be reliable taxonomic characters.

Our new information on black-and-white colobus social systems, combined with this information on vocalizations, provides some fresh insights on the function and evolution of male loud calls in black-and-white colobus. Oates and Trocco (1983) suggested that the main function of roaring is communication over relatively long distances to conspecifics outside the caller's group; they also noted that this communication appears primarily to be between adult males. These conclusions, however, resulted largely from studies on one species, *C. guereza*, and Oates and Trocco acknowledged that the loud call may be used in different contexts by different species.

It is now clear that *C. guereza* is unusual among the black-and-white colobus monkeys in its social structure. Multimale groups are apparently characteristic of *C. satanas* (Sabater Pi, 1973; McKey and Waterman, 1982), *C. angolensis* (Moreno-Black and Bent, 1982; A. Vedder, pers. comm.; this study), and *C. polykomos* (Dasilva, 1989). Several adult males may occur in larger groups of *C. vellerosus* (Olson, 1986 and pers. comm.), but this species remains poorly known. Although two (or more) adult males have been observed in some *C. guereza* groups (Dunbar and Dunbar, 1974; Oates, 1977; von Hippel, 1996; this study), one male is typically dominant to the others; subordinate males are harassed by other adults, are peripheral to the group, and often eventually leave their natal group. A uni-male, social system is apparently common in this species.

Roaring calls are not used in the same way in the different species. Bocian's observations in the Ituri Forest show that, in the same habitat, *C. guereza* males roar more frequently than *C. angolensis* males, and that predawn roaring choruses are very much more frequent in *C. guereza* than in *C. angolensis*. *Colobus guereza* male roars seem to be used much more frequently in long-range communication than are the roars of *Colobus angolensis*.

This evidence suggests that the different acoustic properties of male roars in black-and-white colobus may be related to how the calls function within the social system; these different properties may reflect the path of social evolution within this subgenus. Oates and Trocco (1983) argued that the roar of *C. satanas* (high dominant frequency, fast pulse rate) is primitive relative to that of the other species, and is probably related to the possession of a small larynx, the presumed primitive condition for African colobines. Building on the work of Grubb (1978, 1982) Oates and Trocco suggested that the ancestor of the *C. angolensis–C. polykomos* cluster evolved from a *C. satanas*-like animal, and that *C. vellerosus* and *C. guereza* evolved in a stepwise fashion from *C. polykomos* (or its ancestor) (Fig. 16.5).

Beginning as an alarm call in an animal like *C. satanas*, roaring could have been modified into a call used also in male–male competition within a multimale-group social system (as seen in *C. angolensis*). In this process, sexual selection could have acted to produce a louder, lower-frequency call, likely to indicate greater development and strength in the musculature operating the vocal apparatus (and therefore potentially providing information on the physical condition of the individual). Once a louder, lower-frequency call exists, it can function more effectively in long-range communication; such communication may be both with individuals (especially males) in other bisexual groups and with solitary males. In *C. guereza* a loud low-frequency call is commonly used in long-range communication; in this species, strong male–male competition (and high intolerance of other adult males within a group) often results in a one-male group structure.

Colobus polykomos appears to be intermediate between *C. angolensis* and *C. guereza* in its patterns of male relationships. Predawn choruses occur more frequently in *C. polykomos* than in *C. angolensis*, but less frequently than in *C. guereza* (JFO, unpublished observations from Tiwai Island, Sierra Leone). In Dasilva's *C. polykomos* study group at Tiwai, the dominant male did not have exclusive breeding access to females, but male–male interactions (which were infrequent) tended to be agonistic; between-group interactions were more antagonistic than in Ituri *C. angolensis*, and there appeared to be a positive correlation between the number of adult males in a group and the group's success in displacing other groups (Dasilva, 1989).

These species differences in social behavior are probably correlated with

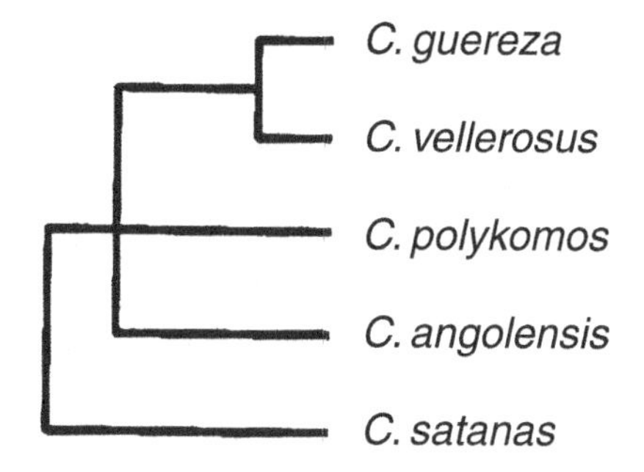

Fig. 16.5. Cladogram of inferred relationships among black-and-white colobus monkey species.

ecological differences, as suggested by the very different ranging patterns of *C. angolensis* and *C. guereza* in the Ituri Forest. The ranges of *C. angolensis* groups are much larger than those of *C. guereza* groups, and are shared with other groups. *Colobus satanas* groups also seem to use relatively large ranges (60 ha: McKey and Waterman, 1982; 84 ha: Harrison, cited in Oates, 1994), whereas *C. polykomos* home ranges are more like those of *C. guereza* (24 ha: Dasilva, 1989).

In general, then, there appears to be an evolutionary sequence (related to features of ecology and social organization) in how black-and-white colobus males use their loud calls, with roars used least in long-range male–male competition in *C. satanas*, and most in *C. guereza*. Such changes in the function of the calls have influenced the evolution of their acoustic properties, with the calls becoming progressively slower-paced and lower-pitched. These acoustic changes are presumably the result of changes in the anatomical and neurological structures involved in call production.

Further insights on the evolution of this calling system could be gained both by anatomical studies, and by more careful field studies on behavior and call use in *C. satanas* and *C. vellerosus*, two species whose social behavior remains relatively poorly known. Based on the great similarity of their male loud calls, we predict that *C. vellerosus* will be more similar to *C. guereza* than to *C. polykomos* in the details of its social organization.

Summary

We used new recordings and analytical equipment to re-examine the conclusions reached by Oates and Trocco (1983) in their study of variation in the loud calls (roars) of male black-and-white colobus monkeys. Our reanalysis supported Oates and Trocco's main finding, that the acoustic structure of *C. satanas* roars can be readily distinguished from those of other species, whose calls group into two clusters, *C. angolensis* with *C. polykomos*, and *C. guereza* with *C. vellerosus*. Within these clusters, some commonly-recognized subspecies of *C. guereza* and *C. satanas* can be distinguished by loud call features. These findings support the view that male loud calls are indeed reliable taxonomic characters, and that *C. vellerosus* is reasonably regarded as a distinct species, rather than a subspecies of *C. polykomos*.

We also report new observations on the behavior of *C. angolensis* and *C. guereza* living sympatrically in the Ituri Forest of D.R. Congo. *Colobus angolensis* was found living in larger groups with more adult males than was *C. guereza*, and a *C. angolensis* study group ranged over a much larger area and had more frequent intergroup interactions than did a *C. guereza* group. Males of both species roar in contexts suggestive of arousal or alarm, and *C. angolensis* roars are also used in close-range male–male interactions. Roaring in *C. guereza* is used more frequently in long-range communication than in *C. angolensis*. We suggest that differences in the acoustic properties of male roars within the black-and-white colobus monkeys are related to different patterns of social organization among the species.

Acknowledgements

The Institut Zairois pour la Conservation de la Nature, and its Director, Mankoto ma Mbaelele, gave permission for the Ituri Forest study. Field work in Ituri and elsewhere was supported by grants from the National Science Foundation, the Research Foundation of CUNY, the Wildlife Conservation Society, the Chicago Zoological Society, the Department of Biology at City College, and an NIMH Training Grant. C.T. was supported by a fellowship at CUNY from the New York Consortium in Evolutionary Primatology (NYCEP). C.B. also thanks Frances Bocian for financial assistance, and John and Terese Hart and Karl and Rosemarie Ruf for logistical support. We thank Peter Marler, Dietrich Schaaf and Thomas Struhsaker for allowing us to use their recordings, and Elena Cunningham and Katherine Gonder for their help in digitizing and measuring many recordings. J.M. Whitehead is thanked for his constructive comments on the manuscript.

References

Dasilva, G.L. (1989). *The Ecology of the Western Black and White Colobus (*Colobus polykomos polykomos *Zimmerman 1780) on a Riverine*

Island in Southeastern Sierra Leone. DPhil. thesis, University of Oxford.
Dunbar, R.I.M. & Dunbar, E.P. (1974). Ecology and population dynamics of *Colobus guereza* in Ethiopia. *Folia primatol.* **21**, 188–208.
Gautier, J.-P. (1988). Interspecific affinities among guenons as deduced from vocalizations. In *A Primate Radiation: Evolutionary Biology of the African Guenons*, ed. A. Gautier-Hion, F. Bourliere, J.-P. Gautier & J. Kingdon, pp. 194–226. Cambridge: Cambridge University Press.
Groves, C.P. , Angst, R. & Westwood, C. (1993). The status of *Colobus polykomos dollmani. Int. J. Primatol.* **14**, 573–86.
Grubb, P. (1978). Patterns of speciation in African mammals. *Bull. Carnegie Mus. Nat. Hist.* **6**, 152–67.
Grubb, P. (1982). Refuges and dispersal in the speciation of African forest mammals. In *Biological Diversification in the Tropics*, ed. G. Prance, pp. 537–53. New York: Columbia University Press.
Hill, W.C.O. & Booth, A.H. 1957. Voice and larynx in African and Asian Colobidae. *J. Bombay Nat. Hist. Soc.* **54**, 309–21.
Hull, D.B. (1979). A craniometric study of the black and white colobus Illiger 1811 (Primates: Cercopithecoidea). *Am. J. phys. Anthrop.* **51**, 163–82.
Marler, P. (1957). Specific distinctiveness in the communication signals of birds. *Behaviour* **11**, 13–39.
Marler, P. (1972). Vocalizations of East African monkeys. II. Black and white colobus. *Behaviour* **42**,175–97.
Marshall, J.T. & Marshall, E.R. (1976). Gibbons and their territorial songs. *Science* **193**, 235–7.
Masters, J.C. (1991). Loud calls of *Galago crassicaudatus* and *G. garnettii* and their relation to habitat structure. *Primates* **32**, 153–67.
McKey, D.B. & Waterman, P.G. (1982). Ranging behavior of a group of black colobus (*Colobus satanas*) in the Douala-Edea Reserve, Cameroon. *Folia primatol.* **39**, 264–304.
Mitani, J.C. (1987). Species discrimination of male song in gibbons. *Am. J. Primatol.* **13**, 413–23.
Moreno-Black, G.S. & Bent, E.F. (1982). Secondary compounds in the diet of *Colobus angolensis*. *Afr. J. Ecol.* **20**, 29–36.
Oates, J.F. (1977). The social life of a black-and-white colobus monkey, *Colobus guereza. Z. Tierpsychol.* **45**, 1–60.
Oates, J.F. (1994). The natural history of African colobines. In *Colobine Monkeys: Their Ecology, Behaviour and Evolution*, ed. A.G. Davies & J.F. Oates, pp. 75–128. Cambridge: Cambridge University Press.
Oates, J.F. & Trocco, T. F. (1983). Taxonomy and phylogeny of black-and-white colobus monkeys: inferences from an analysis of loud call variation. *Folia primatol.* **40**, 83–113.
Olson, D.K. (1986). Determining range size for arboreal monkeys: methods, assumptions, and accuracy. In *Current Perspectives in Primate Social Dynamics*, ed. D.M. Taub & F.A. King, pp. 212–27. New York: Van Nostrand Reinhold.
Rahm, U.H. (1970). Ecology, zoogeography, and systematics of some African forest monkeys. In *Old World Monkeys: Evolution, Systematics, and Behavior*, ed. J. R. Napier & P. H. Napier, pp. 589–626. New York: Academic Press.
Sabater Pi, J. (1973). Contribution to the ecology of *Colobus polykomos satanas* (Waterhouse, 1838) of Rio Muni (Republic of Equatorial Guinea). *Folia primatol.* **19**, 193–207.

Snowdon, C.T., Hodun, A., Rosenberger, A.L. & Coimbra-Filho, A.F. (1986). Long-call structure and its relation to taxonomy in lion tamarins. *Am. J. Primatol.* **11**, 253–61.

Sokal, R.R. & Rohlf, J. (1981). *Biometry*. San Francisco: W.H. Freeman.

Struhsaker, T.T. (1970). Phylogenetic implications of some vocalizations of *Cercopithecus* monkeys. In *Old World Monkeys: Evolution, Systematics, and Behavior*, ed. J.R. Napier & P.H. Napier, pp. 365–444. New York: Academic Press.

von Hippel, F.A. (1996). Interactions between overlapping multimale groups of black and white colobus monkeys (*Colobus guereza*) in the Kakamega Forest, Kenya. *Am. J. Primatol.* **38**, 193–209.

Whitehead, J.M. (1995). Vox Alouattinae: A preliminary survey of the acoustic characteristics of long-distance calls of howling monkeys. *Int. J. Primatol.* **16**, 121–44.

Wilkinson, L. (1990). *Systat: The System for Statistics*. Evanston, Illinois.

Wilson, W.L. & Wilson, C.C. (1975). Species-specific vocalizations and the determination of phylogenetic affinities of the *Presbytis aygula-melalophos* group in Sumatra. In *Contemporary Primatology*, ed. S. Kondo, M. Kawai & A. Ehara, pp. 459–63. Basel: S. Karger.

Zimmermann, E. (1990). Differentiation of vocalizations in bushbabies (Galaginae, Prosimiae, Primates) and the significance for assessing phylogenetic relationships. *Z. zool. Syst. Evolut.-forsch.* **28**, 217–39.

17

Agonistic and affiliative relationships in a blue monkey group

MARINA CORDS

Introduction

The literature often contrasts features of the social organization of cercopithecines with that of other taxa, including colobines (McKenna, 1979; Borries *et al.*, 1994; Newton and Dunbar 1994), platyrhines (O'Brien, 1993), and apes (Watts, 1994). Most of what we know about "cercopithecine" social organization, however, derives from studies of only a few species, namely baboons (*Papio* and *Theropithecus*), some macaques (*Macaca*), and vervet monkeys and their relatives (*Chlorocebus*) (Erwin and Zucker, 1987; Strier, 1990), even though these species do not provide a representative sample of cercopithecine genera or of habitats in which cercopithecine monkeys live. In particular, the social organization of the African forest-dwelling cercopithecins (*Cercopithecus, Miopithecus*, and *Allenopithecus*) is poorly known. The forested habitat and arboreal habits of these monkeys make study in the wild difficult, and captive groups of naturalistic size do not exist.

Some recent publications have begun to provide information on cercopithecine species previously overlooked, especially lesser-known papionins, mostly studied in captivity (e.g. Baker and Estep, 1985; Ehardt, 1988; Oi, 1990; Thierry *et al.*, 1990; Gust and Gordon, 1993, 1994; Gust, 1994, 1995). In an attempt to correct prevalent generalizations about cercopithecine social organization by including cercopithecins, Rowell (1988) contrasted their social organization with that of the better-known papionins. Noting that most members of both tribes live in female-bonded, matrilocal groups, she identified four main behavioral differences: (1) cercopithecins maintain group cohesion by monitoring the behavior of other group members, and adjusting their own behavior accordingly, whereas papionins emphasize more the exchange of specialized communicative signals; (2) cercopithecin groups are territorial except where populations occur at low density; (3)

adult male cercopithecins are mutually intolerant in the presence of females, whereas papionin males can at least tolerate each other, and can even cooperate; and (4) adult male cercopithecins seldom interact with non-receptive females, whereas papionin males maintain affiliative bonds with females throughout their reproductive cycle, and thus contribute to a more unified group social organization. Vervets, swamp monkeys (*Allenopithecus*), and talapoins (*Miopithecus*) provided some exceptions to these generalizations.

Conspicuously absent from Rowell's discussion of cercopithecins, especially in light of the literature on macaques, baboons, and vervets, is the within-group organization of social relationships, especially among females. It is this particular gap in our knowledge that the present contribution attempts partially to redress. Using focal animal data from a group of individually recognized wild blue monkeys (*C. mitis stuhlmanni*), I focus on the social relationships among adult and subadult females, with special attention to agonism. An earlier report on the same study group was made by Rowell *et al.* (1991), using only *ad libitum* sampling techniques; the focal animal data collected in the present study are able to address a greater variety of questions.

Methods

Subjects and general sampling methods

Data were collected from the habituated T_w group of blue monkeys in the Kakamega Forest, Kenya (Cords, 1987). This group has been monitored since 1979. Data reported here were collected during an annual two-to-four-month period between June and September, 1992–1996. During this period, group size varied from 31 to 35 individuals, excluding adult males. The density of blue monkeys at this site has been estimated at 169 individuals/km^2 (Cords, 1987).

Social behavior was sampled in two ways. In all five years, I collected *ad libitum* records of all observed agonism, noting the identities of opponents as well as the context and nature of assertive, aggressive, or submissive signals exchanged. Three types of agonistic interactions were scored. In *approach–retreat interactions*, one animal approached a second showing no aggression but the animal approached showed submissive signals, including fleeing. If it fled when the approacher came within 1 m, the interaction was termed a "supplant"; otherwise it was termed an "avoid". In *spontaneous submission*, one animal directed submissive signals (geckering, trilling, cowering, fleeing) to the other, even though the latter had neither approached nor shown aggressive signals. In *aggressive interactions* one animal chased, threatened, or attacked the other, which responded by ignoring or returning aggression, or by showing submission.

In 1992 and 1993, focal animal samples were also conducted on all adult (parous) females, as well as those nulliparous, sub-adult females who had attained full body size, and whose behavior generally resembled that of adults rather than juveniles. Specifically, they were not observed to play (see also Bramblett and Coelho, 1987), and they engaged in mutual grooming with several adult partners, whereas smaller juveniles focus grooming on their own mothers. Female blue monkeys first give birth at about six to seven years of age, and all sub-adult subjects gave birth within two years of being sampled. The subjects were thus 12 adults and three sub-adults in 1992, and 14 adults and one sub-adult in 1993. All subjects were sampled in both years, except for one 1992 adult who did not survive to 1993, and one 1993 primipara, who was too young to be included in 1992.

Using 10×40 binoculars, I observed each female for five hours in 1992, and 11 hours in 1993. A grid of paths spaced at 50–100 m intervals (Cords, 1987) facilitated tracking of individuals, but I left the paths when necessary to maintain visual contact. In 1996, one minute focal samples were conducted on adult females feeding on four different species of fruiting tree (*Harungana madagascariensis*, *Maesopsis eminii*, *Solanum mauritianum*, and *Bischoffia javanica*), to count the fruit intake rate. No female was sampled more than once in a given tree species. These data were then related to agonistic rank.

Focal sampling protocol (1992–1993)

Focal samples were scheduled to last one hour, but sometimes the subject disappeared from view. If she remained out of sight for 15 minutes, the sample was terminated at the moment of disappearance, and the remainder of the hour was collected at the next opportunity. If the subject disappeared for less than 15 minutes, the sample was extended until she had been in view for a total of 60 minutes. A monkey usually disappeared because she moved into dense vegetation, not because she engaged in behavior that was difficult to follow.

Because a blue monkey group is typically dispersed irregularly over several hundred meters in a visually dense environment, a predetermined sampling schedule was impractical. Instead, I tried to spread observations of individuals across each sampling period by completing a round of samples on all subjects before any one was re-sampled. To ensure that all females' samples were similarly distributed throughout the day, the latter was broken into three periods (8:00–11:00 hrs, 11:00–15:00 hrs, and 15:00–18:00 hrs) within which I gathered nearly identical numbers of samples for each subject. Finally, I tried to ensure that different habitat types used by the group were equally represented in each female's samples. Habitat types included high-canopy forest (Cords, 1987) and woodland. In the latter, there was a less continuous canopy of smaller trees, and a well developed ground layer of grass and *Pavetta* shrubs. In 1993, the monkeys spent some hours in the woodland about every third day. Because their use of space was potentially different in the woodland, each female was sampled there for an equal and representative time (2 of 11 hours).

During focal samples, I continuously recorded all the subject's activities and social interactions, and noted transition times (to the nearest second) between activity classes. As well as actual ingestion, "feeding" included brief movements (≤ 6 steps, usually lasting ≤ 3 seconds) between ingestion sites. It also included foraging for insects by manual searching, and by visual searching when the subject obviously gazed at nearby substrates. I also noted transition times between different plant parts consumed. "Moving" included all locomotion that lasted ≥ 2 seconds, unless it satisfied the conditions for "feeding". "Resting" included sitting, standing, and lying; the subject was immobile, but was neither feeding nor engaged in grooming. "Socializing" included active grooming, receiving grooming, soliciting grooming and sitting in contact. Whenever the subject moved from one plant to another, the time and the plant's identity were noted.

When the subject fed in a tree, the number of other individuals in the same tree was continuously monitored, to assess the potential for feeding competition. Where an exact count was impossible, I noted whether the tree contained no, one, some (≥ 1) or many (≥ 5) other monkeys. A host tree and its associated liana(s) were treated as separate, unless the liana(s) covered the tree's entire crown. Some lianas connected and thus unified the crowns of several trees; in such cases, the liana itself was identified as one tree, regardless of how many trunks or crowns it connected. Where several lianas grew on the same supporting tree(s), the entire liana tangle was considered one plant. When a monkey fed from one tree while sitting in another, the monkeys in the tree that was the food source were counted.

Feeding trees were further characterized by size, which was expressed in terms of how many feeding monkeys the tree could hold. This metric was adopted because it incorporated the monkeys' perspective better than arbitrary measurements such as crown diameter. Trees with deep crowns might have many potential feeding sites, despite a small crown diameter. On the basis of 15 years experience of how blue monkeys at this site typically space themselves within different trees, three classes were distinguished: "small" trees could be occupied by no more than two animals, and typically had crown diameters (CD) ≤ 2 m; "medium" trees (2–8 m CD) could be occupied by three to eight monkeys; "large" trees could be occupied by at least nine individuals, and typically had CD > 8 m. In assessing how individuals were using trees relative to the space available in them, I noted when trees were "full", with at least oneother blue monkey in a small tree, more than two monkeys in a medium tree, more than five monkeys in a large tree, or "many" monkeys in any tree.

Data analysis

To measure rates of aggression, only focal samples were used. Both focal and *ad libitum* data (1992–1995) were used to analyze types and contexts of agonism and to evaluate agonistic relationships. Rare episodes of polyadic aggression were broken down into constituent dyads. The 1996 *ad libitum* data were included only to evaluate inter-annual consistency in agonistic relationships.

Decided agonistic interactions, in which one and only one animal showed submission, were distinguished from undecided interactions. Approach–retreat interactions and spontaneous submission were by definition decided. Aggressive interactions in which the recipient initially fought back but then fled were also included. The only undecided interactions sampled involved recipients of aggression who either ignored it or fought back, but neither partner showed submission.

To assess linearity in dominance hierarchies, Appleby's (1983) test was applied initially. Because this test is overly conservative when there are many unknown relationships, I followed deVries's (1995) recommendation to apply his randomization test to hierarchies yielding $0.15 > p > 0.01$ by Appleby's test. The randomization tests were performed using version 3.2 of MatMan (deVries *et al.* 1995), an earlier version of which was described in deVries (1993).

In tallying social interactions from focal samples to get frequencies, ≥ 2 minutes had to elapse in which the partners were not within 1 m and did not interact before a new interaction was scored.

All statistical tests were two-tailed, and an alpha of 0.05 was accepted as a criterion for statistical significance. Correlations with agonistic rank were

Table 17.1. *Rates of agonism*

Type of aggression	Mean rate freq/hr*	SEM
Showing aggression (attacking, chasing, threatening)	0.085	0.025
Receiving aggression (being attacked, chased, threatened)	0.055	0.016
Receiving submission or deference** (supplanting, being avoided, trilled or geckered at)	0.119	0.027
Showing submission or deference** (being supplanted, avoiding, trilling or geckering)	0.195	0.041

Notes:
*n = 16 adult and subadult females, sampled as focal subjects in 1992, 1993 or both.
** By definition, submission or deference was not associated with received aggression.

measured using the Spearman correlation coefficient. Means are reported with associated standard errors.

Results

General patterns of agonism

Data combined across the two years of focal sampling (in which general patterns were very similar) indicated that an average female participated in an agonistic encounter every 2.2 hours. Rates of showing and receiving overtly aggressive behavior were lower than rates of showing or receiving spatial displacement and spontaneous submission (Table 17.1).

Of 596 agonistic interactions recorded (484 from *ad libitum*, 112 from focal observations), 14 *ad libitum* observations were incomplete, though the interactions were known to be "decided" in outcome. For the remaining 582 cases, half involved overt aggression, while slightly less than half were approach–retreat interactions (Table 17.2). The few remaining cases involved spontaneous submission. The proportions of interactions in these categories are somewhat different in the focal data, with a higher percentage of approach–retreat than aggressive interactions. This difference probably arises because approach–retreat interactions are less conspicuous than aggression, and hence more likely to be overlooked in *ad libitum* sampling.

Table 17.2. *Types of agonism*

Type of Agonism	All data ($n=582$)	Focal data only ($n=112$)
Approach retreat interactions	264 (45%)	67 (60%)
Avoids (no submission except retreat)	49	12
Supplants (no submission except retreat)	172	41
Retreat with submissive	43	14
Aggressive interactions	290 (50%)	41 (37%)
Recipient responds with submission only	274	40
Recipient returns aggression, but then submits	4	1
Recipient returns aggression, no clear outcome	5	0
Recipient ignores received aggression	3	0
Recipient's response not noted	4	0
Spontaneous submission	28 (5%)	4 (3%)

Few recipients of aggression returned or ignored aggression, resulting in few "undecided" interactions (8/596).

Aggressive interactions usually involved just two animals. Only 5% (14/290) of such episodes in the total sample were polyadic: in nine of these, a third animal joined the aggressor, and in five, it intervened to defend the victim. Recipients of aggression redirected onto third parties only twice (<1% of interactions). The dyadic nature of agonism is even more pronounced in the focal data: only 1% of agonistic episodes involved defensive coalitions; and no coalitions with aggressors or cases of redirected aggression were observed.

The occurrence of reconciliation after aggression was not investigated formally by comparing rates of post-conflict affiliation to baseline rates (Veenema *et al.*, 1994). I noted, however, whenever opponents had a friendly reunion within 4 minutes of their original aggressive exchange. Given the general infrequency of friendly interactions in blue monkeys (see below), it is likely that any reunions following so closely after aggression would indicate statistically elevated rates, and so would satisfy a formal, operational definition of reconciliation. Reconciliation rates may be underestimated, however, by ignoring reunions occurring at longer intervals. Four percent (12/290) of aggressive interactions in the total sample (2/41 in focal samples) were followed by friendly reunion within 4 minutes. Of 12 post-conflict reunions, nine involved grooming, two involved brief touching, and one involved presenting the hindquarters. The loser of the preceding aggressive encounter initiated all these reunions except two grooming bouts.

Table 17.3. *Contexts of agonism (focal and* ad lib *data combined)*

Context	Approach–retreat interactions	Aggressive interactions	Spontaneous submission	All interactions
Feeding/ drinking	168	152	10	330
Space	13	10	1	24
Access to social partners	14	4	1	19
Access to infants	1	3	4	8
Coalition formation	–	14	–	14
Other*	0	7	2	9
Unclear	68	100	10	178

Notes:
*Other contexts included redirected aggression, nursing conflicts, and apparent confusion or fear when an adult male suddenly appeared nearby.

Most agonism (82%) occurred in the context of feeding, or drinking at arboreal water holes, regardless of the interaction type (Table 17.3). Typically, the loser moved to another location in the tree, or left it entirely. The next largest categories, each accounting for about 5% of agonistic interactions with known contexts, included spatial competition (e.g. A attacked B who passed by too closely, not in a feeding tree) and access to social or sexual partners.

The focal sample data indicated even more strongly that competition for food and drinking holes was the most frequent context of agonism – 93% of focal agonism occurred in this context. The amount of agonism related to particular food types was distributed differently from expectations based on the time monkeys spent feeding on each food type ($\chi^2 = 16.18$, $df = 2$, $p < 0.001$, focal data). Agonism over fruit occurred 1.7 times more often than expected, while agonism during insect and foliage feeding occurred respectively at 0.3 and 0.75 times the expected rates. These results suggest that blue monkey agonism is related primarily to competition for fruit.

Agonistic relationships and hierarchy

Data on agonism were used to assess agonistic relationships in each year. Only 12–33% of all possible dyads were seen to interact agonistically in a given year. Of these dyads, 62–69% interacted only once, and fewer than 14% were observed interacting more than twice.

For each year, hierarchies were constructed in which an individual's rank was based on the number of other individuals against which it usually prevailed. It was possible to arrange individuals so that there were few or no reversals (Fig. 17.1), but the resultant hierarchies were not significantly linear by Appleby's test (χ^2 test, $p>0.10$, $p>0.25$, $p>0.995$, $p>0.75$, respectively, in chronological order from 1992–1995), though one (1992) showed significant linearity by the randomization test ($p=0.024$). Adult and sub-adult females were sampled disproportionately because they were focal subjects and were easy to recognize; to see whether linearity could be detected with a greater density of data, analyses were repeated for hierarchies including only these females. In this subset, data were available for 27–50% of possible dyads across the four years, but again, none of the hierarchies was significantly linear according to Appleby's test ($p>0.25$, $p>0.10$, $p>0.975$, $p>0.05$ in chronological order), while the hierarchies of 1993 and 1995 showed significant linearity ($p=0.020$, $p=0.023$ respectively) when reanalyzed with the randomization test.

Although data on agonism were sparse, and a hierarchical structure to agonistic relationships was not discernable within every year, the fact that agonistic relationships were consistent across years suggests that they were not artifacts of sampling. Table 17.4 shows that those dyadic relationships that could be assessed in two or more years were overwhelmingly consistent in the direction of dominance. This consistency suggests that it is reasonable to look for hierarchical structure in data pooled across the four years (Fig. 17.2). With the greater density of data (70% of dyads were observed to interact, and 49% of these interacted more than twice), linearity was statistically significant even by Appleby's test ($\chi^2=89.17$, $df=26$, $p<0.001$). There were reversals in only four dyads (of 124 with data), and one triangular set of relationships. The rankings derived from this hierarchy were used in subsequent analyses.

Rank did not appear to be a function of a female's age, measured as the time since the birth of her first infant (Fig. 17.2). Age was not linearly related to rank ($r_s=0.251$, $n=19$), and there was no indication of any other kind of systematic relationship between these variables.

There was some indication that juveniles matched their mothers' dominance status in agonistic encounters. Across the years, agonism occurred in 23 juvenile–adult dyads in which the agonistic relationship between the adult and the juvenile's mother was documented in the same year as the agonism. In 22 dyads, the juvenile's status with respect to the opponent

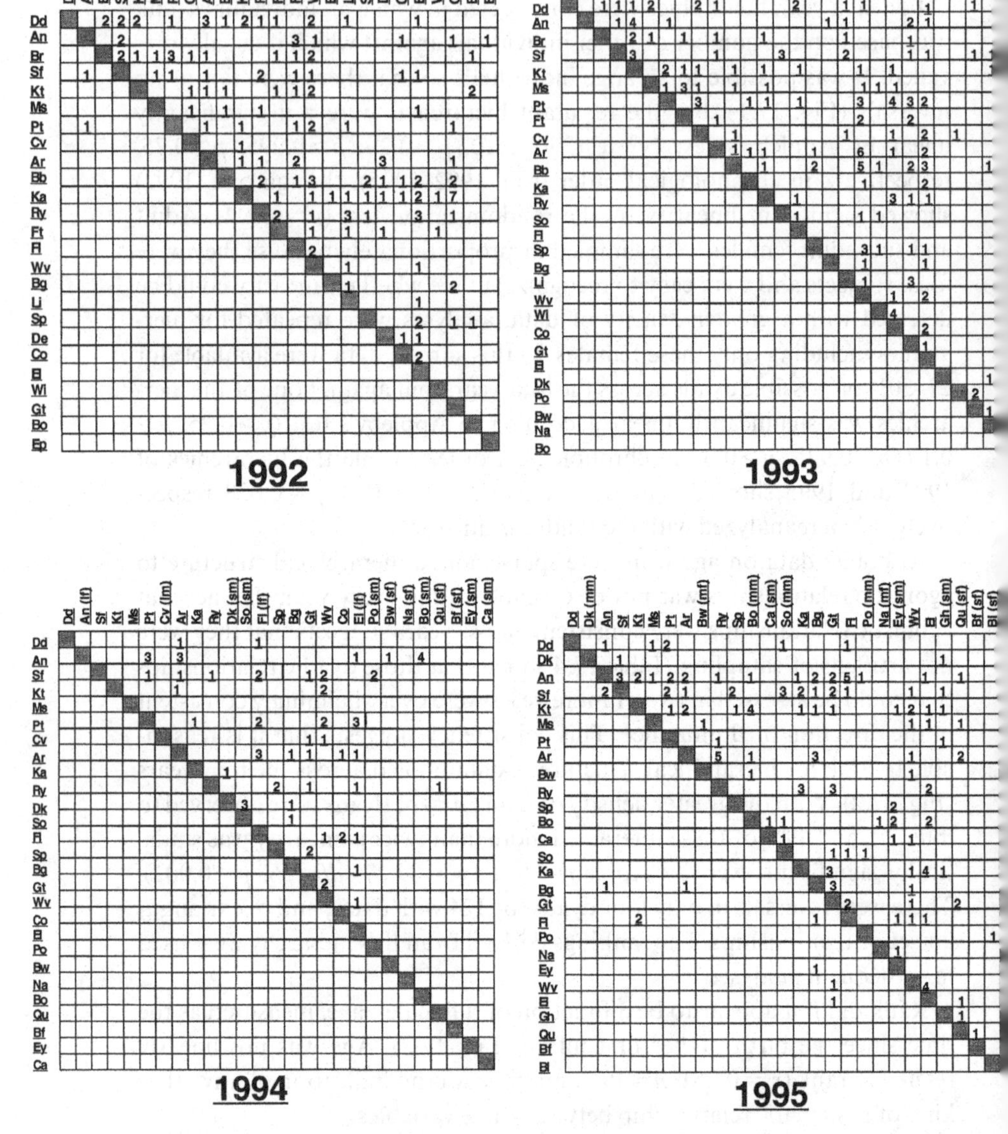

Fig. 17.1. Dominance hierarchies in T_w group for each year between 1992–95. Adult males and infants were excluded, but all other group members are included. Juveniles' names at top of each grid are followed by letters indicating their size (small, medium, large) and sex. Because of the sparse data, these particular depictions are not the only possible hierarchical arrangements.

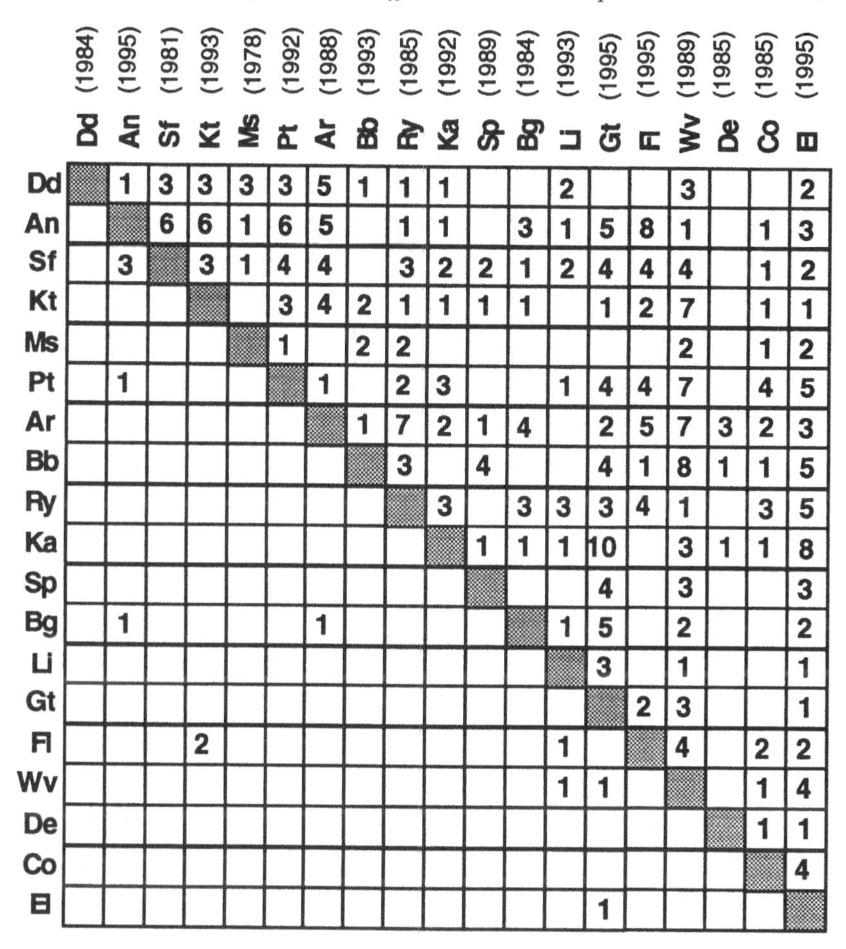

	Dd (1984)	An (1995)	Sf (1981)	Kt (1993)	Ms (1978)	Pt (1992)	Ar (1988)	Bb (1993)	Ry (1985)	Ka (1992)	Sp (1989)	Bg (1984)	Li (1993)	Gt (1995)	Fl (1995)	Wv (1989)	De (1985)	Co (1985)	El (1995)
Dd		1	3	3	3	3	5	1	1	1			2			3			2
An			6	6	1	6	5		1	1		3	1	5	8	1		1	3
Sf		3		3	1	4	4		3	2	2	1	2	4	4	4		1	2
Kt						3	4	2	1	1	1	1		1	2	7		1	1
Ms						1		2	2							2		1	2
Pt		1					1		2	3			1	4	4	7		4	5
Ar								1	7	2	1	4		2	5	7	3	2	3
Bb									3		4			4	1	8	1	1	5
Ry										3		3	3	3	4	1		3	5
Ka											1	1	1	10		3	1	1	8
Sp														4		3			3
Bg		1					1						1	5		2			2
Li														3		1			1
Gt															2	3			1
Fl				2									1			4		2	2
Wv													1	1				1	4
De																		1	1
Co																			4
El														1					

Fig. 17.2. Female dominance hierarchy constructed using 1992–95 data pooled across all 4 years. Individuals were included if they were adults for at least one of the four years. The year in which each female first gave birth is indicated above the top of the figure.

matched its mother's. In juvenile–juvenile dyads, this pattern was less marked. In four of six dyads in which the relative ranks of the juveniles' mothers were known, the relative status of the juveniles matched that of their mothers. More dyads could be evaluated if juveniles' agonistic relationships were deduced from data pooled over four years; results were qualitatively similar. Most all-juvenile dyads consisted of equal-sized peers born in the same year, but in two cases, a younger, smaller individual prevailed against its opponent. The two dyads in which juvenile and maternal dominance did not correspond involved a particular male, just beginning

Table 17.4. *Consistency of dyadic dominance relations across years of the study*

	Number of dyads with data for these years	Percentage of dyads showing the same dominance relationship over time
Comparisons of two consecutive years		
1992–1993	45	96
1993–1994	18	100
1994–1995	19	100
1995–1996	19	100
Comparisons of three years		
1992, 1993, 1994	7	100
1992, 1993, 1995	6	100
1992, 1993, 1996	14	92
1992, 1994, 1995	7	100
1992, 1994, 1996	8	100
1992, 1995, 1996	10	100
1993, 1994, 1995	3	100
1993, 1994, 1996	9	100
1993, 1995, 1996	8	80
1994, 1995, 1996	4	100
Comparisons of four years		
1992–1995 inclusive	3	100
1992, 1994, 1995, 1996	4	100
1992, 1993, 1995, 1996	5	100
1992, 1993, 1994, 1996	6	100
1993–1996 inclusive	2	100
Comparisons of five years		
1992–1996 inclusive	2	100%

Notes:
A particular dyad may have been a datum in more than one comparison: for example, the two dyads evaluated over a five-year period are included in the data for every other comparison listed in the table.

his adolescent growth spurt, dominating smaller animals at least 12 months younger. Once blue monkey males reach sub-adulthood, they are able to supplant and chase many adult females and most (if not all) juveniles.

Relationship between agonistic rank, agonism, feeding competition, and reproductive output

The relationships between agonistic rank (Fig. 17.2) and other aspects of behavior were evaluated using data from the 16 focal subjects. Data from

1992 and 1993 were combined for those individuals sampled in both years.

High-ranking females showed aggression ($r_s = 0.511$) and received submission and deference (i.e. supplanted, were avoided, received submissive vocalizations; $r_s = 0.519$) at higher rates per unit time than lower-ranking females ($p < 0.05$). The rate of receiving aggression was not significantly correlated with rank ($r_s = 0.404$, $p < 0.20$), but low-ranking females showed higher rates of submission and deference ($r_s = -0.614$, $p < 0.02$). If data on all types of agonism are combined, it is clear that high-ranking females engaged in agonistic encounters in which they prevailed at a higher rate ($r_s = 0.559$, $p < 0.05$) and engaged in agonistic encounters in which they yielded at a lower rate ($r_s = -0.695$, $p < 0.005$) than lower-ranking females. Higher-ranking females also prevailed in a larger proportion of their agonistic encounters ($r_s = 0.719$, $p < 0.005$), and had higher ratios of "wins" to "losses" ($r_s = 0.654$, $p < 0.02$). Although these results might appear self-evident, the way in which dyadic agonistic relationships were hierarchically ordered depends only on the prevailing *direction* of agonism within dyads; the way in which agonistic rank relates to rates of agonistic behavior, however, depends on the frequency as well as the direction of agonism in different dyads.

Agonistic rank was not correlated with any aspect of feeding behavior that I examined, regardless of whether I used individual ranks or rank classes in the analysis. Females of different rank did not differ systematically with respect to time spent feeding, nor did they differ with respect to the fraction of feeding time devoted to major dietary constituents, including fruits (averaging 30% of feeding time, $n = 16$), young leaves and shoots (30%), mature leaves (19%), and invertebrates (15%). There were also no rank differences among females when foods were classified as clumped (fruits, young leaves) versus dispersed (mature leaves, invertebrates, gum, blossoms). Rank was also uncorrelated with time spent feeding in trees of different size, or with the number of other individuals in the tree. Similarly, rank was not related to time spent in trees that were "full" of group mates.

It is conceivable that even without rank-related differences in feeding *time*, females of different ranks have different feeding *rates*, and thus differ ultimately in the amount of food consumed. Because agonism occurred disproportionately often when monkeys fed on fruit, rank differences in food intake are expected to be greatest during fruit feeding. The 1996 data on feeding rates in four species of fruiting tree, however, did not substantiate any rank-related differences in fruit intake rate: for all four tree species, correlations between rank and ingestion rate were not significant.

Table 17.5. *Rank and reproductive output*

Female*	Rank	Rate of producing offspring that survived at least 12 months (infants/year)**	Dates
Dd	1	0.300	1981–95
Sf	3	0.286	1981–95
Ms	5	0.063	1979–95
Ar	7	0.286	1988–95
Ry	9	0.300	1985–95
Sp	11	0.333	1989–95
Bg	12	0.364	1984–95
Wv	16	0.400	1989–95
De	17	0.429	1985–92
Co	18	0.111	1985–94

Notes:
* Analysis limited to females included in Table 17.2 for whom at least three years of reproductive records were available. All females except Ms gave birth for the first time during the initial year of monitoring; Ms was parous when she was first identified in 1979.
** A 12 month survival criterion was used because most juvenile mortality occurs in the first year in this population; because of intermittent monitoring, it is possible that some infants were born and died without our knowing it.

Finally, I examined the relationship between agonistic rank and reproductive output. Because the interbirth interval in this population averaged 34 months when the first of two infants survived, the analysis included only those females whose reproduction had been monitored at least three years. Much of this monitoring occurred before 1992, and the analysis assumes that relative ranks were stable before and after this time. There was no correlation between rank and the production of offspring that survived to one year (Table 17.5). If anything, lower-ranking females tended to produce viable offspring at a higher rate than higher-ranking females, but two females (Ms, Co) with exceptionally low rates obscure any statistically discernable pattern.

Friendly interactions

Grooming was the most frequent type of explicitly friendly interaction in which adult and sub-adult females engaged (mean $= 0.53 \pm 0.07$ encounters/hour, $n = 16$). Nose-to-mouth inspections, solicitations for grooming which did not ensue, brief touches and sitting in contact occurred less than one third as often, even when grouped together (mean $= 0.15 \pm 0.03$

bouts/hour). Coming into proximity (<1 m), without engaging in any exchange of signals, occurred slightly less frequently (mean = 0.54 ± 0.05 encounters/hour) than all explicitly friendly interactions considered together (mean = 0.68 ± 0.07 encounters/hour), but was more frequent than grooming.

Adult and sub-adult females spent an average of 7% (± 1%, n = 16) of their time engaged in overtly social activity. Only grooming, sitting in contact and play were interactions whose durations were timed, and grooming accounted for 96% (± 1.5%, n = 16) of all social time. The remainder consisted of sitting in contact, plus one bout of play.

On average, blue monkey females spent 11.6% (± 1.5%, n = 16) of their time in proximity to other monkeys, excluding infants. Of this proximity time, 40% did not include overt socializing (grooming or contact), occurring mostly as animals rested or fed. Females' rank and age did not correlate with the time spent grooming with or in proximity to one another.

Adult and sub-adult females slightly preferred one another as social partners. They spent more time close to and grooming with each other than would be expected from their numbers in the group. In 1992, an average of 67% of grooming time and 78% of proximity time (which included grooming) involved other adult and sub-adult females, who constituted only 60% of possible partners. In 1993, an average of 74% of grooming time and 61% of proximity time was spent with adult and sub-adult females, who accounted for only 50% of possible partners.

Among adult and sub-adult females, grooming was not equally distributed among all possible partners (Fig. 17.3). In combined data from the 1992 and 1993 focal samples, when each dyad was monitored for 32 hours, an average female groomed with 6.6 of 14 available partners (n = 14 females sampled in both years, range = 5–9). Typically, time spent grooming with the most favored partner (mean = 29.1 mins) was at least 10 times greater than time spent grooming with the least favored partner (mean = 1.2 mins). The number of grooming partners was not related to rank, even in dyads with high total grooming times.

Similarly, females did not spend equal amounts of time in proximity to one another. In combined data from 1992 and 1993, an average female had 9.9 proximity partners (of 14 available, range = 6–12, n = 14 females sampled in both years). The range of values for proximity time was even more extreme than for grooming time – proximity time with the most favored partner was at least 10 (7 females) and often 100 times (7 females) greater than proximity time with the least favored partner (mean = 48.3 vs. 0.6 minutes, n = 14). The number of proximity partners was not related to

Recipient

	Dd	Sf	Kt	Ms	Pt	Ar	Bb	Ry	Ka	Sp	Bg	Li	Wv	De	Co	El
Dd		16.4	.4			.4					1.2				7.4	
Sf	12.4		1.4		4.9				1.0						1.9	
Kt	.3				5.1	2.4		9.5	5.7	.3						
Ms					6.9			.1				3.5			.4	
Pt			1.2	5.0		4.5			7.4		4.3					3.1
Ar			3.0	11.8	11.5			1.5			3.1		9.4		23.9	
Bb								6.8		1.2		8.9			.7	
Ry			21.0			11.5	32.9		29.5	2.8	1.2				9.3	
Ka			3.4		15.6			6.5			26.9		5.6			
Sp			3.0	2.0				5.3			1.0					
Bg					2.3	4.4			3.8			2.8	4.3			1.7
Li				4.3			11.5						1.2	1.2	8.1	
Wv						3.1			4.8	1.0	4.5			2.5		
De													.4			
Co	.3					10.9		.5				2.4				11.7
El					39.5	5.3			12.6	1.6	4.7				16.4	

Fig. 17.3. Distribution of grooming among 16 adult and subadult females. Females are arranged from top to bottom and from left to right in order of dominance rank. Numbers in cells represent minutes in which the row animal groomed the column animal. Total sampling time for each dyad is the sum of sampling times for its two members. De and Bb were sampled only in 1992 (5 h) and 1993 (11 hs) respectively; all other individuals were sampled in both years (16 h).

rank, even when analysis was limited to dyads with high total proximity times.

Rank-related attractiveness of social partners did not seem to explain the distribution of grooming or proximity among these females. There were no correlations between rank and amount of grooming given ($r_s = 0.213$) or received ($r_s = 0.015$, $n = 14$ females sampled in both years). Females did not groom or receive grooming from higher (or lower) ranking individuals at rates exceeding a random expectation – in 20 of 37 dyads that groomed and for which there were also data on agonism, the lower-ranking partner groomed more, while in 17 dyads, the higher-ranked partner groomed more. When grooming times were summed over all 37 dyads, the amount of grooming directed up the hierarchy (283 minutes, or 53%) was similar to the amount directed down the hierarchy (252 minutes).

Table 17.6. *Rank distance and grooming among the 16 adult and large juvenile females who were focal subjects in 1992 and 1993*

Rank distance	Number of dyads that groomed/number of dyads in this rank distance category (%)	Average number of grooming (in mins) in dyads belonging to this rank distance category
1	10/15(67)	10.80
2	8/14 (57)	1.00
3	5/13 (38)	2.17
4	3/12 (25)	2.78
5	5/11 (45)	5.91
6	4/10 (40)	1.93
7	5/9 (56)	4.36
8	2/8 (25)	1.06
9	1/7 (14)	0.76
10	2/6 (33)	5.98
11	2/5 (40)	10.17
12	0/4 (0)	–
13	1/3 (33)	0.62
14	1/2 (50)	3.84
15	0/1 (0)	–

Grooming was observed in dyads whose members' ranks were adjacent as well as those whose members' ranks were widely separated (Table 17.6). The proportion of dyads in which grooming was observed was higher for dyads whose members were closer in rank ($r_s = -0.525$ for rank distance and proportion of possible dyads that groomed, $p = 0.05$, $n = 15$ rank distance classes). Rank distance was not generally correlated with the average amount of grooming per dyad in those dyads that groomed ($r_s = -0.424$), but dyads of adjacently ranked individuals accounted for a larger proportion of total grooming time (30%) than would be expected by their representation in the group (12.5%); thus there was some indication that adjacently ranked animals were particularly favored grooming partners.

Proximity time (including grooming) showed a similar pattern. Rank was not correlated with the total time spent in proximity to other adult and subadult females ($r_s = 0.248$, $n = 14$). Females did not consistently spend more time with higher (or lower) ranking females than expected from their numbers. Dyads of similarly ranked animals were not more likely to spend time in proximity, or to spend more time in proximity than dyads whose members' rank positions were more dissimilar. Again, however, adjacently ranked partners seemed to be special – they accounted for twice as much proximity time (28.4%) as would be expected based on their representation in the group (12.5%).

Studies of many other cercopithecines suggest that kinship is relevant in explaining social preferences, but kinship among the adult females in this study was not known. In some cases mother–daughter relationships could be ruled out if females were too close in age, or if no appropriately aged daughter had been born to a particular mother. Among dyads that groomed for long periods (≥5 minutes) in either 1992 or 1993, seven of nine (1992) and five of nine (1993) were certainly *not* mother–offspring pairs. Another way to assess the importance of kinship is to compare grooming and proximity time of mothers and their own juvenile offspring versus other juveniles. Four of seven females who were observed to groom with juveniles spent more time grooming with their own offspring than with all other juveniles combined. These numbers do not suggest a strong kinship effect, but they are small. A stronger kinship effect is apparent in the proximity data: six of seven females who spent any time in proximity to juveniles spent more of it near their own offspring than near any other juvenile, and all seven females spent more time with their own offspring than with an average non-descendant juvenile.

The distribution of grooming among individuals was quite different in 1992 and 1993. In each year there were nine dyads characterized by ≥5 minutes of grooming, but only one of these dyads was the same in both years. Thus the identity of partners with the strongest grooming relationships differed greatly from year to year. (In contrast to this general pattern, the one exceptional dyad had been reported to groom often, with each partner as the other's top choice, in previous years; Rowell *et al.*, 1991.) The attractiveness of infants, and hence their mothers, could not account for the changes. Similar differences between years were found in patterns of general proximity as well. If these discrepancies do not reflect the difficulty of sampling rare behavior, they may reflect real inter-annual changes in affiliative relationships. Such change, however, would be in marked contrast to the consistency observed in agonistic relationships (Table 17.4).

Discussion

Agonism and affiliation in blue monkeys

Agonism in blue monkeys is dominated by approach–retreat interactions. Aggression also constitutes a large proportion of agonistic interactions, but spontaneous submission is rare. Although agonism occurs infrequently, it is possible to detect dominance relationships among group mates, in that most agonistic interactions are unidirectional and decided, and the direction of multiple interactions within a given dyad is consistent, even over years. Whether these dyadic dominance relationships can be ordered into a hierarchy is less clear. Significant linearity emerged when data from all four years were combined, but even then, about two-thirds of possible dyads were observed to interact two or fewer times. The hierarchy is thus abstracted from minimal characterizations of its constituent relationships.

The biological significance of this hierarchy is even less clear. Compared to lower-ranking females, high-ranking animals participated in agonism as winners at higher rates, and won more of their agonistic interactions. Because most agonism occurred during feeding, and the season in which data were collected is one when availability of preferred fruit is low, one might expect rank to relate to the outcome of feeding competition. There was no indication, however, that lower-ranking females, who suffered more frequent interruptions, were less efficient when feeding on fruit, or that they compensated by spending more time feeding. Their diet was qualitatively similar to that of higher-ranking females. Nor did lower-ranking females feed in trees of different size or with different numbers of group mates. Lower-ranking females did not spend less time in popular trees, full of other individuals. Data were not collected on the efficiency of feeding on non-fruit items, or on the biochemical make-up of different foods; such data could in principle reveal rank-related differences in diet (e.g. Barton and Whiten, 1993).

The conclusion that rank has little or no impact upon diet must be viewed as provisional. Nevertheless, it is consistent with the finding that rank was not related to the rate of producing viable offspring, which is presumably related to nutritional status. This lack of correlation also suggests that in this population, female reproductive competition does not operate directly through harassment and associated stress-induced infertility, inversely proportional to rank. It remains possible that rank has an effect on lifetime reproductive success by influencing survival rather than fecundity; three of four adult females (of various ages) that disappeared and presumably died during the study were from the bottom half of the hierarchy. How rank might influence longevity is not immediately obvious. Van Noordwijk and van Schaik (1987) showed that in long-tailed macaques (*M. fascicularis*), lower-ranking females were more often spatially peripheral, and thus presumably at greater risk of predation. In this group of blue monkeys, the lowest ranking female (El) was often the last animal in directed group progressions, but two other females from the top (Dd) and middle (Ka) of the hierarchy were also generally peripheral. More systematic study would be needed to explore this possibility thoroughly.

Rank was also not related to the distribution of grooming, the most frequent form of friendly interaction. Across dyads, grooming was as likely to go down as up the hierarchy. Dyads whose members had similar ranks were slightly more likely to groom, but only those of adjacent rank spent extraordinary amounts of time grooming each other. Seyfarth (1977) proposed that grooming in monkeys is a way to cultivate alliances, but it could be interpreted more generally as a way to elicit benefits (like tolerance) from powerful individuals. Because blue monkeys hardly ever form alliances, and because access to food sources does not appear to be related significantly to rank, it is entirely consistent that grooming is not distributed as predicted by the model.

This study confirms some results from our earlier report on the same study group (Rowell *et al.*, 1991), even though data were collected in different years and in different ways. In particular, rates of agonism and grooming were low, hierarchical relationships were weak but detectable, and, among adults, mother–daughter kinship seemed not to be a primary determinant of grooming frequency. In other respects, however, the two reports differ. The present analysis, based upon time spent grooming rather than frequency of bouts, showed greater differentiation among partners – grooming and proximity time were distributed quite unevenly across possible partners. The way in which agonism related to partner choice also differed – the rank-related clusters of grooming partners reported previously were not detectable with the present data on grooming durations, and the number of partners per female was not related to rank. Also, whereas middle-aged females had been most socially active in terms of grooming bouts, age was not systematically related to the amount of time spent grooming (or being in proximity). To understand these discrepancies, further study is called for, concentrating especially on the extent, basis, and function of differentiation among affiliative relationships.

Comparison with better-known cercopithecine monkeys

The agonism and affiliation of female blue monkeys differs in several respects from that seen in baboons, macaques, and vervets. First, rates of agonism appear to be low. Strictly comparable data from the wild are hard to find, as studies differ in (and do not always specify) how agonism is defined, which animals are used to calculate rates, and how many potential partners there are. Among forest-dwelling macaques, adult females engaged in agonistic interactions at rates that were at least three times higher than those reported here (*M. fascicularis*, van Noordwijk and van Schaik, 1987; *M. fuscata*, Hill and Okayasu, 1995). Gelada (*Theropithecus gelada*) females received attacks from higher-ranking females (Dunbar, 1980) at rates 3–10 times greater than blue monkeys (who were receiving aggression from all group members). Among anubis baboons (*Papio hamadryas anubis*) living in a savanna-woodland habitat, adult females interacted agonistically 1.5 times as often as female blue monkeys (Barton, 1993; Barton *et al.*, 1996). In chacma baboons (*P. h. ursinus*) in rocky, mountainous terrain, females interacted agonistically with other females at rates 20 times greater than those reported here for blue monkeys interacting with all possible partners (Seyfarth, 1976). These comparisons of wild forest cercopithecins and papionins support previous comparisons of captive or provisioned groups (Rowell, 1971; Kaplan, 1987).

Another aspect of agonism that has been reported as characterizing captive and provisioned cercopithecins is the high proportion of agonistic encounters that include aggression, even extreme and apparently uncontrolled aggression, as opposed to approach–retreat encounters or spontaneous submission (Rowell, 1971; Kaplan, 1987). This feature may be related to the fact that cercopithecins do not use many ritualized signals of submission or appeasement (Rowell, 1971; Kaplan, 1987; Loy *et al.*, 1993). In this study, signals of submission other than fleeing were indeed rare (the geckering vocalization was the next most common submissive response), but the physical, uncontrolled and cascading nature of aggression stressed by Rowell (1971) in captive Sykes' monkeys (*C. m. kolbi*) was not apparent. The proportion of agonistic encounters that included aggression (37%) was the same as that reported by Rowell (1971) for Sykes' monkeys, and centered within the range reported by Bernstein *et al.* (1983) for five papionin species studied in captivity (13–62%). In wild papionin populations, the proportion of aggressive interactions has been variously reported as higher (46% in *M. fascicularis* females; van Noordwijk and van Schaik, 1987) and lower (4% in *P. h. ursinus* female–female interactions; Seyfarth, 1976, table III) than that found in this study. It seems likely that cercopithecin aggression is more conspicuous and extreme in captivity, where escalation cannot be avoided by spreading out (Kaplan, 1987). Data from wild populations, however, suggest that there may be little difference both in the degree to which aggression is escalated and in the proportion of agonism that includes aggression relative to other cercopithecines.

Like the better studied cercopithecines, blue monkeys show agonistic interactions and relationships that are asymmetrical; however it is less clear for blue monkeys that these dyadic relationships can be arranged into a dominance hierarchy. Because rates of agonism are low, data accumulate slowly, and many dyads are seen to interact rarely if at all. Finding a hierarchy among wild blue monkeys is therefore something of a struggle, whereas hierarchies are obvious in the better-known cercopithecine species (Walters and Seyfarth, 1987). Hierarchies have been reported for other cercopithecins studied in captivity (Rowell, 1971, 1978; Kirkevold and Crockett, 1987; Loy and Loy, 1987; Zucker, 1987; Loy and Harnois, 1988; Erhart, 1993; Loy *et al.*, 1993), though sometimes they are based on criteria other than asymmetric pairwise relationships. Until the 1990s, however, there were no reports of hierarchies in wild cercopithecins, apart from vervet monkeys. Nakagawa (1992) alluded to stable linear hierarchies among patas (*Erythrocebus patas*) females that were occasionally provisioned, but published data come from only two studies (Rowell *et al.*, 1991, on the focal group of this study; Isbell and Pruetz, 1998). Most earlier reports of hierarchies, whether from captive or wild monkeys, did not assess linearity statistically; Erhart's (1993) Sykes' monkeys showed an unstable hierarchy that was not always linear.

Although this and other studies have found evidence of linear dominance hierarchies in cercopithecins other than vervets, the significance of these hierarchies, and of attaining a particular position in them, is often reported as unclear or inconsistent compared with baboons, macaques, and vervets (Rowell, 1971; Kaplan, 1987; Zucker, 1987). In the latter species, numerous reports document positive correlations between rank and success in reproduction or in competition for resources or spatial positions thought to influence reproductive output (van Noordwijk, 1987; Barton and Whiten, 1993; Silk, 1993). Positive correlations between rank and measures of inter-female competition do not occur in every population, probably because the effectiveness of agonistic competition depends on ecological conditions (Harcourt, 1989; Silk, 1993). Nevertheless, the fact that such correlations are observed and are positive when observed, that females seem to strive for high agonistic rank, and that rank correlates with many aspects of social behavior (e.g. Cheney and Seyfarth, 1990) suggests that achieving high rank has real functional consequences. Whatever those consequences may be, most researchers would surely agree that an adequate description of social organization in these species must include a consideration of rank. In blue monkeys and patas, however, it is not so clear that dominance rank is a variable with much or consistent explanatory power, in terms of its relationship both to competitive success and to the patterning of other social behavior (Kaplan, 1987; Loy and Harnois, 1988; York and

Rowell, 1988; Nakagawa, 1992). Particularly noteworthy contrasts with baboons, macaques and vervets concern differences in the frequency of coalition formation, and in the degree to which agonistic rank structures affiliative relationships.

Unanswered questions

There remain at least two important questions. First, why do blue monkeys have hierarchical relationships at all? While predictable agonistic relations at the dyadic level are understandable as conventions that obviate escalated aggression (Bernstein, 1981), and thus may benefit animals that meet regularly, this advantage of dyadic dominance cannot explain why a set of dyadic relationships is organized hierarchically. Linearity in hierarchies suggests that some underlying quality of individuals, usually thought to relate to their power in agonistic interactions, is the basis for the group-wide pattern. In blue monkeys, however, where rank apparently does not correlate with the presumed prerogatives of power, such as access to food, linearity in the hierarchy is difficult to understand. Perhaps the material advantages of high rank are apparent only at infrequent 'crunch' times, which did not occur during the periods reported here. Alternatively, the advantage of high rank may simply be the psychological reward of being less often frustrated in achieving one's goals.

Second, how can we explain the differences in female social relationships among cercopithecine monkeys? Variation in female agonistic relationships has been related to modes of competition that depend on ecological factors, especially the abundance and distribution of food and the degree to which cohesiveness protects against predation (Wrangham, 1980; van Schaik, 1989). These schemata fail to fit the forest cercopithecins well, for they lack female relationships that are well-differentiated according to status even though they rely on fruit (Cords, 1986), the quintessentially patchy resource that can be monopolized aggressively. Future research must evaluate more carefully the ways in which monkeys interact with their resources – blue monkeys, for example, may avoid direct competition by spreading out and by switching to more abundant or dispersed foods when fruit is scarce (Isbell, 1991; Cords, 1993).

It is also possible that heritage, or phylogeny, explains some resemblances among cercopithecine monkeys (Struhsaker, 1969). It is remarkable how much published accounts of patas monkeys agree with the findings of this study, even though patas live under very different ecological conditions. On the other hand, the social organization of vervets, which are generally considered to be even more closely related to blue monkeys than are patas, differs substantially from that of blue monkeys.

To test either of these ideas satisfactorily, comparable data will be required for more populations and species. Meanwhile, however, we should recognize that social relationships and organization among cercopithecine species do vary, and we should be wary of overgeneralizing from the few species that have been well studied.

Acknowledgements

I am grateful to the Office of the President, Government of Kenya, for permission to study blue monkeys in the Kakamega Forest, to the Zoology Department, University of Nairobi, for local sponsorship, and to the Kakamega Forest Station staff for their cooperation. Field work has been supported by the National Science Foundation (SBR 95-23623), the L.S.B. Leakey Foundation, the Wenner-Gren Foundation, and Columbia University. Thanks to my Kenyan and American student-interns for contributing to the data and analysis, to Han de Vries for running his program for me, and to the reviewers and editors for comments on the developing manuscript.

References

Appleby, M.C. (1983). The probability of linearity in hierarchies. *Anim. Behav.* **31**, 600–8.

Baker, S.C. & Estep, D.Q. (1985). Kinship and affiliative behavior patterns in a captive group of Celebes black apes *(Macaca nigra)*. *J. comp. Psychol.* **99**, 356–60.

Barton, R.A. (1993). Sociospatial mechanisms of feeding competition in female olive baboons, *Papio anubis*. *Anim. Behav.* **46**, 791–802.

Barton, R.A. & Whiten, A. (1993). Feeding competition among female olive baboons, *Papio anubis*. *Anim. Behav.* **46**, 777–89.

Barton, R.A., Byrne, R.W. & Whiten, A. (1996). Ecology, feeding competition and social structure in baboons. *Behav. Ecol. Sociobiol.* **38**, 321–9.

Bernstein, I.S. (1981). Dominance: the baby and the bathwater. *Behav. Brain. Sci.* **4**, 419–57.

Bernstein, I., Williams, L. & Ramsay, M. (1983). The expression of aggression in old world monkeys. *Am. J. Primatol.* **4**, 113–25.

Borries, C., Sommer, V. & Srivastava, A. (1994). Weaving a tight social net: allogrooming in free-ranging female langurs *(Presbytis entellus)*. *Int. J. Primatol.* **15**, 421–43.

Bramblett, C.A. & Coelho, A.M. (1987). Development of social behavior in vervet monkeys, Sykes' monkeys, and baboons. In *Comparative Behavior of African Monkeys*, ed.E. L. Zucker, pp. 67–79. New York: Alan R. Liss.

Cheney, D.L. & Seyfarth, R.M. (1990). *How Monkeys See the World*. Chicago: University of Chicago Press.

Cords, M. (1986). Forest guenons and patas monkeys: male–male competition in one-male groups. In *Primate Societies*, ed. B.B. Smuts, D.L. Cheney, R.M. Seyfarth, R.W. Wrangham & T.T. Struhsaker, pp. 98–111. Chicago: University of Chicago Press.
Cords, M. (1987). Mixed-species association of *Cercopithecus* monkeys in the Kakamega Forest, Kenya. *Univ. Calif. Pub. Zool.* **109**, 1–109.
Cords, M. (1993). The behavior of adult female blue monkeys during a period of seasonal food scarcity. *Am. J. Primatol.* **30**, 304–5.
deVries, H. (1993). Improved test of linearity in dominance hierarchies containing unknown or tied relationships. *Anim. Behav.* **50**, 1375–89.
deVries, H., Netto, W.J. & Hanegraaf, P.L.H. (1995). MatMan: a program for the analysis of sociometric matrices and behavioural transition matrices. *Behaviour* **125**, 157–75.
Dunbar, R.I.M. (1980). Determinants and evolutionary consequences of dominance among female gelada baboons. *Behav. Ecol. Sociobiol.* **7**, 253–65.
Ehardt, C. L. (1988). Absence of strongly kin-preferential behavior by adult female sooty mangabeys (*Cercocebus atys*). *Am. J. phys. Anthrop.* **76**, 233–43.
Erhart, E.M. (1993). Diachronic changes in the dominance relations of adult females in a Sykes' monkey matriline. *Am. J. Primatol.* **30**, 308.
Erwin, J. & Zucker, E.L. (1987). African monkeys in behavioral research: a 5-year retrospective analysis. In *Comparative Behavior of African Monkeys*, ed. E.L. Zucker, pp. 1–21. New York: Alan R. Liss.
Gust, D.A. (1994). A brief report on the social behavior of the crested mangabey (*Cercocebus galeritus galeritus*) with a comparison to the sooty mangabey (*C. torquatus atys*). *Primates* **35**, 375–83.
Gust, D.A. (1995). Moving up the dominance hierarchy in young sooty mangabeys. *Anim. Behav.* **50**, 15–21.
Gust, D.A. & Gordon, T.P. (1993). Conflict resolution in sooty mangabeys. *Anim. Behav.* **46**, 685–94.
Gust, D.A. & Gordon, T.P. (1994). The absence of a matrilineally based dominance system in sooty mangabeys, *Cercocebus torquatus atys*. *Anim. Behav.* **47**, 589–94.
Harcourt, A.H. (1989). Social influences on competitive ability: alliances and their consequences. In *Comparative Socioecology* ed. V. Standen & R.A. Foley, pp. 223–42. Oxford: Blackwell Scientific Publications.
Hill, D.A. & Okayasu, N. (1995). Absence of 'youngest ascendancy' in the dominance relations of sisters in wild Japanese macaques (*Macaca fuscata yakui*). *Behaviour* **132**, 367–79.
Isbell, L.A. (1991). Contest and scramble competition: patterns of female aggression and ranging behavior among primates. *Behav. Ecol.* **2**, 143–55.
Isbell, L.A. & Pruetz, J.D. (1998). Differences between vervets (*Cercopithecus aethiops*) and patas monkeys (*Erythrocebus patas*) in agonistic interactions between adult females. *Int. J. Primatol.* **19**, 837–55.
Kaplan, J.R. (1987). Dominance and affiliation in the Cercopithecini and Papionini: a comparative examination. In *Comparative Behavior of African Monkeys*, ed. E.L. Zucker, pp. 127–50. New York: Alan R. Liss.
Kirkevold, B.C. & Crockett, C.M. (1987). Behavioral development and proximity patterns in captive DeBrazza's monkeys. In *Comparative Behavior of African Monkeys*, ed. E.L. Zucker, pp. 39–65. New York: Alan R. Liss, Inc.
Loy, J., Argo, B., Nestell, G., Vallett, S. & Wanamaker, G. (1993). A reanalysis of patas monkeys' "grimace and gecker" display and a discussion of their lack of formal dominance. *Int. J. Primatol.* **14**, 879–93.

Loy, J. & Harnois, M. (1988). An assessment of dominance and kinship among patas monkeys. *Primates* **29**, 331–42.

Loy, K.M. & Loy, J. (1987). Sexual differences in early social development among captive patas monkeys. In *Comparative Behavior of African Monkeys*, ed. E.L. Zucker, pp. 23–37. New York: Alan R. Liss.

McKenna, J.J. (1979). The evolution of allomothering behavior among colobine monkeys: function and opportunism in evolution. *Am. Anthropol.* **81**, 818–40.

Nakagawa, N. (1992). Distribution of affiliative behaviors among adult females within a group of wild patas monkeys in a nonmating, nonbirth season. *Int. J. Primatol.* **13**, 73–96.

Newton, P.N. & Dunbar, R.I.M. (1994). Colobine monkey society. In *Colobine Monkeys: Their Ecology, Behaviour and Evolution*, ed. A. G. Davies & J. F. Oates, pp. 311–46. Cambridge: Cambridge University Press.

O'Brien, T.G. (1993). Allogrooming behaviour among adult female wedge-capped capuchin monkeys. *Anim. Behav.* **46**, 499–510.

Oi, T. (1990). Patterns of dominance and affiliation in wild pig-tailed macaques (*Macaca nemestrina*). *Int. J. Primatol.* **11**, 339–56.

Rowell, T.E. (1971). Organization of caged groups of *Cercopithecus* monkeys. *Anim. Behav.* **19**, 625–45.

Rowell, T.E. (1978). How female reproductive cycles affect interaction patterns in groups of patas monkeys. In *Recent Advances in Primatology*, ed. D.J. Chivers & J. Herbert, pp. 489–90. New York: Academic Press.

Rowell, T.E. (1988). The social system of guenons, compared with baboons, macaques and mangabeys. In *A Primate Radiation: Evolutionary Biology of the African Guenons*, ed. A. Gautier-Hion, F. Bourliere, J. P. Gautier & J. Kingdon, pp. 439–51. Cambridge: Cambridge University Press.

Rowell, T.E., Wilson, C. & Cords, M. (1991). Reciprocity and partner preference in grooming of female blue monkeys. *Int. J. Primatol.* **12**, 319–36.

Seyfarth, R.M. (1976). Social relationships among adult female baboons. *Anim. Behav.* **24**, 917–38.

Seyfarth, R.M. (1977). A model of social grooming among adult female monkeys. *J. theoret. Biol.* **65**, 671–98.

Silk, J.B. (1993). The evolution of social conflict among female primates. In *Primate Social Conflict*, ed. W.A. Mason & S.P. Mendoza, pp. 49–83. Albany: State University of New York Press.

Strier, K.B. (1990). New World primates, new frontiers: insights from the woolly spider monkey (*Brachyteles arachnoides*). *Int. J. Primatol.* **11**, 7–19.

Struhsaker, T.T. (1969). Correlates of ecology and social organization among African Cercopithecines. *Folia primatol.* **11**, 80–118.

Thierry, B., Gauthier, C. & Peignot, P. (1990). Social grooming in Tonkean macaques. *Int. J. Primatol.* **11**, 357–75.

van Noordwijk, M.A. & van Schaik, C.P. (1987). Competition among female long-tailed macaques, *Macaca fascicularis*. *Anim. Behav.* **35**, 577–89.

van Schaik, C.P. (1989). The ecology of social relationships amongst female primates. In *Comparative Socioecology*, ed. V. Standen & R.A. Foley, pp. 195–218. Oxford: Blackwell Scientific Publications.

Veenema, H.C., Das, M. & Aureli, F. (1994). Methodological improvements for the study of reconciliation. *Behav. Proc.* **31**, 29–38.

Walters, J.R. & Seyfarth, R.M. (1987). Conflict and cooperation. In *Primate Societies*, ed. B.B. Smuts, D.L. Cheney, R.M. Seyfarth, R.W. Wrangham & T.T. Struhsaker, pp. 306–17. Chicago: University of Chicago Press.

Watts, D.P. (1994). Agonistic relationships between female mountain gorillas (*Gorilla gorilla berengei*). *Behav. Ecol. Sociobiol.* **34**, 439–51.
Wrangham, R.W. (1980). An ecological model of female-bonded primate groups. *Behaviour* **75**, 262–300.
York, A.D. & Rowell, T.E. (1988). Reconciliation following aggression in patas monkeys, *Erythrocebus patas*. *Anim. Behav.* **36**, 502–9.
Zucker, E.L. (1987). Social status and the distribution of social behavior by adult female patas monkeys: a comparative perspective. In *Comparative Behavior of African Monkeys*, ed. E.L. Zucker, pp. 151–73. New York: Alan R. Liss.

18

Locomotor behavior in Ugandan monkeys

DANIEL L. GEBO AND COLIN A. CHAPMAN

Introduction

Studies of positional behavior have helped our understanding of postcranial adaptation in primates and this in turn has contributed to discussions concerning how and why particular directions in primate evolution occurred. At first, anatomists observed positional behavior, whether in captivity or in the wild, in order to describe what primates actually do. These observations led to simple categorization of primate species into, for example, brachiators or arboreal quadrupeds, as well as evolutionary scenarios reconstructing adaptive pathways in primate and human locomotor evolution (e.g. Keith, 1923; Clark, 1959; Napier and Walker, 1967). With the proliferation of field studies, many of the early categories proved less than useful and the association between particular anatomical features and specific behaviors came under closer scrutiny (e.g. Stern and Oxnard, 1973; Mittermeier and Fleagle, 1976; Morbeck *et al.*, 1979). These works, which used more sophisticated methods, specifically the quantification of primate positional behavior, helped to direct studies of positional behavior towards ecology as well as morphology. Despite these efforts made in the 1970s, surprisingly few species have been adequately sampled quantitatively in the wild, and, perhaps more importantly, very few studies have focused upon a particular research problem (for example, changes in body size and its effect on arboreal locomotion; Napier, 1967; Cartmill, 1974; Fleagle and Mittermeier, 1980; Fleagle, 1985; Jungers, 1985). Thus, how and why primates make the day-to-day choices they do, as well as why species are adapted to particular environments, are particularly central questions today, and few answers are to be found in the literature on positional behavior.

With this in mind, we began a field project in 1990 in Kibale Forest, Uganda, to examine positional behavior, body size, and habitat use in five sympatric cercopithecid monkey species (Gebo and Chapman, 1995a). We wanted first to identify the distinctive locomotor abilities of each species and second, to test relationships between positional behavior, body size, and habitat use. Fleagle and Mittermeier (1980) had noted several trends in South American monkey locomotion and we wondered whether African monkeys might conform to these same tendencies (e.g. smaller primates showing higher leaping frequencies and less use of the mid- and upper canopy). The five cercopithecid species (*Cercopithecus ascanius, C. mitis, Lophocebus albigena, Colobus badius*, and *C. guereza*) are sympatric within the primary forest. Their sympatry and close taxonomic relationship minimize problems associated with comparing animals with very different anatomy, or living in different habitats. We also recorded locomotor behavior of one species, the red colobus monkey (*C. badius*), in a variety of ecological contexts, thus providing an overall assessment of its locomotor variation. We sampled behavior within the same season in different years, within different seasons of the same year, and within three different forest settings – primary, secondary, and pine forests (Gebo and Chapman, 1995b). We also sampled locomotor behavior of red colobus monkeys in crisis situations (i.e. when responding to predators) (Gebo *et al.*, 1994).

Methods

The Kibale Forest Reserve (560 km^2) is situated in western Uganda, near the base of the Ruwenzori Mountains. It is a moist, evergreen forest with the canopy generally 25–30 m in height (Struhsaker, 1975; Kasenene, 1980; Skorupa, 1986, 1988; Kalina, 1988; Butynski, 1990). Parts of the reserve are comprised of swamp, grassland, plantations of pine, thicket, and colonizing forest (Butynski, 1990). The study site, Kanyawara, is situated at an elevation of 1500 m. Besides the species included in this study, the forest is inhabited by three strepsirhine primates (*Galagoides demidovii, Galago inustus*, and *Perodicticus potto*), chimpanzees (*Pan troglodytes*) and three other cercopithecid monkeys (*Cercopithecus aethiops, Cercopithecus lhoesti*, and *Papio hamadryas anubis*).

The positional behavior of the five species was observed in primary forest during the dry seasons (May–August) of 1990 and 1991. Additional observations on *Colobus badius* were made in secondary and pine forests, and during the wet season of 1990. All study populations were habituated to observers, although no animals could be approached closer than three meters. When approached, most individuals settled down to their normal regime after an initial moment of uncertainty. A focal animal technique of

Table 18.1. *Definitions of positional activities*

Locomotion

Quadrupedalism: all four limbs move in a regular pattern above a support or on the ground; includes walking, running, and galloping.

Leaping: the hindlimbs propel an animal across a gap. A leap included quadrupedal standing then leaping, or pumping the body up and down before leaping, vertical clinging and then leaping, and quadrupedal running and leaping. Dropping down from a branch was not scored as a leap.

Climbing: a movement up or down a vertical or steeply inclined support or through irregular and intertwined small supports; all four limbs move in an often irregular pattern with abducted arms and knees and with variable hand and foot positions; the arms are used to pull the animal while the legs alternately push the body upward/forward.

Other: includes, *quadrupedal suspensory movements,* in which the body is progressing below a support using three or four limbs; *bridging,* where spatial gaps are crossed by body stretching; *bimanualism,* in which the hands grasp a support and are used to pull the body up to a support from below (bimanual pull-up); *bipedalism,* in which only the hind feet are used to take a short walk; and *vertical bounding,* a succession of short jump-clings up a vertical support.

Postures

Sitting: animal supports weight on its haunches; feet may or may not be in contact with the support, above or below the body, legs splayed or close to midline.

Standing: animal stands on all four limbs.

Reclining: animal lies on its belly, side, or back.

Other: includes, *vertical clinging:* animal clings to a vertical support without sitting; *quadrupedal suspension:* animal hangs underneath a support by all four limbs; *bimanual suspension:* animal hangs from hands, usually with bent elbows; *hindlimb suspension:* animal hangs from feet; *bipedal stand:* animal stands on hindfeet, usually with the heel elevated above the support.

continuous sampling (Altmann, 1974) was used, and more than 20 adult individuals of each species were sampled. Positional behaviors were defined as shown in Table 18.1, and were recorded as a series of bouts. Each bout included a single behavior, bounded by a different posture or movement (see Fleagle, 1976; Fleagle and Mittermeier, 1980; Susman, 1984; Gebo, 1992). Animals were observed continuously from first contact until approximately 500 positional bouts were recorded for the day (between 7 and 9 hours of observation). Most types of locomotion and postures used by cercopithecid monkeys are described and illustrated in Ripley (1967), Morbeck (1975), Mittermeier and Fleagle (1976), Fleagle (1978, 1980), and Rose (1979).

The behavioral context of each observation was recorded as *travel* (long distance movements between trees, usually between a series of trees, from or to feeding or resting sites), *feeding* (movements within a single tree), or *resting* (periods of inactivity). The circumference of the support used was recorded as *large* (>25 cm), *medium* (6–25 cm), or *small* (<5 cm); its location within the canopy as within the *upper* (>16 m above ground level), *middle* (5–15 m) or *lower* (<5 m) *zone*, and its orientation as *horizontal* (0–15° from horizontal), *oblique* (15–75°) or *vertical* (75–90°).

Results

Positional behavior (Fig. 18.1)

Cercopithecus ascanius preferred to move by climbing and quadrupedalism, while *C. mitis* was predominately quadrupedal. Neither of the guenons frequently used quadrupedal suspensory movements, bridging, bimanualism, bipedalism, or vertical bounding ("Other" in Fig. 18.1). Their frequencies of sitting and standing were similar. *Lophocebus albigena* had a locomotor and postural profile similar to that of blue monkeys, but leapt more frequently. Both *Colobus* species leapt more often than any of the cercopithecines. *Colobus guereza* was the most frequent leaper and the least frequent climber. Bounding and galloping along horizontal, usually large diameter supports was observed more often in the guereza than in the red colobus (see also Morbeck, 1975, 1976, 1977, 1979; Mittermeier and Fleagle, 1976; Rose, 1978, 1979). When stationary, *C. badius* preferred to sit or stand, while *C. guereza* preferred sitting and reclining postures.

The smallest species, *Cercopithecus ascanius*, climbed most frequently and leapt rather infrequently compared to the other four species, while *Colobus guereza*, the largest species, leapt the most and climbed least often. *Cercopithecus mitis* climbed at approximately the same frequency as the larger *L. albigena.* Both colobines leapt more frequently than the similarly sized mangabey, which leapt as much as the smaller *C. ascanius.* It is evident that differences in body size among these species do not correspond closely to differences in locomotor frequency, although there is a tendency for size to correlate negatively with frequency of climbing, and positively with that of leaping.

Figure 18.2 compares male and female locomotor frequencies. Quadrupedalism differs between the sexes by no more than four percentage points in all five species. The frequency of climbing typically shows a difference of less than two percentage points, while leaping varies by four

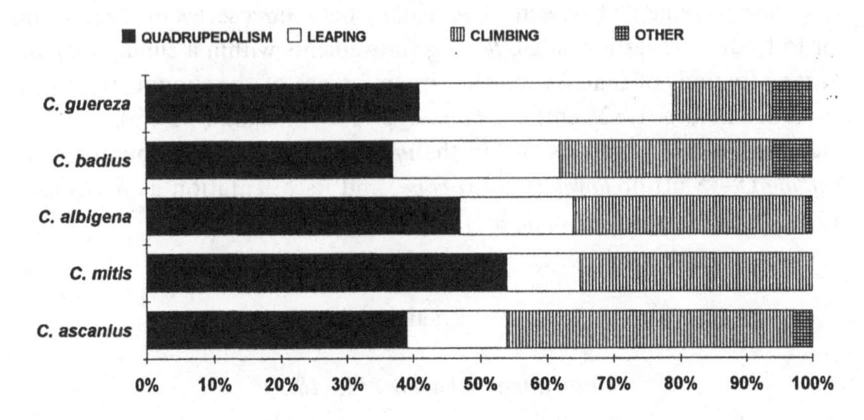

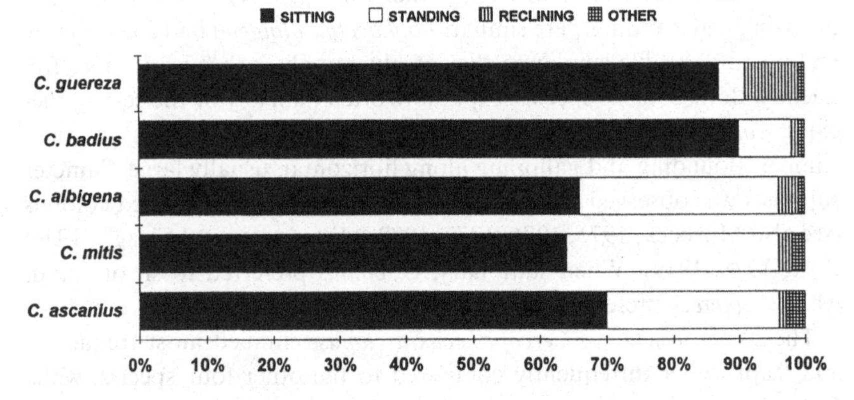

Fig. 18.1. Locomotor and postural frequencies.

percentage points or less. Thus, no appreciable locomotor differences associated with sex can be documented in these sexually dimorphic species.

Habitat and support use

All five species were observed in each of the three height zones (Fig. 18.3). With the exception of *Colobus guereza*, which used the upper zone most often, all species preferred the middle zone. *Colobus badius*, and to a lesser extent *L. albigena*, was observed in the middle and upper zones about equally often. Guenons clearly preferred the middle zone over the upper,

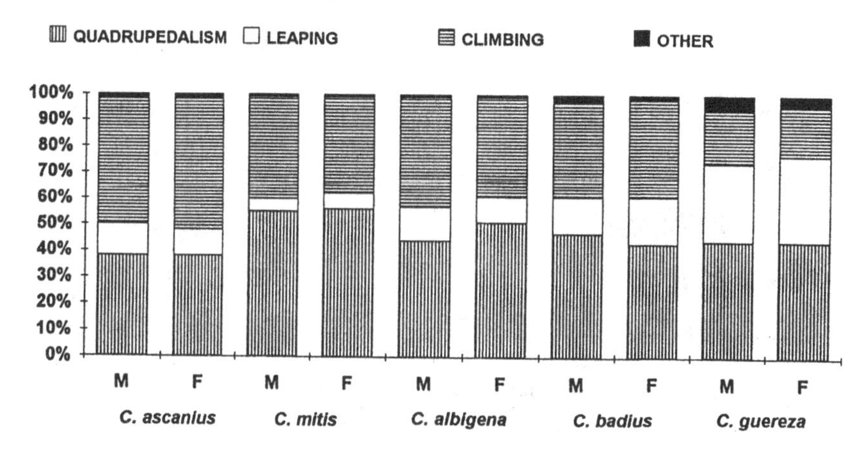

Fig. 18.2 Locomotion by sex and species (M : males; F : females).

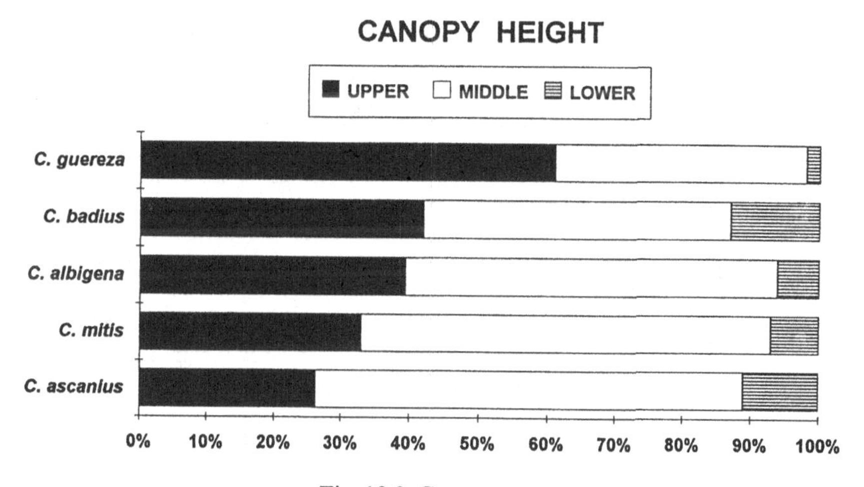

Fig. 18.3. Canopy use.

while *Cercopithecus ascanius* and *Colobus badius* were observed in the lower zone more frequently than the other species. All five species were observed to come to the ground occasionally.

All five species used medium sized supports approximately half to two-thirds of the time. *Cercopithecus ascanius*, the lightest species, used the smallest supports most often, while *Colobus guereza*, the heaviest, used the largest supports most frequently. Beyond this simple correspondence, associations with body weight are less clear. *Lophocebus albigena*, for example,

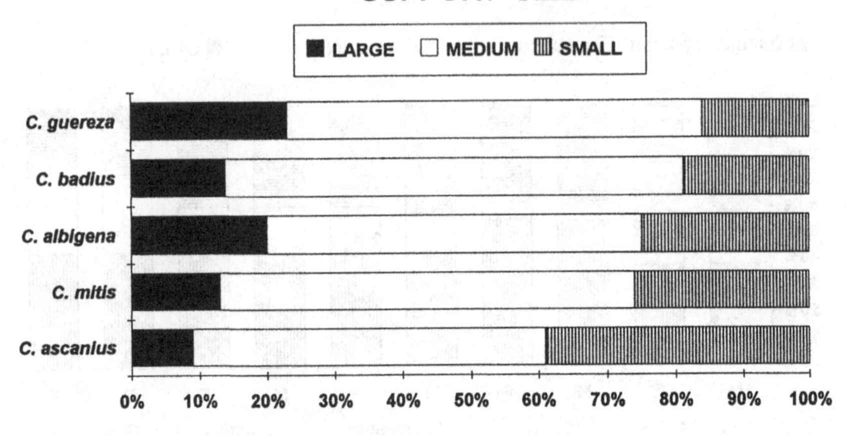

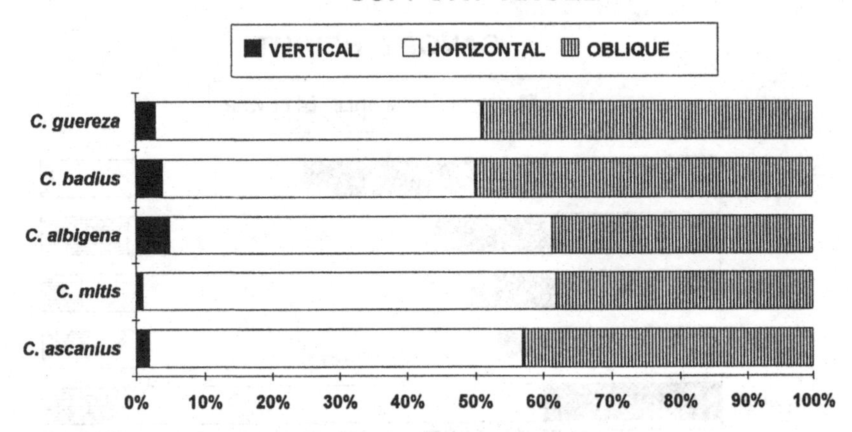

Fig. 18.4. Support use by size and orientation.

used the largest supports more often than the heavier *C. badius*, and all five species preferred medium sized supports.

All five species were observed on all three types of supports (horizontal, oblique, and vertical), and all used vertical supports least often (Fig. 18.4). The three largest species, *L. albigena, C. badius*, and *C. guereza*, used vertical supports more often than did the smaller guenons. The cercopithecines preferred horizontal supports while the colobines used horizontal and oblique supports about equally often. Separating movements from postures showed little variation in support use across species.

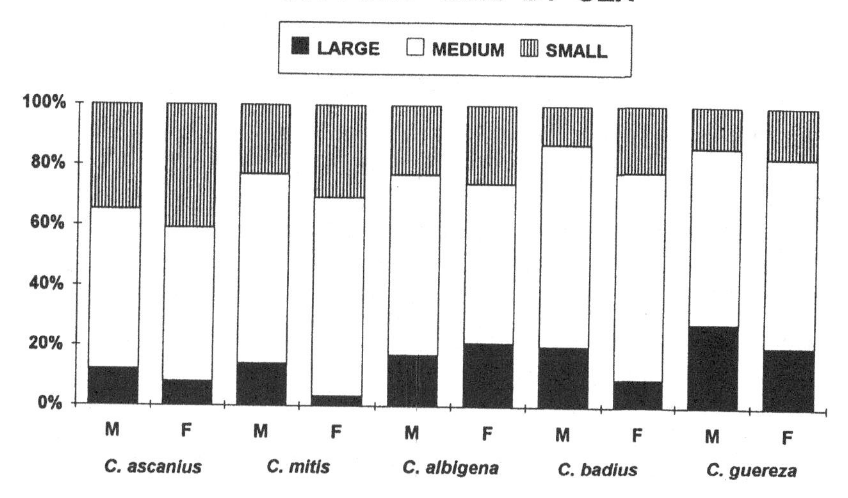

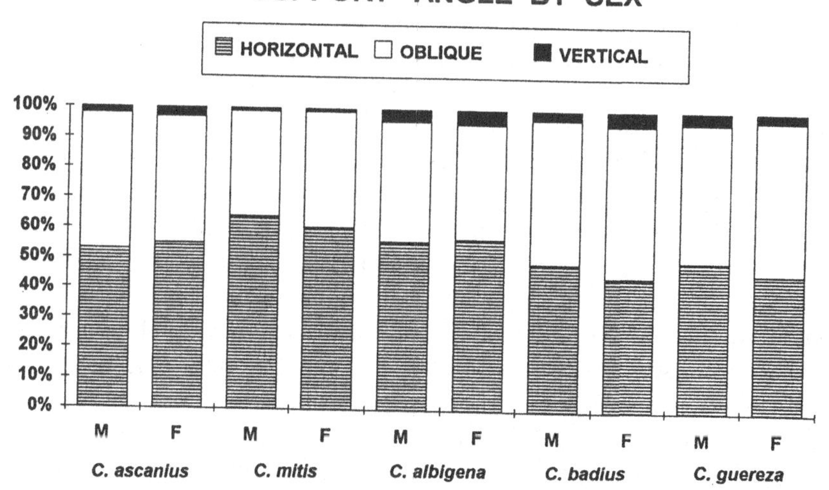

Fig. 18.5. Use of different support sizes and orientations by sex and species.

As shown in Figure 18.5, there is little or no evidence of intraspecific variation in support use associated with sex in these species, and the slight differences that are seen are not associated with the species' degree of sexual dimorphism in body size. The smaller-sized females do use large supports slightly less often and small supports slightly more often than do males of the same species (but see *L. albigena*, Fig. 18.5). Sex differences are most marked in *C. badius*, the least dimorphic species in terms of its body size (Table 18.2), but in no species are the sex differences very dramatic. The use of supports of different orientation likewise shows little association with sex within species. The biggest difference is observed in *C. guereza* where females use oblique supports more frequently than do males.

From data derived by scanning of the spatial position of individuals at 10-minute intervals (Gebo and Chapman, 1995a), it is evident that the crown of the tree was used more often than the major branches or the trunk, with two exceptions. *Lophocebus albigena* utilized the crown only slightly more frequently than the major branches, while *C. guereza*, on the other hand, showed a decided preference for major branches.

Habitat and seasonal effects on positional behavior of red colobus monkeys

We compared the positional behavior of *C. badius* during the dry season in primary and secondary forest. Secondary forest is distinguished from primary forest in being less continuous. Large open stretches between trees are common. During travel, quadrupedalism was more frequently observed in secondary forest than in primary forest. During feeding, quadrupedalism was also more often observed in secondary forest, as is the case for leaping, while climbing decreases dramatically within secondary forest (Fig. 18.6). We also observed *C. badius* in a mature pine plantation, a very different type of forest. Here, trees were spaced very close together and small branches tended to break when red colobus monkeys walk out away from the trunk. Locomotor frequencies during travel differed by no more than five percentage points between pine and secondary forests. In feeding, quadrupedalism and leaping were more frequent in the pine forest, while climbing decreased substantially. Overall, red colobus monkeys utilized quadrupedalism, leaping, and climbing more equally in the pine plantation than in primary and secondary forests. Quadrupedalism is used most extensively in secondary forests. The largest observed differences in locomotor frequencies among the different forests occurred during feeding (Fig. 18.6).

We also compared positional behavior of red colobus monkeys in primary forest during the dry and wet seasons of 1990. The data show a five and six percentage point difference in frequencies of quadrupedalism and leaping during travel (Fig. 18.6). Locomotion associated with feeding, however, shows a twenty-one percentage point decrease in quadrupedalism in the wet season, and a compensatory increase of nine and eight percentage points for leaping and climbing, respectively. Clearly, seasonal effects

Table 18.2. *Body weights (kg) and number of bouts observed for each species*

	Cercopithecus ascanius	*Cercopithecus mitis*	*Lophocebus albigena*	*Colobus badius*	*Colobus guereza*
Male body weight	4.2	6.0	9.0	8.3	10.1
Female body weight	3.0	3.5	6.4	8.2	8.0
Male:female weight ratio	1.4	1.7	1.4	1.0	1.3
Mean female and male	3.6	4.8	8.0	8.25	9.1
Total bouts	6,450	6,444	6,165	7,515	6,452

Source: Body weights from Waser (1987) and Fleagle (1988).

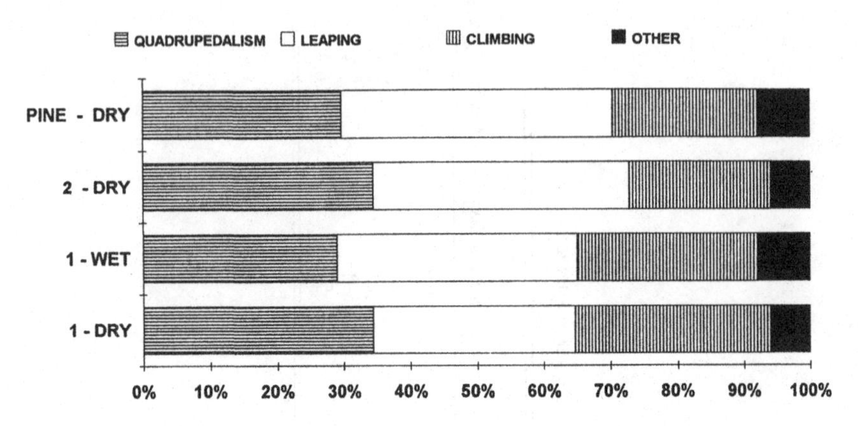

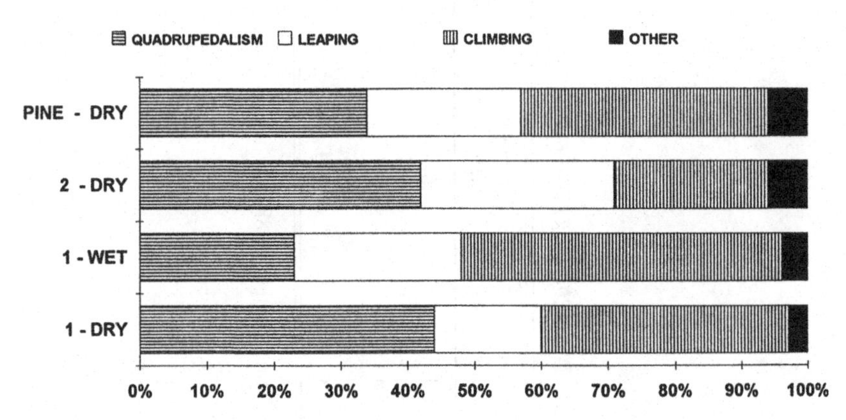

Fig. 18.6. Use of forest types (primary, 1; secondary, 2; and pine) for travel and feeding by red colobus monkeys, in wet and dry seasons.

on locomotor behavior are greater than the effects of forest type, and are especially marked during feeding.

The effects of predation threat upon locomotion in red colobus monkeys

Two approaches were utilized to study the effects of the apparent presence of a predator upon locomotor frequencies. We simulated the approach of a potential terrestrial predator by moving towards unhabituated groups of

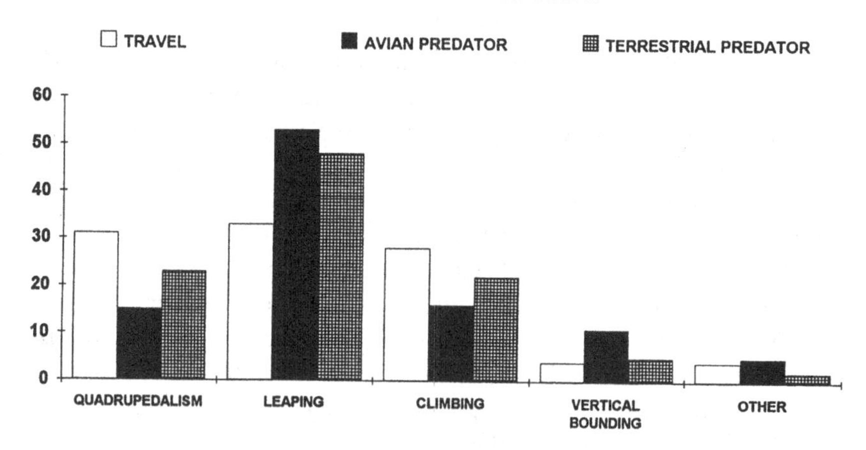

Fig. 18.7. Locomotor responses of red colobus to different predator situations, compared to normal locomotion during travel.

red colobus monkeys and scored their fleeing responses. Second, we played calls of the crowned hawk eagle (*Stephanoaetus coronatus*) to red colobus groups and again scored their fleeing movements (see Gebo *et al.*, 1994, for details). Comparing the escape movements in the two situations (terrestrial and avian "predator") to "normal" travel movements shows that quadrupedalism and climbing decreased, while leaping increased dramatically (Fig. 18.7). Of the rarer locomotor behaviors, only vertical bounding increased substantially, during playback of the crowned hawk-eagle call (Fig. 18.7).

The linear distance moved during a bout was also affected by the fleeing responses provoked by the crowned hawk-eagle call. Longer mean distances, higher maximum distances, and more frequent use of long distance movements occurred (Gebo *et al.*, 1994). In contrast, responses to the "terrestrial predator" showed distance measures similar to those observed during "normal" travel.

In summary, leaping increased in frequency in times of crisis, and obviously represents a strategy for rapidly crossing large distances. Of the two simulated predators, the hawk-eagle provoked more substantial changes in locomotor behavior, perhaps indicating that these predators pose a greater threat (see Leland and Struhsaker, 1993).

Discussion and conclusion

The results of the Kibale Forest project show that all five species commonly utilize five varieties of positional behavior (quadrupedalism, climbing, leaping, sitting, standing), and that each species displays a distinctive pattern of use for these five behaviors. Canopy, crown, and support use are fairly similar across the five species. Positional behavior shows much greater interspecific difference during feeding than during travel. The smaller species tend to climb more and leap less often than the larger species – the exact reverse of the trend documented for sympatric platyrrhines (Fleagle and Mittermeier, 1980). On the other hand, intraspecific body size differences due to sexual dimorphism are not manifested in obvious differences in the frequencies of positional behavior, or in the use of supports of different size and orientation.

The study of red colobus monkeys in different forests, in different seasons, and in the predator experiments shows that substantial changes in locomotor frequencies can occur, especially during feeding or times of crisis. This series of observations demonstrates that this primate species, and probably primates in general, are ecologically flexible in their use of positional behavior (see also Crompton, 1984; Boinski, 1989; Doran and Hunt, 1994; Dagosto, 1995; Doran, 1996; Dagosto and Yamashita, 1998) and that intraspecific variation is an important consideration when attempting to sort out interspecific comparisons (in contrast to Garber and Preutz, 1995; McGraw, 1996). We believe that individual primates are making movement choices according to a number of current and past factors. Anatomical design and body size affect movement possibilities while food and predators change seasonally throughout the life of a primate. Thus to truly understand primate positional behavior, we need to quantify intraspecific variation. This means better sampling over longer time periods, in different seasons, and if possible, in different habitats. Increased sampling will help to ensure that the "movement sample" adequately addresses the particular problem at hand. What are the major problems? First, positional studies need to determine how movements and arboreal pathways are linked and why. Second, how do primate activities (e.g. feeding behavior) affect locomotor abilities? Third, how does size affect positional behavior? In the end, we need to understand cause and effect.

Traditionally, positional studies have played a central role in determining locomotor adaptation among fossil primates. Given the number of fossil discoveries over the past decade, positional studies will continue to play an important role in these evolutionary debates. If we truly want to understand how and why living and extinct primates move, we believe that we need to renew our emphasis on field work to help alleviate the many behavioral, ecological, and anatomical concerns that exist in the field today.

Acknowledgements

First, we would like to thank Paul Whitehead and Cliff Jolly for inviting us to participate in this volume. Second, we thank Dr. Andrew Johns and Dr. G. Isabirye-Basuta at the Makerere University Biological Field Station (Kibale Forest, Uganda) for all their help, as well as the support from the Office of the President, Uganda, the National Research Council, and the Ugandan Forest Department. We also thank Lauren Chapman, Marian Dagosto, and Joanna Lambert for their help in the field. This research project was supported by a LSB Leakey Grant in Aid as well as by a Northern Illinois University Summer Research Grant (DLG).

References

Altmann, J. (1974). Observational study of behavior: sampling methods. *Behaviour* **49**, 227–67.

Boinski, S. (1989). The positional behavior and substrate use of squirrel monkeys: ecological implications. *J. hum. Evol.* **18**, 659–78.

Butynski, T.M. (1990). Comparative ecology of blue monkeys (*Cercopithecus mitis)* in high- and low-density subpopulations. *Ecol. Monogr.* **60**, 1–26.

Cartmill, M. (1974). Pads and claws in arboreal locomotion. In *Primate Locomotion*, ed. F. Jenkins, pp. 45–83. New York: Academic Press.

Clark, W.E.L. (1959). *The Antecedents of Man.* Edinburgh: Edinburgh Universtiy Press.

Crompton, R.H. (1984). Foraging, habitat structure, and locomotion in two species of *Galago.* In *Adaptations for Foraging in Nonhuman Primates*, ed. P. S. Rodman & J.G.H. Cant, pp. 73–109. New York: Columbia University Press.

Dagosto, M. (1995). Seasonal variation in positional behavior of Malagasy lemurs. *Int. J. Primatol.* **16**, 807–33.

Dagosto, M. & Yamashita, N. (1998). Effect of habitat structure on positional behavior and support use in three species of lemur. *Primates* **39**, 459–72.

Doran, D. (1996). Comparative positional behavior of the African apes. In *Great Ape Societies*, ed. W.C. McGrew, L.F. Marchant & T. Nishida, pp. 213–24. Cambridge: Cambridge University Press.

Doran, D. & Hunt, K.D. (1994). The comparative locomotor behavior of chimpanzees and bonobos: species and habitat differences. In *Chimpanzee Cultures*, ed. R.W. Wrangham, W.C. McGrew, F.B.M. de Waal & P.G. Heltne, pp. 93–108. Cambridge, MA: Harvard University Press.

Fleagle, J.F. (1976). Locomotion and posture of the Malayan siamang and implications for hominoid evolution. *Folia primatol.* **26**, 245–69.

Fleagle, J.F. (1978). Locomotion, posture, and habitat utilization in two sympatric Malaysian leaf-monkeys (*Presbytis obscura* and *Presbytis melalophos)*. In *The Ecology of Arboreal Folivores*, ed. G.G. Montgomery, pp. 243–51. Washington, DC: Smithsonian Press.

Fleagle, J.F. (1980). Locomotion and posture. In *Malayan Forest Primates*, ed. D.J. Chivers, pp. 191–207. New York: Plenum Press.

Fleagle, J.F. (1985). Size and adaptation in primates. In *Size and Scaling in Primate Biology*, ed. W.L. Jungers, pp. 1–19. New York: Plenum Press.

Fleagle, J.G. (1988). *Primate Adaptation and Evolution.* New York: Academic Press.

Fleagle, J.G. & Mittermeier, R.A. (1980). Locomotor behavior, body size, and comparative ecology of seven Surinam monkeys. *Am. J. phys. Anthrop.* **52**, 301–14.

Garber, P.A. & Preutz, J.D. (1995). Positional behavior in moustached tamarin monkeys: effects of habitat on locomotor variability and locomotor stability. *J. Hum. Evol.* **28**, 411–26.

Gebo, D.L. (1992). Locomotor and postural behavior in *Alouatta palliata* and *Cebus capucinus. Am. J. Primatol.* **26**, 277–90.

Gebo, D.L. & Chapman, C.A. (1995a). Positional behavior in five sympatric Old World monkeys. *Am. J. phys. Anthrop.* **97**, 49–76.

Gebo, D.L. & Chapman, C.A. (1995b). Habitat, annual, and seasonal effects on positional behavior in red colobus monkeys. *Am. J. phys. Anthrop.* **96**, 73–82.

Gebo, D. L., Chapman, C. A., Chapman, L.J. & Lambert, J. (1994). Locomotor response to predator threat in red colobus monkeys. *Primates* **35**, 219–23.

Jungers, W.L. (1985). Body size and scaling of limb proportions in primates. In *Size and Scaling in Primate Biology*, ed. W.L. Jungers, pp. 345–81. New York: Plenum Press.

Kalina, J. (1988). *Ecology and Behavior of the Black-and-White Casqued Hornbill* (Bycanistes subcylindricus subquadratus) *in Kibale Forest, Uganda.* PhD Dissertation, Michigan State University, East Lansing.

Kasenene, J.M. (1980). *Plant Regeneration and Rodent Populations in Selectively Felled and Unfelled Areas of Kibale Forest, Uganda.* MSc Thesis, Makerere University, Kampala (Uganda).

Keith, A. (1923). Man's posture: its evolution and disorders. *Br. Med. J.* **1**, 451–4.

Leland, L. & Struhsaker, T.T. (1993). Teamwork tactics. *Nat. Hist.* **4**, 42–8.

McGraw, W.S. (1996). Cercopithecid locomotion, support use, and support availability in the Tai Forest, Ivory Coast. *Am. J. phys. Anthrop.* **100**, 507–22.

Mittermeier, R.A. & Fleagle, J.F. (1976). The locomotor and postural repertoires of *Ateles geoffroyi* and *Colobus guereza*, and a reevaluation of the locomotor category semibrachiation. *Am. J. phys. Anthrop.* **45**, 235–56.

Morbeck, M.E. (1975). Positional behavior of *Colobus guereza:* a preliminary quantitative analysis. In *Symp. 5th Cong. Int. Primatol. Soc.*, ed. S. Kondo, M. Kawai, A. Ehmara & S. Kawamura, pp. 331–43. Tokyo: Japan Science Press.

Morbeck, M.E. (1976). Leaping, bounding, and bipedalism in *Colobus guereza:* a spectrum of positional behavior. *Yrbk. phys. Anthropol.* **20**, 408–20.

Morbeck, M.E. (1977). Positional behavior, selective use of habitat substrate and associated non-positional behavior in free-ranging *Colobus guereza* (Ruppel, 1835). *Primates* **18**, 35–58.

Morbeck, M.E. (1979). Forelimb use and positional adaptation in *Colobus guereza:* integration of behavioral, ecological, and anatomical data. In *Environment, Behavior, and Morphology: Dynamic Interactions in Primates*, ed. M.E. Morbeck, H. Preuschoft & N. Gomberg, pp. 95–117, New York: Gustav Fischer.

Morbeck M.E., Preuschoft, H. & Gomberg, N. (1979). *Environment, Behavior, and Morphology: Dynamic Interactions in Primates.* New York: Gustav Fischer.

Napier, J.R. (1967). Evolutionary aspects of primate locomotion. *Am. J. phys. Anthrop.* **27**, 333–42.

Napier, J.R. & Walker, A.C. (1967). Vertical clinging and leaping: a newly recognized category of primate locomotion. *Folia primatol.* **6**, 204–19.

Ripley, S. (1967). The leaping of langurs: a problem in the study of locomotor adaptation. *Am. J. phys. Anthrop.* **26**, 149–70.

Rose, M.D. (1978). Feeding and associated positional behavior of black and white colobus monkeys (*Colobus guereza*). In *The Ecology of Arboreal Folivores*, ed. G.G. Montgomery, pp. 253– 62. Washington, DC: Smithsonian Press.

Rose, M.D. (1979). Positional behavior of natural populations: some quantitative results of a field study of *Colobus guereza* and *Cercopithecus aethiops.* In *Environment, Behavior, and Morphology: Dynamic Interactions in Primates*, ed. M.E. Morbeck, H. Preuschoft & N. Gomberg, pp. 95–117. Gustav Fisher: New York.

Skorupa, J.P. (1986). Responses of rainforest primates to selective logging in Kibale Forest, Uganda: a summary report. In *Primates: The Road to Self-Sustaining Populations*, ed. K. Benirschke, pp. 57–70. New York: Springer-Verlag.

Skorupa, J.P. (1988). *The Effect of Selective Timber Harvesting on Rain-forest Primates in Kibale Forest, Uganda.* PhD Dissertation, University of California, Davis.

Stern, J.T. & Oxnard, C.E. (1973). Primate locomotion: some links with evolution and morphology. *Primatologia* **4**, 1–93.

Struhsaker, T.T. (1975). *The Red Colobus Monkey.* Chicago: University of Chicago Press.

Susman, R.L. (1984). The locomotor behavior of *Pan paniscus* in the Lomako Forest. In *The Pygmy Chimpanzee*, ed. R. L. Susman, pp. 369–93. New York: Plenum Press.

Waser, P. (1987). Interactions among primate species. In *Primate Societies*, ed. B.B Smuts, D.L. Cheney, R.M. Seyfarth, R.W. Wrangham & T.T. Struhsaker, pp. 210–26. Chicago: University of Chicago Press.

19

The behavioral ecology of Asian colobines

CAREY P. YEAGER AND KAREN KOOL

Introduction

This chapter will focus on the relationships among physiology, behavior, and ecology in the Asian colobine monkeys. Colobines are known for their specialized digestive physiology, including, especially, their unique sacculated stomach containing anaerobic cellulytic bacteria (Bauchop and Martucci, 1968). This physiological specialization allows them to extract nutrients from foliage more efficiently, but digestive efficiency via microbial fermentation has a cost: a slower rate of digestion. This, combined with a small size (when compared to other animals utilizing microbial symbionts), limits colobines' gross intake of food, and forces them to balance the quality and the quantity of food ingested. The costs and benefits of the colobine digestive system have a profound impact on social structure and ecology, underlying social relationships, home range size, population density, activity patterns, and intergroup interactions as well as diet. For example, the ability to digest low quality food may widen the resource base, directly affecting home range size and intergroup interactions.

In the following review, we emphasize detailed, longer-term studies, many of them only recently completed, which in many cases have clarified our perceptions of this group. The tables summarize information from this literature. We follow the taxonomic classifications of Oates, Davies and Delson (1994) with one exception; we retain the separation of *Rhinopithecus* and *Pygathrix* (Jablonski and Peng, 1993; Jablonski, 1995), thus recognizing seven genera.

Social structure

Asian colobines are typically organized into one-male social groups (one male, several females, and offspring). Of the 24 species listed in Table 19.1,

17 usually live in one-male social groups. Three species appear to be both one-male and multi-male – *Semnopithecus entellus, Trachypithecus obscurus*, and *Pygathrix nemaeus.* Two are found in both monogamous and one-male social groups – *Simias concolor* and *Presbytis comata* – and one is monogamous, *P. potenziani.* Insufficient data are available to classify *P. frontata*, but it is reported to live in small groups that may be one-male.

The advantage of group living is usually said to be predation avoidance (Alexander, 1974; van Schaik, 1983) or access to food resources (Wrangham, 1980). (See also Struhsaker, Chapter 15.) Both factors may be important to Asian colobines, depending on the level of predation risk and the distribution and abundance of resources (Newton and Dunbar, 1994). Almost all Asian colobines form one-male, rather than multi-male groups. The number of males in a group appears to be a function of the number of adult females (Mitani *et al.*, 1996). It is in a male's best reproductive interest to exclude other males from the group if possible (Goss-Custard *et al.*, 1972). If the number of associating females is small enough to be defensible, then the group should have only one male.

Female association patterns in primates are generally attributed to kinship relationships and the spatial distribution of resources (Wrangham, 1980). Food resources may limit the number of females that can associate. The average number of females in an Asian colobine group is relatively small. Given their ability to exploit mature foliage – a relatively abundant, evenly distributed food resource – it would appear that these females should form larger groups. However, preferred food resources (fruit/seeds) are more limited and more patchily distributed, and the availability of these may be limiting female associations to a size defensible by a single male. An alternative explanation is that social factors are responsible. Colobine aggressive behavior increases with density and crowding (Poirier, 1974). Home range size does not appear to increase with group size in the Asian colobines (see below), thus, as group size increases, so does population density. The resultant, higher rates of intragroup aggression, especially among females, may cause females to emigrate, thus limiting group size. Another possibility is that extra-group males may be attracted by groups with many females, resulting in increased harassment of females and perhaps increased risk of infanticide (as in *Alouatta seniculus*, Crockett and Janson, 1993).

Fission–fusion of relatively stable one-male groups within bands is common in *Nasalis* and *Rhinopithecus.* Early studies of short duration reported proboscis and snub-nosed monkeys to be organized into multi-male groups (Bai *et al.*, 1986; Kawabe and Mano, 1972; Kern, 1964; Li *et*

Table 19.1. *Behavior and social structure of the Asian colobines*

Species (conservation status)	%MON	%OMG	%MM	AMG	SOL	B	AP	I	D	HRO	IGA	IGT	FI	CP	Reference
Presbytis comata (HE)		100		?	x		x		m	9	m				Ruhiyat (1983) (UD)
	80		20	?	x				m	high	m				Ruhiyat (1983) (HD)
Presbytis frontata	small groups														Wilson & Wilson (1975)
Presbytis hosei		100							m,f?	20.1	DO	x			Mitchell (1994)
Presbytis melalophos		100		x	x		x		m,f	79	DO	x			Bennett (1983)
		66.7	33.3							23.5	DO				Johns (1983) (UD)
		100			x				m	3	DO	x			Johns (1983) (HD)
P. melalophos femoralis		85.7	14.3	x	x				m,f	24.5	m		x	mm	Megantara (1989)
Presbytis potenziani (E)	100				x		x		m,f	~40	m,f	x			Fuentes (1994)
Presbytis rubicunda	11.1	88.9			x				m,f	~25					Supriatna *et al.* (1986)
		100		x			x		m	10	m	x			Davies (1984)
Presbytis thomasi (V)		100		x					m	28.4	m	x			Assink & van Dijk (1990)
		82.6	17.4	x	x				m	mod.	m				Gurmaya (1986)
		92	8	x	x		x		m,f	x	m		x	mm	Steenbeck (1994)
							x	x					x	mm	Sterck (1992, 1995)
Trachypithecus auratus (V)		77.8	22.2	x	x		x		m,f	23	m	x	x		Kool (1989)
Trachypithecus cristatus (V)		80	20				x			low	m	x	x	mm	Bernstein (1968)
				x	x			x	m						Wolf & Fleagle (1977)
										mod.		x			Hock & Sasekumar (1979)
Trachypithecus francoisi (E)		100			x		x			x			x	sm	Li Z-Y 1993, pc
Trachypithecus geei (V)	6.3	81.3	12.5	x	x		x		m	x		x			Mukherjee (1978)
															Mukherjee & Saja (1974)
Trachypithecus johnii (HV)		62.5	37.5	x	x		x		m,f	low	m	x		mm	Poirier (1968a,b, 1969)
Trachypithecus obscuras			100		x		x		m	3	m				Curtin (1980)
		most	x	x	x				m	low	DO				MacKinnon & MacKinnon (1980)
Trachypithecus phayrei (V)		100													Stanford (1988)
Trachypithecus pileatus (V)		100		x	x		x		m,f	~100	m	x	x	sm	Stanford (1991)
	17.4	69.6	13		x		x		m,f	x	DO	x	x		Islam & Husain (1982)
		100		x	x				m	x	DO				Green (1981)

Trachypithecus vetulus monticola (HV)		100		x			x		m,f	high	m	x	x		Rudran (1973a,b) (UD)
T. vetulus senex		93.1	6.9	x			x	x	m,f	low	m		x		Rudran (1973a,b) (HD)
Semnopithecus entellus		92.9	7.1	x	x			x	m,f	50.7	m,f	x	x		Newton (1987, 1992)
		86	14	x	x		x	x	m,f	high	m,f	x	x	sm,mm	Hrdy (1977)
		42.6	57.4	x				x							Newton (1988) (UD)
		65.2	34.8	x				x							Newton (1988)* (HD)
Pygathrix nemaeus (E)		50	50										x		Gochfeld (1974)
		67	33		x		x		m,f						Lippold (1977)
Rhinopithecus avunculus (HE)		100		x		x			m	~100		x			Boonratana & Canh (1994)
Rhinopithecus bieti (HE)		100		x		x	x		m	~100	m	x		sm?	Kirkpatrick (1994, 1996)
Rhinopithecus brelechi (HE)		100		x		x	x		m	~100	m	x			Bleisch *et al.* (1993, pc)
Rhinopithecus roxellana (E)		100		x	x	x	x		m	~100		x			Poirier & Hu (1983)
													x	sm	Clarke (1991) (captive)
Nasalis larvatus (HV)		100		x	x	x			m,f	~100		x			Bennett & Sebastian (1988)
	10	90		x	x	x	x		m,f	95.9	m	x	x	sm	Yeager (1989)
Simias concolor (HE)	45	50	5	x	x				m,f	x	DO				Teneza & Fuentes (1995) (HD)
	100			x	x				m,f		DO				Tilson (1977) (HD)
	100									low	DO				Watanabe (1981) (high HD)
	33.3	66.7			x					8	DO	x			Watanabe (1981) (low HD)

Notes:
If a range was given we used the midpoint. Conservation status based on Action Plan for Asian Primate Conservation (IUCN) (Eudey, 1987).
HE = highly endangered, E = endangered, HV = highly vulnerable, V = vulnerable, %MON = percent groups monogamous, %OMG = percent one-male groups, %MM = percent muti-male groups, AMG = all-male or non-reproductive groups, SOL = solitary individuals, B = band comprised of omgs and amgs, AP = alloparental care, I = infanticide, D = dispersal, HRO = average percent home range overlap, IGA = physical intergroup aggression (chase, fight), DO = display only, IGT = intergroup tolerance, FI = female initiation of sexual behavior, CP = copulation pattern, x = present, ? = unclear from report, low = little overlap based on report or figures (<10%), mod. = moderate overlap based on report or figures (>10%), high = high overlap based on report or figures (>25%), m = male, f = female, sm = single mount to ejaculation, mm = multiple mount to ejaculation pattern, * average taken from long term studies only listed in table 1 of Newton (1988) (undisturbed had disturbance ratings less than 2.0). pc = personal communication, UD = undisturbed, HD = human disturbance, may include logging and hunting.

al., 1982; Yang, 1988). However, more recent, longer term studies (Bennett and Sebastian, 1988; Boonratana, 1994; Bleisch *et al.*, 1993; Kirkpatrick, 1994, 1996; Yeager, 1989, 1995) have found bands composed of one-male and all-male groups, associations akin to those reported for gelada baboons. In proboscis monkeys, fission–fusion of relatively stable one-male groups probably serves at least two functions. It avoids displacement at feeding sites (two or more small groups may equal a large group in size, thereby reducing the risk of displacement), and temporary associations may allow groups to coordinate river crossings, thereby reducing individual risk of predation by crocodilians (i.e., dilution effect; Sweeney and Vannote, 1982) (Yeager, 1992). In *Rhinopithecus bieti*, associations between groups may be the result of past predation pressure by large cats now locally extinct (R.C. Kirkpatrick, pers. com.). The primary food resource (lichens) appears to be evenly distributed and abundant, thereby allowing large numbers of individuals to congregate (Kirkpatrick, 1994, 1996). Current associations may simply be the result of inertia, or perhaps the benefits obtained from social groups (grooming, access to potential mates) more than outweigh the small potential cost in food competition. *Rhinopithecus bieti* bands occasionally fission temporarily, but this is rare (R.C. Kirkpatrick, pers. com.). Food-tree patchiness may lead to fission–fusion of groups of *R. brelechi* and *R. roxellana* (Bleisch *et al.*, 1993; Bennett and Davies, 1994). However, as even the smallest resource patches are sufficient for more than a single one-male group, these species are still able to maintain associations between groups. It is not known, however, whether specific groups associate preferentially, as in *Nasalis larvatus.*

The stability of groups appears to vary, and does not appear to be controlled by male tenure, since group composition does not change simply due to a male take-over. By stability, we mean the degree of change in female membership due to immigration or emigration (or fission-fusion), apart from male take-overs. In *Presbytis thomasi* and *N. larvatus*, some groups remain stable for a year or more, while others change composition radically from year to year (Sterk, 1995; Steenbeck, pers. com.; Ranjanathan and Bennett, 1990). Group stability appears to vary, at least in part, with environmental degradation (Johns, 1983). For example, following severe habitat degradation due to fire, logging, and mining, one female proboscis monkey with an infant was observed to change groups at least three times during a single year, living in groups ranging in size from 5 to 21 individuals (C. Yeager, unpub. data)

All-male groups, or groups consisting of males and non-reproductive females, have been observed in 17 (possibly 18) of the 24 species of Asian colobines (Table 19.1). The exceptions are *Presbytis comata, P. frontata, P. hosei, P. potenziani, Trackypithecus francoisi, T. phayrei* and *Pygathrix nemaeus.* Apart from *Presbytis potenziani* which is monogamous, and *P. frontata* which has small groups, these are one-male group species. As they are also among the least studied, it may simply be that all-male groups have not yet been sighted. It has been suggested that all-male groups are more common amongst terrestrial species, and that this grouping behavior is an adaptation to increased predation pressures faced by terrestrial species (Struhsaker, 1969). If this is so, the prevalence of all-male groups in colobines suggests that predation may be just as important a pressure for arboreal species (see Isbell, 1994). The formation of all-male groups may also be a response to mate competition; males may be able to form coalitions and alliances that improve their chance of success during a group takeover, as reported for *Semnopithecus entellus* (Hrdy, 1977).

Female intragroup relationships

Female intragroup relationships have been poorly documented for the majority of the Asian colobines, presumably because of poor visibility of arboreal species, a paucity of studies of well habituated groups, and apparently low rates of interaction (see also Cords, Chapter 17). Linear dominance hierarchies among females have been reported in *S. entellus, T. johnii* and *P. comata* (Hrdy, 1977; Poirier, 1969; Ruhiyat, 1983), but reports are conflicting for *S. entellus* (see Newton, 1987 for a review). Hollihn (1972) reports dominance hierarchies among captive *N. larvatus* and *Pygathrix nemaeus.* Female–female displacement has been reported in *T. cristatus* (Furaya, 1961–62), *S. entellus* (Hrdy, 1977), *Presbytis melalophos* (Bennett, 1983) and *N. larvatus* (Yeager, 1989). Social interactions are reported to be rare in *P. melalophos* (Bennett, 1983; Curtin, 1980), *P. comata* (Ruhiyat, 1983) and *T. geei* (Mukherjee and Saha, 1974). Female–female affiliative behaviors (allogrooming, sitting in proximity) been reported in *P. comata* (Ruhiyat, 1983), *P. melalophos* (Bennett, 1983; Curtin, 1980), *P. rubicunda* (Supriatna *et al.*, 1986), *P. thomasi* (Gurmaya, 1986), *T. auratus* (K. Kool, pers. obs.), *T. cristatus* (Bernstein, 1968), *T. francoisi* (Z. Li, pers. com.), *T. obscurus* (Curtin, 1980), *T. johnii* (Tanaka, 1965), *T. pileatus* (Stanford, 1991), *S. entellus* (Hrdy, 1977), *Pygathrix nemaeus* (Lippold, 1977), *R. bieti* (R.C. Kirkpatrick, pers. com.), captive *R. roxellana* (Clarke, 1991), and *N. larvatus* (Yeager, 1989).

Affiliative relationships are often considered to be the social "glue" holding groups together (Mason, 1976; Yeager, 1990). In one-male groups it is the relationships among females that help maintain group stability and cohesiveness. Colobines appear to have lower rates of female–female interaction than cercopithecines (Oates, 1987; Newton and Dunbar, 1994), which may partially explain the group instability observed in *Presbytis thomasi* and *N. larvatus* (see section Social structure) and the presence of both male and female dispersal in colobines (see section Dispersal). Colobine groups may simply be more loosely bonded socially than those of cercopithecines.

Alloparental care

Alloparental care has been reported in the majority of Asian colobine species for which there are data (19 species, Table 19.1). Adult and juvenile females are typically involved, although juvenile and adult males occasionally participate. Alloparental care has been hypothesized to serve a variety of functions. These include, providing release time for mothers to forage (Poirier, 1968b), ensuring social integration of new infants into the group and thus improving their probability of adoption if the mother should become disabled or die (Lancaster, 1971), and improving parenting skills of the alloparent ("learning to mother") (Hrdy, 1977). Abusive handling of new infants has been suggested to reduce resource competition for the alloparents' own offspring (Wasser and Barash, 1981). Alternatively, alloparenting may simply be misdirected parental behavior that has no selective advantage (Quiatt, 1979).

Support for two of these explanations is provided by observations of proboscis monkeys. Females have been observed on several occasions to leave their offspring in a "play group" with one adult female, while the others actively forage (C. Yeager, pers. obs.). Juvenile alloparents actively seek out opportunities to interact with infants, and they have extremely poor infant handling skills, frequently holding infants upside-down or dragging them by a limb (C. Yeager, pers. obs.).

Sexual behavior

Female initiation is the norm among Asian colobines (Table 19.1). Female proceptive behavior has been observed in all 10 species for which there are data. In addition, a characteristic head shake during solicitation has been observed in *P. cristata* (Bernstein, 1968), *T. auratus* (Kool, 1989), *T. francoisi* (Z. Li, pers. com.), *T. francoisi* (Z. Li, pers. com.) *T. vetulus* (Rudran, 1973a), *T. pileatus* (Islam and Husain, 1982; Stanford, 1991), *N. larvatus*

(Yeager, 1989) and *S. entellus* (Hrdy, 1977). Of the 10 species for which data are available (Table 19.1), five appear to be single mount ejaculators and four are multiple mount ejaculators, with both ejaculation patterns reported for *S. entellus* (Hrdy, 1977). In cercopithecines, single mount ejaculation is believed to correlate with a low degree of direct, regular male–male competition (Caldecott, 1986; Shively *et al.*, 1982). Given the predominance of the one-male groups, and thus low rates of intragroup, male–male competition, Asian colobines should exhibit a single mount ejaculation pattern. The above data appear to conflict with this prediction. However, if we examine the relationship between mount pattern and proportion of one-male groups found in the population, single mount patterns were significantly more common in species with a greater proportion of one-male groups (Mann-Whitney U test, $z = 2.68$, $p = 0.0073$, based on nine species for which there was a clear pattern).

Infanticide

This phenomenon has been well documented in some populations of *S. entellus*, but its significance is still contested (Sussman *et al.*, 1995; Hrdy *et al.*, 1995). The two major explanations are that infanticide represents a reproductive strategy of males during group takeovers, accelerating the mother's next estrus (Hrdy, 1977), or, alternatively, that infanticide is a behavioral aberration, brought about by social stress (Curtin, 1977). Hrdy and Hausfater, (1984) discuss additional theoretical explanations. Newton (1987) has suggested that infanticide is associated with one-male group structure, but infanticide has rarely been reported in other Asian colobines, the majority of which are organized in one-male groups, and has only been infrequently observed in *P. thomasi* (Sterck, 1995), *T. cristatus* (Wolf and Fleagle, 1977), and *T. vetulus* (Rudran, 1973b). In *T. cristatus* and *T. vetulus* the infanticides ocurred in disturbed, high density populations but within the context of a male takeover of the group. These data do not clarify the picture. Moreover, *N. larvatus* females have been documented to transfer to new groups with their infants (Bennett and Sebastian, 1988; Rajanathan and Bennett, 1990), a behavior which puts them at some risk. The circumstances under which females transferred is not clear.

Dispersal

Male dispersal is observed in all species of Asian colobines for which there are data (Table 19.1). Female dispersal has been observed in 12 (possibly

13) species. Female dispersal is not the typical cercopithecoid pattern. Why do female colobines disperse? Possible advantages are incest avoidance, to obtain better quality habitat, to increase access to food resources and reduce the level of aggression received, to gain access to a higher quality male (better genes hypothesis), and to increase mating opportunities (see Moore, 1984, for a review). Incest avoidance and increasing mating opportunities probably do not apply to these colobine females, as all males are dispersing from their natal group, adult male tenure is probably shorter than the time it takes an infant female to mature, and each group has an adult male with whom the females can breed. As the home ranges of Asian colobine groups often overlap, dispersing females would probably not benefit from increased habitat quality unless they dispersed a significant distance. In *N. larvatus*, in those instances in which the identity of the dispersing female was known, she joined a group in the same area (Rajanathan and Bennett, 1990; Yeager, 1989, unpub. data). Colobine females may leave groups to reduce intragroup aggression and reduce competition, as female–female aggression, though rare, can be of high intensity. Gaining access to a higher quality male cannot be ruled out, however, and a female's decision to disperse is probably influenced by multiple factors.

Isbell and Van Vuren (1996) suggest that costs differ between social versus locational dispersal for female primates. Social dispersal involves changing groups, but not general area, and females attempting to transfer may risk intergroup aggression from other females. Locational dispersal involves changing areas, thus risking loss of knowledge concerning location of food resources and increased vulnerability to predation. They predict frequent (in multi-male groups) or occasional (in one-male groups) female transfer when both social and locational costs are low (i.e. female intergroup aggression absent; weak or indiscernable female dominance hierarchies; highly overlapping home ranges). The pattern of female Asian colobine dispersal seems generally to match Isbell and Van Vuren's (1996) prediction, although the degree of overlap is lower than predicted. That is, female Asian colobines rarely participate in intergroup aggression (Table 19.1), have weak or indiscernable dominance hierarchies (see section Female intragroup relationships), and groups generally have overlapping home ranges (Table 19.1).

Both natal and breeding dispersal are observed in Asian colobines. Juvenile males disperse and typically join all-male or non-reproductive groups, in which juvenile females are also occasionally observed (e.g. *N. larvatus*, C. Yeager, unpub. data; *T. auratus*, Kool, 1989; *T. johnii*, Ali *et al.*, 1985). Juveniles may be peripheralized in their groups prior to dispersal. In some instances, adult group members may force immatures to disperse through aggression (Poirier, 1974; Moore, 1984). In other cases, immature individuals disperse spontaneously (perhaps as a side effect of maturation, or to find receptive mates) (Pusey and Packer, 1987). Adult females have been known to leave groups in which they have previously bred and to join new groups (Rudran, 1973b, Rajanathan and Bennett, 1990; Isbell and Van Vuren, 1996).

Home ranges, territoriality and intergroup encounters

Are Asian colobines territorial? Although intergroup aggression does occur, group ranges overlap considerably in many species and the majority are tolerant of at least some other groups in close proximity (Table 19.1). The term "territorial" may be problematic, as its operational definition varies (Maher and Lott, 1995; van Schaik *et al.*, 1992, Kaufmann, 1983). Maher and Lott, (1995, p. 1589) suggest the following, " a fixed space from which . . . a group of mutually tolerant individuals actively excludes competitors for a specific resource or resources". This definition implies intergroup aggressive behavior and space-dependent dominance resulting in exclusive, defended areas. If this strict definition of territoriality is used, the majority of Asian colobines are not territorial.

Whittenberger (1981:562) has stated that "If food is relatively abundant but spatiotemporally variable, it should be exploited by social groups with overlapping home ranges". This appears to describe the situation with Asian colobines, which are folivore/frugivores with overlapping home ranges; mature foliage is relatively abundant and available year round, while fruit and young leaves are spatiotemporally variable.

In a recent review of territoriality in Southeast Asian langurs, van Schaik *et al.* (1992) concluded that intergroup aggression was due to mate defense by males rather than food resource defense by females and males. If animals are defending resources, both the ability to defend an area (measured by Mitani and Rodman's, 1979, index of defensibility) and population density should be negatively correlated with spatial overlap. van Schaik *et al.*, (1992), however, found the opposite; defensibility was positively correlated with the percentage of home range overlap, and, moreover, as population density increased, the percentage range overlap also increased. Using a larger data set, we found that home range overlap was not significantly correlated with either defensibility or density (Tables 19.2 and 19.3).

However, mate and resource defense explanations are not necessarily mutually exclusive. Perhaps the important factor may be the necessity for resource defense, rather than the ability to defend resources. Butynski (1990:1) emphasized the importance of "investigating the density of primate populations relative to the carrying capacities of their environments, and the influence of this relationship on behavior and ecology". If colobines are below carrying capacity in an area, there may be no need for resource defense. Density may thus be partially uncoupled from territoriality, so that density and overlap may be positively correlated when the population is below carrying capacity, and negatively correlated when the population approaches or exceeds it (see Figure 19.1). If the relationship between density and carrying capacity, rather than density *per se*, is the important factor, this could explain the apparent conflicts in the literature.

Feeding ecology

Although previously considered to be primarily folivorous (e.g. by Oates, 1987), more recent evidence shows that most Asian colobines include a substantial proportion of fruit and/or seeds in their diet. Although colobines can probably sustain themselves on a diet of lower quality (Gaulin, 1979) than monogastric primates, their small size and corresponding gut capacity make it important to feed selectively. Typical Asian colobine diets include young leaves and unripe fruits and seeds, which are preferred over mature foliage. Ripe fruit and seeds are usually not eaten. As they mature, leaves become higher in fiber content and often in secondary compounds (Choo *et al.*, 1981; McKey *et al.*, 1981; Milton, 1979). The distribution of nutrients and defense compounds may vary throughout a plant part (Waterman and Kool, 1994). Colobines may preferentially eat the petioles of mature leaves, perhaps to increase their intake of certain minerals (e.g. Baranga, 1983). Unripe fruit is lower in simple sugars than ripe fruit, which can lead to bloat in colobines (Collins and Roberts, 1978). Seeds generally harden as they mature, requiring greater force to masticate, and at least one species, *N. larvatus*, chooses softer seeds (Yeager and Dierenfeld, 1994).

Colobines need to select food high in certain limiting essential nutrients and low in digestion-inhibiting compounds (particularly lignin, a non-digestible fiber component). Stanford (1990) has suggested that Asian colobines vary in their ability to break down digestion inhibitors, citing intergeneric variation in diet amongst the Asian colobines. He states that *Trachypithecus* include less fruit and more foliage in their diet than *Presbytis*, and that *Presbytis* have been reported to eat ripe fruit occasionally. Bennett and Davies (1994) also note that *Trachypithecus* eat a higher proportion of foliage, particularly mature leaves, than *Presbytis*. At

Table 19.2. *Asian colobine feeding ecology, home range size and day range, group size and individual density*

Species	Habitat	YL	ML	LV	SD	FF	FR	FL	O	HR	DR	GS	ID	Reference
Presbytis comata	montane forest	59.1	5.6	64.7	0.7	13.5	14.2	7.0	12.2	38	500	8	11.5	Ruhiyat (1983) (UD)
	montane/sec.for.									14		6	35	Ruhiyat (1983) (HD)
Presbytis frontata	lowland dipterocarp												5.7	Wilson & Wilson (1975)
Presbytis hosei	lowland dipterocarp	45.4	5.1	77.8	16.7		19.4	2.8		34.5	691	9.0		Mitchell (1994) (UD)
	logged lowland dip.	42.3	1.3	59.7	21.3		40.1	0.2		44.8	794	7.0		Mitchell (1994) (HD)
Presbytis melalophos	lowland dipterocarp	26.5	7.3	32.8	26.2		46.4	16.9	1.3	29.5	780	15.0	105	Bennett (1983)
	logged lowland for.	30	4	34			60	4	1	14.5	614	14	47.6	Johns (1983) (UD)
	lowland rainforest	58		58			42			18.5	495	8.7	47.5	Johns (1983) (HD)
	lowland rainforest	24	11	35	8		56	6	2	21	754	15.9		Curtin (1980)
	lowland rainforest			42.6			42.8	14.6		21	1150	9.3	74	MacKinnon & MaKinnon (1980)
Presbytis melalophos femoralis	plantation/sec. for.	26.3	2.5	28.8	35.7	17.9	57.7	.4	13.3	19.3	936	12.5	52.9	Megantara (1989)
Presbytis potenziani	lowland dipterocarp	36.9	18.1	55.0	x		32.0	x	13.0	34.3	540	3.3	11.0	Fuentes (1994)
Presbytis rubicunda	lowland dipterocarp	36.5	1.1	37.6	30.1		49.3	11.1	1.1	84.8	890	7.0	18.9	Davies (1984)
Presbytis thomasi	lowland rainforest	30.4	13.4	44.8			36.4	3.6	16.2		1218	7.9		Sterck (pc, 1995)
	lowland rainforest											10.2	63.6	Steenbeck (1994 pc)
	plantation/garden			32.1	x		57.7	7.5	2.7	14	684	8.0	21	Gurmaya (1986)
	lowland rainforest									37.7	1073	8.9	37.9	Assink & van Dijk (1990)
Trachypithecus auratus	plantation/sec. for.	45.4	0.7	50.2			32.3	13.7	4.1	5.5	550	12.4	267.6	Kool (1989)
Trachypithecus cristatus	mangrove			91.0			9.0							Hock & Sasekumar (1979)
	mangrove/sec. for.									20	350	31.8	~160	Bernstein (1968)
Trachypithecus francoisi	rainforest/sec. for.									157	1000	10.0	<45	Li (1993, pc)
				x			x	x				10.5		Burton *et al.* (1995)
Trachypithecus geei	moist deciduous			x			x	x	x		750	10.5		Mukherjee (1978)
														Mukherjee & Saja (1974)
Trachypithecus johnii	moist deciduous									~6.5		8.9		Poirier (1968a, 1969)
	tropical evergreen	25.5	26.8	59.4	x		26.3	13.4	1.0	24		17		Oates *et al.* (1980)
Trachypithecus obscuras	lowland rainforest	36	22	58	3		35	7		33	559	17.0		Curtin (1980)
	lowland rainforest			48.2			47.3	4.5		28.5	950	10.3	31	MacKinnon & MacKinnon (1980)

Table 19.2. (*cont.*)

Species	Habitat	YL	ML	LV	SD	FF	FR	FL	O	HR	DR	GS	ID	Reference
Trachypithecus phayrei	moist deciduous			68.1			24.4		7.5	27.5	500	9.0		Stanford (1988, 1991)
Trachypithecus pileatus	moist deciduous	10.9	42.0	66.8	9.3		33.7	7	x	21.6	324.5	8.2	52	Stanford (1991)
Trachypithecus vetulus	secondary forest	20	40	60			28	12						Hladik (1977)
T. vetulus monticola	cloud forest ec. for.									6.8		8.9	92.6	Rudran (1973) (UD)
T. vetulus senex	secondary forest									2.5		8.4	215	Rudran (1973) (HD)
Semnopithecus entellus	moist decid./meadow	14.2	34.9	49.1			24.4	9.5	4.0	74.5	1083	21.7	46.2	Newton (1987, 1992)
	mosit/dry deciduous	34.1	23.0	62.0	12.0		12.9	17.9	2.5			148		Kar-Gupta & Kumar (1994)
	secondary forest	27	21	48	x		45	7	x	12.5		25		Hladik (1977)
	moist/dry deciduous											31.6	32.1	Newton (1988)* (UD)
	village/farm/sec. for.											21	41.2	Newton (1988)* (HD)
Pygathrix nemaeus		x		x			x	x				9.3		Lippold (1977)
												6		Gochfeld (1974)
Rhinopithecus avunculus	montane limestone	38	<1	38.5	15		62				>1000	14.8	7.0	Boonratana & Canh (1994)
Rhinopithecus bieti	montane forest									2525	~1250	<15	~7.0	Kirkpatrick (pc, 1996)
Rhinopithecus brelechi	montane forest			74	x		13	12.4	x	3500	>1000	6.2	<4.5	Bleisch *et al.* (1993, 1999)
Rhinopithecus roxellana	montane forest			x	x		x		x			~11	11.7	Poirier & Hu (1983)
	montane forest			x	x		x		x	2600	710			Su *et al.* (1999)
Nasalis larvatus	mangrove/rainforest	38	3	41	15		56	3		900	706	11.4	5.9	Bennett & Sebastian (1988)
	riverine swamp	41.2	5.2	51.9	37.0		40.3	3.0	4.8	130	~800	12.6	62.6	Yeager (1989, unpub. data)
Simias concolor	lowland dip./swamp			x			x			27.5		3.6	9.75	Tilson (1977)
	lowland dip./swamp									~13.5		4.1	21	Teneza & Fuentes (1995)
	lowland dip./sec. for.									13.5		3.2	8	Watanabe (1981) (high HD)
	secondary forest									3.5		7.9	220	Watanabe (1981) (low HD)

Notes:
If a range was given we used the midpoint. YL = percent young leaves in diet, ML = percent mature leaves, LV = total percent leaves, SD = percent seeds, FF = percent fruit flesh, FR = total percent fruit, FL = percent flower, O = percent other or unknown, HR = average home range in hectares, DR = average day range in meters, GS = average group size of reproductive units, ID = individual density per km^2, x = present in diet, * = mean taken from long term studies only listed in table 1 Newton (1988) (undisturbed had disturbance ratings less than 2.0), UD = undisturbed, HD = human disturbance, may include logging and hunting, pc = personal communication.

Table 19.3. *Nonparametric correlations (Spearman's rho) among the defensibility index (DI), home range size (HR), percent home range overlap (HRO), individual density (ID), and group size (GS) for Asian colobines*

	DI	HR	HRO	ID	GS
Presbytis and *Trachypithecus*					
DI					
HR	0.03				
HRO	0.16	−0.24			
ID	0.24	−0.85*	0.05		
GS	0.26	−0.51	−0.37	0.63	
All Asian colobines					
DI					
HR	−0.41				
HRO	−0.47	0.34			
ID	0.50	−0.71*	−0.41		
GS	0.20	−0.07	−0.09	0.29	

Notes:
*$p<0.05$; n for *Presbytis* and *Trachypithecus* varied between 8 and 13; n for all Asian colobines varied between 12 and 18.

present, there are no experimental data to test Stanford's (1990) hypothesis, although there are some physiological differences between the genera. *Trachypithecus* have larger stomachs than *Presbytis* (Chivers, 1994), implying that they have an increased capacity to process foliage. The diets of most species have not been phytochemically analyzed, but it is possible that the individual ripe fruits chosen may be low in glucose, or that the overall level of glucose in the diet is acceptable, given other dietary components. However, the available data on diet in these two genera (Table 19.2) do not support the assertion that they differ in percentage foliage consumed; they showed no significant differences for total foliage in diet (t (11) = 1.5, $p>$ 0.05), nor proportion of total fruit in diet ((t (11) = 1.0, $p>0.05$). It does, however, appear that the proportion of mature foliage in their diet differs significantly (t (9) = 2.5, $p<0.05$), with *Trachypithecus* eating more mature leaves.

Are food resources limiting?

Davies (1994) has suggested that colobines are limited by their food supply, citing as evidence the fact that in three species, population densities

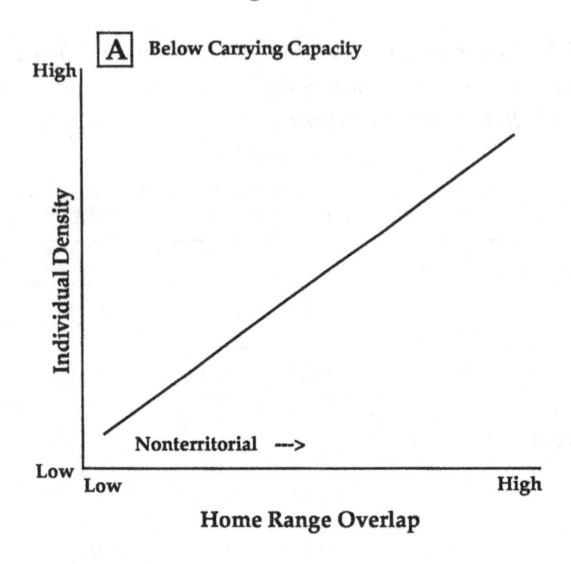

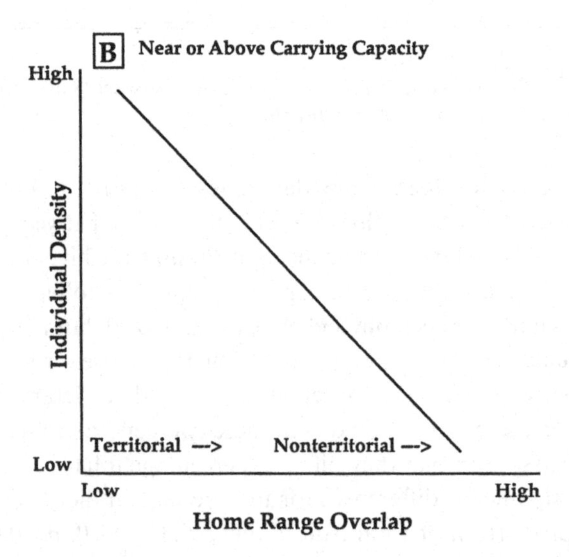

Fig. 19.1. The hypothesized theoretical relationship among home range overlap, individual density, and carrying capacity, with respect to the expression of territorial behavior. In (A) individual density and home range overlap are positively correlated when the population is below carrying capacity and groups are nonterritorial. In (B) individual density and home range overlap are negatively correlated when the population is near or above carrying capacity and groups are territorial at high density.

declined following catastrophic loss of forest cover. All populations are, of course, ultimately limited by access to food resources. The question is, therefore, whether Asian colobine populations are normally at carrying capacity. Home range size and individual density are significantly negatively correlated (Table 19.3), indicating that many populations may be below carrying capacity. Population density is generally higher in disturbed than undisturbed habitats (Table 19.2). Disturbance can increase the productivity of individual trees by increasing light exposure. Ganzhorn (1995) suggested that low-level disturbance may increase an area's carrying capacity. He found that leaf nutritional quality increased following light logging, but productivity increases only partially compensated for loss of leaf biomass. Compared with unlogged forest, the edges but not the interior of logged forest produced more fruit. This suggests that increased productivity and nutritional quality may be able to maintain or perhaps raise carrying capacity slightly after light logging, but are unlikely to raise carrying capacity significantly in moderately or severely logged forests. The fact that some non-human primate species doubled or even tripled population densities in disturbed habitats, even though food almost certainly did not double or triple in abundance, suggests that these species may normally be below carrying capacity in undisturbed areas, and that they are not normally limited by food resources.

If Asian colobines are not normally food limited, the explanation may lie in their ability to exploit mature foliage when necessary. Such foliage is generally abundant and evenly distributed, but has relatively poor nutritional content, with low protein and high fiber. It is usually eaten in small amounts, and only becomes a substantial portion of the diet during times of food scarcity. Asian colobine populations may be able to sustain themselves for long time periods on a low quality diet of mature foliage, and are thus somewhat more buffered against fluctuations in food supply than primates with simple stomachs. However, the impact of a low quality diet will vary by sex and age. Infants, juveniles, and pregnant or lactating females have higher nutritional needs that in the long-term cannot be met by mature foliage alone (e.g. Lee, 1987). Physiological stress may increase juvenile or infant mortality, and reduce birth rates. Asian colobine populations thus appear to be *partially* limited by the abundance and availability of *preferred* food resources, but may be better able to cope with short- to moderate-term food shortages than other primates.

Their ability to exploit mature foliage may explain why colobines appear to be generally less aggressive than non-colobine species, and more tolerant of other groups sharing spatial resources. As they are less at risk of losing access to at least some food resources, the costs of defending food resources may be too high. Why risk physical harm defending a food resource when an alternative food resource is available? The cost of accepting an alternative food is much lower than the cost of an injury from direct competition for resources. Cercopithecines do not have the same option; being unable to exploit mature foliage, they may not eat at all if they do not compete directly for more limited fruits and young leaves.

So what limits Asian colobine populations? Disease is one possibility, but there are few data suggesting that disease has severely diminished primate populations (four instances, Young, 1994). Predation is another, but in her recent review of predation, Isbell (1994) found only one primate species (red colobus) possibly limited by predation. Human predation is currently limiting some colobine populations (e.g. *S. concolor*). Social factors are another possibility. Poirier (1974) noted that colobine aggression appeared to be more common at high densities. Dunbar (1987) found that intragroup aggression inhibited reproduction in large groups of guerezas (*Colobus abyssinicus*) at high densities. Increased density can lead to increased social stress, as individuals must negotiate a greater number of relationships. Higher stress levels have a variety of potential effects, including increased agonism (Fedigan, 1982), immunosuppression (Scott, 1988), and decreased reproductive success (Lee, 1987). These effects can have a significant impact on population density, group size, and composition.

Resource competition, resource quality, and territoriality

Isbell (1991) suggests that for female primates, food distribution affects intragroup competition, while food abundance affects intergroup competition. Competition is presumed to occur, at least in part, because of the potential impact on individual reproductive success via nutritional factors. She (Isbell, 1991) hypothesizes that clumped food resources lead to intragroup competition, since clumped or patchily distributed food resources can be more easily defended and monopolized. Food scarcity is hypothesized to lead to intergroup competition. In contrast to other models (e.g. Wrangham, 1980; van Schaik, 1989), the pattern of food distribution is hypothesized to have no effect on intergroup competition (Isbell, 1991). Many Asian colobines may not normally be food limited, thus fitting Isbell's model for non-food limited species; i.e., home range size and day range length do not appear to increase with group size (*N. larvatus*, Yeager, 1989; *R. bieti*, Kirkpatrick, 1994, pers. com.; *R. brelechi*, Bleisch *et al.*, 1993; *T. pileatus*, Stanford, 1991; *P. melalophos*, Bennett, 1983). Most species do not exhibit dominance hierarchies, and females do not participate in inter-group aggression.

Food quality affects the energy available to an individual. Agonistic behavior requires considerable energy expenditure above average basal metabolic rates; if available energy is insufficient there may be a decrease in agonistic behavior. Thus, food quality may indirectly affect agonism and direct competition. Food quality has only recently been directly incorporated as a variable in theoretical models concerning competition, resource distribution and social structure. Janson and Goldsmith (1995) found an apparent lack of foraging competition in folivores (defined by having more than 40% foliage in diet) and suggest that group sizes of folivorous species may be explained by the "energetic constraints of a leafy diet". It is generally assumed that abundant, non-patchily distributed resources (i.e., leaves) are of lower nutritional quality (Wrangham, 1980). Evidence suggests that this assumption may be in error (Oftedal, 1991) because it neglects differences in digestion inhibitors (fiber, secondary toxins), plus protein levels and other nutrients, found among leaves of different ages and species. Additionally, the "quality" of a particular food item may vary among exploiting species, depending upon their ability to use available nutrients and deal with digestion inhibitors. A particular food's quality for a colobine is probably quite different from its quality for a cercopithecine.

Food quality, abundance, and distribution in space and time probably all interact in their impact on competitive agonism. If food is extremely low-quality, requiring extended gut passage time, the benefits derived from defending a resource both within and between groups may not outweigh the energy expenditure required. In fact, the energy obtained from low quality food may be less than that required to find, process, and digest the item, plus maintain basal metabolic rate, resulting in a net deficit. If food is scarce, as well as of low quality, there may be insufficient energy both to locate food and to engage in resource defense. Johns (1986) has suggested that energetic constraints may force *P. melalophos* groups to switch from territoriality to non-territoriality. Dependent upon the amount of energy reserves (body fat) available, there may be a lag time to onset of nutritional or energetic deficiencies. Thus the proximate mechanisms underlying the exhibition of aggressive behavior in Asian colobines may be quite complex.

Conservation status

Asian colobines include some of the most threatened primate species and subspecies (Table 19.1), confined to limited, fragmented, and shrinking habitats. Their vulnerability stems from their dependence on forest habitats that are increasingly being cleared for commercial logging, commercial agricultural development, and subsistence farming (Marsh *et al.*, 1987).

Hunting is also a major threat in some regions (Oates and Davies, 1994). Selective logging may affect species differently, but it generally leads to a population decline with recovery some years later (Johns, 1983, 1986). However, the combination of selective logging with hunting facilitated by the opening up of forest patches by roads and tracks, can have a disastrous effect on colobine populations (Oates and Davies, 1994).

Colobines do not generally survive well in captivity (Collins and Roberts, 1978), and therefore captive breeding as a conservation measure is of doubtful merit. Continued survival for many species will depend upon a combination of increased habitat protection and elimination of hunting. This will only be achieved through better education, the development of sustainable economic alternatives for the local community, and sufficient baseline information regarding population and environmental trends to make informed management decisions.

Summary

Asian colobines are folivore/frugivores commonly organized into one-male groups. The majority of species appear to be non-territorial, with overlapping home ranges and inter-group tolerance. Resource defense may only occur when species are at or near carrying capacity. Infanticide does not appear to be a generally defining characteristic of the Asian colobines, apart from *S. entellus.* Given their specialized digestive physiology, Asian colobines may be less food-limited than other primates. The ability to include mature leaves in their diet may buffer them at least partially from food limitations.

Acknowledgements

The authors wish to thank Lynne Isbell, Charles Janson, Craig Kirkpatrick, Jim Moore, Scott Silver, and Paul Whitehead for helpful comments and discussions. Due to page limitations, we cited dissertations and theses when possible, as opposed to the numerous publications resulting from these. Partial support for this work was provided to Carey Yeager by a Clare Booth Luce Assistant Professorship in Biological Sciences at Fordham University, the Douroucouli Foundation and the International Scientific Support Trust.

References

Alexander, R.D. (1974). The evolution of social behavior. *Ann. Rev. Ecol. Syst.* **5**, 325–83.

Ali, R., Johnson, J.M. & Moore, J. (1985). Female emigration in *Presbytis johnii*: a life-history strategy. *J. Bombay Nat. Hist. Soc.* **82**, 249–52.

Assink, P.R. & van Dijk, I.F. (1990). *Social Organization, Ranging and Density of* Presbytis thomasi *at Ketambe (Sumatra) and a Comparison with Other* Presbytis *species at Several South-east Asian Locations.* MS thesis, University of Utrecht, Netherlands.

Bai, S.C., Zou, S.Q., Lin, S., Two, D., Tu, Z. & Zhong, T. (1986).The preliminary observation on the distribution, the number and the population structure of *Rhinopithecus bieti* in Biama Xueshan Natural Reserve, Yunnan, China. *Primate Rep.* **14**, 265.

Baranga, D. (1983). Changes in chemical composition of food plants in the diet of colobus monkeys. *Ecology* **64**, 668–73.

Bauchop, T. & Martucci, R.W. (1968). Ruminant-like digestion of the langur monkey. *Science* **161**, 698–700.

Bennett, E.L. (1983). *The Banded Langur – Ecology of a Colobine in West Malaysian Rain Forest.* PhD thesis. University of Cambridge.

Bennett, E.L. & Davies, A.G. (1994). The ecology of Asian colobines. In *Colobine Monkeys: Their Ecology, Behaviour and Evolution*, ed. A.G. Davies & J.F. Oates, pp. 129–71. Cambridge: Cambridge University Press.

Bennett, E.L. & Sebastian, T. (1988). Social organization and ecology of proboscis monkeys (*Nasalis larvatus*) in mixed coastal forest in Sarawak. *Int. J. Primatol.* **9**, 233–56.

Bernstein, I.S. (1968). The lutong of Kuala Selangor. *Behaviour* **32**, 1–16.

Bleisch, W., Cheng, A.-S., Ren, X.-D., Xie, J.-H. (1993). Preliminary results from a field study of wild Guizhou snub-nosed monkeys (*Rhinopithecus brelichi*). *Folia primatol.* **60**, 72–82.

Bleisch, W.V. & Xie, J.H. (1999). Ecology and behavior of the Guizhou golden monkey, *Rhinopithecus brelichi*. In *The Natural History of the Dourc's and Snub-nosed Langurs*, ed. N. Jablonski, Singapore: World Scientific Publications.

Boonratana, R. & Canh, L.X. (1994). *A report on the ecology, status and conservation of the Tonkin Snub-nosed monkey (Rhinopithecus avunculus) in Northern Vietnam.* Unpublished report to the Wildlife Conservation Society, New York.

Burton, F.D., Snarr, K.A. & Harrison, S.E. (1995). Preliminary report on *Presbytis francoisi leucocephalus*. *Int. J. Primatol.* **16**, 311–27.

Butynski, T.M. (1990). Comparative ecology of blue monkeys (*Cercopithecus mitis*) in high- and low-density subpopulations. *Ecol. Monogr.* **60**, 1–26.

Caldecott, J.O. (1986). Mating patterns, societies and the ecogeography of macaques. *Anim. Behav.* **34**, 208–20.

Chen, F., Min, Z., Luo, S.& Xie, W. (1983). An observation on the behavior and some ecological habits of the golden monkey (*Rhinopithecus roxellanae*) in Qing mountains (China). *Acta Ther. Sin.* **3**, 141–6.

Chivers, D.J. (1994). Functional anatomy of the gastrointestinal tract. In *Colobine Monkeys: Their Ecology, Behaviour and Evolution*, ed. A.G. Davies & J.F. Oates, pp. 205–27. Cambridge: Cambridge University Press.

Choo, G.M., Waterman, P.G., McKey, D.B. & Gartlan, J.S. (1981). A simple enzyme assay for dry matter digestibility and its value in studying food selection by generalist herbivores. *Oecologia* **49**, 170–8.

Clarke, A.S. (1991). Sociosexual behavior of captive Sichuan golden monkeys (*Rhinopithecus roxellanae*). *Zoo Biology* **10**, 369–74.

Collins, L. & Roberts, M. (1978). Arboreal folivores in captivity – maintenance of a delicate minority. In *The Ecology of Arboreal Folivores*, ed. G.G. Montgomery, pp. 5–12. Washington: Smithsonian Institute.

Crockett, C.M. & Janson, C.H. (1993). The costs of sociality in red howler monkeys: infanticide or food competition? *Am. J. Primatol.* **30**, 306.

Curtin, S.H. (1977). Langur social behavior and infant mortality. *Kroeber Anthro. Soc. Papers* **50**, 27–36.

Curtin, S.H. (1980). Dusky and banded leaf monkeys. In *Malaysian Forest Primates: Ten Years' Study in Tropical Rain Forest*, ed. D.J. Chivers, pp. 107–45. London: Plenum Press.

Davies, A.G. (1984). *An Ecological Study of the Red Leaf Monkey (Presbytis rubicunda) in the Dipterocarp Forest of Northern Borneo*. PhD dissertation, University of Cambridge.

Davies, A.G. (1994). Colobine populations. In *Colobine Monkeys: Their Ecology, Behaviour and Evolution*, ed. A.G. Davies & J.F. Oates, pp. 285–310. Cambridge: Cambridge University Press.

Dunbar, R.I.M. (1987). Habitat quality, population dynamics, and group composition in colobus monkeys (*Colobus guereza*). *Int. J. Primatol.* **8**, 299–330.

Eudey, A.A. (1987). *Action Plan for Asian Primate Conservation: 1987–91*. Stony Brook, NY: IUCN/SSC Primate Specialist Group.

Fedigan, L.M. (1982). *Primate Paradigms*. Montreal: Eden Press.

Fuentes, A. (1994). *The Socioecology of the Mentawai Langur Presbytis potenziani*. PhD Dissertation, University of California, Berkeley.

Furuya, Y. (1961–62). The social life of silvered leaf monkeys, *Trachypithecus cristatus*. *Primates* **3**, 41–60.

Ganzhorn, J.U. (1995). Low-level forest disturbance effects on primary production, leaf chemistry, and lemur populations. *Ecology* **76**, 2084–96.

Gaulin, S.J. (1979). A Jarman/Bell model of primate feeding niches. *Hum. Ecol.* **7**, 1–20.

Gochfeld, M. (1974). Douc langurs. *Nature* **247**, 167.

Goss-Custard, J.D., Dunbar, R.I.M. & Aldrich-Blake, F.P.G. (1972). Survival, mating, and rearing in the evolution of primate social structure. *Folia. Primatol.* **17**, 1–19.

Green, K.M. (1981). Preliminary observations on the ecology and behaviour of the capped langur, *Presbytis pileatus*, in the Madhupur Forest of Bangladesh. *Int. J. Primatol.* **2**, 131–51.

Gurmaya, K.J. (1986). Ecology and behaviour of *Presbytis thomasi* in North Sumatra. *Primates* **27**, 151–72.

Hladik, C.M. (1977). A comparative study of the feeding strategies of two sympatric species of leaf monkey: *Presbytis senex* and *Presbytis entellus*. In *Primate Ecology*, ed. T.H. Clutton-Brock, pp. 31–7. New York: Academic Press.

Hock, L.B. & Sasekumar, A. (1979). A preliminary study on the feeding biology of mangrove forest primates, Kuala Selangor. *Malay. Nat. J.* **33**, 105–12.

Hollihn, U. (1972). Remarks on the breeding and maintenance of Colobus monkeys, proboscis monkeys and douc langurs in zoos. *Int. Zoo Ybk.* 185–7.

Hrdy, S.B. (1977). *The Langurs of Abu: Female and Male Strategies of Reproduction*. Cambridge, MA: Harvard University Press.

Hrdy, S.B. & Hausfater, G. (1984). Comparative and evolutionary perspectives on

infanticide: introduction and overview. In *Infanticide: Comparative and Evolutionary Perspectives*, ed. S.B. Hrdy & G. Hausfater, pp. xiii–xxxv. New York: Aldine.
Hrdy, S.B., Janson, C. & van Schaik, C. (1995). Infanticide: let's not throw out the baby with the bath water. *Evol. Anthropol.* **3**, 151–4.
Isbell, L.A. (1991). Contest and scramble competition: patterns of female aggression and ranging behavior among primates. *Behav. Ecol.* **2**, 143–55.
Isbell, L.A. (1994). Predation on primates: ecological patterns and evolutionary consequences. *Evol. Anthropol.* **3**, 61–71.
Isbell, L.A. & Van Vuren, D. (1996). Differential costs of locational and social dispersal and their consequences for female group-living primates. *Behaviour.* 133.
Islam, M.A. & Husain, K.Z. (1982). A preliminary study on the ecology of the capped langur. *Folia primatol.* **39**, 145–59.
Jablonski, N.G. (1995). The phyletic position and systematics of the douc langurs of Southeast Asia. *Am. J. Primatol.* **35**, 185–205.
Jablonski, N.G. & Peng, Y.-Z. (1993). The phylogenetic relationships and classification of the doucs and snub-nosed langurs of China and Vietnam. *Folia primatol.* **60**, 36–55.
Janson, C.H. & Goldsmith, M.L. (1995). Predicting group size in primates: foraging costs and predation risks. *Behav. Ecol.* **6**, 326–36.
Johns, A.D. (1983). *Ecological Effects of Selective Logging in a West Malaysian Rain-forest*. PhD thesis, University of Cambridge.
Johns, A.D. (1986). Effects of selective logging on the behavioral ecology of West Malaysian primates. *Ecology* **67**, 684–94.
Kar-Gupta, K. & Kumar, A. (1994). Leaf chemistry and food selection by common langurs (*Presbytis entellus*) in Rajaji National Park, Uttar Pradesh, India. *Int. J. Primatol.* **15**, 75–93.
Kaufman, J.H. (1983). On the definitions and functions of dominance and territoriality. *Biol. Rev.* **58**, 1–20.
Kawabe, M. & Mano, T. (1972). Ecology and behaviour of the wild proboscis monkey, *Nasalis larvatus* (Wermb.) in Sabah, Malaysia. *Primates* 213–27.
Kern, J.A. (1964). Observations on the habits of the proboscis monkey, *Nasalis larvatus* (Wurmb), made in the Brunai Bay area, Borneo. *Zoologica* **49**, 183–92.
Kirkpatrick, R.C. (1994). The natural history of the Yunnan snub-nosed monkey (Colobinae: *Rhinopithecus bieti*). Paper presented at the *XIVth Congress of the International Primatological Society,Bali, Indonesia.* [Oral presentation.]
Kirkpatrick, R.C. (1996). *Socioecology of the Yunnan Snub-nosed Monkey (Rhinopithecus bieti)*. PhD dissertation, University of California, Davis.
Kool, K.M. (1989). *Behavioural Ecology of the Silver Leaf Monkey, Trachypithecus auratus sondiacus, in the Pangandaran Nature Reserve, West Java, Indonesia.* PhD dissertation, University of New South Wales.
Lancaster, J. (1971). Play-mothering: the relations between juvenile females and young infants among free-ranging vervet monkeys (*Cercopithecus aethiops*). *Folia Primatol.* **15**, 161–82.
Lee, P.C. (1987). Nutrition, fertility and maternal investment in primates. *J.Zool.* **213**, 409–22.
Li, Z.X., Ma, S.L., Hua, C.H. & Wang, Y.X. (1982). The distribution and habits of the Yunnan golden monkey, *Rhinopithecus bieti*. *J. Hum. Evol.* **11**, 633–8.
Li, Z.-Y. (1993). Preliminary investigation of the habitats of *Presbytis francoisi*

and *Presbytis leucocephalus*, with notes on the activity pattern of *Presbytis leucocephalus. Folia primatol.* **60**, 83–93.
Lippold, L.K. (1977). The douc langur: a time for conservation. In *Primate Conservation*, ed. H.S.H. Prince Rainier III of Monaco & G.H.Bourne, pp. 513–38. New York: Academic Press.
MacKinnon, J.R. & MacKinnon, K.S. (1980). Niche differentiation in a primate community. In *Malaysian Forest Primates: Ten Years' Study in Tropical Rain Forest*, ed. D.J. Chivers, pp. 167–90 . London: Plenum Press.
Maher, C.R. & Lott, D.F. (1995). Definitions of territoriality used in the study of variation in vertebrate spacing systems. *Anim. Behav.* **49**, 1581–97.
Marsh, C.W., Johns, A.D. & Ayres, J.M. (1987). Effects of habitat disturbance on rain forest primates. In *Primate Conservation in the Tropical Rain Forest*, ed. C.W. Marsh & R. A. Mittermeier, pp. 83–107. New York: Alan Liss, Inc.
Mason, W.A. (1976). Primate social behavior: pattern and process. In *Evolution of Brain and Behavior in Vertebrates*, ed. R.B. Masterton, C.B.G. Campbell, M.E. Bitterman & N. Hotton, pp. 425–55. Hillsdale, NJ: Lawrence Erlbaum Associates.
McKey, D.B., Gartlan, J.S., Waterman, P.G. & Choo, G.M. (1981). Food selection by black colobus monkeys (*Colobus satanas*) in relation to plant chemistry. *Biol. J. Linn. Soc.* **16**, 115–46.
Megantara, E.N. (1989). Ecology, behavior and sociality of *Presbytis femoralis* in Eastcentral Sumatra. *Comp. Primatol. Mon.* **2**, 171–301.
Milton, K. (1979). Factors influencing leaf choice by howler monkeys: a test of some hypothesis of food selection by generalist herbivores. *Am. Nat.* **114**, 362–78.
Mitani, J.C., Gros-Louis, J. & Manson, J.H. (1996). The number of males in primate groups: comparative tests of competing hypothesis. *Am.J. Primatol.* **38**, 315–32.
Mitani, J.C. & Rodman, P.S. (1979). Territoriality: the relation of ranging pattern and home range size to defendability, with an analysis of territoriality among primate species. *Behav. Ecol. Sociobiol.* **5**, 241–51.
Moore, J. (1984). Female transfer in primates. *Int. J. Primatol.* **5**, 537–89.
Mitchell, A.H. (1994). *Ecology of Hose's Langur,* Presbytis hosei, *in Mixed Logged and Unlogged Dipterocarp Forest of North Borneo*. PhD dissertation, Yale University.
Mukherjee, R.P. (1978). Further observations on the golden langur (*Presbytis geei* Khajuria 1956), with a note on the capped langur (*Presbytis pileatus* Blyth 1843) of Assam. *Primates* **19**, 737–47.
Mukherjee, R.P. & Saha, S.S. (1974). The golden langurs (*Presbytis geei* Khajuria, 1956) of Assam. *Primates* **15**, 327–40.
Newton, P.N. (1987). The social organization of forest hanuman langurs (*Presbytis entellus*). *Int. J. Primatol.* **8**, 199–232.
Newton, P.N. (1988). The variable social organization of the hanuman langur (*Presbytis entellus*), infanticide and the monopolization of females. *Int. J. Primatol.* **9**, 59–77.
Newton, P.N. (1992). Feeding and ranging patterns of forest hanuman langurs (*Presbytis entellus*). *Int. J. Primatol.* **13**, 245–85.
Newton, P.N. & Dunbar, R.I.M. (1994). Colobine monkey society. In *Colobine Monkeys: Their Ecology, Behaviour and Evolution*, ed. A.G. Davies & J.F. Oates, pp. 311–46. Cambridge: Cambridge University Press.
Oates, J.F. (1987). Food distribution and foraging behavior. In *Primate Societies*, ed. B.B. Smuts, D.L. Cheney, R.M.Seyfarth, R.W.Wrangham & T.T.Struhsaker, pp. 197–209. Chicago:University of Chicago Press.

Oates, J.F. & Davies, A.G. (1994). Conclusions: the past, present and future of the colobines. In *Colobine Monkeys: Their Ecology, Behaviour and Evolution*, ed. A.G. Davies & J.F. Oates, pp. 347–58. Cambridge: Cambridge University Press.

Oates, J.F., Davies, A.G., Delson, E. (1994). The diversity of living colobines. In *Colobine Monkeys: Their Ecology, Behaviour and Evolution*, ed. A.G. Davies & J.F. Oates, pp. 45–73. Cambridge: Cambridge University Press.

Oates, J.F., Waterman, P.G. & Choo, G.M. (1980). Food selection by the South Indian leaf-monkey, *Presbytis johnii*, in relation to leaf chemistry. *Oecologia*, **45**, 45–56.

Oftedal, O.T. (1991). The nutritional consequences of foraging in primates: the relationship of nutrient intakes to nutrient requirements. *Phil. Trans. R. Soc. Lond.* **334**, 161–70.

Parra, R. (1978). Comparison of foregut and hindgut fermentation in herbivores. In *The Ecology of Arboreal Folivores*, ed. G.G. Montgomery, pp. 205–30. Washington, DC: Smithsonian Institute.

Poirier, F.E. (1974). Colobine aggression: a review. In *Primate Aggression, Territoriality and Xenophobia*, ed. R.L. Holloway, pp. 123–57. New York: Academic Press.

Poirier, F.E. (1968a). Analysis of a Nilgiri langur (*Presbytis johnii*) home range change. *Primates* **9**, 29–43.

Poirier, F.E. (1968b). The Nilgiri langur (*Presbytis johnii*) mother–infant dyad. *Primates* **9**, 45–68.

Poirier, F.E. (1969). The Nilgiri langur (*Presbytis johnii*) troop: its composition, structure, function and change. *Folia primatol.* **10**, 20–47.

Poirier, F.E. & Hu, H.X. (1983). *Macaca mulatta* and *Rhinopithecus* in China: preliminary research results. *Curr. Anthropol.* **24**, 387–8.

Pusey, A.E. & Packer, C. (1987). Dispersal and philopatry. In *Primate Societies*, ed. B.B. Smuts, D.L. Cheney, R.M. Seyfarth, R.W. Wrangham & T.T. Struhsaker, pp. 250–66. Chicago:University of Chicago Press.

Quiatt, D. (1979). Aunts and mothers: adaptive implications of allomaternal behavior of non-human primates. *Am. Anthropol.* **81**, 310–19.

Rajanathan, R. & Bennett, E.L. (1990). Notes on the social behaviour of wild proboscis monkeys (*Nasalis larvatus*). *Malay. Nat. J.* **44**, 35–44.

Rudran, R. (1973a). The reproductive cycles of two subspecies of purple-faced langurs (*Presbytis senex*) with relation to environmental factors. *Folia primatol.* **19**, 41–60.

Rudran, R. (1973b). Adult male replacement in one-male troop of purple-faced langurs (*Presbytis senex senex*) and its effect on population structure. *Folia Primatol.* **19**, 166–92.

Ruhiyat, Y. (1983). Socio-ecological study of *Presbytis aygula* in West Java. *Primates* **42**, 344–59.

Scott, M.E. (1988). The impact of infection and disease on animal populations: implications for conservation biology. *Con. Bio.* **2**, 40–56.

Shively, C., Clarke, S., King, N., Schapiro, S. & Mitchell, G. (1982). Patterns of sexual behavior in male macaques. *Am. J. Primatol.* **2**, 373–84.

Stanford, C.B. (1988). Ecology of the capped langur and Phayre's leaf monkey in Bangladesh. *Primate Con.* **9**, 125–8.

Stanford, C.B. (1990). Colobine socioecology and female-bonded models of primate social structure. *Kroeber Anthro. Soc. Papers* **71–72**, 21–8.

Stanford, C.B. (1991). *The Capped Langur in Bangladesh:Behavioral Ecology and Reproductive Tactics.* Basel:Karger.

Sterck, E.H.M. (1992). Timing of female migrations in the Thomas langur

(*Presbytis thomasi*). Paper presented at the XIVth Congress of the International Primatological Society, Strasbourg, France.

Sterck, E.H.M. (1995). *Females, Foods, and Fights*. PhD dissertation, University of Utrecht.

Steenbeck, R. (1994). Constraints for female migration in Thomas langurs (*Presbytis thomasi*): A natural experiment. Paper presented at the *XIVth Congress of the International Primatological Society, Bali, Indonesia*. [Oral presentation.]

Struhsaker, T.T. (1969). Correlates of ecology and social organization among African Cercopithecines. *Folia primatol.* **11**, 80–118.

Su, Y., Ren, R., Yan, K., Li, J. & Zhou, Y. (1999). Preliminary survey of the home range and ranging behavior of golden monkeys (*Rhinopithecus roxellanae*) in Shennongjia National Natural Reserve, Hubei, China. In *The Natural History of the Doucs and Snub-nosed Monkeys*, ed. N. Jablonski, Singapore: World Scientific Publications.

Supriatna, J., Manullang, B.O. & Soekara, E. (1986). Group composition, home range and diet of the maroon leaf monkey (*Presbytis rubicunda*) at Tanjung Puting, Central Kalimantan, Indonesia. *Primates* **27**, 185–90.

Sussman, R.W., Cheverud, J.M. & Barlett, T.Q. (1995). Infant killing as an evolutionary strategy: reality or myth. *Evol. Anthropol.* **3**, 149–51.

Sweeney, B.W. & Vannote, R.L. (1982). Population synchrony in mayflies: a predator satiation hypothesis. *Evolution* **36**, 810–21.

Tanaka, J. (1965). Social structure of Nilgiri langurs. *Primates* **6**, 107–22.

Teneza, R. & Fuentes, A. (1995). Monandrous social organization of pigtailed langurs (*Simias concolor*) in the Pagai islands, Indonesia. *Int. J. Primatol.* **16**, 295–310.

Tilson, R. (1977). Social organization of Simakobu monkeys (*Nasalis concolor*) in Siberut Island, Indonesia. *J. Mammol.* **58**, 202–12.

van Schaik, C.P., Assink, P.R. & Salafsky, N. (1992). Territorial behavior in Southeast Asian langurs: resource defense or mate defense? *Am. J. Primatol.* **26**, 233–42.

van Schaik, C.P. (1983). Why are diurnal primates living in groups? *Behaviour* **87**, 120–44.

van Schaik, C.P. (1989). The ecology of social relationships amongst female primates. *Comparative Socioecology*, ed. V. Standen & R. Foley, pp. 195–218. Oxford: Blackwell.

Wasser, S.K. & Barash, P.P. (1981). The selfish "allomother": a comment on Scollay and DeBold (1980). *Ethol. Sociobiol.* **2**, 91–3.

Watanabe, K. (1981). Variation in group composition and population density of the two sympatric Mentawaian leaf monkeys. *Primates* **22**, 145–60.

Waterman, P.G. & Kool, K.M. (1994). Colobine food selection and plant chemistry. In *Colobine Monkeys: Their Ecology, Behaviour and Evolution*, ed. A.G. Davies & J.F. Oates, pp. 251–84. Cambridge: Cambridge University Press.

Wilson, C.C. & Wilson, W.L. (1975). The influence of selective logging on primates and some other animals in East Kalimantan. *Folia primatol.* **23**, 245–74.

Whittenberger, J.F. (1981). *Animal Social Behavior*. Boston: Duxbury Press.

Wolf, K. & Fleagle, J. (1977). Adult male replacement in a group of silvered leaf-monkeys (*Presbytis cristata*) at Kuala Selangor, Malaysia. *Primates* **18**, 949–55.

Wrangham, R.W. (1980). An ecological model of female bonded primate groups. *Behaviour* **75**, 262–300.

Yang, D.H. (1988). Black snub-nosed monkeys in China. *Oryx* **22**, 41–3.
Yeager, C.P. (1989). *Proboscis Monkey* (Nasalis larvatus) *Social Organization and Ecology*. PhD thesis, University of California, Davis.
Yeager, C.P. (1992). Proboscis monkey (*Nasalis larvatus*) social organization: the nature and possible functions of intergroup patterns of association. *Am. J. Primatol.* **26**, 133–8.
Yeager, C.P. (1995). Does intraspecific variation in social systems explain reported differences in the social structure of the proboscis monkey (*Nasalis larvatus*)? *Primates* **36**, 577–84.
Yeager, C.P. & Dierenfeld, E. (1994). Phytochemical composition and physical characteristics of seeds and fruits used as food resources by proboscis monkeys (*Nasalis larvatus*). Invited presentation at the *XVth Congress of the International Primatological Society, Bali, Indonesia.*
Young, T.P. (1994). Natural die-offs of large mammals: implications for conservation. *Con. Bio.* **8**, 410–18.

Index